FIFTH EDITION

FUNDAMENTALS OF
PHYSICS

FIFTH EDITION

FUNDAMENTALS OF
PHYSICS

PART 3

DAVID HALLIDAY
University of Pittsburgh

ROBERT RESNICK
Rensselaer Polytechnic Institute

JEARL WALKER
Cleveland State University

JOHN WILEY & SONS, INC.

New York • Chichester • Brisbane • Toronto • Singapore

ACQUISITIONS EDITOR Stuart Johnson
DEVELOPMENTAL EDITOR Rachel Nelson
SENIOR PRODUCTION SUPERVISOR Cathy Ronda
PRODUCTION ASSISTANT Raymond Alvarez
MARKETING MANAGER Catherine Faduska
ASSISTANT MARKETING MANAGER Ethan Goodman
DESIGNER Dawn L. Stanley
MANUFACTURING MANAGER Mark Cirillo
PHOTO EDITOR Hilary Newman
COVER PHOTO William Warren/Westlight
ILLUSTRATION EDITOR Edward Starr
ILLUSTRATION Radiant/Precision Graphics

This book was set in Times Roman by Progressive Information Technologies, and printed and bound by Von Hofmann Press. The cover was printed by Phoenix Color Corp.

Recognizing the importance of preserving what has been written, it is a policy of John Wiley & Sons, Inc. to have books of enduring value published in the United States printed on acid-free paper, and we exert our best efforts to that end.

The paper on this book was manufactured by a mill whose forest management programs include sustained yield harvesting of its timberlands. Sustained yield harvesting principles ensure that the number of trees cut each year does not exceed the amount of new growth.

ISBN 0-471-14855-5

Printed in the United States of America

10 9 8 7 6 5 4

Hello There!

You are about to begin your first college level physics course. You may have heard from friends and fellow students that physics is a difficult course, especially if you don't plan to go on to a career in the hard sciences. But that doesn't mean it has to be difficult for you. The key to success in this course is to have a good understanding of each chapter before moving on to the next. When learned a little bit at a time, physics is straightforward and simple. Here are some ideas that can make this text and this class work for you:

- Read through the **Sample Problems** and solutions carefully. These problems are similar to many of the end-of-chapter exercises, so reading them will help you solve homework problems. In addition, they offer a look at how an experienced physicist would approach solving the problem.

- Try to answer the **Checkpoint** questions as you read through the chapter. Most of these can be answered by thinking through the problem, but it helps if you have some scratch paper and a pencil nearby to work out some of the harder questions. Hold the answer page at the back of the book with your thumb (or a tab or bookmark) so you can refer to it easily. The end-of-chapter **Questions** are very similar— you can use them to quiz yourself after you've read the whole chapter.

- Use the **Review & Summary** sections at the end of each chapter as the first place to look for formulas you might neeed to solve homework problems. These sections are also helpful in making study sheets for exams.

- The biggest tip I can give you is pretty obvious. Do your homework! Understanding the homework problems is the best way to master the material and do well in the course. Doing a lot of homework problems is also the best way to review for exams. And by all means, consult your classmates whenever you are stuck. Working in groups will make your studying more effective.

The study of basic physics is required for degrees in Engineering, Physics, Biology, Chemistry, Medicine, and many other sciences because the fundamentals of physics are the framework on which every other science is built. Therefore, a solid grasp of basic physical principles will help you understand upper-level science courses and make your study of these courses easier.

I took introductory physics because it was a prerequisite for my B.S. in Mechanical Engineering. Even though engineering and pure physics are worlds apart, I find myself using this book as a reference almost every day. I urge you to keep it after you have finished your course. The knowledge you will gain from this book and your introductory physics course is the foundation for all other sciences. This is the primary reason to take this class seriously and be successful in it.

Best of luck!

Josh Kane

Josh Kane

Preface

For four editions, *Fundamentals of Physics* has been successful in preparing physics students for careers in science and engineering. The first three editions were coauthored by the highly regarded team of David Halliday and Robert Resnick, who developed a groundbreaking text replete with conceptual structure and applications. In the fourth edition, the insights provided by new coauthor, Jearl Walker, took the text into the 1990s and met the challenge of guiding students through a time of tremendous advances and a ferment of activity in the science of physics. Now, in the fifth edition, we have expanded on the conventional strengths of the earlier editions and enhanced the applications that help students forge a bridge between concepts and reasoning. We not only *tell* students how physics works, we *show* them, and we give them the opportunity to show us what they have learned by testing their understanding of the concepts and applying them to real-world scenarios. Concept checkpoints, problem solving tactics, sample problems, electronic computations, exercises and problems—all of these skill-building signposts have been developed to help students establish a connection between conceptual theories and application. The students reading this text today are the scientists and engineers of tomorrow. It is our hope that the fifth edition of *Fundamentals of Physics* will help prepare these students for future endeavors by contributing to the enhancement of physics education.

CHANGES IN THE FIFTH EDITION

Although we have retained the basic framework of the fourth edition of *Fundamentals of Physics,* we have made extensive changes in portions of the book. Each chapter and element has been scrutinized to ensure clarity, currency, and accuracy, reflecting the needs of today's science and engineering students.

Content Changes

Mindful that textbooks have grown large and that they tend to increase in length from edition to edition, we have reduced the length of the fifth edition by combining several chapters and pruning their contents. In doing so, six chapters have been rewritten completely, while the remaining chapters have been carefully edited and revised, often extensively, to enhance their clarity, incorporating ideas and suggestions from dozens of reviewers.

- *Chapters 7 and 8 on energy* (and sections of later chapters dealing with energy) have been rewritten to provide a more careful treatment of energy, work, and the work–kinetic energy theorem. As the same time, the text material and problems at the end of each chapter still allow the instructor to present the more traditional treatment of these subjects.

- *Temperature, heat, and the first law of thermodynamics* have been condensed from two chapters to one chapter (*Chapter 19*).

- *Chapter 21 on entropy* now includes a statistical mechanical presentation of entropy that is tied to the traditional thermodynamical presentation.

- *Chapters on Faraday's law and inductance* have been combined into one new chapter (*Chapter 31*).

- *Treatment of Maxwell's equations* has been streamlined and moved up earlier into the chapter on magnetism and matter (*Chapter 32*).

- *Coverage of electromagnetic oscillations and alternating currents* has been combined into one chapter (*Chapter 33*).

- *Chapters 39, 40, and 41 on quantum mechanics* have been rewritten to modernize the subject. They now include experimental and theoretical results of the last few years. In addition, quantum physics and special relativity are introduced in some of the early chapters in short sections that can be covered quickly. These early sections lay some of the groundwork for the ''modern physics'' topics that appear later in the extended version of the text and add an element of suspense about the subject.

New Pedagogy

In the interest of addressing the needs of science and engineering students, we have added a number of new pedagogical features intended to help students forge a bridge between concepts and reasoning and to marry theory with practice. These new features are designed to help students test their understanding of the material. They were also developed to help students prepare to apply the information to exam questions and real-world scenarios.

- To provide opportunities for students to check their understanding of the physics concepts they have just read, we have placed **Checkpoint** questions within the chapter presentations. Nearly 300 Checkpoints have been added to help guide the student away from common errors and misconceptions. All of the Checkpoints require decision making and reasoning on the part of the student (rather than computations requiring calculators) and focus on the key points of the physics that students need to understand in order to tackle the exercises and problems at the end of each chapter. Answers to all of the Checkpoints are found in the back of the book, sometimes with extra guidance to the student.

- Continuing our focus on the key points of the physics, we have included additional **Checkpoint-type questions** in the Questions section at the end of each chapter. These new questions require decision making and reasoning on the part of the student; they ask the student to organize the physics concepts rather than just plug numbers into equations. Answers to the odd-numbered questions are now provided in the back of the book.

- To encourage the use of computer math packages and graphing calculators, we have added an **Electronic Computation problem section** to the Exercises and Problems sections of many of the chapters.

These new features are just a few of the pedagogical elements available to enhance the student's study of physics. A number of tried-and-true features of the previous edition have been retained and refined in the fifth edition, as described below.

CHAPTER FEATURES

The pedagogical elements that have been retained from previous editions have been carefully planned and crafted to motivate students and guide their reasoning process.

- *Puzzlers* Each chapter opens with an intriguing photograph and a "puzzler" that is designed to motivate the student to read the chapter. The answer to each puzzler is provided within the chapter, but it is not identified as such to ensure that the student reads the entire chapter.

- *Sample Problems* Throughout each chapter, sample problems provide a bridge from the concepts of the chapter to the exercises and problems at the end of the chapter. Many of the nearly 400 sample problems featured in the text have been replaced with new ones that more sharply focus on the common difficulties students experience in solving the exercises and problems. We have been especially mindful of the mathematical difficulties students face. The sample problems also provide

an opportunity for the student to see how a physicist thinks through a problem.

- *Problem Solving Tactics* To help further bridge concepts and applications and to add focus to the key physics concepts, we have refined and expanded the number of problem solving tactics that are placed within the chapters, particularly in the earlier chapters. These tactics provide guidance to the students about how to organize the physics concepts, how to tackle mathematical requirements in the exercises and problems, and how to prepare for exams.

- *Illustrations* Because the illustrations in a physics textbook are so important to an understanding of the concepts, we have altered nearly 30 percent of the illustrations to improve their clarity. We have also removed some of the less effective illustrations and added many new ones.

- *Review & Summary* A review and summary section is found at the end of each chapter, providing a quick review of the key definitions and physics concepts *without* being a replacement for reading the chapter.

- *Questions* Approximately 700 thought-provoking questions emphasizing the conceptual aspects of physics appear at the ends of the chapters. Many of these questions relate back to the checkpoints found throughout the chapters, requiring decision making and reasoning on the part of the student. Answers to the odd-numbered questions are provided in the back of the book.

- *Exercises & Problems* There are approximately 3400 end-of-chapter exercises and problems in the text, arranged in order of difficulty, starting with the exercises (labeled "E"), followed by the problems (labeled "P"). Particularly challenging problems are identified with an asterisk (*). Those exercises and problems that have been retained from previous editions have been edited for greater clarity; many have been replaced. Answers to the odd-numbered exercises and problems are provided in the back of the book.

VERSIONS OF THE TEXT

The fifth edition of *Fundamentals of Physics* is available in a number of different versions, to accommodate the individual needs of instructors and students alike. The Regular Edition consists of Chapters 1 through 38 (ISBN 0-471-10558-9). The Extended Edition contains seven additional chapters on quantum physics and cosmology (Chapters 1–45) (ISBN 0-471-10559-7). Both editions are available as single, hardcover books, or in the alternative versions listed on page ix:

- Volume 1—Chapters 1–21 (Mechanics/Thermodynamics), cloth, 0-471-15662-0
- Volume 2—Chapters 22–45 (E&M and Modern Physics), cloth, 0-471-15663-9
- Part 1—Chapters 1–12, paperback, 0-471-14561-0
- Part 2—Chapters 13–21, paperback, 0-471-14854-7
- Part 3—Chapters 22–33, paperback, 0-471-14855-5
- Part 4—Chapters 34–38, paperback, 0-471-14856-3
- Part 5—Chapters 39–45, paperback, 0-471-15719-8

The Extended edition of the text is also available on CD ROM.

SUPPLEMENTS

The fifth edition of *Fundamentals of Physics* is supplemented by a comprehensive ancillary package carefully developed to help teachers teach and students learn.

Instructor's Supplements

- *Instructor's Manual* by J. RICHARD CHRISTMAN, U.S. Coast Guard Academy. This manual contains lecture notes outlining the most important topics of each chapter, as well as demonstration experiments, and laboratory and computer exercises; film and video sources are also included. Separate sections contain articles that have appeared recently in the *American Journal of Physics* and *The Physics Teacher.*

- *Instructor's Solutions Manual* by JERRY J. SHI, Pasadena City College. This manual provides worked-out solutions for all the exercises and problems found at the end of each chapter within the text. *This supplement is available only to instructors.*

- *Solutions Disk.* An electronic version of the Instructor's Solutions Manual, for instructors only, available in TeX for Macintosh and Windows™.

- *Test Bank* by J. RICHARD CHRISTMAN, U.S. Coast Guard Academy. More than 2200 multiple-choice questions are included in the Test Bank for *Fundamentals of Physics.*

- *Computerized Test Bank*. IBM and Macintosh versions of the entire Test Bank are available with full editing features to help you customize tests.

- *Animated Illustrations.* Approximately 85 text illustrations are animated for enhanced lecture demonstrations.

- *Transparencies.* More than 200 four-color illustrations from the text are provided in a form suitable for projection in the classroom.

Student's Supplements

- *A Student's Companion* by J. RICHARD CHRISTMAN, U.S. Coast Guard Academy. Much more than a traditional study guide, this student manual is designed to be used in close conjunction with the text. The Student's Companion is divided into four parts, each of which corresponds to a major section of the text, beginning with an overview "chapter." These overviews are designed to help students understand how the important topics are integrated and how the text is organized. For each chapter of the text, the corresponding Companion chapter offers: Basic Concepts, Problem Solving, Notes, Mathematical Skills, and Computer Projects and Notes.

- *Solutions Manual* by J. RICHARD CHRISTMAN, U.S. Coast Guard Academy and EDWARD DERRINGH, Wentworth Institute. This manual provides students with complete worked-out solutions to 30 percent of the exercises and problems found at the end of each chapter within the text.

- **Interactive Learningware** by JAMES TANNER, Georgia Institute of Technology, with the assistance of GARY LEWIS, Kennesaw State College. This software contains 200 problems from the end-of-chapter exercises and problems, presented in an interactive format, providing detailed feedback for the student. Problems from Chapter 1 to 21 are included in Part 1, from Chapters 22 to 38 in Part 2. The accompanying workbooks allow the student to keep a record of the worked-out problems. The Learningware is available in IBM 3.5″ and Macintosh formats.

- **CD Physics.** The entire Extended Version of the text (Chapters 1–45) is available on CD ROM, along with the student solutions manual, study guide, animated illustrations, and Interactive Learningware.

Acknowledgments

A textbook contains far more contributions to the elucidation of a subject than those made by the authors alone. J. Richard Christman, of the U.S. Coast Guard Academy, has once again created many fine supplements for us; his knowledge of our book and his recommendations to students and faculty are invaluable. James Tanner, of Georgia Institute of Technology, and Gary Lewis, of Kennesaw State College, have provided us with innovative software, closely tied to the text's exercises and problems. J. Richard Christman, of the U.S. Coast Guard Academy, and Glen Terrell, of the University of Texas at Arlington, contributed problems to the Electronic Computation sections of the text. Jerry Shi, of Pasadena City College, performed the Herculean task of working out solutions for every one of the Exercises and Problems in the text. We thank John Merrill, of Brigham Young University, and Edward Derringh, of the Wentworth Institute of Technology for their many contributions in the past. We also thank George W. Hukle of Oxnard, California, for his check of the answers at the back of the book.

At John Wiley, publishers, we have been fortunate to receive strong coordination and support from our former editor, Cliff Mills. Cliff guided our efforts and encouraged us along the way. When Cliff moved on to other responsibilities at Wiley, we were ably guided to completion by his successor, Stuart Johnson. Rachel Nelson has coordinated the developmental editing and multilayered preproduction process. Catherine Faduska, our senior marketing manager, and Ethan Goodman, assistant marketing manager, have been tireless in their efforts on behalf of this edition. Jennifer Bruer has built a fine supporting package of ancillary materials. Monica Stipanov and Julia Salsbury managed the review and administrative duties admirably.

We thank Lucille Buonocore, our able production manager, and Cathy Ronda, our production editor, for pulling all the pieces together and guiding us through the complex production process. We also thank Dawn Stanley, for her design; Brenda Griffing, for her copy editing; Edward Starr, for managing the line art program; Lilian Brady, for her proofreading; and all other members of the production team.

Stella Kupferburg and her team of photo researchers, particularly Hilary Newman and Pat Cadley, were inspired in their search for unusual and interesting photographs that communicate physics principles beautifully. We thank Boris Starosta and Irene Nunes for their careful development of a full-color line art program, for which they scrutinized and suggested revisions of every piece. We also owe a debt of gratitude for the line art to the late John Balbalis, whose careful hand and understanding of physics can still be seen in every diagram.

We especially thank Edward Millman for his developmental work on the manuscript. With us, he has read every word, asking many questions from the point of view of a student. Many of his questions and suggested changes have added to the clarity of this volume. Irene Nunes added a final, valuable developmental check in the last stages of the book.

We owe a particular debt of gratitude to the numerous students who used the fourth edition of *Fundamentals of Physics* and took the time to fill out the response cards and return them to us. As the ultimate consumers of this text, students are extremely important to us. By sharing their opinions with us, your students help us ensure that we are providing the best possible product and the most value for their textbook dollars. We encourage the users of this book to contact us with their thoughts and concerns so that we can continue to improve this text in the years to come. In particular, we owe a special debt of gratitude to the students who participated in a final focus group at Union College in Schenecdaty, New York: Matthew Glogowski, Josh Kane, Lauren Papa, Phil Tavernier, Suzanne Weldon, and Rebecca Willis.

Finally, our external reviewers have been outstanding and we acknowledge here our debt to each member of that team:

MARIS A. ABOLINS
Michigan State University

BARBARA ANDERECK
Ohio Wesleyan University

ALBERT BARTLETT
University of Colorado

MICHAEL E. BROWNE
University of Idaho

TIMOTHY J. BURNS
Leeward Community College

JOSEPH BUSCHI
Manhattan College

PHILIP A. CASABELLA
Rensselaer Polytechnic Institute

RANDALL CATON
Christopher Newport College

J. RICHARD CHRISTMAN
U.S. Coast Guard Academy

ROGER CLAPP
University of South Florida

W. R. CONKIE
Queen's University

PETER CROOKER
University of Hawaii at Manoa

WILLIAM P. CRUMMETT
Montana College of Mineral Science and Technology

EUGENE DUNNAM
University of Florida

ROBERT ENDORF
University of Cincinnati

F. PAUL ESPOSITO
University of Cincinnati

JERRY FINKELSTEIN
San Jose State University

ALEXANDER FIRESTONE
Iowa State University

ALEXANDER GARDNER
Howard University

ANDREW L. GARDNER
Brigham Young University

JOHN GIENIEC
Central Missouri State University

JOHN B. GRUBER
San Jose State University

ANN HANKS
American River College

SAMUEL HARRIS
Purdue University

EMILY HAUGHT
Georgia Institute of Technology

LAURENT HODGES
Iowa State University

JOHN HUBISZ
North Carolina State University

JOEY HUSTON
Michigan State University

DARRELL HUWE
Ohio University

CLAUDE KACSER
University of Maryland

LEONARD KLEINMAN
University of Texas at Austin

EARL KOLLER
Stevens Institute of Technology

ARTHUR Z. KOVACS
Rochester Institute of Technology

KENNETH KRANE
Oregon State University

SOL KRASNER
University of Illinois at Chicago

PETER LOLY
University of Manitoba

ROBERT R. MARCHINI
Memphis State University

DAVID MARKOWITZ
University of Connecticut

HOWARD C. MCALLISTER
University of Hawaii at Manoa

W. SCOTT MCCULLOUGH
Oklahoma State University

JAMES H. MCGUIRE
Tulane University

DAVID M. MCKINSTRY
Eastern Washington University

JOE P. MEYER
Georgia Institute of Technology

ROY MIDDLETON
University of Pennsylvania

IRVIN A. MILLER
Drexel University

EUGENE MOSCA
United States Naval Academy

MICHAEL O'SHEA
Kansas State University

PATRICK PAPIN
San Diego State University

GEORGE PARKER
North Carolina State University

ROBERT PELCOVITS
Brown University

OREN P. QUIST
South Dakota State University

JONATHAN REICHART
SUNY—Buffalo

MANUEL SCHWARTZ
University of Louisville

DARRELL SEELEY
Milwaukee School of Engineering

BRUCE ARNE SHERWOOD
Carnegie Mellon University

JOHN SPANGLER
St. Norbert College

ROSS L. SPENCER
Brigham Young University

HAROLD STOKES
Brigham Young University

JAY D. STRIEB
Villanova University

DAVID TOOT
Alfred University

J. S. TURNER
University of Texas at Austin

T. S. VENKATARAMAN
Drexel University

GIANFRANCO VIDALI
Syracuse University

FRED WANG
Prairie View A&M

ROBERT C. WEBB
Texas A&M University

GEORGE WILLIAMS
University of Utah

DAVID WOLFE
University of New Mexico

We hope that our words here reveal at least some of the wonder of physics, the fundamental clockwork of the universe. And, hopefully, those words might also reveal some of our awe of that clockwork.

DAVID HALLIDAY
6563 NE Windermere Road
Seattle, WA 98105

ROBERT RESNICK
Rensselaer Polytechnic Institute
Troy, NY 12181

JEARL WALKER
Cleveland State University
Cleveland, OH 44115

HOW TO USE THIS BOOK:

You are about to begin what could be one the most exciting course that you will undertake in college. It offers you the opportunity to learn what makes our world "tick" and to gain insight into the role physics plays in our everyday lives. This knowledge will not come without some effort, however, and this book has been carefully designed and written with an awareness of the kinds of difficulties and challenges you may face. Therefore, before you begin, we have provided a visual overview of some of the key features of the book that will aid in your studies.

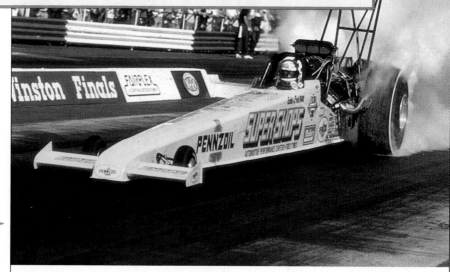

Chapter Opening Puzzlers

Each chapter opens with an intriguing example of physics in action. By presenting high-interest applications of each chapters concepts, the puzzlers are intended to peak your interest and motivate you to read the chapter.

In 1977, Kitty O'Neil set a dragster record by reaching 392.54 mi/h in a sizzling time of 3.72 s. In 1958, Eli Beeding Jr. rode a rocket sled from a standstill to a speed of 72.5 mi/h in an elapsed time of 0.04 s (less than an eye blink). How can we compare these two rides to see which was more exciting (or more frightening)—by final speeds, by elapsed times, or by some other quantity?

SAMPLE PROBLEM 2-6

(a) When Kitty O'Neil set the dragster records for the greatest speed and least elapsed time, she reached 392.54 mi/h in 3.72 s. What was her average acceleration?

SOLUTION: From Eq. 2-7, O'Neil's average acceleration was

$$\bar{a} = \frac{\Delta v}{\Delta t} = \frac{392.54 \text{ mi/h} - 0}{3.72 \text{ s} - 0}$$

$$= +106 \frac{\text{mi}}{\text{h} \cdot \text{s}}, \qquad \text{(Answer)}$$

where the motion is taken to be in the positive x direction. In

Answers to Puzzlers

All chapter-opening puzzlers are answered later in the chapter, either in text discussion or in a sample problem.

If the car
r, the bob
oninertial

CHECKPOINT **1:** In the figure, two perpendicular forces F_1 and F_2 are combined in six different ways. Which ways may be used to correctly determine the net force ΣF?

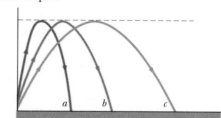

w wish to
accelera-
the stan-
use) the
been as-

ictionless
at by trial
accelera-
definition,
dy has a

body by

Checkpoints

Checkpoints appear throughout the text, focusing on the key points of physics you will need to tackle the exercises and problems found at the end of each chapter. These checkpoints help guide you away from common errors and misconceptions.

Checkpoint Questions

Checkpoint-type questions at the end of each chapter ask you to organize the physics concepts rather than plug numbers into equations. Answers to the odd-numbered questions are provided in the back of the book.

ey puck in

$v = -2t\mathbf{i}$

ponents of
ion vector
nd and t is
-2 and 3?

cal projec-
t identical
n the same
nal speeds

(c)

peed (a) a

$4.9\mathbf{j}$ (x is
). Has the

9. Figure 4-25 shows three paths for a kicked football. Ignoring the effects of air on the flight, rank the paths according to (a) time of flight, (b) initial vertical velocity component, (c) initial horizontal velocity component, and (d) initial speed. Place the greatest first in each part.

FIGURE 4-25 Question 9.

10. Figure 4-26 shows the velocity and acceleration of a particle at a particular instant in three situations. In which situation, and at that instant, is (a) the speed increasing, (b) the speed decreasing, (c) the speed not changing, (d) $\mathbf{v} \cdot \mathbf{a}$ positive, (e) $\mathbf{v} \cdot \mathbf{a}$ negative, and (f) $\mathbf{v} \cdot \mathbf{a} = 0$?

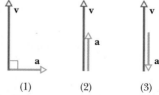

FIGURE 4-26 Question 10.

Sample Problems

The sample problems offer you the opportunity to work through the physics concepts just presented. Often built around real-world applications, they are closely coordinated with the end-of-chapter Questions, Exercises, and Problems.

SAMPLE PROBLEM 4-1

The position vector for a particle is initially

$$\mathbf{r}_1 = -3\mathbf{i} + 2\mathbf{j} + 5\mathbf{k}$$

and then later is

$$\mathbf{r}_2 = 9\mathbf{i} + 2\mathbf{j} + 8\mathbf{k}$$

(see Fig. 4-2). What is the displacement from $\mathbf{r}_1$ to $\mathbf{r}_2$?

SOLUTION: Recall from Chapter 3 that we add (or subtract) two vectors in unit-vector notation by combining the components, axis by axis. So Eq. 4-2 becomes

$$\mathbf{\Delta r} = (9\mathbf{i} + 2\mathbf{j} + 8\mathbf{k}) - (-3\mathbf{i} + 2\mathbf{j} + 5\mathbf{k})$$
$$= 12\mathbf{i} + 3\mathbf{k}. \qquad \text{(Answer)}$$

The displacement vector is parallel to the xz plane, because it lacks any y component, a fact that is easier to pick out in the numerical result than in Fig. 4-2.

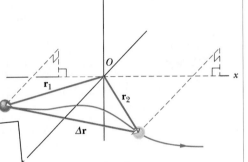

Sample Problem 4-1. The displacement $\mathbf{\Delta r} = $ ...d of $\mathbf{r}_1$ to the head of $\mathbf{r}_2$.

PROBLEM SOLVING TACTICS

TACTIC 1: *Reading Force Problems*

Read the problem statement several times until you have a good mental picture of what the situation is, what data are given, and what is requested. In Sample Problem 5-1, for example, you should tell yourself: "Someone is pushing a sled. Its speed changes, so acceleration is involved. The motion is along a straight line. A force is given in one part and asked for in the other, and so the situation looks like Newton's second law applied to one-dimensional motion."

If you know what the problem is about but don't know what to do next, put the problem aside and reread the text. If you are hazy about Newton's second law, reread that section. Study the sample problems. The one-dimensional-motion parts of Sample Problem 5-1 and the constant acceleration should send you back to Chapter 2 and especially to Table 2-1, which displays all the equations you are likely to need.

TACTIC 2: *Draw Two Types of Figures*

You may need two figures. One is a rough sketch of the actual real-world situation. When you draw the forces on it, place the tail of each force vector either on the boundary of or within the body feeling that force. The other figure is a free-body diagram in which the forces on a *single* body are drawn, with the body represented with a dot or a sketch. Place the tail of each force vector on the dot or sketch.

TACTIC 3: *What l... ...em?*

Problem Solving Tactics

Careful attention has been paid to helping you develop your problem-solving skills. Problem-solving tactics are closely related to the sample problems and can be found throughout the text, though most fall within the first half. The tactics are designed to help you work through assigned homework problems and prepare for exams. Collectively, they represent the stock in trade of experienced problem solvers and practicing scientists and engineers.

Review and Summary

Review & Summary sections at the end of each chapter review the most important concepts and equations.

REVIEW & SUMMARY

Conservative Forces

A force is a **conservative force** if the net work it does on a particle moving along a closed path from an initial point and then back to that point is zero. Or, equivalently, it is conservative if its work on a particle moving between two points does not depend on the path taken by the particle. The gravitational force (weight) and the spring force are conservative forces; the kinetic frictional force is a **nonconservative force.**

Potential Energy

A **potential energy** is energy that is associated with the configuration of a system in which a conservative force acts. When the conservative force does work W on a particle within the system, the change ΔU in the potential energy of the system is

$$\Delta U = -W. \tag{8-1}$$

If the particle moves from point x_i to point x_f, the change in potential energy of the system is

$$\Delta U = -\int_{x_i}^{x_f} F(x)\, dx. \tag{8-6}$$

Gravitational Potential Energy

The potential energy associated with a system consisting of the Earth and a nearby particle is the **gravitational potential energy.** If the particle moves from height y_i to height y_f, the change in gravitational potential energy of the particle–Earth system is

$$\Delta U = mg(y_f - y_i) = mg\,\Delta y. \tag{8-7}$$

If the **reference position** of the particle is set as $y_i = 0$ and the corresponding gravitational potential energy of the system is set as $U_i = 0$, then the gravitational potential energy U when the particle is at any position y is

$$U = mgy. \tag{8-9}$$

in which the subscripts refer to different instants during an transfer process. This conservation can also be written as

$$\Delta E = \Delta K + \Delta U = 0.$$

Potential Energy Curves

If we know the **potential energy function** $U(x)$ for a sy which a force F acts on a particle, we can find the force

$$F(x) = -\frac{dU(x)}{dx}.$$

If $U(x)$ is given on a graph, then at any value of x, the fo the negative of the slope of the curve there and the kinetic of the particle is given by

$$K(x) = E - U(x),$$

where E is the mechanical energy of the system. A **turnin** is a point x where the particle reverses its motion (there, The particle is in **equilibrium** at points where the slope $U(x)$ curve is zero (there, $F(x) = 0$).

Work by Nonconservative Forces

If a nonconservative applied force F does work on particl part of a system having a potential energy, then the wo done on the system by F is equal to the change ΔE in the m ical energy of the system:

$$W_{app} = \Delta K + \Delta U = \Delta E. \tag{8-24}$$

If a kinetic frictional force f_k does work on an obj change ΔE in the total mechanical energy of the object system containing it is given by

$$\Delta E = -f_k d,$$

in which d is the displacement of the object during the wo

Exercises and Problems

A hallmark of this text, nearly 3400 end-of-chapter exercises and problems are arranged in order of difficulty, starting with the exercises (labeled "E"), followed by the problems (labeled "P"). Particularly difficult problems are identified with an asterisk (*). Answers to all the odd-numbered exercises and problems are provided in the back of the book. New electronic computation problems, which require the use of math packages and graphing calculators, have been added to many of the chapters.

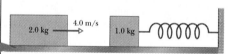

FIGURE 10-44 Problem 56.

57P. Two 22.7 kg ice sleds are placed a short distance apart, one directly behind the other, as shown in Fig. 10-45. A 3.63 kg cat, standing on one sled, jumps across to the other and immediately back to the first. Both jumps are made at a speed of 3.05 m/s relative to the ice. Find the final speeds of the two sleds.

FIGURE 10-45 Problem 57.

58P. The bumper of a 1200 kg car is designed so that it can just absorb all the energy when the car runs head-on into a solid wall at 5.00 km/h. The car is involved in a collision in which it runs at 70.0 km/h into the rear of a 900 kg car moving at 60.0 km/h in the same direction. The 900 kg car is accelerated to 70.0 km/h as a result of the collision. (a) What is the speed of the 1200 kg car immediately after impact? (b) What is the ratio of the kinetic energy absorbed in the collision to that which can be absorbed by the bumper of the 1200 kg car?

59P. A railroad freight car weighing 32 tons and traveling at 5.0

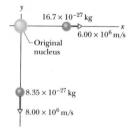

FIGURE 10-46 Exercise 62.

63E. In a game of pool, the cue ball strikes anothe at rest. After the collision, the cue ball moves at 3.5 line making an angle of 22.0° with its original dir tion, and the second ball has a speed of 2.00 m/s angle between the direction of motion of the secon original direction of motion of the cue ball and (b speed of the cue ball. (c) Is kinetic energy conserv

64E. Two vehicles A and B are traveling west and tively, toward the same intersection, where they co together. Before the collision, A (total weight 2700 with a speed of 40 mi/h and B (total weight 3600 lb of 60 mi/h. Find the magnitude and direction of the v (interlocked) vehicles immediately after the collisi

65E. In a game of billiards, the cue ball is given a V and strikes the pack of 15 stationary balls. All engage in nume___ ___ and ball–cushion col time lat___ ___ ___ accident) all

Brief Contents

Contents

FIFTH EDITION

FUNDAMENTALS OF
PHYSICS

22
Electric Charge

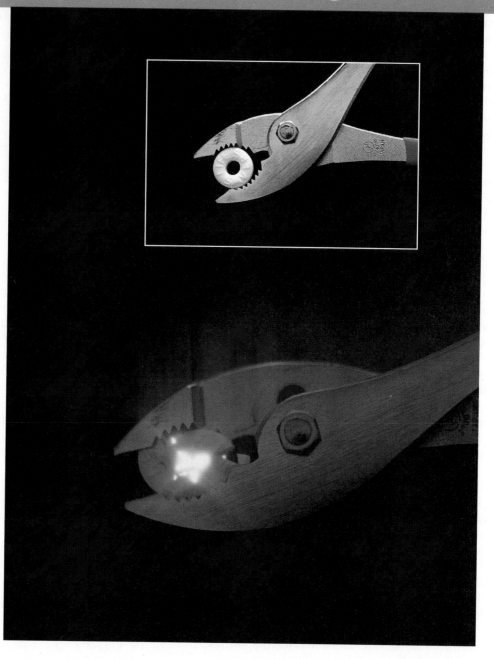

If you adapt your eyes to darkness for about 15 minutes and then have a friend chew a wintergreen LifeSaver, you will see a faint flash of blue light from your friend's mouth with each chomp. (To avoid wear on the teeth, you might crush the candy with pliers, as in the photograph.) What causes this display of light, commonly called "sparking"?

22-1 ELECTROMAGNETISM

The early Greek philosophers knew that if you rubbed a piece of amber, it would attract bits of straw. This ancient observation can be traced down directly to the electronic age in which we live. (The strength of the connection is indicated by our word *electron,* which is derived from the Greek word for amber.) The Greeks also recorded the observation that some naturally occurring "stones," known today as the mineral magnetite, would attract iron.

From these modest origins, the sciences of electricity and magnetism developed separately for centuries—until 1820, in fact, when Hans Christian Oersted found a connection between them: an electric current in a wire can deflect a magnetic compass needle. Interestingly enough, Oersted made this discovery while preparing a lecture demonstration for his physics students.

The new science of *electromagnetism* (the combination of electrical and magnetic phenomena) was developed further by workers in many countries. One of the best was Michael Faraday, a truly gifted experimenter with a talent for physical intuition and visualization. That talent is attested to by the fact that his collected laboratory notebooks do not contain a single equation. In the mid-19th century, James Clerk Maxwell put Faraday's ideas into mathematical form, introduced many new ideas of his own, and put electromagnetism on a sound theoretical basis.

Table 32-1 shows the basic laws of electromagnetism, now called Maxwell's equations. We plan to work our way through them in the chapters between here and there, but you might want to glance at them now, to see our goal.

22-2 ELECTRIC CHARGE

If you walk across a carpet in dry weather, you can produce a spark by bringing your finger close to a metal doorknob. Television advertising has alerted us to the problem of "static cling" in clothing (Fig. 22-1). On a grander scale, lightning is familiar to everyone. Each of these phenomena represents a tiny glimpse of the vast amount of *electric charge* that is stored in the familiar objects that surround us and—indeed—in our own bodies. **Electric charge** is an intrinsic characteristic of the fundamental particles making up those objects; that is, it is a characteristic that automatically accompanies those particles wherever they exist.

The vast amount of charge in an everyday object is usually hidden because the object contains equal amounts of two kinds of charge: *positive charge* and *negative charge*. With such an equality—or *balance*—of charge, the object is said to be *electrically neutral*; that is, it contains no *net* charge to interact with other objects. If the two types of charge are not in balance, then there *is* a net charge

FIGURE 22-1 Static cling, an electrical phenomenon that accompanies dry weather, causes these pieces of paper to stick to one another and to the plastic comb, and your clothing to stick to your body.

that *can* interact with other objects, and we become aware of the existence of the net charge. We say that an object is *charged* to indicate that it has a charge imbalance, or net charge. The imbalance is always very small compared to the total amounts of positive charge and negative charge contained in the object.

Charged objects interact by exerting forces on one another. To show this, we first charge a glass rod by rubbing one end with silk. At points of contact between the rod and the silk, tiny amounts of charge are transferred from one to the other, slightly upsetting the electrical neutrality of each. (We *rub* the silk over the rod to increase the number of contact points and thus the amount, still tiny, of transferred charge.)

Suppose we now suspend the charged rod from a thread to *electrically isolate* it from its surroundings so that its charge cannot change. If we bring a second, similarly charged, glass rod nearby (Fig. 22-2a), the two rods *repel*

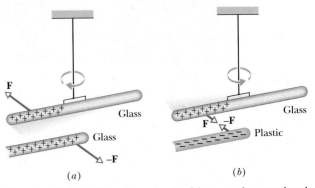

FIGURE 22-2 (a) Two charged rods of the same signs repel each other. (b) Two charged rods of opposite signs attract each other.

each other. That is, each rod experiences a force directed away from the other rod. However, if we rub a plastic rod with fur and bring it near the suspended glass rod (Fig. 22-2b), the two rods *attract* each other. That is, each rod experiences a force directed toward the other rod.

We can understand these two demonstrations in terms of positive and negative charges. When a glass rod is rubbed with silk, the glass loses some of its negative charge and then has a small unbalanced positive charge (represented by the plus signs in Fig. 22-2a). When the plastic rod is rubbed with fur, the plastic gains a small unbalanced negative charge (represented by the minus signs in Fig. 22-2b). Our two demonstrations reveal the following:

> Charges with the same electrical sign repel each other, and charges with opposite electrical signs attract each other.

In Section 22-4, we shall put this rule into quantitative form as Coulomb's law of *electrostatic force* (or *electric force*) between charges. The term *electrostatic* is used to emphasize that, relative to each other, the charges are either stationary or moving only very slowly.

The ''positive'' and ''negative'' labels and signs for electric charge were chosen arbitrarily by Benjamin Franklin. He could easily have interchanged the labels or used some other pair of opposites to distinguish the two kinds of charge. (Franklin was a scientist of international reputation. It has even been said that Franklin's triumphs in diplomacy in France during the American War of Independence were facilitated, and perhaps even made possible, because he was so highly regarded as a scientist.)

The attraction and repulsion between charged bodies have many industrial applications, including electrostatic paint spraying and powder coating, fly-ash collection in chimneys, nonimpact ink-jet printing, and photocopying. Figure 22-3 shows a tiny carrier bead in a Xerox copying machine, covered with particles of black powder called *toner,* that stick to it by means of electrostatic forces. The negatively charged toner particles are eventually attracted from the carrier bead to a rotating drum, where a positively charged image of the document being copied has formed. A charged sheet of paper then attracts the toner particles from the drum to itself, after which they are heat-fused in place to produce the copy.

22-3 CONDUCTORS AND INSULATORS

In some materials, such as metals, tap water, and the human body, some of the negative charge can move rather freely. We call such materials **conductors.** In other materials, such as glass, chemically pure water, and plastic, none of the charge can move freely. We call these materials **nonconductors** or **insulators.**

If you rub a copper rod with wool while holding the rod in your hand, you will not be able to charge the rod, because both you and the rod are conductors. The rubbing will cause a charge imbalance on the rod, but the excess charge will immediately move from the rod through you to the floor (which is connected to Earth's surface), and the rod will quickly be neutralized.

In thus setting up a pathway of conductors between an object and Earth's surface, we are said to *ground* the object. And in neutralizing the object (by eliminating an unbalanced positive or negative charge), we are said to *discharge* the object. (See Fig. 22-4 for a somewhat bizarre example of discharge.) If instead of holding the rod in your hand, you hold it via an insulating handle, you eliminate the conducting path to Earth, and the rod can then be charged by rubbing, as long as you do not touch it directly with your hand.

The properties of conductors and insulators are due to the structure and electrical nature of atoms. Atoms consist of positively charged *protons*, negatively charged *electrons*, and electrically neutral *neutrons*. The protons and neutrons are packed tightly together in a central *nucleus*; in a simple model of an atom, the electrons orbit the nucleus.

The charge of a single electron and that of a single proton have the same magnitude but are opposite in sign. Hence an electrically neutral atom contains equal numbers of electrons and protons. Electrons are held near the nu-

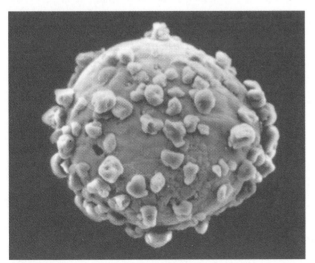

FIGURE 22-3 A carrier bead from a Xerox copying machine; it is covered with toner particles that cling to it by electrostatic attraction. The diameter of the bead is about 0.3 mm.

FIGURE 22-4 Not a parlor stunt but a serious experiment carried out in 1774 to prove that the human body is a conductor of electricity. The etching shows a person suspended by nonconducting ropes while being charged by a charged rod (which probably touched flesh instead of the trousers). When the person brought his face, left hand, or the conducting ball and rod in his right hand near one of the metallic plates, electric sparks flew through the intermediate air, discharging him.

cleus because they have the electrical sign opposite that of the protons in the nucleus and thus are attracted to the nucleus.

When atoms of a conductor like copper come together to form the solid, some of their outermost (and so most loosely held) electrons do not remain attached to the individual atoms but become free to wander about within the solid, leaving behind positively charged atoms (*positive ions*). We call the mobile electrons *conduction electrons*. There are few (if any) free electrons in a nonconductor.

The experiment of Fig. 22-5 demonstrates the mobility of charge in a conductor. A negatively charged plastic rod will attract either end of an isolated neutral copper rod.

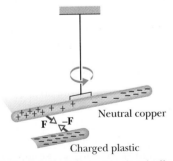

Neutral copper

Charged plastic

FIGURE 22-5 A neutral copper rod is electrically isolated from its surroundings by being suspended on a nonconducting thread. Either end of the copper rod will be attracted by a charged rod. Here, conduction electrons in the copper rod are repelled to the far end of that rod by the negative charge on the plastic rod. Then that negative charge attracts the remaining positive charge on the near end of the copper rod, rotating the copper rod to bring that near end closer to the plastic rod.

What happens is that many of the conduction electrons in the closer end of the copper rod are repelled by the negative charge on the plastic rod. They move to the far end of the copper rod, leaving the near end depleted in electrons and thus with an unbalanced positive charge. This positive charge is attracted to the negative charge in the plastic rod. Although the copper rod is still neutral, it is said to have an *induced charge*, which means that some of its positive and negative charges have been separated owing to the presence of a nearby charge.

Similarly, if a positively charged glass rod is brought near one end of a neutral copper rod, conduction electrons in the copper rod are attracted to that end. That end becomes negatively charged and the other end positively charged, so again an induced charge is set up in the copper rod. Although the copper rod is still neutral, it and the glass rod attract each other.

Note that it is only conduction electrons, with their negative charges, that can move; positive ions are fixed in place. Thus, an object becomes positively charged only through the *removal of negative charges*.

Semiconductors, such as silicon and germanium, are materials that are intermediate between conductors and insulators. The microelectronic revolution that has transformed our lives in so many ways is due to devices constructed of semiconducting materials.

Finally, there are **superconductors,** so called because they present no resistance to the movement of electric charge through them. When charge moves through a material, we say that an **electric current** exists in the material. Ordinary materials, even good conductors, tend to resist the flow of charge through them. In a superconductor, however, the resistance is not just small; it is precisely zero. If you set up a current in a superconducting ring, the flow of electrons persists without change for as long as you care to watch it, with no battery or other source of energy needed to maintain the current.

CHECKPOINT **1:** The figure shows five pairs of plates: *A*, *B*, and *D* are charged plastic plates and *C* is an electrically neutral copper plate. The electrostatic forces between the pairs of plates are shown for three of the pairs. For the remaining two pairs, do the plates repel or attract each other?

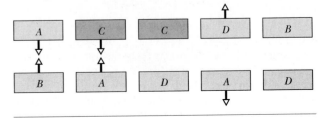

22-4 COULOMB'S LAW

Let two charged particles (also called *point charges*) have charge magnitudes q_1 and q_2 and be separated by a distance r. The **electrostatic force** of attraction or repulsion between them has the magnitude

$$F = k\,\frac{|q_1||q_2|}{r^2} \qquad \text{(Coulomb's law),} \quad (22\text{-}1)$$

in which k is a constant. Each particle exerts a force of this magnitude on the other particle; the two forces form an action-reaction pair. If the particles *repel* each other, the force on each particle points *away from* the other particle (as in Figs. 22-6a and b). If the particles *attract* each other, the force on each particle points *toward* the other particle (as in Fig. 22-6c).

Equation 22-1 is called **Coulomb's law** after Charles Augustin Coulomb, whose experiments in 1785 led him to it. Curiously, the form of Eq. 22-1 is the same as that of Newton's equation for the gravitational force between two particles with masses m_1 and m_2 that are separated by a distance r:

$$F = G\,\frac{m_1 m_2}{r^2}, \qquad\qquad (22\text{-}2)$$

in which G is the gravitational constant.

The constant k in Eq. 22-1, by analogy with the gravitational constant G in Eq. 22-2, may be called the *electrostatic constant*. Both equations describe inverse square laws that involve a property of the interacting particles—the mass in one case and the charge in the other. The laws differ in that gravitational forces are always attractive but electrostatic forces may be either attractive or repulsive,

depending on the signs of the two charges. This difference arises from the fact that, although there is only one kind of mass, there are two kinds of charge (and that is why absolute signs are needed in Eq. 22-1 but not in Eq. 22-2).

Coulomb's law has survived every experimental test; no exceptions to it have ever been found. It holds even within the atom, correctly describing the force between the positively charged nucleus and each of the negatively charged electrons, even though classical Newtonian mechanics fails in that realm and is replaced there by quantum physics. This simple law also correctly accounts for the forces that bind atoms together to form molecules, and for the forces that bind atoms and molecules together to form solids and liquids.

For practical reasons having to do with the accuracy of measurements, the SI unit of charge is derived from the SI unit of electric current, the ampere (A). The SI unit of charge is the **coulomb** (C): *One coulomb is the amount of charge that is transferred through the cross section of a wire in 1 second when there is a current of 1 ampere in the wire.* In Section 30-2 we shall describe how the ampere is defined experimentally. In general, we can write

$$dq = i\,dt, \qquad\qquad (22\text{-}3)$$

in which dq (in coulombs) is the charge transferred by a current i (in amperes) during the time interval dt (in seconds).

For historical reasons (and because doing so simplifies many other formulas), the electrostatic constant k of Eq. 22-1 is usually written $1/4\pi\epsilon_0$. Then Coulomb's law becomes

$$F = \frac{1}{4\pi\epsilon_0}\,\frac{|q_1||q_2|}{r^2} \qquad \text{(Coulomb's law).} \quad (22\text{-}4)$$

The constants in Eqs. 22-1 and 22-4 have the value

$$k = \frac{1}{4\pi\epsilon_0} = 8.99 \times 10^9 \text{ N}\cdot\text{m}^2/\text{C}^2. \quad (22\text{-}5)$$

The quantity ϵ_0, called the **permittivity constant,** sometimes appears separately in equations and is

$$\epsilon_0 = 8.85 \times 10^{-12} \text{ C}^2/\text{N}\cdot\text{m}^2. \quad (22\text{-}6)$$

Still another parallel between the gravitational force and the electrostatic force is that both obey the principle of superposition. If we have n charged particles, they interact independently in pairs, and the force on any one of them, let us say particle 1, is given by the vector sum

$$\mathbf{F}_1 = \mathbf{F}_{12} + \mathbf{F}_{13} + \mathbf{F}_{14} + \mathbf{F}_{15} + \cdots + \mathbf{F}_{1n}, \quad (22\text{-}7)$$

in which, for example, $\mathbf{F}_{14}$ is the force acting on particle 1

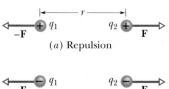

(a) Repulsion

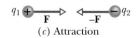

(b) Repulsion

(c) Attraction

FIGURE 22-6 Two charged particles, separated by distance r, repel each other if their charges are (a) both positive and (b) both negative. (c) They attract each other if their charges are of opposite signs. In each of the three situations, the force acting on one particle is equal in magnitude to the force acting on the other particle but points in the opposite direction.

owing to the presence of particle 4. An identical formula holds for the gravitational force.

Finally, the two shell theorems that we found so useful in our study of gravitation have analogs in electrostatics:

> A shell of uniform charge attracts or repels a charged particle that is outside the shell as if all the shell's charge were concentrated at its center.
>
> A shell of uniform charge exerts no electrostatic force on a charged particle that is located inside the shell.

Spherical Conductors

If excess charge is placed on a spherical shell that is made of conducting material, the excess charge spreads uniformly over the (external) surface. For example, if we place excess electrons on a spherical metal shell, those electrons repel one another and tend to move apart, spreading over the available surface until they are uniformly distributed. That arrangement maximizes the distances between all pairs of the excess electrons. According to the first shell theorem, the shell then will attract or repel an external charge as if all the excess charge on the shell were concentrated at its center.

If we remove negative charge from a spherical metal shell, the resulting positive charge of the shell is also spread uniformly over the surface of the shell. For example, if we remove n electrons, there are then n sites of positive charge (sites missing an electron) that are spread uniformly over the shell. According to the first shell theorem, the shell will again attract or repel an external charge as if all the shell's excess charge were concentrated at its center.

PROBLEM SOLVING TACTICS

TACTIC 1: *Symbols Representing Charge*

Here is a general guide to the symbols representing charge. If the symbol q, with or without a subscript, is used in a sentence when no electrical sign has been specified, the charge can be either positive or negative. Sometimes the sign is explicitly shown, as in the notation $+q$ or $-q$.

When more than one charged object is being considered, their charges might be given as multiples of a charge magnitude. As examples, the notation $+2q$ means a positive charge with magnitude twice that of some reference charge magnitude q, and $-3q$ means a negative charge with magnitude three times that of the reference charge magnitude q.

CHECKPOINT 2: The figure shows two protons (symbol p) and one electron (symbol e) on an axis. What are the directions of (a) the electrostatic force on the central proton due to the electron, (b) the electrostatic force on the central proton due to the other proton, and (c) the net electrostatic force on the central proton?

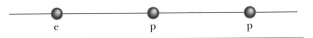

SAMPLE PROBLEM 22-1

Figure 22-7a shows two particles fixed in place: a particle of charge $q_1 = +8q$ at the origin of an x axis and a particle of charge $q_2 = -2q$ at $x = L$. At what point (other than infinitely far away) can a proton be placed so that it is in *equilibrium* (meaning that the net force on it is zero)? Is that equilibrium *stable* or *unstable*?

SOLUTION: If $\mathbf{F}_1$ is the force on the proton due to charge q_1 and $\mathbf{F}_2$ is the force on the proton due to charge q_2, then the point we seek is where $\mathbf{F}_1 + \mathbf{F}_2 = 0$, which requires that

$$\mathbf{F}_1 = -\mathbf{F}_2. \qquad (22\text{-}8)$$

This tell us that at the point we seek, the forces acting on the proton due to the other two particles must be of equal magnitudes,

$$F_1 = F_2, \qquad (22\text{-}9)$$

and that the forces must have opposite directions.

A proton has a positive charge. Thus the proton and the particle of charge q_1 are of the same sign, and force $\mathbf{F}_1$ on the proton must point away from q_1. Also the proton and the particle of charge q_2 are of opposite signs, so force $\mathbf{F}_2$ on the proton must point toward q_2. "Away from q_1" and "toward q_2" can be in opposite directions only if the proton is located on the x axis.

If the proton is on the x axis at any point between q_1 and q_2, such as P in Fig. 22-7b, then $\mathbf{F}_1$ and $\mathbf{F}_2$ are in the same direction and not in opposite directions as required. If the proton is at any point on the x axis to the left of q_1, such as point S in Fig. 22-7b, then $\mathbf{F}_1$ and $\mathbf{F}_2$ are in opposite directions. However, Eq. 22-4 tells us that $\mathbf{F}_1$ and $\mathbf{F}_2$ cannot have equal magnitudes there: F_1 must be larger than F_2, because F_1 is produced by a closer charge (with smaller r) of larger magnitude ($8q$ versus $2q$).

Finally, if the proton is at any point on the x axis to the right of q_2, such as point R, then $\mathbf{F}_1$ and $\mathbf{F}_2$ are again in opposite directions. However, because now the charge of larger magnitude (q_1) is *farther* away from the proton than the charge of smaller magnitude, there is a point at which F_1 is equal to

F_2. Let x be the coordinate of this point, and let q_p be the charge of the proton. Then with the aid of Eq. 22-4, we can rewrite Eq. 22-9 as

$$\frac{1}{4\pi\epsilon_0} \frac{8qq_p}{x^2} = \frac{1}{4\pi\epsilon_0} \frac{2qq_p}{(x-L)^2}. \qquad (22\text{-}10)$$

(Note that only the magnitudes of the charges appear in Eq. 22-10.) Rearranging Eq. 22-10 gives us

$$\left(\frac{x-L}{x}\right)^2 = \frac{1}{4}.$$

After taking the square roots of both sides, we have

$$\frac{x-L}{x} = \frac{1}{2},$$

which gives us

$$x = 2L. \qquad \text{(Answer)}$$

The equilibrium at $x = 2L$ is unstable. That is, if the proton is displaced leftward from point R, then F_1 and F_2 both increase but F_2 increases more (because q_2 is closer than q_1), and a net force will drive the proton farther leftward. And if the proton is displaced rightward, both F_1 and F_2 decrease but F_2 decreases more, and thus a net force will then drive the proton farther rightward. In a stable equilibrium, each time the proton was displaced slightly, it would return to the equilibrium position.

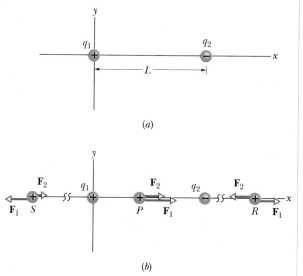

(a)

(b)

FIGURE 22-7 Sample Problem 22-1. (a) Two particles of charges q_1 and q_2 are fixed in place on an x axis, with separation L. (b) Three possible locations S, P, and R for a proton. At each location, the proton experiences electrostatic force $\mathbf{F}_1$ due to q_1 and electrostatic force $\mathbf{F}_2$ due to q_2.

SAMPLE PROBLEM 22-2

Figure 22-8a shows an arrangement of six fixed charged particles, where $a = 2.0$ cm and $\theta = 30°$. All six particles have the same magnitude of charge, $q = 3.0 \times 10^{-6}$ C; their electrical signs are as indicated. What is the net electrostatic force $\mathbf{F}_1$ acting on q_1 due to the other charges?

SOLUTION: From Eq. 22-7 we know that $\mathbf{F}_1$ is the vector sum of forces $\mathbf{F}_{12}$, $\mathbf{F}_{13}$, $\mathbf{F}_{14}$, $\mathbf{F}_{15}$, and $\mathbf{F}_{16}$, which are the electrostatic forces acting on q_1 due to the other charges. Because q_2 and q_4 are equal in magnitude and are both a distance $r = 2a$ from q_1, we have from Eq. 22-4

$$F_{12} = F_{14} = \frac{1}{4\pi\epsilon_0} \frac{|q_1||q_2|}{(2a)^2}. \qquad (22\text{-}11)$$

Similarly, since q_3, q_5, and q_6 are equal in magnitude and are each a distance $r = a$ from q_1, we have

$$F_{13} = F_{15} = F_{16} = \frac{1}{4\pi\epsilon_0} \frac{|q_1||q_3|}{a^2}. \qquad (22\text{-}12)$$

Figure 22-8b is a free-body diagram for q_1. It and Eq. 22-11 show that $\mathbf{F}_{12}$ and $\mathbf{F}_{14}$ are equal in magnitude but opposite in direction; thus those forces cancel. Inspection of Fig. 22-8b and Eq. 22-12 reveals that the y components of $\mathbf{F}_{13}$ and $\mathbf{F}_{15}$ also cancel, and that their x components are identical in magnitude and both point in the direction of decreasing x. Figure

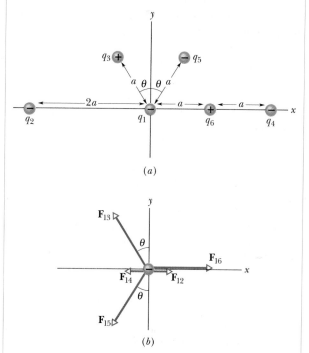

(a)

(b)

FIGURE 22-8 Sample Problem 22-2. (a) An arrangement of six charged particles. (b) The electrostatic forces acting on q_1 due to the other five charges.

22-8b also shows us that $\mathbf{F}_{16}$ points in the direction of increasing x. Thus $\mathbf{F}_1$ must be parallel to the x axis; its magnitude is the difference between F_{16} and twice the x component of $\mathbf{F}_{13}$:

$$F_1 = F_{16} - 2 F_{13} \sin \theta$$
$$= \frac{1}{4\pi\epsilon_0} \frac{|q_1||q_6|}{a^2} - \frac{2}{4\pi\epsilon_0} \frac{|q_1||q_3|}{a^2} \sin \theta.$$

Setting $q_3 = q_6$ and $\theta = 30°$, we find

$$F_1 = \frac{1}{4\pi\epsilon_0} \frac{|q_1||q_6|}{a^2} - \frac{2}{4\pi\epsilon_0} \frac{|q_1||q_6|}{a^2} \sin 30° = 0. \quad \text{(Answer)}$$

Note that the presence of q_6 along the line between q_1 and q_4 does not alter the electrostatic force exerted by q_4 on q_1.

PROBLEM SOLVING TACTICS

TACTIC 2: *Symmetry*

In Sample Problem 22-2 we used the symmetry of the situation to reduce the time and amount of calculation involved in the solution. By realizing that q_2 and q_4 are positioned symmetrically about q_1, and thus that $\mathbf{F}_{12}$ and $\mathbf{F}_{14}$ cancel, we avoided calculating either force. And by realizing that the y components of $\mathbf{F}_{13}$ and $\mathbf{F}_{15}$ cancel and that their x components are identical and add, we saved even more effort. In fact, by using symmetry and by setting up the solution in symbols, we never had to substitute the charge magnitude 3.0×10^{-6} C given in the problem.

TACTIC 3: *Drawing Electrostatic Force Vectors*

When you are given a diagram of charged particles, such as Fig. 22-8a, and are asked to find the net electrostatic force on one of them, you should usually draw a free-body diagram showing only the particle of concern and the forces *it* experiences, as in Fig. 22-8b. If, instead, you choose to superimpose those forces on the given diagram showing all the particles, be sure to draw the force vectors with either their tails (preferably) or their heads on the particle of concern. If you draw the vectors elsewhere in the diagram, you invite confusion. And confusion is guaranteed if you draw the vectors on the particles *causing* the forces on the particle of concern.

$\mathbf{C}$HECKPOINT **3:** The figure shows three arrangements of an electron e and two protons p. (a) Rank the arrangements according to the magnitude of the net electrostatic force on the electron due to the protons, largest first. (b) In situation *c*, is the angle between the

net force on the electron and the line labeled d less than or more than 45°?

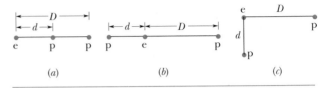

(a)   (b)   (c)

SAMPLE PROBLEM 22-3

In Fig. 22-9a, two identical, electrically isolated conducting spheres A and B are separated by a (center-to-center) distance a that is large compared to the spheres. Sphere A has a positive charge of $+Q$; sphere B is electrically neutral; and initially, there is no electrostatic force between the spheres.

(a) Suppose the spheres are connected for a moment by a conducting wire. The wire is thin enough so that any net charge on it is negligible. What is the electrostatic force between the spheres after the wire is removed?

SOLUTION: When the spheres are wired together, conduction electrons of sphere B are attracted to positively charged sphere A (Fig. 22-9b). As sphere B loses negative charge, it becomes positively charged. And as A gains negative charge, it becomes *less* positively charged. The spheres must end up with the same charge because they are identical. Thus the transfer of charge stops when the excess charge on B has increased to $+Q/2$ and the excess charge on A has decreased to $+Q/2$ (Fig. 22-9c). This condition occurs when a charge of $-Q/2$ has been transferred.

After the wire has been removed, we can assume that the charge on either sphere does not disturb the uniformity of the charge distribution on the other sphere, because the spheres are small relative to their separation. Thus we can apply the first shell theorem to each sphere. By Eq. 22-4 with $q_1 = q_2 = Q/2$ and $r = a$, the electrostatic force between the spheres has a magnitude of

$$F = \frac{1}{4\pi\epsilon_0} \frac{(Q/2)(Q/2)}{a^2} = \frac{1}{16\pi\epsilon_0} \left(\frac{Q}{a}\right)^2. \quad \text{(Answer)}$$

Since both spheres are now positively charged, they repel each other.

(b) Next, suppose sphere A is grounded momentarily, and then the ground connection is removed. What now is the electrostatic force between the spheres?

SOLUTION: The ground connection allows electrons, with a total charge of $-Q/2$, to move from the ground to sphere A (Fig. 22-9d), neutralizing that sphere (Fig. 22-9e). With no charge on sphere A, there is no electrostatic force between the two spheres (just as initially, in Fig. 22-9a).

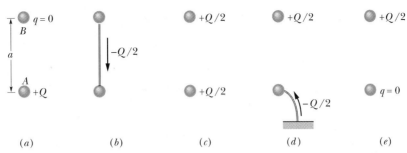

FIGURE 22-9 Sample Problem 22-3. Two small conducting spheres A and B. (a) To start, sphere A is charged positively. (b) Negative charge is transferred between the spheres through a connecting wire. (c) Both spheres are then charged positively. (d) Negative charge is transferred through a grounding wire to sphere A. (e) Sphere A is then neutral.

22-5 CHARGE IS QUANTIZED

In Benjamin Franklin's day, electric charge was thought to be a continuous fluid—an idea that was useful for many purposes. However, we now know that fluids themselves, such as air and water, are not continuous but are made up of atoms and molecules; matter is discrete. Experiment shows that "electrical fluid" is also not continuous but is made up of multiples of a certain elementary charge. That is, any positive or negative charge q that can be detected can be written as

$$q = ne, \qquad n = \pm 1, \pm 2, \pm 3, \cdots, \quad (22\text{-}13)$$

in which e, the **elementary charge,** has the value

$$e = 1.60 \times 10^{-19} \text{ C.} \quad (22\text{-}14)$$

The elementary charge e is one of the important constants of nature. The electron and proton both have a charge of magnitude e (Table 22-1). (Quarks, the constituent particles of protons and neutrons, have charges of $\pm e/3$ or $\pm 2e/3$, but they apparently cannot be detected individually. Hence, we do not take their charges to be the elementary charge.)

You often see phrases—such as "the charge on a sphere," "the amount of charge transferred," and "the charge carried by the electron"—that suggest that charge

is a substance. (Indeed, such statements have already appeared in this chapter.) You should, however, keep in mind what is intended: *particles* are the substance and charge happens to be one of their properties, just as mass is.

When a physical quantity such as charge can have only discrete values rather than any value, we say that the quantity is **quantized**. We have already seen that matter, energy, and angular momentum are quantized; charge adds one more important physical quantity to the list. It is possible, for example, to find a particle that has no charge at all or a charge of $+10e$ or $-6e$, but not a particle with a charge of, say, 3.57e.

The quantum of charge is small. In an ordinary 100 W lightbulb, for example, about 10^{19} elementary charges enter the bulb every second and just as many leave. However, the graininess of electricity does not show up in such large-scale phenomena, just as you cannot feel the individual molecules of water with your hand.

The graininess of electricity is responsible for the blue glow that is emitted by a wintergreen LifeSaver while it is being crushed. When the sugar (sucrose) crystals in the candy rupture, one part of each ruptured crystal has excess electrons while the other part has excess positive ions. Almost immediately, electrons and ions jump across the gap of the rupture to neutralize the two sides. During the jumps, the electrons and positive ions collide with nitrogen molecules in the air that is then flowing into the gap.

The collisions cause the nitrogen to emit ultraviolet light that you cannot see, as well as blue light (from the visible region of the spectrum) that is, however, too dim to see. Oil of wintergreen in the crystals absorbs the ultraviolet light and immediately emits enough blue light to light up a mouth or a pair of pliers. However, if the candy is wet with saliva, the demonstration fails, because the conducting saliva neutralizes the two parts of a fractured crystal before sparking can occur.

TABLE 22-1
THE CHARGES OF THREE PARTICLES

PARTICLE	SYMBOL	CHARGE
Electron	e or e$^-$	$-e$
Proton	p	$+e$
Neutron	n	0

CHECKPOINT 4: Initially, sphere A has a charge of $-50e$ and sphere B has a charge of $+20e$. The spheres are made of conducting material and are identical in size. If the spheres then touch, what is the resulting charge on sphere A?

This is about 2×10^{12} tons! Even if the charges were separated by one Earth diameter, the attractive force would still be huge, about 120 tons. Actually, it is impossible to disturb the electrical neutrality of ordinary matter very much. If we try to remove any sizable fraction of the charge of one electrical sign from a body, a large electrostatic force appears automatically, tending to pull it back.

SAMPLE PROBLEM 22-4

An electrically neutral penny, of mass $m = 3.11$ g, contains equal amounts of positive and negative charge.

(a) Assuming that the penny is made entirely of copper, what is the magnitude q of the total positive (or negative) charge in the coin?

SOLUTION: A neutral atom has a negative charge of magnitude Ze associated with its electrons and a positive charge of the same magnitude associated with the protons in its nucleus, where Z is the *atomic number* of the element in question. For copper, Appendix F tells us that Z is 29, which means that an atom of copper has 29 protons and, when electrically neutral, 29 electrons.

The charge magnitude q we seek is equal to NZe, in which N is the number of atoms in the penny. To find N, we multiply the number of moles of copper in the penny by the number of atoms in a mole (Avogadro's number, $N_A = 6.02 \times 10^{23}$ atoms/mol). The number of moles of copper in the penny is m/M, where M is the molar mass of copper, 63.5 g/mol (from Appendix F). Thus we have

$$N = N_A \frac{m}{M} = 6.02 \times 10^{23} \text{ atoms/mol} \frac{3.11 \text{ g}}{63.5 \text{ g/mol}}$$

$$= 2.95 \times 10^{22} \text{ atoms.}$$

We then find the magnitude of the total positive or negative charge in the penny to be

$$q = NZe$$
$$= (2.95 \times 10^{22})(29)(1.60 \times 10^{-19} \text{ C})$$
$$= 137,000 \text{ C.} \qquad \text{(Answer)}$$

This is an enormous charge. (For comparison, if you rub a plastic rod with fur, you will be lucky to deposit any more than 10^{-9} C on the rod.)

(b) Suppose that the positive charge and the negative charge in a penny could be concentrated into two separate bundles, 100 m apart. What attractive force would act on each bundle?

SOLUTION: From Eq. 22-4 we have

$$F = \frac{1}{4\pi\epsilon_0} \frac{q^2}{r^2}$$

$$= \frac{(8.99 \times 10^9 \text{ N} \cdot \text{m}^2/\text{C}^2)(1.37 \times 10^5 \text{ C})^2}{(100 \text{ m})^2}$$

$$= 1.69 \times 10^{16} \text{ N.} \qquad \text{(Answer)}$$

SAMPLE PROBLEM 22-5

The nucleus in an iron atom has a radius of about 4.0×10^{-15} m and contains 26 protons.

(a) What is the magnitude of the repulsive electrostatic force between two of these protons that happen to be separated by 4.0×10^{-15} m?

SOLUTION: From Eq. 22-4 and Table 22-1 we can write

$$F = \frac{1}{4\pi\epsilon_0} \frac{e^2}{r^2}$$

$$= \frac{(8.99 \times 10^9 \text{ N} \cdot \text{m}^2/\text{C}^2)(1.60 \times 10^{-19} \text{ C})^2}{(4.0 \times 10^{-15} \text{ m})^2}$$

$$= 14 \text{ N.} \qquad \text{(Answer)}$$

This is a small force to be acting on a macroscopic object like a cantaloupe, but an enormous force to be acting on a proton. Such forces should blow apart the nucleus of any element but hydrogen (which has only one proton in its nucleus). But they don't, not even in nuclei with a great many protons. So there must be some attractive nuclear force to counter this enormous repulsive electrostatic force.

(b) What is the magnitude of the gravitational force between those same two protons?

SOLUTION: With m_p ($= 1.67 \times 10^{-27}$ kg) representing the mass of a proton, we write Eq. 22-2 for the gravitational force, finding

$$F = G \frac{m_p^2}{r^2}$$

$$= \frac{(6.67 \times 10^{-11} \text{ N} \cdot \text{m}^2/\text{kg}^2)(1.67 \times 10^{-27} \text{ kg})^2}{(4.0 \times 10^{-15} \text{ m})^2}$$

$$= 1.2 \times 10^{-35} \text{ N.} \qquad \text{(Answer)}$$

This result tells us that the (attractive) gravitational force is far too weak to counter the repulsive electrostatic forces between protons in a nucleus. Instead, the protons are bound together by an enormous force called (aptly) the *strong nuclear force* — a force that acts between protons (and neutrons) when they are close together, as in a nucleus.

Although the gravitational force is many, many times weaker than the electrostatic force, it is more important in large-scale situations because it is always attractive. This means that it can collect many small bodies into huge masses, such as planets and stars, that then exert large gravitational forces. The electrostatic force, on the other hand, is repulsive for charges of the same sign, so it is unable to collect either positive charge or negative charge into large concentrations that would then exert large electrostatic forces.

22-6 CHARGE IS CONSERVED

If you rub a glass rod with silk, a positive charge appears on the rod. Measurement shows that a negative charge of equal magnitude appears on the silk. This suggests that rubbing does not create charge but only transfers it from one body to another, upsetting the electrical neutrality of each body during the process. This hypothesis of **conservation of charge,** first put forward by Benjamin Franklin, has stood up under close examination, both for large-scale charged bodies and for atoms, nuclei, and elementary particles. No exceptions have ever been found. Thus we add electric charge to our list of quantities —including energy and both linear and angular momentum—that obey a conservation law.

Radioactive decay of nuclei, in which a nucleus spontaneously transforms into a different type of nucleus, gives us many instances of charge conservation at the nuclear level. For example, uranium-238, or ^{238}U, which is found in common uranium ore, can decay by emitting an alpha particle (which is a helium nucleus, ^{4}He) and transforming to thorium, ^{234}Th:

$$^{238}\text{U} \rightarrow {}^{234}\text{Th} + {}^{4}\text{He} \qquad \begin{matrix}\text{(radioactive} \\ \text{decay).}\end{matrix} \qquad (22\text{-}15)$$

The atomic number Z of the radioactive *parent* nucleus

^{238}U is 92, which tells us that this nucleus contains 92 protons and has a charge of $92e$. The emitted alpha particle has $Z = 2$, and the *daughter* nucleus ^{234}Th has $Z = 90$. Thus the amount of charge present before the decay, $92e$, is equal to the total amount present after the decay, $90e + 2e$. Charge is conserved.

Another example of charge conservation occurs when an electron e$^-$ (whose charge is $-e$) and its antiparticle, the *positron* e$^+$ (whose charge is $+e$), undergo an *annihilation process* in which they transform into two *gamma rays* (high-energy, chargeless particles of light):

$$\text{e}^- + \text{e}^+ \rightarrow \gamma + \gamma \qquad \text{(annihilation).} \quad (22\text{-}16)$$

In applying the conservation-of-charge principle, we must add the charges algebraically, with due regard for their signs. In the annihilation process of Eq. 22-16 then, the net charge of the system is zero both before and after the event. Charge is conserved.

In *pair production,* the converse of annihilation, charge is also conserved. In this process a gamma ray transforms into an electron and a positron:

$$\gamma \rightarrow \text{e}^- + \text{e}^+ \qquad \text{(pair production).} \quad (22\text{-}17)$$

Figure 22-10 shows such a pair-production event that occurred in a bubble chamber. A gamma ray entered the chamber directly from the left and at one point transformed into an electron and a positron. Because those new particles were charged and moving, each left a trail of tiny bubbles. (The trails were curved because a magnetic field had been set up in the chamber.) The gamma ray, being chargeless, left no trail. Still, you can tell exactly where it underwent pair production—at the tip of the curved V, where the trails of the electron and positron begin.

FIGURE 22-10 A photograph of trails of bubbles left in a bubble chamber by an electron and a positron. The pair of particles was produced by a gamma ray that entered the chamber directly from the left. Being chargeless, the gamma ray did not generate a telltale trail of bubbles along its path, as the electron and positron did.

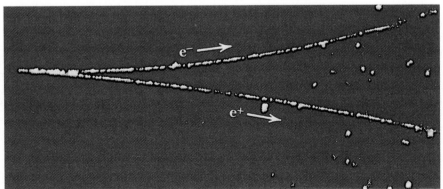

REVIEW & SUMMARY

Electric Charge

The strength of a particle's electric interaction with objects around it depends on its **electric charge,** which can be either positive or negative. Charges with the same sign repel each other and charges with opposite signs attract each other. An object with equal amounts of the two kinds of charge is electrically neutral, whereas one with an imbalance is electrically charged.

Conductors are materials in which a significant number of charged particles (electrons in metals) are free to move. The charged particles in **nonconductors,** or **insulators,** are not free to move. When charge moves through a material, we say that an **electric current** exists in the material.

The Coulomb and Ampere

The SI unit of charge is the **coulomb** (C). It is defined in terms of the unit of current, the ampere (A), as the charge passing a particular point in 1 second when there is a current of 1 ampere at that point.

Coulomb's Law

Coulomb's law describes the **electrostatic force** between small (point) electric charges q_1 and q_2 at rest (or nearly at rest) and separated by a distance r:

$$F = \frac{1}{4\pi\epsilon_0}\frac{|q_1||q_2|}{r^2} \qquad \text{(Coulomb's law).} \qquad (22\text{-}4)$$

Here $\epsilon_0 = 8.85 \times 10^{-12}$ C^2/N·m^2 is the **permittivity constant;** $1/4\pi\epsilon_0 = 8.99 \times 10^9$ N·m^2/C^2.

The force of attraction or repulsion between point charges at rest acts along the line joining the two charges. If more than two charges are present, Eq. 22-4 holds for each pair of charges. The net force on each charge is then found, using the superposition principle, as the vector sum of the forces exerted on the charge by each of the others.

The two shell theorems for electrostatics are

A shell of uniform charge attracts or repels a charged particle that is outside the shell as if all the shell's charge were concentrated at its center.

A shell of uniform charge exerts no electrostatic force on a charged particle that is located inside the shell.

The Elementary Charge

Electric charge is **quantized:** any charge can be written as ne, where n is a positive or negative integer, and e is a constant of nature called the **elementary charge** (approximately 1.60×10^{-19} C). Electric charge is conserved: the (algebraic) net charge of any isolated system cannot change.

QUESTIONS

1. Does Coulomb's law hold for all charged objects?

2. A particle of charge q is to be placed, in turn, outside four metal objects, each of uniform charge Q: (1) a large solid sphere, (2) a large spherical shell, (3) a small solid sphere, and (4) a small spherical shell. The distance between the particle and the center of the object is the same in all four cases, and q is small enough not to alter significantly the uniform distribution of Q. Rank the objects according to the electrostatic force they exert on the particle, greatest first.

3. Figure 22-11 shows four situations in which charged particles are fixed in place on an axis. In which situations is there a point to the left of the particles where an electron will be in equilibrium?

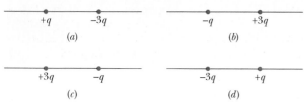

FIGURE 22-11   Question 3.

4. Figure 22-12 shows two charged particles on an axis. The charges are free to move. At one point, however, a third charged particle can be placed such that all three particles are in equilibrium. (a) Is that point to the left of the first two particles, to their right, or between them? (b) Should the third particle be positively or negatively charged? (c) Is the equilibrium stable or unstable?

FIGURE 22-12   Question 4.

5. In the figure for Checkpoint 2, two protons and one electron are fixed in place on an axis. Where on the axis could a fourth charged particle be placed so that the net electrostatic force on it due to the first three particles is zero: to the left of the first three particles; to their right; between the protons; or between the electron and the proton closer to it?

6. In Fig. 22-13, a central particle of charge $-q$ is surrounded by two circular rings of charged particles, of radii r and R, with $R > r$. What are the magnitude and direction of the net electrostatic force on the central particle due to the other particles?

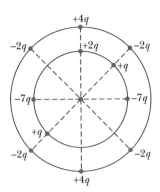

FIGURE 22-13 Question 6.

7. In Fig. 22-14, a central particle of charge $-2q$ is surrounded by a square array of charged particles, separated by either distance d or $d/2$ along the perimeter of the square. What are the magnitude and direction of the net electrostatic force on the central particle due to the other particles?

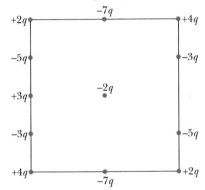

FIGURE 22-14 Question 7.

8. Figure 22-15 shows four arrangements of charged particles. Rank the arrangements according to the magnitude of the net electrostatic force on the particle with charge $+Q$, greatest first.

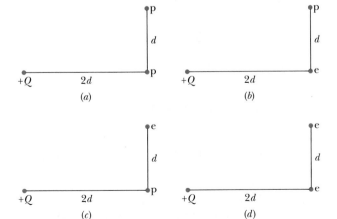

FIGURE 22-15 Question 8.

9. Figure 22-16 shows four situations in which particles of charge $+q$ or $-q$ are fixed in place. In each, the particles on the x axis are equidistant from the y axis. First, consider the middle particle in situation 1; the middle particle experiences an electrostatic force from each of the other two particles. (a) Are the magnitudes F of those forces the same or different? (b) Is the magnitude of the net force on the middle particle equal to, greater than, or less than $2F$? (c) Do the x components of the two forces add or cancel? (d) Do their y components add or cancel? (e) Is the direction of the net force on the middle particle that of the canceling components or the adding components? (f) What is the direction of that net force? Now consider the remaining situations: What is the direction of the net force on the middle particle in (g) situation 2, (h) situation 3, and (i) situation 4?

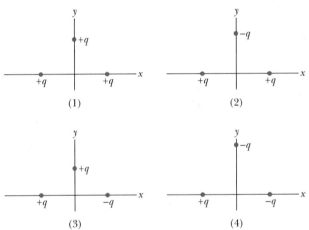

FIGURE 22-16 Question 9.

10. Figure 22-17 shows a pair of particles of charge Q and another pair of particles of charge q. The particle at the origin is free to move; the others are fixed in place. Should q be positive or negative if the net force on the free particle is to be zero and Q is (a) positive and (b) negative?

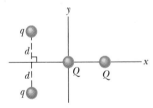

FIGURE 22-17 Question 10.

11. Four identical conducting spheres A, B, C, and D have charges of $-8.0Q$, $-6.0Q$, $-4.0Q$, and $8.0Q$, respectively. Which should be connected together (by thin wire) to produce two or more spheres with charges of (a) $-2.0Q$ and (b) $-2.5Q$? (c) What sequence of connections will produce two spheres with charges of $-3.0Q$?

12. A positively charged ball is brought close to a neutral isolated conductor. The conductor is then grounded while the ball is kept

close. Is the conductor charged positively or negatively, or is it neutral, if (a) the ball is first taken away and then the ground connection is removed and (b) the ground connection is first removed and then the ball is taken away?

13. (a) A positively charged glass rod attracts an object suspended by a nonconducting thread. Is the object definitely negatively charged or only possibly negatively charged? (b) A positively charged glass rod repels a similarly suspended object. Is the object definitely positively charged or only possibly?

14. You are given two identical neutral metal spheres A and B mounted on portable insulating supports, as well as a thin conducting wire and a glass rod that you can rub with silk. You can attach the wire between the spheres or between a sphere and the ground. You cannot touch the rod to a sphere. How can you give the spheres charges of (a) equal magnitudes and the same signs and (b) equal magnitudes and opposite signs?

15. In a simple model of a helium atom, two electrons orbit a nucleus consisting of two protons. Is the magnitude of the force exerted on the nucleus by one of the electrons greater than, less than, or the same as the magnitude of the force exerted on that electron by the nucleus?

16. In Fig. 22-5, the nearby (negatively charged) plastic rod causes some of the conduction electrons in the copper rod to move to the far end of the copper rod. Why does the flow of the conduction electrons quickly cease? After all, a huge number of them are free to move to that far end.

17. Figure 22-18 shows three small spheres that have charges of equal magnitudes and rest on a frictionless surface. Spheres y and z are fixed in place and are equally distant from sphere x. If sphere x is released from rest, which of the five paths shown will it take?

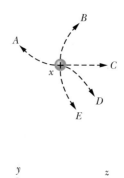

FIGURE 22-18 Question 17.

18. A person standing on an electrically insulated platform touches a charged, electrically isolated conductor. Does this discharge the conductor completely?

EXERCISES & PROBLEMS

SECTION 22-4 Coulomb's Law

1E. In the return stroke of a typical lightning bolt, a current of 2.5×10^4 A exists for 20 μs. How much charge is transferred in this event?

2E. What would be the electrostatic force between two 1.00 C charges separated by a distance of (a) 1.00 m and (b) 1.00 km if such a configuration could be set up?

3E. A point charge of $+3.00 \times 10^{-6}$ C is 12.0 cm distant from a second point charge of -1.50×10^{-6} C. Calculate the magnitude of the force on each charge.

4E. What must be the distance between point charge $q_1 = 26.0$ μC and point charge $q_2 = -47.0$ μC for the electrostatic force between them to have a magnitude of 5.70 N?

5E. Two equally charged particles, held 3.2×10^{-3} m apart, are released from rest. The initial acceleration of the first particle is observed to be 7.0 m/s^2 and that of the second to be 9.0 m/s^2. If the mass of the first particle is 6.3×10^{-7} kg, what are (a) the mass of the second particle and (b) the magnitude of the charge of each particle?

6E. In Figure 22-19, three identical conducting spheres A, B, and C form an equilateral triangle of side length d and have initial charges of $-2Q$, $-4Q$, and $8Q$, respectively. (a) What is the magnitude of the electrostatic force between spheres A and C?

The following steps are then taken: A and B are connected by a thin wire and then disconnected; B is grounded by the wire and the wire is then removed; B and C are connected by the wire and then disconnected. What now are the magnitudes of the electrostatic force (b) between spheres A and C and (c) between spheres B and C?

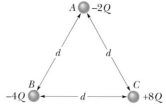

FIGURE 22-19 Exercise 6.

7E. Identical isolated conducting spheres 1 and 2 have equal amounts of charge and are separated by a distance large compared with their diameters (Fig. 22-20a). The electrostatic force acting on sphere 2 due to sphere 1 is $\mathbf{F}$. Suppose now that a third identical sphere 3, having an insulating handle and initially neutral, is touched first to sphere 1 (Fig. 22-20b), then to sphere 2 (Fig. 22-20c), and finally removed (Fig. 22-20d). In terms of $\mathbf{F}$, what is the electrostatic force $\mathbf{F}'$ that now acts on sphere 2?

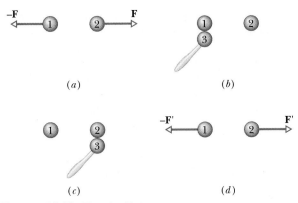

(a)   (b)

(c)   (d)

FIGURE 22-20   Exercise 7.

8P. In Fig. 22-21, three charged particles lie on a straight line and are separated by a distance d. Charges q_1 and q_2 are held fixed. Charge q_3 is free to move but happens to be in equilibrium (no net electrostatic force acts on it). Find q_1 in terms of q_2.

FIGURE 22-21   Problem 8.

9P. Figure 22-22a shows two charges, q_1 and q_2, held a fixed distance d apart. (a) What is the magnitude of the electrostatic force that acts on q_1? Assume that $q_1 = q_2 = 20.0\ \mu C$ and $d = 1.50$ m. (b) A third charge $q_3 = 20.0\ \mu C$ is brought in and placed as shown in Fig. 22-22b. What now is the magnitude of the electrostatic force on q_1?

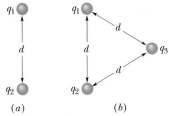

(a)   (b)

FIGURE 22-22   Problem 9.

10P. In Fig. 22-23, what are the horizontal and vertical components of the resultant electrostatic force on the charge in the lower left corner of the square if $q = 1.0 \times 10^{-7}$ C and $a = 5.0$ cm?

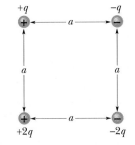

FIGURE 22-23
Problem 10.

11P. Charges q_1 and q_2 lie on the x axis at points $x = -a$ and $x = +a$, respectively. (a) How must q_1 and q_2 be related for the net electrostatic force on charge $+Q$, placed at $x = +a/2$, to

be zero? (b) Repeat (a) but with the $+Q$ charge now placed at $x = +3a/2$.

12P. Two small, positively charged spheres have a combined charge of 5.0×10^{-5} C. If each sphere is repelled from the other by an electrostatic force of 1.0 N when the spheres are 2.0 m apart, what is the charge on each sphere?

13P. Two identical conducting spheres, fixed in place, attract each other with an electrostatic force of 0.108 N when separated by 50.0 cm. The spheres are then connected by a thin conducting wire. When the wire is removed, the spheres repel each other with an electrostatic force of 0.0360 N. What were the initial charges on the spheres?

14P. Two fixed particles, of charges $q_1 = +1.0\ \mu C$ and $q_2 = -3.0\ \mu C$, are 10 cm apart. How far from each should a third charge be located so that no net electrostatic force acts on it?

15P. The charges and coordinates of two charged particles held fixed in the xy plane are: $q_1 = +3.0\ \mu C$, $x_1 = 3.5$ cm, $y_1 = 0.50$ cm, and $q_2 = -4.0\ \mu C$, $x_2 = -2.0$ cm, $y_2 = 1.5$ cm. (a) Find the magnitude and direction of the electrostatic force on q_2. (b) Where could you locate a third charge $q_3 = +4.0\ \mu C$ such that the net electrostatic force on q_2 is zero?

16P. Two *free* point charges $+q$ and $+4q$ are a distance L apart. A third charge is placed so that the entire system is in equilibrium. (a) Find the location, magnitude, and sign of the third charge. (b) Show that the equilibrium of the system is unstable.

17P. (a) What equal positive charges would have to be placed on Earth and on the Moon to neutralize their gravitational attraction? Do you need to know the lunar distance to solve this problem? Why or why not? (b) How many thousand kilograms of hydrogen would be needed to provide the positive charge calculated in (a)?

18P. A certain charge Q is divided into two parts q and $Q - q$, which are then separated by a certain distance. What must q be in terms of Q to maximize the electrostatic repulsion between the two charges?

19P. A charge Q is fixed at each of two opposite corners of a square. A charge q is placed at each of the other two corners. (a) If the net electrostatic force on each Q is zero, what is Q in terms of q? (b) Is there any value of q that makes the net electrostatic force on each of the four charges zero? Explain.

20P. In Fig. 22-24, two tiny conducting balls of identical mass m and identical charge q hang from nonconducting threads of length L. Assume that θ is so small that $\tan\theta$ can be replaced by its

FIGURE 22-24
Problem 20.

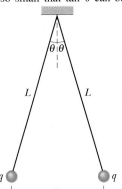

approximate equal, sin θ. (a) Show that, for equilibrium,

$$x = \left(\frac{q^2 L}{2\pi\epsilon_0 mg} \right)^{1/3},$$

where x is the separation between the balls. (b) If $L = 120$ cm, $m = 10$ g, and $x = 5.0$ cm, what is q?

21P. Explain what happens to the balls of Problem 20b if one of them is discharged, and find the new equilibrium separation x, using the given values of L and m and the computed value of q.

22P. Figure 22-25 shows a long, nonconducting, massless rod of length L, pivoted at its center and balanced with a weight W at a distance x from the left end. At the left and right ends of the rod are attached small conducting spheres with positive charges q and $2q$, respectively. A distance h directly beneath each of these spheres is a fixed sphere with positive charge Q. (a) Find the distance x when the rod is horizontal and balanced. (b) What value should h have so that the rod exerts no vertical force on the bearing when the rod is horizontal and balanced?

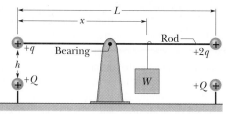

FIGURE 22-25 Problem 22.

SECTION 22-5 Charge Is Quantized

23E. What is the magnitude of the electrostatic force between a singly charged sodium ion (Na^+, of charge $+e$) and an adjacent singly charged chlorine ion (Cl^-, of charge $-e$) in a salt crystal if their separation is 2.82×10^{-10} m?

24E. A neutron consists of one "up" quark of charge $+2e/3$ and two "down" quarks each having charge $-e/3$. If the down quarks are 2.6×10^{-15} m apart inside the neutron, what is the magnitude of the electrostatic force between them?

25E. What is the total charge in coulombs of 75.0 kg of electrons?

26E. How many megacoulombs of positive (or negative) charge are in 1.00 mol of neutral molecular-hydrogen gas (H_2)?

27E. The magnitude of the electrostatic force between two identical ions that are separated by a distance of 5.0×10^{-10} m is 3.7×10^{-9} N. (a) What is the charge of each ion? (b) How many electrons are "missing" from each ion (thus giving the ion its charge imbalance)?

28E. (a) How many electrons would have to be removed from a penny to leave it with a charge of $+1.0 \times 10^{-7}$ C? (b) To what fraction of the electrons in the penny does this correspond? (See Sample Problem 22-4.)

29E. Two tiny, spherical water drops, with identical charges of -1.00×10^{-16} C, have a center-to-center separation of 1.00 cm.

(a) What is the magnitude of the electrostatic force acting between them? (b) How many excess electrons are on each drop, giving it its charge imbalance?

30E. How far apart must two protons be if the magnitude of the electrostatic force acting on either one is equal to the proton's weight at Earth's surface?

31E. An electron is in a vacuum near the surface of Earth. Where should a second electron be placed so that the electrostatic force it exerts on the first electron balances the weight of the first electron?

32P. Earth's atmosphere is constantly bombarded by *cosmic ray protons* that originate somewhere in space. If the protons were all to pass through the atmosphere, each square meter of Earth's surface would intercept protons at the average rate of 1500 protons per second. What would be the corresponding current intercepted by the total surface area of the planet?

33P. A 100 W lamp operated on a 120 V circuit has a current (assumed steady) of 0.83 A in its filament. How long does it take for 1 mol of electrons to pass through the lamp?

34P. Calculate the number of coulombs of positive charge in 250 cm³ of (neutral) water (about a glass full).

35P. In the basic CsCl (cesium chloride) crystal structure, Cs^+ ions form the corners of a cube and a Cl^- ion is at the cube's center (Fig. 22-26). The edge length of the cube is 0.40 nm. The Cs^+ ions are each deficient by one electron (and thus each has a charge of $+e$), and the Cl^- ion has one excess electron (and thus has a charge of $-e$). (a) What is the magnitude of the net electrostatic force exerted on the Cl^- ion by the eight Cs^+ ions at the corners of the cube? (b) If one of the Cs^+ ions is missing, the crystal is said to have a *defect*; what is the magnitude of the net electrostatic force exerted on the Cl^- ion by the seven remaining Cs^+ ions?

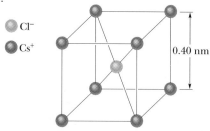

FIGURE 22-26 Problem 35.

36P. We know that, within the limits of measurement, the magnitudes of the negative charge on the electron and the positive charge on the proton are equal. Suppose, however, that these magnitudes differ from each other by 0.00010%. With what force would two copper pennies, placed 1.0 m apart, repel each other? What do you conclude? (*Hint:* See Sample Problem 22-4.)

37P. Two engineering students, John with a weight of 200 lb and Mary with a weight of 100 lb, are 100 ft apart. Suppose each has a 0.01% imbalance in the amount of positive and negative charge, one student being positive and the other negative. Estimate *roughly* the electrostatic force of attraction between them by replacing each student with a sphere of water having the same mass as the student.

SECTION 22-6   Charge Is Conserved

38E. In *beta decay* a massive fundamental particle changes to another massive particle, and either an electron or a positron is emitted. (a) If a proton undergoes beta decay to become a neutron, which particle is emitted? (b) If a neutron undergoes beta decay to become a proton, which particle is emitted?

39E. Using Appendix F, identify X in the following nuclear reactions:

(a) $^1\text{H} + {}^9\text{Be} \rightarrow \text{X} + \text{n}$;

(b) $^{12}\text{C} + {}^1\text{H} \rightarrow \text{X}$;

(c) $^{15}\text{N} + {}^1\text{H} \rightarrow {}^4\text{He} + \text{X}$.

40E. In the radioactive decay of ^{238}U (see Eq. 22-15), the center of the emerging ^4He particle is, at a certain instant, 9.0×10^{-15} m from the center of the daughter nucleus ^{234}Th. At this instant, (a) what is the magnitude of the electrostatic force on the ^4He particle, and (b) what is that particle's acceleration?

ELECTRONIC COMPUTATION

41. In Problem 18, let $q = \alpha Q$. (a) Write an expression for the magnitude F of the force between the charges in terms of α, Q, and the charge separation d. (b) Graph F as a function of α. Graphically find the values of α that give (c) the maximum value of F and (d) half the maximum value of F.

42. Two particles, each of positive charge q, are fixed in place on an x axis, one at $x = 0$ and the other at $x = d$. A particle of charge Q is to be placed along that axis at locations given by $x = \alpha d$. (a) Write expressions, in terms of α, that give the net electrostatic force $\mathbf{F}$ acting on the third particle when it is in the three regions $x < 0$, $0 < x < d$, and $d < x$. The expressions should give a positive result when $\mathbf{F}$ is in the positive direction of the x axis and a negative result when $\mathbf{F}$ is in the negative direction. (b) Graph $\mathbf{F}$ versus α for the range $-2 < \alpha < 3$.

23
Electric Fields

Water heats so well in a microwave oven that you might be able to heat a cup of water as much as 8 C° above the normal boiling temperature of water <u>without causing it to boil</u>. If you then pour coffee powder, or even chips of ice, into the water, it will erupt into a furious boil like that in the photograph, scattering water that could quickly scald you. Why do microwaves heat water❓

23-1 CHARGES AND FORCES: A CLOSER LOOK

Suppose we fix a positively charged particle q_1 in place and then put a second positively charged particle q_2 near it. From Coulomb's law we know that q_1 exerts a repulsive electrostatic force on q_2 and, given enough data, we could determine the magnitude and direction of that force. Still, a nagging question remains: How does q_1 "know" of the presence of q_2? That is, since the charges do not touch, how can q_1 exert a force on q_2?

This question about *action at a distance* can be answered by saying that q_1 sets up an **electric field** in the space surrounding it. At any given point P in that space, the field has both magnitude and direction. The magnitude depends on the magnitude of q_1 and the distance between P and q_1. The direction depends on the direction from q_1 to P and the electrical sign of q_1. Thus when we place q_2 at P, q_1 interacts with q_2 through the electric field at P. The magnitude and direction of that electric field determine the magnitude and direction of the force acting on q_2.

Another action-at-a-distance problem arises if we move q_1, say, toward q_2. Coulomb's law tells us that when q_1 is closer to q_2, the repulsive electrostatic force acting on q_2 must be greater. And it is. But here the nagging question is: Does the electric field at q_2, and thus the force acting on q_2, change immediately?

The answer is no. Instead, the information about the move by q_1 travels outward from q_1 (in all directions) as an electromagnetic wave at the speed of light c. The change in the electric field at q_2, and thus the change in the force acting on q_2, occurs when the wave finally reaches q_2.

23-2 THE ELECTRIC FIELD

The temperature at every point in a room has a definite value. You can measure the temperature at any given point or combination of points by putting a thermometer there. We call the resulting distribution of temperatures a *temperature field*. In much the same way, you can imagine a *pressure field* in the atmosphere: it consists of the distribution of air pressure values, one for each point in the atmosphere. These two examples are of *scalar fields,* because temperature and air pressure are scalar quantities.

The electric field is a *vector field*: it consists of a distribution of *vectors*, one for each point in the region around a charged object, such as a charged rod. In principle, we can define the electric field at some point near the charged object, such as point P in Fig. 23-1a, by placing a *positive* charge q_0, called a *test charge*, at the point. We then measure the electrostatic force F that acts on the test charge.

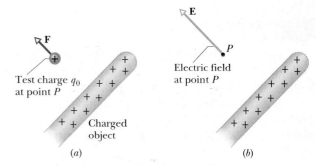

FIGURE 23-1 (a) A positive test charge q_0 placed at point P near a charged object. An electrostatic force **F** acts on the test charge. (b) The electric field **E** at point P produced by the charged object.

The electric field **E** at point P due to the charged object is defined as

$$\mathbf{E} = \frac{\mathbf{F}}{q_0} \qquad \text{(electric field).} \qquad (23\text{-}1)$$

Thus the magnitude of the electric field **E** at point P is $E = F/q_0$, and the direction of **E** is that of the force **F** that acts on the *positive* test charge. As shown in Fig. 23-1b, we represent the electric field at P with a vector whose tail is at P. To define the electric field within some region, we must similarly measure it at all points in the region. The SI unit for the electric field is the newton per coulomb (N/C). Table 23-1 shows the electric fields that occur in a few physical situations.

Although we use a positive test charge to define the electric field of a charged object, that field exists independently of the test charge. The field at point P in Figure 23-1b existed both before and after the test charge of Fig. 23-1a was put there. (We assume that in our defining procedure, the presence of the test charge does not affect the charge distribution on the charged object, and thus does not alter the electric field we are defining.)

TABLE 23-1 SOME ELECTRIC FIELDS

FIELD LOCATION OR SITUATION	VALUE (N/C)
At the surface of a uranium nucleus	3×10^{21}
Within a hydrogen atom, at a radius of 5.29×10^{-11} m	5×10^{11}
Electric breakdown occurs in air	3×10^{6}
Near the charged drum of a photocopier	10^{5}
Near a charged plastic comb	10^{3}
In the lower atmosphere	10^{2}
Inside the copper wire of household circuits	10^{-2}

To examine the role of an electric field in the interaction between charged objects, we have two tasks: (1) calculating the electric field produced by a given distribution of charge, and (2) calculating the force that a given field exerts on a charge placed in it. We perform the first task in Sections 23-4 through 23-7 for several charge distributions. We perform the second task in Sections 23-8 and 23-9 by considering a point charge and a pair of point charges in an electric field. But first, we discuss a way to visualize electric fields.

23-3 ELECTRIC FIELD LINES

Michael Faraday, who introduced the idea of electric fields in the 19th century, thought of the space around a charged body as filled with *lines of force.* Although we no longer attach much reality to these lines, now usually called **electric field lines,** they still provide a nice way to visualize patterns in electric fields.

The relation between the field lines and electric field vectors is this: (1) at any point, the direction of a straight field line or the direction of the tangent to a curved field line gives the direction of **E** at that point, and (2) the field lines are drawn so that the number of lines per unit area, measured in a plane that is perpendicular to the lines, is proportional to the *magnitude* of **E**. This second relation means that where the field lines are close together, E is large; and where they are far apart, E is small.

Figure 23-2*a* shows a sphere of uniform negative charge. If we place a *positive* test charge anywhere near the sphere, an electrostatic force pointing *toward* the center of the sphere will act on the test charge as shown. In other words, the electric field vectors at all points near the sphere are directed radially toward the sphere. This pattern of vectors is neatly displayed by the field lines in Fig. 23-2*b*, which point in the same directions as the force and field vectors. Moreover, the spreading of the field lines with distance from the sphere tells us that the magnitude of the electric field decreases with distance from the sphere.

If the sphere of Fig. 23-2 were of uniform *positive* charge, the electric field vectors at all points near the sphere would be directed radially *away from* the sphere. Thus the electric field lines would also extend radially away from the sphere. We then have the following rule:

> Electric field lines extend away from positive charge and toward negative charge.

Figure 23-3*a* shows part of an infinitely large, nonconducting *sheet* (or plane) with a uniform distribution of positive charge on one side. If we were to place a positive

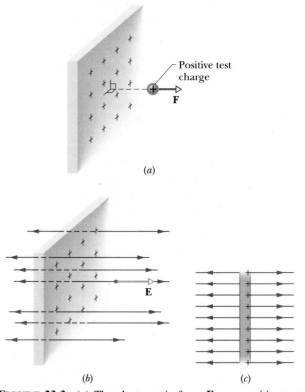

(a)

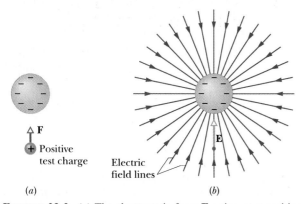

(a)   *(b)*

FIGURE 23-2 (*a*) The electrostatic force **F** acting on a positive test charge near a sphere of uniform negative charge. (*b*) The electric field vector **E** at the location of the test charge, and the electric field lines in the space near the sphere. The field lines extend *toward* the negatively charged sphere. (They originate on distant positive charges.)

(b)   *(c)*

FIGURE 23-3 (*a*) The electrostatic force **F** on a positive test charge near a very large, nonconducting sheet with uniformly distributed positive charge on one side. (*b*) The electric field vector **E** at the location of the test charge, and the electric field lines in the space near the sheet. The field lines extend *away from* the positively charged sheet. (*c*) Side view of (*b*).

test charge at any point near the sheet of Fig. 23-3a, the net electrostatic force acting on the test charge would be perpendicular to the sheet, because forces acting in all other directions would cancel one another as a result of the symmetry. Moreover, the net force on the test charge would point away from the sheet as shown. Thus the electric field vector at any point in the space on either side of the sheet is also perpendicular to the sheet and directed away from it (Figs. 23-3b and c). Since the charge is uniformly distributed along the sheet, all the field vectors have the same magnitude. Such an electric field, with the same magnitude and direction at every point, is a *uniform electric field*.

Of course, no real nonconducting sheet (such as a flat expanse of plastic) is infinitely large, but if we consider a region that is near the middle of a real sheet and not near its edges, the field lines through that region are arranged as in Fig. 23-3b and c.

Figure 23-4 shows the field lines for two equal positive charges. Figure 23-5 shows the pattern for two charges that are equal in magnitude but of opposite sign, a configuration that we call an **electric dipole.** Although we do not often use field lines quantitatively, they are very useful to visualize what is going on. Can you not almost ''see'' the charges being pushed apart in Fig. 23-4 and pulled together in Fig. 23-5?

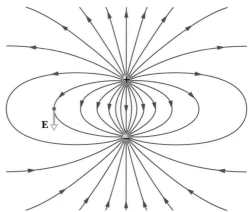

FIGURE 23-5 Field lines for a positive and a nearby negative point charge that are equal in magnitude. The charges attract each other. The pattern of field lines and the electric field it represents have rotational symmetry about an axis passing through both charges. The electric field vector at one point is shown; the vector is tangent to the field line through the point.

SAMPLE PROBLEM 23-1

In Fig. 23-2, how does the magnitude of the electric field vary with distance from the center of the uniformly charged sphere?

SOLUTION: Suppose that N field lines terminate on the sphere of Fig. 23-2. Imagine a concentric sphere of radius r surrounding the charged sphere. The number of lines per unit area on the imaginary sphere is $N/4\pi r^2$. Because E is proportional to this quantity, we can write $E \propto 1/r^2$. Thus the electric field set up by a uniform sphere of charge varies as the inverse square of the distance from the center of the sphere.

23-4 THE ELECTRIC FIELD DUE TO A POINT CHARGE

To find the electric field due to a point charge q (or charged particle), we put a positive test charge q_0 at any point a distance r from the point charge. From Coulomb's law (Eq. 22-4), the magnitude of the electrostatic force acting on q_0 is

$$F = \frac{1}{4\pi\epsilon_0}\frac{|q||q_0|}{r^2}. \tag{23-2}$$

The direction of **F** is directly away from the point charge if q is positive and directly toward the point charge if q is negative. The magnitude of the electric field vector is, from Eq. 23-1,

$$E = \frac{F}{q_0} = \frac{1}{4\pi\epsilon_0}\frac{|q|}{r^2} \quad \text{(point charge).} \tag{23-3}$$

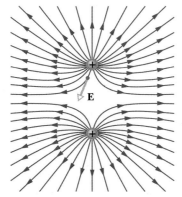

FIGURE 23-4 Field lines for two equal positive point charges. The charges repel each other. (The lines terminate on distant negative charges.) To ''see'' the actual three-dimensional pattern of field lines, mentally rotate the pattern shown here about an axis passing through both charges in the plane of the page. The three-dimensional pattern and the electric field it represents are said to have *rotational symmetry* about that axis. The electric field vector at one point is shown; note that it is tangent to the field line through that point.

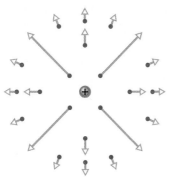

FIGURE 23-6 The electric field vectors at several points around a positive point charge.

The direction of **E** is the same as that of the force on the positive test charge: directly away from the point charge if q is positive, and toward it if q is negative.

We find the electric field in the space around a point charge by moving the test charge around in that space. The field for a positive point charge is shown in Fig. 23-6 in vector form (not as field lines).

We can find the net, or resultant, electric field due to more than one point charge with the aid of the principle of superposition. If we place a positive test charge q_0 near n point charges $q_1, q_2, \cdots, q_n$, then, from Eq. 22-7, the net force $\mathbf{F}_0$ from the n point charges acting on the test charge is

$$\mathbf{F}_0 = \mathbf{F}_{01} + \mathbf{F}_{02} + \cdots + \mathbf{F}_{0n}.$$

So, from Eq. 23-1, the net electric field at the position of the test charge is

$$\mathbf{E} = \frac{\mathbf{F}_0}{q_0} = \frac{\mathbf{F}_{01}}{q_0} + \frac{\mathbf{F}_{02}}{q_0} + \cdots + \frac{\mathbf{F}_{0n}}{q_0}$$
$$= \mathbf{E}_1 + \mathbf{E}_2 + \cdots + \mathbf{E}_n. \qquad (23\text{-}4)$$

Here $\mathbf{E}_i$ is the electric field that would be set up by point charge i acting alone. Equation 23-4 shows us that the principle of superposition applies to electric fields as well as to electrostatic forces.

CHECKPOINT 1: The figure shows a proton p and an electron e on an x axis. What is the direction of the electric field due to the electron at (a) point S and (b) point R? What is the direction of the net electric field at (c) point R and (d) point S?

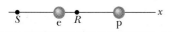

Figure 23-7a shows three particles with charges $q_1 = +2Q$, $q_2 = -2Q$, and $q_3 = -4Q$, each a distance d from the origin. What net electric field **E** is produced at the origin?

SOLUTION: Charges q_1, q_2, and q_3 produce electric field vectors $\mathbf{E}_1$, $\mathbf{E}_2$, and $\mathbf{E}_3$, respectively, at the origin. We seek the vector sum $\mathbf{E} = \mathbf{E}_1 + \mathbf{E}_2 + \mathbf{E}_3$. For this, we first must find the magnitudes and orientations of the three field vectors. To find the magnitude of $\mathbf{E}_1$, which is due to q_1, we use Eq. 23-3, substituting d for r and $2Q$ for $|q|$ and obtaining

$$E_1 = \frac{1}{4\pi\epsilon_0} \frac{2Q}{d^2}.$$

Similarly, we find the magnitudes of the fields $\mathbf{E}_2$ and $\mathbf{E}_3$ to be

$$E_2 = \frac{1}{4\pi\epsilon_0} \frac{2Q}{d^2} \quad \text{and} \quad E_3 = \frac{1}{4\pi\epsilon_0} \frac{4Q}{d^2}.$$

We next must find the orientations of the three electric field vectors at the origin. Because q_1 is a positive charge, the field vector it produces points directly *away* from it. And because q_2 and q_3 are both negative, the field vectors they produce point directly *toward* each of them. Thus the three electric fields produced at the origin by the three charged particles are oriented as in Fig. 23-7b. (Note that we have placed the tails of the vectors at this point where the fields are to be evaluated; doing so decreases the chance of error.)

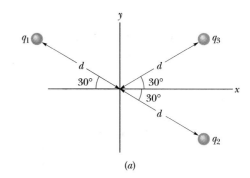

(a)

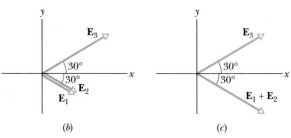

(b) (c)

FIGURE 23-7 Sample Problem 23-2. (a) Three particles with charges q_1, q_2, and q_3 are at the same distance d from the origin. (b) The electric field vectors $\mathbf{E}_1$, $\mathbf{E}_2$, and $\mathbf{E}_3$ at the origin due to the three particles. (c) The electric field vector $\mathbf{E}_3$ and the vector sum $\mathbf{E}_1 + \mathbf{E}_2$ at the origin.

We can now add the fields vectorially as usual, by finding the x and y components of each vector, and then the net x component E_x and the net y component E_y. To get the magnitude E we would use the Pythagorean theorem, and to find the orientation of $\mathbf{E}$ we would use the definition of the tangent of an angle.

However, here we can use symmetry to simplify the procedure. From Fig. 23-7b, we see that $\mathbf{E}_1$ and $\mathbf{E}_2$ point in the same direction. Hence their vector sum points in that direction and has the magnitude

$$E_1 + E_2 = \frac{1}{4\pi\epsilon_0}\frac{2Q}{d^2} + \frac{1}{4\pi\epsilon_0}\frac{2Q}{d^2}$$
$$= \frac{1}{4\pi\epsilon_0}\frac{4Q}{d^2},$$

which happens to equal the magnitude of $\mathbf{E}_3$.

We must now combine two vectors, $\mathbf{E}_3$ and the vector sum $\mathbf{E}_1 + \mathbf{E}_2$, that have the same magnitude and that are oriented symmetrically about the x axis, as shown in Fig. 23-7c. From the symmetry of Fig. 23-7c, we realize that the equal y components of our two vectors cancel and the equal x components add. Thus, the net electric field $\mathbf{E}$ at the origin points along the positive direction of x and has the magnitude

$$E = 2E_{3x} = 2E_3 \cos 30°$$
$$= (2)\frac{1}{4\pi\epsilon_0}\frac{4Q}{d^2}(0.866) = \frac{6.93Q}{4\pi\epsilon_0 d^2}. \quad \text{(Answer)}$$

SAMPLE PROBLEM 23-3

The nucleus of a uranium atom has a radius R of 6.8 fm. Assuming that the positive charge of the nucleus is distributed uniformly, determine the electric field at a point on the surface of the nucleus due to that charge.

SOLUTION: The nucleus has a positive charge of Ze, where the atomic number $Z (= 92)$ is the number of protons within the nucleus, and $e (= 1.60 \times 10^{-19}$ C$)$ is the charge of a proton. If this charge is distributed uniformly, then the first shell theorem of Chapter 22 applies. The electrostatic force on a positive test charge placed near the surface of the nucleus is the same as if the nuclear charge were concentrated at the nuclear center.

From Eq. 23-1, we then know that the electric field produced by the nucleus is also the same as if the nuclear charge were concentrated at the nuclear center. Equation 23-3 applies to such a pointlike concentration of charge, and we can write, for the magnitude of the field,

$$E = \frac{1}{4\pi\epsilon_0}\frac{Ze}{R^2}$$
$$= \frac{(8.99 \times 10^9 \text{ N·m}^2/\text{C}^2)(92)(1.60 \times 10^{-19} \text{ C})}{(6.8 \times 10^{-15} \text{ m})^2}$$
$$= 2.9 \times 10^{21} \text{ N/C}. \quad \text{(Answer)}$$

Since the charge of the nucleus is positive, the electric field vector $\mathbf{E}$ points outward, away from the center of the nucleus.

CHECKPOINT 2: The figure shows four situations in which charged particles are at equal distances from the origin. Rank the situations according to the magnitude of the net electric field at the origin, greatest first.

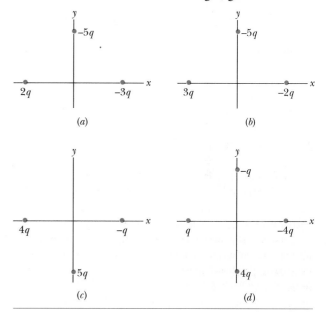

23-5 THE ELECTRIC FIELD DUE TO AN ELECTRIC DIPOLE

Figure 23-8a shows two charges of magnitude q but of opposite sign, separated by a distance d. As was noted in connection with Fig. 23-5, we call this configuration an *electric dipole*. Let us find the electric field due to the dipole of Fig. 23-8a at a point P, a distance z from the midpoint of the dipole and on its central axis, which is called the *dipole axis*.

From symmetry, the electric field $\mathbf{E}$ at point P—and also the fields $\mathbf{E}_{(+)}$ and $\mathbf{E}_{(-)}$ due to the separate charges that make up the dipole—must lie along the dipole axis, which we take to be a z axis. Applying the superposition principle for electric fields, we find that the magnitude E of the electric field at P is

$$E = E_{(+)} - E_{(-)}$$
$$= \frac{1}{4\pi\epsilon_0}\frac{q}{r^2_{(+)}} - \frac{1}{4\pi\epsilon_0}\frac{q}{r^2_{(-)}}$$
$$= \frac{q}{4\pi\epsilon_0(z - \frac{1}{2}d)^2} - \frac{q}{4\pi\epsilon_0(z + \frac{1}{2}d)^2}. \quad (23\text{-}5)$$

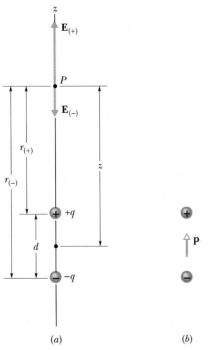

(a)   (b)

FIGURE 23-8 (a) An electric dipole. The electric field vectors $\mathbf{E}_{(+)}$ and $\mathbf{E}_{(-)}$ at point P on the dipole axis resulting from the two charges are shown. P is at distances $r_{(+)}$ and $r_{(-)}$ from the individual charges that make up the dipole. (b) The dipole moment $\mathbf{p}$ of the dipole points from the negative charge to the positive charge.

After a little algebra, we can rewrite this equation as

$$E = \frac{q}{4\pi\epsilon_0 z^2}\left[\left(1 - \frac{d}{2z}\right)^{-2} - \left(1 + \frac{d}{2z}\right)^{-2}\right]. \quad (23\text{-}6)$$

We are usually interested in the electrical effect of a dipole only at distances that are large compared with the dimensions of the dipole, that is, at distances such that $z \gg d$. At such large distances, we have $d/2z \ll 1$ in Eq. 23-6. We can then expand the two quantities in the brackets in that equation by the binomial theorem, obtaining for those quantities

$$\left[\left(1 + \frac{2d}{2z(1!)} + \cdots\right) - \left(1 - \frac{2d}{2z(1!)} + \cdots\right)\right].$$

So,

$$E = \frac{q}{4\pi\epsilon_0 z^2}\left[\left(1 + \frac{d}{z} + \cdots\right) - \left(1 - \frac{d}{z} + \cdots\right)\right]. \quad (23\text{-}7)$$

The unwritten terms in the two expansions in Eq. 23-7 involve d/z raised to progressively higher powers. Since $d/z \ll 1$, the contributions of those terms are progressively less, and to approximate E at large distances, we can ne-

glect them. Then, in our approximation, we can rewrite Eq. 23-7 as

$$E = \frac{q}{4\pi\epsilon_0 z^2}\frac{2d}{z} = \frac{1}{2\pi\epsilon_0}\frac{qd}{z^3}. \quad (23\text{-}8)$$

The product qd, which involves the two intrinsic properties q and d of the dipole, is the magnitude p of a vector quantity known as the **electric dipole moment p** of the dipole. Thus we can write Eq. 23-8 as

$$E = \frac{1}{2\pi\epsilon_0}\frac{p}{z^3} \quad \text{(electric dipole).} \quad (23\text{-}9)$$

The direction of $\mathbf{p}$ is taken to be from the negative to the positive end of the dipole, as indicated in Fig. 23-8b. We can use $\mathbf{p}$ to specify the orientation of a dipole.

Equation 23-9 shows that, if we measure the electric field of a dipole only at distant points, we can never find q and d separately, only their product. The field at distant points would be unchanged if, for example, q were doubled and d simultaneously halved. So the dipole moment is a basic property of a dipole.

Although Eq. 23-9 holds only for distant points along the dipole axis, it turns out that E for a dipole varies as $1/r^3$ for *all* distant points, regardless of whether they lie on the dipole axis; here r is the distance between the point in question and the dipole center.

Inspection of Fig. 23-8 and of the field lines in Fig. 23-5 shows that the direction of $\mathbf{E}$ for distant points on the dipole axis is always the direction of the dipole moment vector $\mathbf{p}$. This is true whether point P in Fig. 23-8a is on the upper or the lower part of the dipole axis.

Inspection of Eq. 23-9 shows that if you double the distance of a point from a dipole, the electric field at the point drops by a factor of 8. If you double the distance from a single point charge, however (see Eq. 23-3), the electric field drops only by a factor of 4. Thus the electric field of a dipole decreases more rapidly with distance than does the electric field of a single charge. The physical reason for this rapid decrease in electric field for a dipole is that from distant points a dipole looks like two equal but opposite charges that almost—but not quite—coincide. So their electric fields at distant points almost—but not quite—cancel each other.

SAMPLE PROBLEM 23-4

A molecule of water vapor causes an electric field in the surrounding space as if it were an electric dipole like that of Fig. 23-8. Its dipole moment has a magnitude $p = 6.2 \times 10^{-30}$ C·m. What is the magnitude of the electric field at a distance $z = 1.1$ nm from the molecule on its dipole axis? (This distance is large enough for Eq. 23-9 to apply.)

SOLUTION: From Eq. 23-9

$$E = \frac{1}{2\pi\epsilon_0}\frac{p}{z^3}$$

$$= \frac{6.2 \times 10^{-30}\ \text{C}\cdot\text{m}}{(2\pi)(8.85 \times 10^{-12}\ \text{C}^2/\text{N}\cdot\text{m}^2)(1.1 \times 10^{-9}\ \text{m})^3}$$

$$= 8.4 \times 10^7\ \text{N/C.} \qquad \text{(Answer)}$$

23-6 THE ELECTRIC FIELD DUE TO A LINE OF CHARGE

So far we have considered the electric field that is produced by one or, at most, a few point charges. We now consider charge distributions that consist of a great many closely spaced point charges (perhaps billions) that are spread along a line, over a surface, or within a volume. Such distributions are said to be **continuous** rather than discrete. Since these distributions can include an enormous number of point charges, we find the electric fields that they produce by means of calculus rather than by considering the point charges one by one. In this section we discuss the electric field caused by a line of charge. We consider a charged surface in the next section. A charged volume is the subject of Sample Problem 23-3, where we found the field outside a uniformly charged sphere. In the next chapter, we shall find the field inside such a sphere.

When we deal with continuous charge distributions, it is most convenient to express the charge on an object as a *charge density* rather than as a total charge. For a line of charge, for example, we would report the linear charge density (or charge per length) λ, whose SI unit is the coulomb per meter. Table 23-2 shows the other charge densities we shall be using.

Figure 23-9 shows a thin ring of radius R with a uniform positive linear charge density λ around its circumference. We may imagine the ring to be made of plastic or some other insulator, so that the charges can be regarded as fixed in place. What is the electric field $\mathbf{E}$ at point P, a distance z from the plane of the ring along its central axis?

TABLE 23-2 SOME MEASURES OF ELECTRIC CHARGE

NAME	SYMBOL	SI UNIT
Charge	q	C
Linear charge density	λ	C/m
Surface charge density	σ	C/m²
Volume charge density	ρ	C/m³

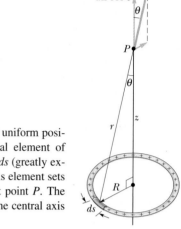

FIGURE 23-9   A ring of uniform positive charge. A differential element of charge occupies a length ds (greatly exaggerated for clarity). This element sets up an electric field $d\mathbf{E}$ at point P. The component of $d\mathbf{E}$ along the central axis of the ring is $dE\cos\theta$.

To answer, we cannot just apply Eq. 23-3, which gives the electric field set up by a point charge, because the ring is obviously not a point charge. However, we can mentally divide the ring into differential elements of charge that are so small that they are like point charges, and then we can apply Eq. 23-3 to each of them. Next, we can add the electric fields set up at P by all the differential elements. Their vector sum gives us the field set up at P by the ring.

Let ds be the (arc) length of any differential element of the ring. Since λ is the charge per unit length, the element has a charge of magnitude

$$dq = \lambda\ ds. \qquad (23\text{-}10)$$

This differential charge sets up a differential electric field $d\mathbf{E}$ at point P, which is a distance r from the element. Treating the element as a point charge, and using Eq. 23-10, we can rewrite Eq. 23-3 to express the magnitude of $d\mathbf{E}$ as

$$dE = \frac{1}{4\pi\epsilon_0}\frac{dq}{r^2} = \frac{1}{4\pi\epsilon_0}\frac{\lambda\ ds}{r^2}. \qquad (23\text{-}11)$$

From Fig. 23-9, we can rewrite Eq. 23-11 as

$$dE = \frac{1}{4\pi\epsilon_0}\frac{\lambda\ ds}{(z^2 + R^2)}. \qquad (23\text{-}12)$$

Figure 23-9 shows us that $d\mathbf{E}$ is at an angle θ to the central axis (which we have taken to be a z axis) and has components perpendicular to and parallel to that axis.

Every element of charge in the ring sets up a differential field $d\mathbf{E}$ at P, with magnitude given by Eq. 23-12. All these $d\mathbf{E}$ vectors have identical components parallel to the central axis, in both magnitude and direction. All these $d\mathbf{E}$ vectors have components perpendicular to the central axis as well; these perpendicular components are identical in magnitude but point in different directions. In fact, for any perpendicular component that points in a given direction, there is another one that points in the opposite direction. The sum of this pair of components, like the sum of all other pairs of oppositely directed components, is zero.

So the perpendicular components cancel and we need not consider them further. This leaves the parallel components; they are all in the same direction, so the net electric field at P is their sum.

From Fig. 23-9, we see that the parallel component of $d\mathbf{E}$ has magnitude $dE \cos \theta$. We also see that

$$\cos \theta = \frac{z}{r} = \frac{z}{(z^2 + R^2)^{1/2}}. \qquad (23\text{-}13)$$

Then Eqs. 23-13 and 23-12 give us, for the parallel component of $d\mathbf{E}$,

$$dE \cos \theta = \frac{z\lambda}{4\pi\epsilon_0(z^2 + R^2)^{3/2}} \, ds. \qquad (23\text{-}14)$$

To add the parallel components $dE \cos \theta$ produced by all the elements, we integrate Eq. 23-14 around the circumference of the ring, from $s = 0$ to $s = 2\pi R$. Since the only quantity in Eq. 23-14 that varies during the integration is s, the other quantities can be moved outside the integral sign. The integration then gives us

$$E = \int dE \cos \theta = \frac{z\lambda}{4\pi\epsilon_0(z^2 + R^2)^{3/2}} \int_0^{2\pi R} ds$$

$$= \frac{z\lambda(2\pi R)}{4\pi\epsilon_0(z^2 + R^2)^{3/2}}. \qquad (23\text{-}15)$$

Since λ is the charge per length of the ring, the term $\lambda(2\pi R)$ in Eq. 23-15 is q, the total charge on the ring. We then can rewrite Eq. 23-15 as

$$E = \frac{qz}{4\pi\epsilon_0(z^2 + R^2)^{3/2}} \qquad \text{(charged ring).} \qquad (23\text{-}16)$$

If the charge on the ring is negative, instead of positive as we have assumed, the magnitude of the field at P is still given by Eq. 23-16. However, the electric field vector then points toward the ring instead of away from it.

Let us check Eq. 23-16 for a point on the central axis that is so far away that $z \gg R$. For such a point, the expression $z^2 + R^2$ in Eq. 23-16 can be approximated as z^2, and Eq. 23-16 becomes

$$E = \frac{1}{4\pi\epsilon_0} \frac{q}{z^2} \qquad \text{(charged ring at large distance).} \qquad (23\text{-}17)$$

This is a reasonable result, because from a large distance, the ring "looks" like a point charge. If we replace z by r in Eq. 23-17, we indeed do have Eq. 23-3, the electric field due to a point charge.

Let us next check Eq. 23-16 for a point at the center of the ring, that is, for $z = 0$. At that point, Eq. 23-16 tells us that $E = 0$. This is a reasonable result, because if we were to place a test charge at the center of the ring, there would be no net electrostatic force acting on it: the force due to any element of the ring would be canceled by the force due to the element on the opposite side of the ring. And, by Eq. 23-1, if the force were zero, the electric field at the center of the ring would have to be zero.

SAMPLE PROBLEM 23-5

Figure 23-10a shows a plastic rod having a uniformly distributed charge $-Q$. The rod has been bent in a 120° circular arc of radius r. We place coordinate axes such that the axis of symmetry of the rod lies along the x axis and the origin is at the center of curvature P of the rod. In terms of Q and r, what is the electric field $\mathbf{E}$ due to the rod at point P?

SOLUTION: Consider a differential element of the rod, having arc length ds and located at an angle θ above the x axis (Fig. 23-10b). If we let λ represent the linear charge density of the rod, our element ds has a differential charge of magnitude

$$dq = \lambda \, ds. \qquad (23\text{-}18)$$

Our element produces a differential electric field $d\mathbf{E}$ at point P, which is a distance r from the element. Treating the element as a point charge, we can rewrite Eq. 23-3 to express the magnitude of $d\mathbf{E}$ as

$$dE = \frac{1}{4\pi\epsilon_0} \frac{dq}{r^2} = \frac{1}{4\pi\epsilon_0} \frac{\lambda \, ds}{r^2}. \qquad (23\text{-}19)$$

The direction of $d\mathbf{E}$ is toward ds, because charge dq is negative.

Our element has a symmetrically located (mirror image) element ds' in the bottom half of the rod. The electric field $d\mathbf{E}'$ set up at P by ds' also has the magnitude given by Eq. 23-19, but the field vector points toward ds' as shown in Fig. 23-10b. If we resolve the electric field vectors of ds and ds' into x and y components as shown in Fig. 23-10b, we see that their y components cancel (because they have equal magnitudes and are in opposite directions). We also see that their x components have equal magnitudes and are in the same direction.

Thus to find the electric field set up by the rod, we need sum (via integration) only the x components of the differential electric fields set up by all the differential elements of the rod. From Fig. 23-10b and Eq. 23-19, we can write the component dE_x set up by ds as

$$dE_x = dE \cos \theta = \frac{1}{4\pi\epsilon_0} \frac{\lambda}{r^2} \cos \theta \, ds. \qquad (23\text{-}20)$$

Equation 23-20 has two variables, θ and s. Before we can integrate it, we must eliminate one variable. We do so by replacing ds, using the relation

$$ds = r \, d\theta,$$

in which $d\theta$ is the angle at P that includes arc length ds (Fig. 23-10c). With this replacement, we can integrate Eq. 23-20 over the angle made by the rod at P, from $\theta = -60°$ to $\theta = 60°$; that will give us the magnitude of the electric field at P due to the rod:

$$E = \int dE_x = \int_{-60°}^{60°} \frac{1}{4\pi\epsilon_0} \frac{\lambda}{r^2} \cos\theta\, r\, d\theta$$

$$= \frac{\lambda}{4\pi\epsilon_0 r} \int_{-60°}^{60°} \cos\theta\, d\theta = \frac{\lambda}{4\pi\epsilon_0 r} \left[\sin\theta \right]_{-60°}^{60°}$$

$$= \frac{\lambda}{4\pi\epsilon_0 r} [\sin 60° - \sin(-60°)]$$

$$= \frac{1.73\lambda}{4\pi\epsilon_0 r}. \qquad (23\text{-}21)$$

(If we had reversed the limits on the integration, we would have gotten the same result but with a minus sign. Since the integration gives only the magnitude of **E**, we would then have discarded the minus sign.)

To evaluate λ, we note that the rod has an angle of 120° and so is one-third of a full circle. Its arc length is then $2\pi r/3$,

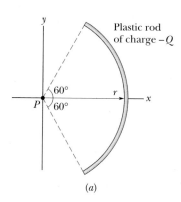

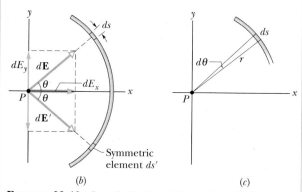

FIGURE 23-10 Sample Problem 23-5. (a) A plastic rod of charge $-Q$ is a circular section of radius r and central angle 120°; point P is the center of curvature of the rod. (b) A differential element in the top half of the rod, at an angle θ to the x axis and of arc length ds, sets up a differential electric field $d\mathbf{E}$ at P. An element ds', symmetric to ds about the x axis, sets up a field $d\mathbf{E}'$ at P with the same magnitude. (c) Arc length ds makes an angle $d\theta$ about point P.

and its linear charge density must be

$$\lambda = \frac{\text{charge}}{\text{length}} = \frac{Q}{2\pi r/3} = \frac{0.477Q}{r}.$$

Substituting this into Eq. 23-21 and simplifying give us

$$E = \frac{(1.73)(0.477Q)}{4\pi\epsilon_0 r^2} = \frac{0.83Q}{4\pi\epsilon_0 r^2}. \qquad \text{(Answer)}$$

The direction of **E** is toward the rod, along the axis of symmetry of the charge distribution.

PROBLEM SOLVING TACTICS

TACTIC 1: *A Field Guide for Lines of Charge*

Here is a generic guide for finding the electric field **E** produced at a point P by a line of uniform charge, either circular or straight. The general strategy is to pick out an element dq of the charge, find $d\mathbf{E}$ due to that element, and integrate $d\mathbf{E}$ over the entire line of charge.

STEP 1. If the line of charge is circular, let ds be the arc length of an element of the distribution. If the line is straight, run an x axis along it and let dx be the length of an element. Mark the element on a sketch.

STEP 2. Relate the charge dq of the element to the length of the element with either $dq = \lambda\, ds$ or $dq = \lambda\, dx$. Consider dq and λ to be positive, even if the charge is actually negative. (The sign of the charge is used in the next step.)

STEP 3. Express the field $d\mathbf{E}$ produced at P by dq with Eq. 23-3, replacing q in that equation with either $\lambda\, ds$ or $\lambda\, dx$. If the charge on the line is positive, then at P draw a vector $d\mathbf{E}$ that points directly away from dq. If the charge is negative, draw the vector pointing directly toward dq.

STEP 4. Always look for any symmetry of the situation. If P is on an axis of symmetry of the charge distribution, resolve the field $d\mathbf{E}$ produced by dq into components that are perpendicular and parallel to the axis of symmetry. Then consider a second element dq' that is located symmetrically to dq about the line of symmetry. At P draw the vector $d\mathbf{E}'$ that this symmetrical element produces, and resolve it into components. One of the components produced by dq is a *canceling component*: it is canceled by the corresponding component produced by dq' and needs no further attention. The other component produced by dq is an *adding component*: it adds to the corresponding component produced by dq'. Add the adding components of all the elements via integration.

STEP 5. Here are four general types of uniform charge distributions, with strategies for simplifying the integral of step 4. Each type can be made more challenging by having the distribution consist of a line of positive charge and a line of negative charge.

Ring, with point P on (central) axis of symmetry, as in Fig. 23-9. In the expression for dE, replace r^2 with $z^2 + R^2$, as in

Eq. 23-12. Express the adding component of $d\mathbf{E}$ in terms of θ. That introduces $\cos\theta$, but θ is identical for all elements and thus is not a variable. Replace $\cos\theta$ as in Eq. 23-13. Integrate over s, around the circumference of the ring.

Circular arc, with point P at the center of curvature, as in Fig. 23-10. Express the adding component of $d\mathbf{E}$ in terms of θ. That introduces either $\sin\theta$ or $\cos\theta$. Reduce the resulting two variables s and θ to one, θ, by replacing ds with $r\,d\theta$. Integrate over θ, as in Sample Problem 23-5, from one end of the arc to the other end.

Straight line, with point P on an extension of the line, as in Fig. 23-11a. In the expression for dE, replace r with x. Integrate over x, from end to end of the line of charge.

Straight line, with point P at perpendicular distance y from the line of charge, as in Fig. 23-11b. In the expression for dE, replace r with an expression involving x and y. If P is on the perpendicular bisector of the line of charge, find an expression for the adding component of $d\mathbf{E}$. That will introduce either $\sin\theta$ or $\cos\theta$. Reduce the resulting two variables x and θ to one, x, by replacing the trigonometric function with an expression (its definition) involving x and y. Integrate over x from end to end of the line of charge. If P is not on a line of symmetry, as in Fig. 23-11c, set up an integral to sum the components dE_x, and integrate over x to find E_x. Also set up an integral to sum the components dE_y, and integrate over x again to find E_y. Use the components E_x and E_y in the usual way to find the magnitude E and the orientation of $\mathbf{E}$.

STEP 6. One arrangement of the integration limits gives a positive result. The reverse arrangement gives the same result with a minus sign; discard the minus sign. If the result is to be stated in terms of the total charge Q of the distribution, replace λ with Q/L, in which L is the length of the distribution. For a ring, L is the ring's circumference.

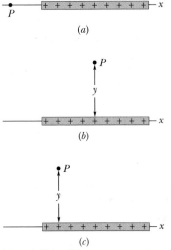

(a)

(b)

(c)

FIGURE 23-11 (a) Point P is on an extension of the line of charge. (b) P is on a line of symmetry of the line of charge, at perpendicular distance y from that line. (c) Same as (b) except that P is not on a line of symmetry.

CHECKPOINT **3:** The figure shows three nonconducting rods, one circular and two straight. Each has a uniform charge of magnitude Q along its top half and another along its bottom half. For each rod, what is the direction of the net electric field at point P?

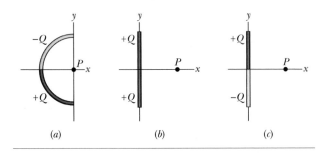

(a)   (b)   (c)

23-7 THE ELECTRIC FIELD DUE TO A CHARGED DISK

Figure 23-12 shows a circular plastic disk of radius R that has a positive surface charge of uniform density σ on its upper surface (see Table 23-2). What is the electric field at point P, a distance z from the disk along its central axis?

Our plan is to divide the disk into concentric flat rings and then to calculate the electric field at point P by adding up (that is, by integrating) the contributions of all the rings. Figure 23-12 shows one such ring, with radius r and radial width dr. Since σ is the charge per unit area, the charge on the ring is

$$dq = \sigma\,dA = \sigma(2\pi r\,dr), \qquad (23\text{-}22)$$

where dA is the differential area of the ring.

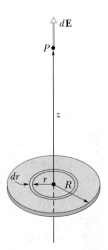

FIGURE 23-12 A disk of radius R and uniform positive charge. The ring shown has radius r and radial width dr. It sets up a differential electric field $d\mathbf{E}$ at point P on its central axis.

We have already solved the problem of the electric field due to a ring of charge. Substituting dq from Eq. 23-22 for q in Eq. 23-16, and replacing R in Eq. 23-16 with r, we obtain an expression for the electric field dE at P due to our flat ring:

$$dE = \frac{z\sigma 2\pi r \, dr}{4\pi\epsilon_0(z^2 + r^2)^{3/2}},$$

which we may write as

$$dE = \frac{\sigma z}{4\epsilon_0} \frac{2r \, dr}{(z^2 + r^2)^{3/2}}.$$

We can now find E by integrating over the surface of the disk, that is, by integrating with respect to the variable r from $r = 0$ to $r = R$. Note that z remains constant during this process. We get

$$E = \int dE = \frac{\sigma z}{4\epsilon_0} \int_0^R (z^2 + r^2)^{-3/2}(2r) \, dr. \quad (23\text{-}23)$$

To solve this integral, we cast it in the form $\int X^m \, dX$ by setting $X = (z^2 + r^2)$, $m = -\frac{3}{2}$, and $dX = (2r) \, dr$. For the recast integral we have

$$\int X^m \, dX = \frac{X^{m+1}}{m + 1},$$

so Eq. 23-23 becomes

$$E = \frac{\sigma z}{4\epsilon_0} \left[\frac{(z^2 + r^2)^{-1/2}}{-\frac{1}{2}} \right]_0^R.$$

Taking the limits and rearranging, we find

$$E = \frac{\sigma}{2\epsilon_0} \left(1 - \frac{z}{\sqrt{z^2 + R^2}} \right) \quad \begin{matrix} \text{(charged} \\ \text{disk)} \end{matrix} \quad (23\text{-}24)$$

as the magnitude of the electric field produced by a flat, circular, charged disk on its central axis. (In carrying out the integration, we assumed that $z \geq 0$.)

If we let $R \rightarrow \infty$ while keeping z finite, the second term in the parentheses in Eq. 23-24 approaches zero, and this equation reduces to

$$E = \frac{\sigma}{2\epsilon_0} \quad \text{(infinite sheet).} \quad (23\text{-}25)$$

This is the electric field produced by an infinite sheet of uniform charge located on one side of a nonconductor such as plastic. The electric field lines for such a situation are shown in Fig. 23-3.

We also get Eq. 23-25 if we let $z \rightarrow 0$ in Eq. 23-24 while keeping R finite. This shows that at points very close to the disk, the electric field set up by the disk is the same as if the disk were infinite in extent.

When the electric field surrounding a charged object (like this charged metal cap) becomes large enough, the air surrounding the object undergoes *electrical breakdown*: air molecules are ionized (electrons are removed from the molecules), and momentary conducting paths appear. The *electric sparks* you see here reveal those paths.

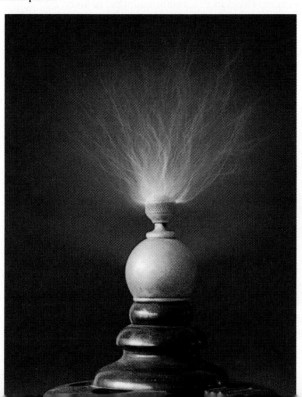

SAMPLE PROBLEM 23-6

The disk of Fig. 23-12 has a surface charge density σ of $+5.3$ $\mu\text{C/m}^2$ on its upper face. (This, incidentally, is a reasonable value for the surface charge density on the photosensitive cylinder of a photocopying machine.)

(a) What is the electric field at the surface of the disk?

SOLUTION: From Eq. 23-25 we have

$$E = \frac{\sigma}{2\epsilon_0} = \frac{5.3 \times 10^{-6} \text{ C/m}^2}{(2)(8.85 \times 10^{-12} \text{ C}^2/\text{N} \cdot \text{m}^2)}$$

$$= 3.0 \times 10^5 \text{ N/C.} \quad \text{(Answer)}$$

This value holds for all points that are close to the surface of the disk but not near its edge.

When the electric field in a material is large enough, the material undergoes *electrical breakdown* in which conducting paths suddenly appear in the material. Electrical breakdown occurs in air (at atmospheric pressure) when the electric field exceeds about 3×10^6 N/C. During breakdown, electrons flow along one or more conducting paths, creating *electric sparks*. Since the computed electric field in this sample problem is only 3×10^5 N/C, the charged disk will not cause sparks in the surrounding air.

(b) Using the binomial theorem, find an expression for the electric field at a point on the central axis far from the disk.

SOLUTION: The phrase *far from the disk* means that the distance z is much greater than the size of the disk, as measured by, say, its radius. As you will see, this allows us to use the binomial theorem to approximate the square-root term in Eq. 23-24.

From Appendix E, we write the general form of the binomial theorem as

$$(1 + x)^n = 1 + \frac{n}{1!} x + \frac{n(n-1)}{2!} x^2 + \cdots , \quad (23\text{-}26)$$

where $|x| \ll 1$. To get ready for the approximation, we rewrite the square-root term as

$$\frac{z}{\sqrt{z^2 + R^2}} = \frac{z}{z\sqrt{1 + \frac{R^2}{z^2}}} = \left(1 + \frac{R^2}{z^2} \right)^{-\frac{1}{2}},$$

which is in the proper form for use of the binomial theorem with $x = R^2/z^2$ and $n = -\frac{1}{2}$. Because z is much greater than R, the condition $|x| \ll 1$ is satisfied.

Applying Eq. 23-26, we now write

$$\left(1 + \frac{R^2}{z^2} \right)^{-\frac{1}{2}} = 1 + \frac{-\frac{1}{2}}{1!} \frac{R^2}{z^2} + \frac{-\frac{1}{2}(-\frac{1}{2} - 1)}{2!} \frac{R^4}{z^4} + \cdots .$$

Successive terms on the right side are progressively less. We can approximate the required result closely enough by discarding terms smaller than R^2/z^2, which leaves us

$$\frac{z}{\sqrt{z^2 + R^2}} = 1 - \frac{R^2}{2z^2}.$$

Substituting this expression in Eq. 23-24 gives us

$$E = \frac{\sigma}{2\epsilon_0} \left[1 - \left(1 - \frac{R^2}{2z^2} \right) \right]$$

$$= \frac{\sigma}{4\epsilon_0} \frac{R^2}{z^2}. \quad \text{(Answer)}$$

We can rewrite this in terms of the charge q on the upper face of the disk by noting that $\sigma = q/A$ and, for the disk, $A = \pi R^2$. Then

$$E = \frac{\sigma}{4\epsilon_0} \frac{R^2}{z^2} = \frac{q}{4\epsilon_0 \pi R^2} \frac{R^2}{z^2}$$

$$= \frac{1}{4\pi\epsilon_0} \frac{q}{z^2}. \quad \text{(Answer)} \quad (23\text{-}27)$$

Equation 23-27 tells us that at points on the central axis where $z \gg R$, the electric field produced by the charge q spread over the face of the disk is the same as that produced by a particle of the same charge q.

23-8 A POINT CHARGE IN AN ELECTRIC FIELD

In the preceding four sections we worked at the first of our two tasks: given a charge distribution, to find the electric field it produces in the surrounding space. Here we begin the second task: to determine what happens to a charged particle that is in an electric field that is produced by other stationary or slowly moving charges.

What happens is that an electrostatic force acts on the particle. This force, a vector quantity, is given by

$$\mathbf{F} = q\mathbf{E}, \quad (23\text{-}28)$$

in which q is the charge of the particle (including its sign)

In an electrostatic precipitator, an electric field exerts a force on charged ash as it ascends a stack, so that much of it is collected in the stack, hence does not enter and pollute the atmosphere. The precipitator is operating in the photograph at the left but not in the photograph at the right.

and **E** is the electric field that other charges have produced at the location of the particle. (The field is *not* the field set up by the particle itself; to distinguish the two fields, the field acting on the particle in Eq. 23-28 is often called the *external field*. A charged particle (or object) is not affected by its own electric field.) Equation 23-28 tells us:

> The electrostatic force **F** acting on a charged particle located in an external electric field **E** points in the direction of **E** if the charge q of the particle is positive and in the opposite direction if q is negative.

CHECKPOINT **4:** (a) In the figure, what is the direction of the electrostatic force on the electron due to the electric field shown? (b) In which direction will the electron accelerate if it is moving parallel to the y axis before it encounters the electric field? (c) If, instead, the electron is initially moving rightward, will its speed increase, decrease, or remain constant?

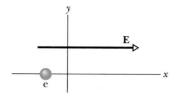

Measuring the Elementary Charge

Equation 23-28 played a role in the measurement of the elementary charge e by American physicist Robert A. Millikan in 1910–1913. Figure 23-13 is a representation of his apparatus. When tiny oil drops are sprayed into chamber A, some of them become charged, either positively or negatively, in the process. Consider a drop that drifts downward through the small hole in plate P_1 and into chamber C. Let us assume that this drop has a negative charge q.

If switch S in Fig. 23-13 is open as shown, battery B has no electrical effect on chamber C. If the switch is closed (the connection between chamber C and the positive terminal of the battery is then complete), the battery causes an excess positive charge on conducting plate P_1 and an excess negative charge on conducting plate P_2. The charged plates set up a downward-pointing electric field **E** in chamber C. According to Eq. 23-28, this field exerts an electrostatic force on any charged drop that happens to be in the chamber and affects its motion. In particular, our negatively charged drop will tend to drift upward.

By timing the motion of oil drops with the switch opened and closed and thus determining the effect of the charge q, Millikan discovered that the values of q were

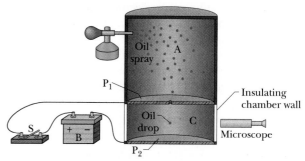

FIGURE 23-13 The Millikan oil-drop apparatus for measuring the elementary charge e. When a charged oil drop drifted into chamber C through the hole in plate P_1, its motion could be controlled by closing and opening switch S and thereby setting up or eliminating an electric field in chamber C. The microscope was used to view the drop, to permit timing of its motion.

always given by

$$q = ne, \qquad n = 0, \pm 1, \pm 2, \pm 3, \cdots, \quad (23\text{-}29)$$

in which e turned out to be the fundamental constant we call the *elementary charge*, 1.60×10^{-19} C. Millikan's experiment is convincing proof that charge is quantized, and he earned the 1923 Nobel Prize in physics in part for this work. Modern measurements of the elementary charge rely on a variety of interlocking experiments, all more precise than the pioneering experiment of Millikan.

Ink-Jet Printing

The need for high-quality, high-speed printing has caused a search for an alternative to impact printing, such as occurs in a standard typewriter. Building up letters by squirting tiny drops of ink at the paper is one such alternative.

Figure 23-14 shows a negatively charged drop moving between two conducting deflecting plates, between which

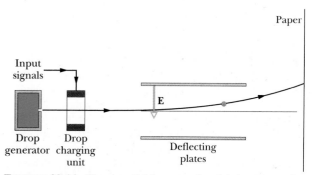

FIGURE 23-14 The essential features of an ink-jet printer. An input signal from a computer controls the charge given to each drop and thus the position on the paper at which the drop lands. About 100 tiny drops are needed to form a single character.

a uniform, downward-pointing electric field **E** has been set up. The drop is deflected upward according to Eq. 23-28 and then strikes the paper at a position that is determined by the magnitudes of **E** and the charge q of the drop.

In practice, E is held constant and the position of the drop is determined by the charge q delivered to the drop in the charging unit, through which the drop must pass before entering the deflecting system. The charging unit, in turn, is activated by electronic signals that encode the material to be printed.

SAMPLE PROBLEM 23-7

In the Millikan oil-drop apparatus of Fig. 23-13, a drop of radius $R = 2.76$ μm has an excess charge of three electrons. What are the magnitude and direction of the electric field that is required to balance the drop so it remains stationary in the apparatus? The density ρ of the oil is 920 kg/m^3.

SOLUTION: To balance the drop, the electrostatic force acting on it must be upward and have a magnitude equal to the weight mg of the drop. From Eqs. 23-28 and 23-29, we can write the *magnitude* of the electrostatic force as $F = (3e)E$. We can also write the mass of the drop as the product of its volume and its density. Thus the balance of forces gives us

$$\tfrac{4}{3}\pi R^3 \rho g = (3e)E.$$

Solving for E yields

$$E = \frac{4\pi R^3 \rho g}{9e}$$

$$= \frac{(4\pi)(2.76 \times 10^{-6} \text{ m})^3 (920 \text{ kg/m}^3)(9.80 \text{ m/s}^2)}{(9)(1.60 \times 10^{-19} \text{ C})}$$

$$= 1.65 \times 10^6 \text{ N/C.} \qquad \text{(Answer)}$$

Because the drop is negatively charged, Eq. 23-28 tells us that **E** and **F** are in opposite directions: $\mathbf{F} = -3e\mathbf{E}$. So the electric field must point downward.

SAMPLE PROBLEM 23-8

Figure 23-15 shows the deflecting plates of an ink-jet printer, with superimposed coordinate axes. An ink drop with a mass m of 1.3×10^{-10} kg and a negative charge of magnitude $Q = 1.5 \times 10^{-13}$ C enters the region between the plates, initially moving along the x axis with speed $v_x = 18$ m/s. The length L of the plates is 1.6 cm. The plates are charged and thus produce an electric field at all points between them. Assume that the downward-pointing field **E** is uniform and has a magnitude of 1.4×10^6 N/C. What is the vertical deflection of the drop at the far edge of the plates? (The weight of the drop is small relative to the electrostatic force acting on the drop and can be neglected.)

SOLUTION: Since the drop is negatively charged and the electric field is downward, Eq. 23-28 tells us that a constant electrostatic force of magnitude QE acts *upward* on the charged drop. Thus as the drop travels parallel to the x axis at constant speed v_x, it accelerates upward with constant acceleration a_y. Applying Newton's second law $F = ma$ along the y axis, we find that

$$a_y = \frac{F}{m} = \frac{QE}{m}. \qquad (23\text{-}30)$$

Let t represent the time required for the drop to pass through the region between the plates. During t the vertical and horizontal displacements of the drop are

$$y = \tfrac{1}{2}a_y t^2 \quad \text{and} \quad L = v_x t, \qquad (23\text{-}31)$$

respectively. Eliminating t between these two equations and substituting Eq. 23-30 for a_y, we find

$$y = \frac{QEL^2}{2mv_x^2}$$

$$= \frac{(1.5 \times 10^{-13} \text{ C})(1.4 \times 10^6 \text{ N/C})(1.6 \times 10^{-2} \text{ m})^2}{(2)(1.3 \times 10^{-10} \text{ kg})(18 \text{ m/s})^2}$$

$$= 6.4 \times 10^{-4} \text{ m} = 0.64 \text{ mm.} \qquad \text{(Answer)}$$

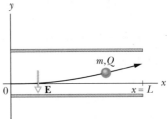

FIGURE 23-15   Sample Problem 23-8. An ink drop of mass m and charge magnitude Q is deflected in the electric field of an ink-jet printer.

23-9 A DIPOLE IN AN ELECTRIC FIELD

We have defined the electric dipole moment **p** of an electric dipole to be a vector that points from the negative to the positive end of the dipole. As you will see, the behavior of a dipole in a uniform external electric field **E** can be described completely in terms of the two vectors **E** and **p**, with no need of any details about the dipole's structure.

As was noted in Sample Problem 23-4, a molecule of water (H_2O) is an electric dipole. Figure 23-16 shows why. There the black dots represent the oxygen nucleus (having eight protons) and the two hydrogen nuclei (having one proton each). The colored enclosed areas represent the region in which the electrons orbit the nuclei.

In a water molecule, the two hydrogen atoms and the oxygen atom do not lie on a straight line but form an angle

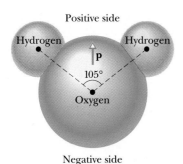

Positive side

Hydrogen   Hydrogen

p

105°

Oxygen

Negative side

FIGURE 23-16 A molecule of H_2O, showing the three nuclei (represented by dots) and the regions in which the electrons orbit the nuclei. The electric dipole moment **p** points from the (negative) oxygen side to the (positive) hydrogen side of the molecule.

of about 105°, as shown in Fig. 23-16. As a result, the molecule has a definite "oxygen side" and "hydrogen side." Moreover, the 10 electrons of the molecule tend to remain closer to the oxygen nucleus than to the hydrogen nuclei. This makes the oxygen side of the molecule slightly more negative than the hydrogen side and creates an electric dipole moment **p** that points along the symmetry axis of the molecule as shown. If the water molecule is placed in an external electric field, it behaves as would be expected of the more abstract electric dipole of Fig. 23-8.

To examine this behavior, we now consider such an abstract dipole in a uniform external electric field **E**, as shown in Fig. 23-17a. We assume that the dipole is a rigid structure (due to internal electrostatic forces) that consists

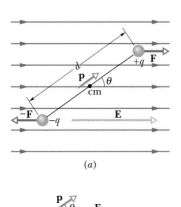

(a)

p

$\tau \otimes$ θ **E**

(b)

FIGURE 23-17 (a) An electric dipole in a uniform electric field **E**. Two centers of equal but opposite charge are separated by distance d. Their center of mass cm is assumed to be midway between them. The bar between them represents their rigid connection. (b) Field **E** causes a torque τ on dipole moment **p**. The direction of the torque vector τ is into the plane of the page, as represented by the symbol $\otimes$.

of two centers of opposite charge, each of magnitude q, separated by a distance d. The dipole moment **p** makes an angle θ with **E**.

At the charged ends of the dipole, electrostatic forces, **F** and $-$**F**, act in opposite directions and with the same magnitude $F = qE$. Thus the net force exerted on the dipole by the field is zero. However, these forces exert a net torque τ on the dipole about its center of mass, which we can take to be midway along the line connecting the charged ends. From Eq. 11-30, with $r = d/2$, we can write the magnitude of this net torque τ as

$$\tau = F\frac{d}{2}\sin\theta + F\frac{d}{2}\sin\theta = Fd\sin\theta. \quad (23\text{-}32)$$

We can also write the magnitude of τ in terms of the magnitudes of the electric field E and the dipole moment qd. To do so, we substitute qE for F and p/q for d in Eq. 23-32, finding that the magnitude of τ is

$$\tau = pE\sin\theta. \quad (23\text{-}33)$$

We can generalize this equation to vector form as

$$\boldsymbol{\tau} = \mathbf{p} \times \mathbf{E} \quad \text{(torque on a dipole).} \quad (23\text{-}34)$$

Vectors **p** and **E** are shown in Fig. 23-17b. The torque acting on a dipole tends to rotate **p** (hence the dipole) into the direction of **E**, thereby reducing θ. In Fig. 23-17, such rotation is clockwise. As we discussed in Chapter 11, we can represent a torque that gives rise to a clockwise rotation by including a minus sign with the magnitude of the torque. With such notation, the torque of Fig. 23-17 is

$$\tau = -pE\sin\theta. \quad (23\text{-}35)$$

Potential Energy of an Electric Dipole

Potential energy can be associated with the orientation of an electric dipole in an electric field. The dipole has its least potential energy when it is in its equilibrium orientation, which is when its moment **p** is lined up with the field **E** (then $\boldsymbol{\tau} = \mathbf{p} \times \mathbf{E} = 0$). It has greater potential energy in all other orientations. Thus the dipole is like a pendulum, which has *its* least gravitational potential energy in *its* equilibrium orientation—at its lowest point. To rotate the dipole or the pendulum to any other orientation requires work by some external agent.

In any situation involving potential energy, we are free to define the zero-potential-energy configuration in a perfectly arbitrary way, because only differences in potential energy have physical meaning. It turns out that the expression for the potential energy of an electric dipole in an external electric field is simplest if we choose the potential energy to be zero when the angle θ in Fig. 23-17 is 90°. We then can find the potential energy U of the dipole at any other value of θ with Eq. 8-1 ($\Delta U = -W$) by calculating the work W done by the field on the dipole when the dipole is rotated to that value of θ from 90°. With the aid of Eq. 11-44 ($W = \int \tau \, d\theta$) and Eq. 23-35, we find that the potential energy U at any angle θ is

$$U = -W = -\int_{90°}^{\theta} \tau \, d\theta$$

$$= \int_{90°}^{\theta} pE \sin \theta \, d\theta. \tag{23-36}$$

Evaluating the integral leads to

$$U = -pE \cos \theta. \tag{23-37}$$

We can generalize this equation to vector form as

$$U = -\mathbf{p} \cdot \mathbf{E} \qquad \begin{array}{l}\text{(potential energy} \\ \text{of a dipole).}\end{array} \tag{23-38}$$

Equations 23-37 and 23-38 show us that the potential energy of the dipole is least ($U = -pE$) when $\theta = 0$, which is when $\mathbf{p}$ and $\mathbf{E}$ are in the same direction; the potential energy is greatest ($U = pE$) when $\theta = 180°$, which is when $\mathbf{p}$ and $\mathbf{E}$ are in opposite directions.

CHECKPOINT **5:** The figure shows four orientations of an electric dipole in an external electric field. Rank the orientations according to (a) the magnitude of the torque on the dipole and (b) the potential energy of the dipole, greatest first.

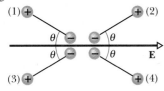

Microwave Cooking

In liquid water, where molecules are relatively free to move around, the electric field produced by each molecular dipole affects the surrounding dipoles. As a result, the molecules bond together in groups of two or three, because the negative (oxygen) end of one dipole and a positive (hydrogen) end of another dipole attract each other. Each time a group is formed, electric potential energy is transferred to the random thermal motion of the group and the surrounding molecules. And each time collisions among the molecules break up a group, the transfer is reversed. The temperature of the water (which is associated with the average thermal motion) does not change because, on the average, the net transfer of energy is zero.

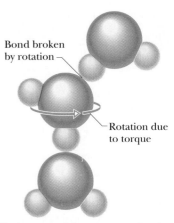

FIGURE 23-18   A group of three water molecules. A torque due to an oscillating electric field in a microwave oven breaks one of the bonds between the molecules and thus breaks up the group.

In a microwave oven, the story differs. When the oven is operated, the microwaves produce (in the oven) an electric field that rapidly oscillates back and forth in direction. If there is water in the oven, the oscillating field exerts oscillating torques on the water molecules, continually rotating them back and forth to align their dipole moments with the field direction. Molecules that are bonded as a pair can twist around their common bond to stay aligned, but molecules that are bonded in a group of three must break at least one of their two bonds (Fig. 23-18).

The energy to break these bonds comes from the electric field, that is, from the microwaves. Then molecules that have broken away from groups can form new groups, transferring the energy they just gained into thermal energy. Thus thermal energy is added to the water when the groups form but is not removed when the groups break apart, and the temperature of the water increases. Foods that contain water can be cooked in a microwave oven because of the heating of that water. If a water molecule were not an electric dipole, this would not be so and microwave ovens would be useless.

SAMPLE PROBLEM 23-9

A neutral water molecule (H_2O) in its vapor state has an electric dipole moment of 6.2×10^{-30} C·m.

(a) How far apart are the molecule's centers of positive and negative charge?

SOLUTION: There are 10 electrons and 10 protons in this molecule. So the magnitude of the dipole moment is

$$p = qd = (10e)(d),$$

in which d is the separation we are seeking and e is the elementary charge. Thus

$$d = \frac{p}{10e} = \frac{6.2 \times 10^{-30} \text{ C} \cdot \text{m}}{(10)(1.60 \times 10^{-19} \text{ C})}$$

$$= 3.9 \times 10^{-12} \text{ m} = 3.9 \text{ pm.} \qquad \text{(Answer)}$$

This distance is not only small, but it is actually smaller than the radius of a hydrogen atom.

(b) If the molecule is placed in an electric field of 1.5×10^4 N/C, what maximum torque can the field exert on it? (Such a field can easily be set up in the laboratory.)

SOLUTION: From Eq. 23-33 we know that the torque is a maximum when $\theta = 90°$. Substituting this value in that equation yields

$$\tau = pE \sin \theta$$

$$= (6.2 \times 10^{-30} \text{ C} \cdot \text{m})(1.5 \times 10^4 \text{ N/C})(\sin 90°)$$

$$= 9.3 \times 10^{-26} \text{ N} \cdot \text{m.} \qquad \text{(Answer)}$$

(c) How much work must an external agent do to turn this molecule end for end in this field, starting from its fully aligned position, for which $\theta = 0$?

SOLUTION: The work is the difference in potential energy between the positions $\theta = 180°$ and $\theta = 0$. Using Eq. 23-37, we get

$$W = U(180°) - U(0)$$

$$= (-pE \cos 180°) - (-pE \cos 0)$$

$$= 2pE = (2)(6.2 \times 10^{-30} \text{ C} \cdot \text{m})(1.5 \times 10^4 \text{ N/C})$$

$$= 1.9 \times 10^{-25} \text{ J.} \qquad \text{(Answer)}$$

REVIEW & SUMMARY

Electric Field

One way to explain the electrostatic force between charges is to assume that each charge sets up an electric field in the space around it. The electrostatic force acting on any one charge is then due to the electric field set up at its location by the other charges.

Definition of Electric Field

The *electric field* **E** at any point is defined in terms of the electrostatic force **F** that would be exerted on a positive test charge q_0 placed there:

$$\mathbf{E} = \frac{\mathbf{F}}{q_0}. \qquad (23\text{-}1)$$

Electric Field Lines

Electric field lines provide a means for visualizing the direction and magnitude of electric fields. The electric field vector at any point is tangent to a field line through that point. The density of field lines in any region is proportional to the magnitude of the electric field in that region. Field lines originate on positive charges and terminate on negative charges.

Field Due to a Point Charge

The magnitude of the electric field **E** set up by a point charge q at a distance r from the charge is

$$E = \frac{1}{4\pi\epsilon_0} \frac{|q|}{r^2}. \qquad (23\text{-}3)$$

The direction of **E** is away from the point charge if the charge is positive and toward the point charge if the charge is negative.

Field Due to an Electric Dipole

An *electric dipole* consists of two particles with charges of equal magnitude q but opposite sign, separated by a small distance d. Their **dipole moment p** has magnitude qd and points from the negative charge to the positive charge. The magnitude of the electric field set up by the dipole at a distant point on the dipole axis (which runs through both charges) is

$$E = \frac{1}{2\pi\epsilon_0} \frac{p}{z^3}, \qquad (23\text{-}9)$$

where z is the distance between the point and the dipole center.

Field Due to a Continuous Charge Distribution

The electric field due to a *continuous charge distribution* is found by treating charge elements as point charges and then summing, via integration, the electric field vectors produced by all the charge elements.

Force on a Point Charge in an Electric Field

When a point charge q is placed in an electric field **E** set up by other charges, the electrostatic force **F** that acts on the point charge is

$$\mathbf{F} = q\mathbf{E}. \qquad (23\text{-}28)$$

Force **F** points in the direction of **E** if q is positive and opposite **E** if q is negative.

Dipole in an Electric Field

When an electric dipole of dipole moment **p** is placed in an electric field **E**, the field exerts a torque τ on the dipole:

$$\boldsymbol{\tau} = \mathbf{p} \times \mathbf{E}. \qquad (23\text{-}34)$$

The dipole has a potential energy U associated with its orientation in the field:

$$U = -\mathbf{p} \cdot \mathbf{E}. \qquad (23\text{-}38)$$

This potential energy is defined to be zero when **p** is perpendicular to **E**; it is least ($U = -pE$) when **p** is aligned with **E**, and most ($U = pE$) when **p** is directed opposite **E**.

QUESTIONS

1. Figure 23-19 shows three electric field lines. What is the direction of the electrostatic force on a positive test charge placed at (a) point A and (b) point B? (c) At which point, A or B, will the acceleration of the test charge be greater if the charge is released?

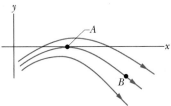

FIGURE 23-19 Question 1.

2. Figure 23-20a shows two charged particles on an axis. (a) Where on the axis (other than at an infinite distance) is there a point at which their net electric field is zero: between the charges, to their left, or to their right? (b) Is there a point of zero electric field off the axis (other than at an infinite distance)?

FIGURE 23-20 Questions 2 and 3.

3. Figure 23-20b shows two protons and an electron that are evenly spaced on an axis. Where on the axis (other than at an infinite distance) is there a point at which their net electric field is zero: to the left of the particles, to their right, between the two protons, or between the electron and the nearer proton?

4. Figure 23-21 shows two square arrays of charged particles. The squares, which are centered on point P, are misaligned. The particles are separated by either d or $d/2$ along the perimeters of the squares. What are the magnitude and direction of the net electric field at P?

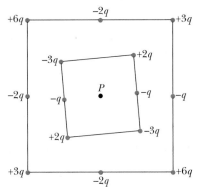

FIGURE 23-21 Question 4.

5. In Fig. 23-22, two particles of charge $-q$ are arranged symmetrically about the y axis; each produces an electric field at point P on that axis. (a) Are the magnitudes of the fields at P equal? (b) Does each electric field point toward or away from the charge producing it? (c) Is the magnitude of the net electric field at P

equal to the sum of the magnitudes of the two field vectors (is it equal to $2E$)? (d) Do the x components of those two field vectors add or cancel? (e) Do their y components add or cancel? (f) Is the direction of the net field at P that of the canceling components or the adding components? (g) What is the direction of the net field?

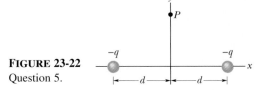

FIGURE 23-22
Question 5.

6. Three circular nonconducting rods of the same radius of curvature have uniform charges. Rod A has charge $+2Q$ and subtends an arc of 30°; rod B has charge $+6Q$ and subtends 90°; rod C has charge $+4Q$ and subtends 60°. Rank the rods according to their linear charge density, greatest first.

7. In Fig. 23-23, two identical circular nonconducting rings are centered on the same line. For three situations, the uniform charges on rings A and B are, respectively, (1) q_0 and q_0, (2) $-q_0$ and $-q_0$, (3) $-q_0$ and q_0. Rank the situations according the magnitude of the net electric field at (a) point P_1 midway between the rings, (b) point P_2 at the center of ring 2, (c) point P_3 to the right of ring 2, greatest first.

FIGURE 23-23
Question 7.

8. In Fig. 23-24a, a circular plastic rod with uniform charge $+Q$ produces an electric field of magnitude E at the center of curvature (at the origin). In Figs. 23-24b, c, and d, more circular rods with identical uniform charges $+Q$ are added until the circle is complete. A fifth arrangement (which would be labeled e) is like that in d except that the rod in the fourth quadrant has charge $-Q$.

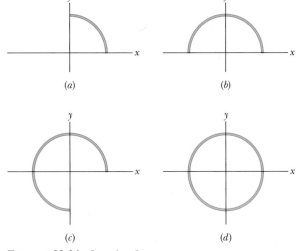

FIGURE 23-24 Question 8.

Rank the five arrangements according to the magnitude of the electric field at the center of curvature, greatest first.

9. In Fig. 23-25, an electron e travels through a small hole in plate A and then toward plate B. A uniform electric field in the region between the plates then slows the electron without deflecting it. (a) What is the direction of the field? (b) Four other particles similarly travel through small holes in either plate A or plate B and then into the region between the plates. Three have charges $+q_1$, $+q_2$, and $-q_3$. The fourth (labeled n) is a neutron, which is electrically neutral. Does the speed of each of those four other particles increase, decrease, or remain the same in the region between the plates?

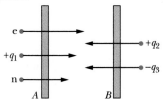

FIGURE 23-25 Question 9.

10. Figure 23-26 shows the path of negatively charged particle 1 through a rectangular region of uniform electric field; the particle is deflected toward the top of the page. (a) Is the field directed leftward, rightward, toward the top of the page, or toward the bottom? (b) Three other charged particles are shown approaching the region of electric field. Which are deflected toward the top of the page and which toward the bottom?

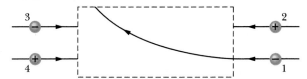

FIGURE 23-26 Question 10.

11. Figure 23-27 shows three arrangements of electric field lines. In each arrangement, a proton is released from rest at point A and is then accelerated through point B by the electric field. Points A and B have equal separations in the three arrangements. Rank the arrangements according to the linear momentum of the proton when it reaches point B, greatest first.

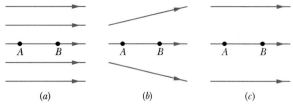

FIGURE 23-27 Question 11.

12. (a) In Checkpoint 5, if the dipole rotates from orientation 1 to orientation 2, is the work done on the dipole by the field positive, negative, or zero? (b) If, instead, the dipole rotates from orientation 1 to orientation 4, is the work done by the field more than, less than, or the same as in (a)?

13. The potential energies associated with four orientations of an electric dipole in an electric field are (1) $-5U_0$, (2) $-7U_0$, (3) $3U_0$, and (4) $5U_0$, where U_0 is positive. Rank the orientations according to (a) the angle between the electric dipole moment $\mathbf{p}$ and the electric field $\mathbf{E}$, and (b) the magnitude of the torque on the electric dipole, greatest first.

14. If you walk across some types of carpet on a dry day and then reach for a metal doorknob or (for more fun) the back of someone's neck, you might produce a spark. Why does the spark occur? (You can increase the brightness and noise of the spark if you reach with a pointed finger or, even better, a metal key with the pointed end forward.)

EXERCISES & PROBLEMS

SECTION 23-3 Electric Field Lines

1E. In Fig. 23-28 the electric field lines on the left have twice the separation as those on the right. (a) If the magnitude of the field at A is 40 N/C, what force acts on a proton at A? (b) What is the magnitude of the field at B?

FIGURE 23-28 Exercise 1.

2E. Sketch qualitatively the electric field lines for two nearby point charges $+q$ and $-2q$.

3E. In Fig. 23-29, three point charges are arranged in an equilateral triangle. Sketch the field lines due to $+Q$ and $-Q$, and from them determine the direction of the force that acts on $+q$ because of the presence of the other two charges. (*Hint:* See Fig. 23-5.)

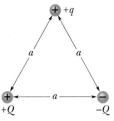

FIGURE 23-29 Exercise 3.

4E. Sketch qualitatively the electric field lines both between and outside two concentric conducting spherical shells when a uniform positive charge q_1 is on the inner shell and a uniform negative charge $-q_2$ is on the outer. Consider the cases $q_1 > q_2$, $q_1 = q_2$, and $q_1 < q_2$.

5E. Sketch qualitatively the electric field lines for a thin, circular, uniformly charged disk of radius R. (*Hint:* Consider as limiting cases points very close to the disk, where the electric field is perpendicular to the surface, and points very far from it, where the electric field is like that of a point charge.)

SECTION 23-4 The Electric Field Due to a Point Charge

6E. What is the magnitude of a point charge that would create an electric field of 1.00 N/C at points 1.00 m away?

7E. What is the magnitude of a point charge whose electric field 50 cm away has the magnitude 2.0 N/C?

8E. Two opposite charges of equal magnitude 2.0×10^{-7} C are held 15 cm apart. What are the magnitude and direction of **E** at the point midway between the charges?

9E. An atom of plutonium-239 has a nuclear radius of 6.64 fm and the atomic number $Z = 94$. Assuming that the positive charge of the nucleus is distributed uniformly, what are the magnitude and direction of the electric field at the surface of the nucleus due to the positive charge?

10P. A particle of charge $-q_1$ is located at the origin of an x axis. (a) At what location should a second particle of charge $-4q_1$ be placed so that the net electric field of the two particles is zero at $x = 2.0$ mm? (b) If, instead, a particle of charge $+4q_1$ is placed at that location, what is the direction of the net electric field at $x = 2.0$ mm?

11P. In Fig. 23-30, two point charges $q_1 = +1.0 \times 10^{-6}$ C and $q_2 = +3.0 \times 10^{-6}$ C are separated by a distance $d = 10$ cm. Plot their net electric field $E(x)$ as a function of x for both positive and negative values of x, taking E to be positive when the vector **E** points to the right and negative when **E** points to the left.

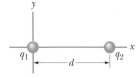

FIGURE 23-30
Problems 11 and 12.

12P. (a) In Fig. 23-30, two point charges $q_1 = -5q$ and $q_2 = +2q$ are separated by distance d. Locate the point (or points) at which the electric field due to the two charges is zero. (b) Sketch the electric field lines qualitatively.

13P. In Fig. 23-31, point charges $+1.0q$ and $-2.0q$ are fixed a distance d apart. (a) Find **E** at points A, B, and C. (b) Sketch the electric field lines.

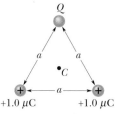

FIGURE 23-31 Problem 13.

14P. Two charges $q_1 = 2.1 \times 10^{-8}$ C and $q_2 = -4.0q_1$ are placed 50 cm apart. Find the point along the straight line passing through the two charges at which the electric field is zero.

15P. In Fig. 23-32, what is the electric field at point P due to the four point charges shown?

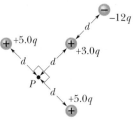

FIGURE 23-32 Problem 15.

16P. A proton and an electron form two corners of an equilateral triangle of side length 2.0×10^{-6} m. What is the magnitude of their net electric field at the third corner?

17P. A clock face has negative point charges $-q$, $-2q$, $-3q$, $\cdots$, $-12q$ fixed at the positions of the corresponding numerals. The clock hands do not perturb the net field due to the point charges. At what time does the hour hand point in the same direction as the electric field vector at the center of the dial? (*Hint:* Use symmetry.)

18P. An electron is placed at each corner of an equilateral triangle having sides 20 cm long. What is the magnitude of the electric field at the midpoint of one of the sides?

19P. Calculate the direction and magnitude of the electric field at point P in Fig. 23-33.

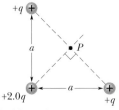

FIGURE 23-33 Problem 19.

20P. In Fig. 23-34, charges are placed at the vertices of an equilateral triangle. For what value of Q (both sign and magnitude) does the total electric field vanish at C, the center of the triangle?

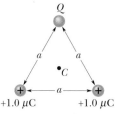

FIGURE 23-34 Problem 20.

21P. In Fig. 23-35, four charges form the corners of a square and four more charges lie at the midpoints of the sides of the square. The distance between adjacent charges on the perimeter of the square is d. What are the magnitude and direction of the electric field at the center of the square?

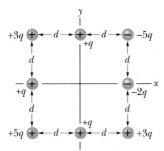

FIGURE 23-35   Problem 21.

22P.   What are the magnitude and direction of the electric field at the center of the square of Fig. 23-36 if $q = 1.0 \times 10^{-8}$ C and $a = 5.0$ cm?

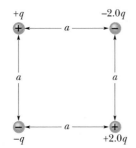

FIGURE 23-36
Problem 22.

SECTION 23-5   The Electric Field Due to an Electric Dipole

23E.   Calculate the electric dipole moment of an electron and a proton 4.30 nm apart.

24E.   In Fig. 23-8, let both charges be positive. Assuming $z \gg d$, show that E at point P in that figure is then given by

$$E = \frac{1}{4\pi\epsilon_0}\frac{2q}{z^2}.$$

25P.   Find the magnitude and direction of the electric field at point P due to the electric dipole in Fig. 23-37. P is located at a distance $r \gg d$ along the perpendicular bisector of the line joining the charges. Express your answer in terms of the magnitude and direction of the electric dipole moment $\mathbf{p}$.

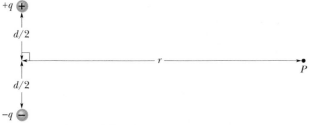

FIGURE 23-37   Problem 25.

26P*.   *Electric quadrupole.* Figure 23-38 shows an electric quadrupole. It consists of two dipoles with dipole moments that are equal in magnitude but opposite in direction. Show that the

value of E on the axis of the quadrupole for points a distance z from its center (assume $z \gg d$) is given by

$$E = \frac{3Q}{4\pi\epsilon_0 z^4},$$

in which $Q \, (= 2qd^2)$ is known as the *quadrupole moment* of the charge distribution.

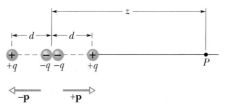

FIGURE 23-38   Problem 26.

SECTION 23-6   The Electric Field Due to a Line of Charge

27E.   Make a quantitative plot of the electric field along the central axis of a charged ring having a diameter of 6.0 cm and a uniformly distributed charge of 1.0×10^{-8} C.

28E.   Figure 23-39 shows two parallel nonconducting rings arranged with their central axes along a common line. Ring 1 has uniform charge q_1 and radius R; ring 2 has uniform charge q_2 and the same radius R. The rings are separated by a distance $3R$. The net electric field at point P on the common line, at distance R from ring 1, is zero. What is the ratio q_1/q_2?

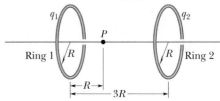

FIGURE 23-39   Exercise 28.

29P.   At what distance along the central axis of a ring of radius R and uniform charge is the magnitude of the electric field due to the ring's charge maximum?

30P.   An electron is constrained to the central axis of the ring of charge of radius R discussed in Section 23-6. Show that the electrostatic force exerted on the electron can cause it to oscillate through the center of the ring with an angular frequency

$$\omega = \sqrt{\frac{eq}{4\pi\epsilon_0 mR^3}},$$

where q is the ring's charge and m is the electron's mass.

31P.   In Fig. 23-40a, two curved plastic rods, one of charge $+q$ and the other of charge $-q$, form a circle of radius R in an xy plane. The x axis passes through their connecting points, and the charge is distributed uniformly on both rods. What are the magnitude and direction of the electric field $\mathbf{E}$ produced at P, the center of the circle?

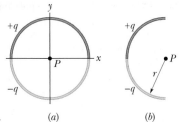

FIGURE 23-40
Problems 31 and 32.   (a)   (b)

32P. A thin glass rod is bent into a semicircle of radius r. A charge $+q$ is uniformly distributed along the upper half, and a charge $-q$ is uniformly distributed along the lower half, as shown in Fig. 23-40b. Find the magnitude and direction of the electric field **E** at P, the center of the semicircle.

33P. A thin nonconducting rod of finite length L has a charge q spread uniformly along it. Show that the magnitude E of the electric field at point P on the perpendicular bisector of the rod (Fig. 23-41) is given by

$$E = \frac{q}{2\pi\epsilon_0 y} \frac{1}{(L^2 + 4y^2)^{1/2}}.$$

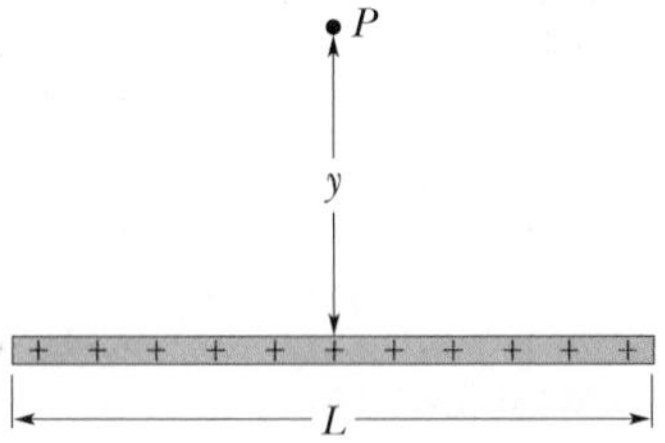

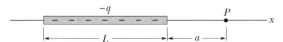

FIGURE 23-41
Problem 33.

34P. In Fig. 23-42, a nonconducting rod of length L has charge $-q$ uniformly distributed along its length. (a) What is the linear charge density of the rod? (b) What is the electric field at point P, a distance a from the end of the rod? (c) If P were very far from the rod compared to L, the rod would look like a point charge. Show that your answer to (b) reduces to the electric field of a point charge for $a \gg L$.

FIGURE 23-42 Problem 34.

35P*. In Fig. 23-43, a "semi-infinite" nonconducting rod (that is, infinite in one direction only) has uniform linear charge density λ. Show that the electric field at point P makes an angle of $45°$ with the rod and that this result is independent of the distance R. (*Hint:* Separately find the parallel and perpendicular (to the rod) components of the electric field at P, and then compare those components.)

FIGURE 23-43 Problem 35.

SECTION 23-7 The Electric Field Due to a Charged Disk

36E. A disk of radius 2.5 cm has a surface charge density of 5.3 μC/m² on its upper face. What is the magnitude of the electric field produced by the disk at a point on its central axis at distance $z = 12$ cm from the disk?

37P. (a) What total (excess) charge q must the disk in Sample Problem 23-6 (Fig. 23-12) have for the electric field on the surface of the disk at its center to equal the value at which air breaks down electrically, producing sparks? Take the disk radius as 2.5 cm, and use the listing for air in Table 23-1. (b) Suppose that each atom at the surface has an effective cross-sectional area of 0.015 nm². How many atoms are needed to make up the disk's surface? (c) The charge in (a) results from some of the surface atoms having one excess electron. What fraction of the surface atoms must be so charged?

38P. At what distance along the central axis of a uniformly charged plastic disk of radius R is the magnitude of the electric field equal to one-half the magnitude of the field at the center of the surface of the disk?

SECTION 23-8 A Point Charge in an Electric Field

39E. An electron is released from rest in a uniform electric field of magnitude 2.00×10^4 N/C. Calculate the acceleration of the electron. (Ignore gravitation.)

40E. An electron is accelerated eastward at 1.80×10^9 m/s² by an electric field. Determine the magnitude and direction of the electric field.

41E. Calculate the magnitude of the force, due to an electric dipole of dipole moment 3.6×10^{-29} C·m, on an electron 25 nm from the center of the dipole, along the dipole axis. Assume that this distance is large relative to the dipole's charge separation.

42E. Humid air breaks down (its molecules become ionized) in an electric field of 3.0×10^6 N/C. In that field, what is the magnitude of the electrostatic force on (a) an electron and (b) an ion with a single electron missing?

43E. An alpha particle (the nucleus of a helium atom) has a mass of 6.64×10^{-27} kg and a charge of $+2e$. What are the magnitude and direction of the electric field that will balance its weight?

44E. A charged cloud system produces an electric field in the air near the Earth's surface. A particle of charge -2.0×10^{-9} C is acted on by a downward electrostatic force of 3.0×10^{-6} N when placed in this field. (a) What is the magnitude of the electric field? (b) What are the magnitude and direction of the electrostatic force exerted on a proton placed in this field? (c) What is the gravitational force on the proton? (d) What is the ratio of the electrostatic force to the gravitational force in this case?

45E. An electric field **E** with an average magnitude of about 150 N/C points downward in the atmosphere near Earth's surface. We wish to "float" a sulfur sphere weighing 4.4 N in this field by charging the sphere. (a) What charge (both sign and magnitude) must be used? (b) Why is the experiment impractical?

46E. (a) What is the acceleration of an electron in a uniform electric field of 1.40×10^6 N/C? (b) How long would it take for the electron, starting from rest, to attain one-tenth the speed of light? (c) How far would it travel in that time? (Use Newtonian mechanics.)

47E. Beams of high-speed protons can be produced in "guns" using electric fields to accelerate the protons. (a) What acceleration would a proton experience if the gun's electric field were 2.00×10^4 N/C? (b) What speed would the proton attain if the field accelerated the proton through a distance of 1.00 cm?

48E. An electron with a speed of 5.00×10^8 cm/s enters an electric field of magnitude 1.00×10^3 N/C, traveling along the field in the direction that retards its motion. (a) How far will the electron travel in the field before stopping momentarily and (b) how much time will have elapsed? (c) If, instead, the region of electric field is only 8.00 mm wide (too small for the electron to stop), what fraction of the electron's initial kinetic energy will be lost in that region?

49E. A spherical water drop 1.20 μm in diameter is suspended in calm air owing to a downward-directed atmospheric electric field $E = 462$ N/C. (a) What is the weight of the drop? (b) How many excess electrons does it have?

50E. In Millikan's experiment, an oil drop of radius 1.64 μm and density 0.851 g/cm^3 is suspended in chamber C when a downward-pointing electric field of 1.92×10^5 N/C is applied. Find the charge on the drop, in terms of e.

51P. In one of his experiments, Millikan observed that the following measured charges, among others, appeared at different times on a single drop:

6.563×10^{-19} C	13.13×10^{-19} C	19.71×10^{-19} C
8.204×10^{-19} C	16.48×10^{-19} C	22.89×10^{-19} C
11.50×10^{-19} C	18.08×10^{-19} C	26.13×10^{-19} C

What value for the elementary charge e can be deduced from these data?

52P. A uniform electric field exists in a region between two oppositely charged plates. An electron is released from rest at the surface of the negatively charged plate and strikes the surface of the opposite plate, 2.0 cm away, in a time 1.5×10^{-8} s. (a) What is the speed of the electron as it strikes the second plate? (b) What is the magnitude of the electric field **E**?

53P. An object having a mass of 10.0 g and a charge of $+8.00 \times 10^{-5}$ C is placed in an electric field **E** with $E_x = 3.00 \times 10^3$ N/C, $E_y = -600$ N/C, and $E_z = 0$. (a) What are the magnitude and direction of the force on the object? (b) If the object is released from rest at the origin, what will be its coordinates after 3.00 s?

54P. At some instant the velocity components of an electron moving between two charged parallel plates are $v_x = 1.5 \times 10^5$ m/s and $v_y = 3.0 \times 10^3$ m/s. Suppose that the electric field between the plates is given by $\mathbf{E} = (120$ N/C$)\mathbf{j}$. (a) What is the acceleration of the electron? (b) What will be the velocity of the electron after its x coordinate has changed by 2.0 cm?

55P. Two large parallel copper plates are 5.0 cm apart and have a uniform electric field between them as depicted in Fig. 23-44. An electron is released from the negative plate at the same time that a proton is released from the positive plate. Neglect the force of the particles on each other and find their distance from the positive plate when they pass each other. (Does it surprise you that you need not know the electric field to solve this problem?)

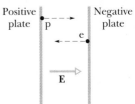

FIGURE 23-44 Problem 55.

56P. In Fig. 23-45, a pendulum is hung from the higher of two large horizontal plates. The pendulum consists of a small nonconducting sphere of mass m and charge $+q$ and an insulating thread of length l. What is the period of the pendulum if a uniform electric field **E** is set up between the plates by (a) charging the top plate negatively and the lower plate positively and (b) vice versa? In both cases, the field points away from one plate and directly toward the other plate.

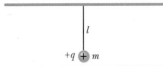

FIGURE 23-45 Problem 56.

57P. In Fig. 23-46, a uniform, upward-pointing electric field **E** of magnitude 2.00×10^3 N/C has been set up between two horizontal plates by charging the lower plate positively and the upper plate negatively. The plates have length $L = 10.0$ cm and separation $d = 2.00$ cm. An electron is then shot between the plates from the left edge of the lower plate. The initial velocity $\mathbf{v}_0$ of the electron makes an angle $\theta = 45.0°$ with the lower plate and has a magnitude of 6.00×10^6 m/s. (a) Will the electron strike one of the plates? (b) If so, which plate and how far horizontally from the left edge?

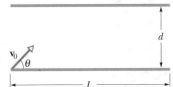

FIGURE 23-46 Problem 57.

SECTION 23-9 A Dipole in an Electric Field

58E. An electric dipole, consisting of charges of magnitude 1.50 nC separated by 6.20 μm, is in an electric field of strength 1100 N/C. (a) What is the magnitude of the electric dipole moment? (b) What is the difference in potential energy corresponding to dipole orientations parallel to and antiparallel to the field?

59E. An electric dipole consists of charges $+2e$ and $-2e$ separated by 0.78 nm. It is in an electric field of strength 3.4×10^6 N/C. Calculate the magnitude of the torque on the dipole when the dipole moment is (a) parallel to, (b) perpendicular to, and (c) antiparallel to the electric field.

60P. Find the work required to turn an electric dipole end for end in a uniform electric field **E**, in terms of the magnitude p of the dipole moment, the magnitude E of the field, and the initial angle θ_0 between **p** and **E**.

61P. Find the angular frequency of oscillation of an electric dipole, of dipole moment p and rotational inertia I, for small amplitudes of oscillation about its equilibrium position in a uniform electric field of magnitude E.

62P. An electric dipole with dipole moment

$$\mathbf{p} = (3.00\mathbf{i} + 4.00\mathbf{j})(1.24 \times 10^{-30} \text{ C·m})$$

is in an electric field $\mathbf{E} = (4000$ N/C$)\mathbf{i}$. (a) What is the potential energy of the electric dipole? (b) What is the torque acting on it?

(c) If an external agent turns the dipole until its electric dipole moment is

$$\mathbf{p} = (-4.00\mathbf{i} + 3.00\mathbf{j})(1.24 \times 10^{-30} \text{ C·m}),$$

how much work is done by the agent?

Electronic Computation

63. Two particles, each of positive charge q, are fixed in place on an y axis, one at $y = d$ and the other at $y = -d$. (a) Write an expression that gives the magnitude E of the net electric field at points on the x axis given by $x = \alpha d$. (b) Graph E versus α for the range $0 < \alpha < 4$. From the graph, determine the values of α that give (c) the maximum value of E and (d) half the maximum value of E.

64. For the data of Problem 51, assume that the charge q on the drop is given by $q = ne$, where n is an integer and e is the elementary charge. (a) Find n for each measurement of q. (b) Do a linear regression fit of the values of q versus the values of n; from it find e.

24
Gauss' Law

Lightning strikes Manhattan in a brilliant display, each strike delivering about 10^{20} electrons from the cloud base to the ground. How wide is a lightning strike? Since it can be seen from kilometers away, is it as wide as, say, a car?

24-1 A NEW LOOK AT COULOMB'S LAW

If you want to find the center of mass of a potato, you can do so by experiment or by laborious calculation, involving the numerical evaluation of a triple integral. However, if the potato happens to be a uniform ellipsoid, you know from its symmetry exactly where the center of mass is without calculation. Such are the advantages of symmetry. Symmetrical situations arise in all areas of physics; when possible, it makes sense to cast the laws of physics in forms that take full advantage of this fact.

Coulomb's law is the governing law in electrostatics, but it is not cast in a form that particularly simplifies the work in situations involving symmetry. In this chapter we introduce a new formulation of Coulomb's law, derived by German mathematician and physicist Carl Friedrich Gauss (1777–1855). This law, called **Gauss' law,** *can* be used to take advantage of special symmetry situations. For electrostatics problems, it is the full equivalent of Coulomb's law; which of them we choose to use depends only on the problem at hand.

Central to Gauss' law is a hypothetical closed surface called a **Gaussian surface.** The Gaussian surface can be of any shape you wish to make it, but the most useful surface is one that mimics the symmetry of the problem at hand. Thus the Gaussian surface will often be a sphere, a cylinder, or some other symmetrical form. It must always be a *closed* surface, so that a clear distinction can be made between points that are inside the surface, on the surface, and outside the surface.

Imagine that you have established a Gaussian surface around a distribution of charges. Then Gauss' law comes into play:

> Gauss' law relates the electric fields at points on a (closed) Gaussian surface and the net charge enclosed by that surface.

Figure 24-1 shows a simple situation in which the Gaussian surface is a sphere. Suppose you know that there is an electric field at every point on the surface and that all the fields have the same magnitude and point radially outward. Without knowing anything about Gauss' law, you can guess that some net positive charge must be inside the Gaussian surface. If you *do* know Gauss' law, you can calculate just how much net positive charge is inside the surface. To make the calculation, you need know only "how much" electric field is intercepted by the surface: this "how much" involves the *flux* of the electric field through the surface.

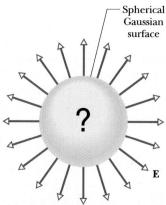

FIGURE 24-1 A spherical Gaussian surface. If the electric field vectors are of uniform magnitude and point radially outward at all surface points, you can conclude that a net positive distribution of charge must lie within the surface and have spherical symmetry.

24-2 FLUX

Suppose, as in Fig. 24-2*a*, that you aim a wide airstream of uniform velocity **v** at a small square loop of area A. Let Φ represent the *volume flow rate* (volume per unit time) at

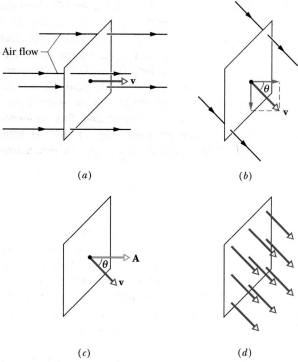

FIGURE 24-2 (*a*) A uniform airstream of velocity **v** is perpendicular to the plane of a square loop of area A. (*b*) The component of **v** perpendicular to the plane of the loop is $v \cos \theta$, where θ is the angle between **v** and a normal to the plane. (*c*) The area vector **A** is perpendicular to the plane of the loop and makes an angle θ with **v**. (*d*) The velocity field intercepted by the area of the loop.

which air flows through the loop. This rate depends on the angle between **v** and the plane of the loop. If **v** is perpendicular to the plane, the rate Φ is equal to vA.

If **v** is parallel to the plane of the loop, no air moves through the loop, so Φ is zero. For an intermediate angle θ, the rate Φ depends on the component of **v** that is normal to the plane (Fig. 24-2b). Since that component is $v \cos \theta$, the rate of volume flow through the loop is

$$\Phi = (v \cos \theta)A. \qquad (24\text{-}1)$$

This rate of flow through an area is an example of a **flux** —a *volume flux* in this situation. Before we discuss a flux that is involved in electrostatics, we need to rewrite Eq. 24-1 in terms of vectors.

To do this, we first define an *area vector* **A** as being a vector whose magnitude is equal to an area (here the area of the loop) and whose direction is normal to the plane of the area (Fig. 24-2c). We then rewrite Eq. 24-1 as the scalar (or dot) product of the velocity vector **v** of the airstream and the area vector **A** of the loop:

$$\Phi = vA \cos \theta = \mathbf{v} \cdot \mathbf{A}, \qquad (24\text{-}2)$$

where θ is the angle between **v** and **A**.

The word "flux" comes from the Latin word meaning "to flow." That meaning makes sense if we talk about the flow of air volume through the loop. However, Eq. 24-2 can be regarded in a more abstract way. To see it, note that we can assign a velocity vector to each point in the airstream passing through the loop (Fig. 24-2d). The composite of all those vectors is a *velocity field*. So we can interpret Eq. 24-2 as giving the *flux of the velocity field through the loop*. With this interpretation, flux no longer means the actual flow of something through an area. Rather it means the product of an area and the field across that area.

24-3 FLUX OF AN ELECTRIC FIELD

To define the flux of an electric field, consider Fig. 24-3a, which shows an arbitrary (asymmetric) Gaussian surface immersed in a nonuniform electric field. Let us divide the surface into small squares of area ΔA, each square being small enough to permit us to neglect any curvature and consider the individual square to be flat. We represent each such element of area with an area vector $\Delta \mathbf{A}$, whose magnitude is the area ΔA. Each vector $\Delta \mathbf{A}$ is perpendicular to the Gaussian surface and directed away from the interior of the surface.

Because the squares have been taken to be arbitrarily small, the electric field **E** may be taken as constant over any given square. The vectors $\Delta \mathbf{A}$ and **E** for each square then make some angle θ with each other. Figure 24-3b

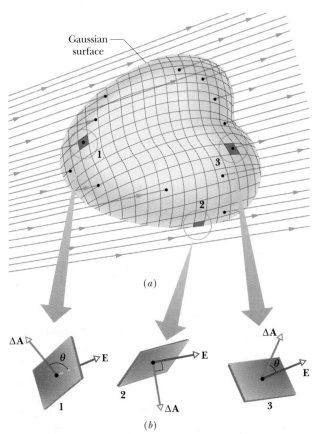

FIGURE 24-3 (a) A Gaussian surface of arbitrary shape immersed in an electric field. The surface is divided into small squares of area ΔA. (b) The electric field vectors **E** and the area vectors $\Delta \mathbf{A}$ for three representative squares, marked 1, 2, and 3.

shows an enlarged view of three squares (1, 2, and 3) on the Gaussian surface, and the angle θ for each.

A provisional definition for the flux of the electric field for the Gaussian surface of Fig. 24-3 is

$$\Phi = \sum \mathbf{E} \cdot \Delta \mathbf{A}. \qquad (24\text{-}3)$$

This equation instructs us to visit each square on the Gaussian surface, to evaluate the scalar product $\mathbf{E} \cdot \Delta \mathbf{A}$ for the two vectors **E** and $\Delta \mathbf{A}$ that we find there, and to sum the results algebraically (that is, with signs included) for all the squares that make up the surface. The sign resulting from each scalar product determines whether the flux through any given square is positive, negative, or zero. As Table 24-1 shows, squares like 1, in which **E** points inward, make a negative contribution to the sum of Eq. 24-3. Squares like 2, in which **E** lies in the surface, make zero contribution. And squares like 3, in which **E** points outward, make a positive contribution.

The exact definition of the flux of the electric field through a closed surface is found by allowing the area of the squares shown in Fig. 24-3a to become smaller and

TABLE 24-1 THREE SQUARES ON THE GAUSSIAN SURFACE OF FIG. 24-3

SQUARE	θ	DIRECTION OF E	SIGN OF $E \cdot \Delta A$
1	$> 90°$	Into the surface	Negative
2	$= 90°$	Parallel to the surface	Zero
3	$< 90°$	Out of the surface	Positive

smaller, approaching a differential limit dA. The area vectors then approach a differential limit dA. The sum of Eq. 24-3 then becomes an integral and we have, for the definition of electric flux,

$$\Phi = \oint \mathbf{E} \cdot d\mathbf{A} \quad \text{(electric flux through a Gaussian surface).} \quad (24\text{-}4)$$

The circle on the integral sign indicates that the integration is to be taken over the entire (closed) surface. The flux of the electric field is a scalar, and its SI unit is the newton–square-meter per coulomb ($N \cdot m^2/C$).

We can interpret Eq. 24-4 in the following way: First recall that we can use the density of electric field lines passing through an area as a measure of an electric field $\mathbf{E}$ there. Specifically, the magnitude E is proportional to the number of electric field lines per unit area. Thus, the dot product $\mathbf{E} \cdot d\mathbf{A}$ in Eq. 24-4 is proportional to the number of electric field lines passing through area dA. Then, because the integration in Eq. 24-4 is carried out over a Gaussian surface, which is closed, we see that

The electric flux Φ through a Gaussian surface is proportional to the net number of electric field lines passing through that surface.

For all points on the left cap, the angle θ between $\mathbf{E}$ and $d\mathbf{A}$ is 180° and the magnitude E of the field is constant. Thus,

$$\int_a \mathbf{E} \cdot d\mathbf{A} = \int E(\cos 180°) \, dA = -E \int dA = -EA,$$

where $\int dA$ gives the cap's area, $A \ (= \pi R^2)$. Similarly, for the right cap, where $\theta = 0$ for all points,

$$\int_c \mathbf{E} \cdot d\mathbf{A} = \int E(\cos 0) \, dA = EA.$$

Finally, for the cylindrical surface, where the angle θ is 90° at all points,

$$\int_b \mathbf{E} \cdot d\mathbf{A} = \int E(\cos 90°) \, dA = 0.$$

Substituting these results into Eq. 24-5 leads us to

$$\Phi = -EA + 0 + EA = 0. \quad \text{(Answer)}$$

This result is perhaps not surprising because the field lines that represent the electric field all pass entirely through the Gaussian surface, entering through the left end cap, leaving through the right end cap, and giving a net flux of zero.

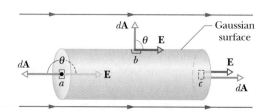

FIGURE 24-4 Sample Problem 24-1. A cylindrical Gaussian surface, closed by end caps, is immersed in a uniform electric field. The cylinder axis is parallel to the field direction.

SAMPLE PROBLEM 24-1

Figure 24-4 shows a Gaussian surface in the form of a cylinder of radius R immersed in a uniform electric field $\mathbf{E}$, with the cylinder axis parallel to the field. What is the flux Φ of the electric field through this closed surface?

SOLUTION: We can write the flux as the sum of three terms: integrals over the left cylinder cap a, the cylindrical surface b, and the right cap c. Thus from Eq. 24-4,

$$\Phi = \oint \mathbf{E} \cdot d\mathbf{A}$$

$$= \int_a \mathbf{E} \cdot d\mathbf{A} + \int_b \mathbf{E} \cdot d\mathbf{A} + \int_c \mathbf{E} \cdot d\mathbf{A}. \quad (24\text{-}5)$$

C**HECKPOINT 1:** The figure shows a Gaussian cube of face area A immersed in a uniform electric field $\mathbf{E}$ that points in the positive direction of the z axis. In terms of E and A, what is the flux through (a) the front face (which is in the xy plane), (b) the rear face, (c) the top face, and (d) the whole cube?

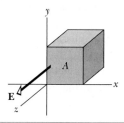

SAMPLE PROBLEM 24-2

A *nonuniform* electric field given by $\mathbf{E} = 3.0x\mathbf{i} + 4.0\mathbf{j}$ pierces the Gaussian cube shown in Fig. 24-5. ($\mathbf{E}$ is in newtons per coulomb and x is in meters.) What is the electric flux through the right face, the left face, and the top face?

SOLUTION: *Right face:* An area vector $\mathbf{A}$ is always perpendicular to its surface and always points away from the interior of a Gaussian surface. Thus, the vector $d\mathbf{A}$ for the right face of the cube must point in the positive x direction. In unit vector notation, then,

$$d\mathbf{A} = dA\mathbf{i}.$$

From Eq. 24-4, the flux Φ_r through the right face is then

$$\Phi_r = \int \mathbf{E} \cdot d\mathbf{A} = \int (3.0x\mathbf{i} + 4.0\mathbf{j}) \cdot (dA\mathbf{i})$$

$$= \int [(3.0x)(dA)\mathbf{i} \cdot \mathbf{i} + (4.0)(dA)\mathbf{j} \cdot \mathbf{i})]$$

$$= \int (3.0x \, dA + 0) = 3.0 \int x \, dA.$$

We are about to integrate over the right face, but we note that x has the same value everywhere on that face, namely $x = 3.0$ m. This means we can substitute that constant value for x. Then

$$\Phi_r = 3.0 \int (3.0) \, dA = 9.0 \int dA.$$

Now the integral merely gives us the area $A = 4.0$ m² of the right face. So,

$$\Phi_r = (9.0 \text{ N/C})(4.0 \text{ m}^2) = 36 \text{ N} \cdot \text{m}^2/\text{C}. \quad \text{(Answer)}$$

Left face: The procedure for finding the flux through the left face is the same as that for the right face. However, two factors change. (1) The differential area vector $d\mathbf{A}$ points in the negative x direction and thus $d\mathbf{A} = -dA\mathbf{i}$. (2) The term x again appears in our integration, and it is again constant over the face being considered. But on the left face, $x = 1.0$ m. With these two changes, we find that the flux Φ_l through the left face is

$$\Phi_l = -12 \text{ N} \cdot \text{m}^2/\text{C}. \quad \text{(Answer)}$$

Top face: The differential area vector $d\mathbf{A}$ points in the positive y direction and thus $d\mathbf{A} = dA\mathbf{j}$. The flux Φ_t through the top face is then

$$\Phi_t = \int (3.0x\mathbf{i} + 4.0\mathbf{j}) \cdot (dA\mathbf{j})$$

$$= \int [(3.0x)(dA)\mathbf{i} \cdot \mathbf{j} + (4.0)(dA)\mathbf{j} \cdot \mathbf{j})]$$

$$= \int (0 + 4.0 \, dA) = 4.0 \int dA$$

$$= 16 \text{ N} \cdot \text{m}^2/\text{C}. \quad \text{(Answer)}$$

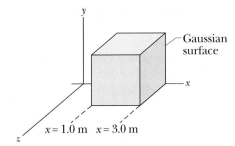

FIGURE 24-5 Sample Problem 24-2. A Gaussian cube with one edge on the x axis lies within a nonuniform electric field.

24-4 GAUSS' LAW

Gauss' law relates the net flux Φ of an electric field through a closed surface (a Gaussian surface) to the *net* charge q_{enc} that is *enclosed* by that surface. It tells us that

$$\epsilon_0 \Phi = q_{enc} \quad \text{(Gauss' law).} \quad (24\text{-}6)$$

By substituting Eq. 24-4, the definition of flux, we can also write Gauss' law as

$$\epsilon_0 \oint \mathbf{E} \cdot d\mathbf{A} = q_{enc} \quad \text{(Gauss' law).} \quad (24\text{-}7)$$

Equations 24-6 and 24-7 hold only when the net charge is located in a vacuum or (what is the same for most practical purposes) in air. In Section 26-8, we modify Gauss' law to include situations in which materials such as mica, oil, or glass are present.

In Eqs. 24-6 and 24-7, the net charge q_{enc} is the algebraic sum of all the *enclosed* positive and negative charges, and it can be positive, negative, or zero. We include the sign, rather than just use the magnitude of the charge, because the sign tells us something about the net flux through the Gaussian surface: if q_{enc} is positive, the net flux is *outward;* if q_{enc} is negative, the net flux is *inward.*

Charge outside the surface, no matter how large or how close it may be, is not included in the term q_{enc} in Gauss' law. The exact form or location of the charges inside the Gaussian surface is also of no concern; the only things that matter, on the right side of Eq. 24-7, are the magnitude and sign of the net enclosed charge. The $\mathbf{E}$ on the left side of Eq. 24-7, however, is the electric field resulting from *all* charges, both those inside and those outside the Gaussian surface. This may seem to be inconsistent, but keep in mind what we saw in Sample Problem 24-1: the electric field due to a charge outside the Gaussian surface contributes zero net flux *through* the surface, because as many field lines due to that charge enter the surface as leave it.

Let us apply these ideas to Fig. 24-6, which shows two charges, equal in magnitude but opposite in sign, and the field lines describing the electric fields that they set up in the surrounding space. Four Gaussian surfaces are also shown, in cross section. Let us consider each in turn.

SURFACE S_1. The electric field is outward for all points on this surface. Thus the flux of the electric field through this surface is positive. So is the net charge within the surface, as Gauss' law requires. (That is, in Eq. 24-6, if Φ is positive, q_{enc} must be also.)

SURFACE S_2. The electric field is inward for all points on this surface. Thus the flux of the electric field is negative and so is the enclosed charge, as Gauss' law requires.

SURFACE S_3. This surface contains no charge, and thus $q_{enc} = 0$. Gauss' law (Eq. 24-6) requires that the net flux of the electric field through this surface be zero. That is reasonable because all the field lines pass entirely through the surface, entering it at the top and leaving at the bottom.

SURFACE S_4. This surface encloses no *net* charge, because the enclosed positive and negative charges have

equal magnitudes. Gauss' law requires that the net flux of the electric field through this surface be zero. That is reasonable because there are as many field lines leaving surface S_4 as entering it.

What would happen if we were to bring an enormous charge Q up close to surface S_4 in Fig. 24-6? The pattern of the field lines would certainly change, but the net flux for the four Gaussian surfaces would not change. We can understand this because the field lines associated with the added Q would pass entirely through each of the four Gaussian surfaces, making no contribution to the net flux through any of them. The value of Q would not enter Gauss' law in any way, because Q lies outside all four of the Gaussian surfaces that we are considering.

SAMPLE PROBLEM 24-3

Figure 24-7 shows five charged lumps of plastic and an electrically neutral coin. The cross section of a Gaussian surface S is indicated. What is the net electric flux through the surface if $q_1 = q_4 = +3.1$ nC, $q_2 = q_5 = -5.9$ nC, and $q_3 = -3.1$ nC.

SOLUTION: The neutral coin makes no contribution to the net charge q_{enc} enclosed by surface S even though the positive and negative charges within the coin may be separated by the electric field in which the coin is immersed. Charges q_4 and q_5 are outside surface S and are therefore not included in q_{enc}. Thus, q_{enc} is $q_1 + q_2 + q_3$ and Eq. 24-6 gives us

$$\Phi = \frac{q_{enc}}{\epsilon_0} = \frac{q_1 + q_2 + q_3}{\epsilon_0}$$

$$= \frac{+3.1 \times 10^{-9}\ C - 5.9 \times 10^{-9}\ C - 3.1 \times 10^{-9}\ C}{8.85 \times 10^{-12}\ C^2/N \cdot m^2}$$

$$= -670\ N \cdot m^2/C. \qquad \text{(Answer)}$$

The minus sign shows that the net charge within the surface is negative and that the net flux through the surface is inward.

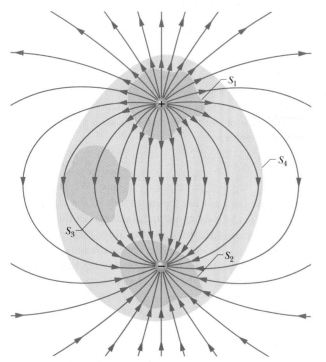

FIGURE 24-6 Two point charges, equal in magnitude but opposite in sign, and the field lines that represent their net electric field. Four Gaussian surfaces are shown in cross section. Surface S_1 encloses the positive charge. Surface S_2 encloses the negative charge. Surface S_3 encloses no charge. And surface S_4 encloses both charges, and thus no net charge.

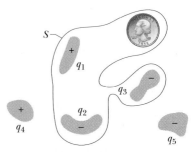

FIGURE 24-7 Sample Problem 24-3. Five plastic objects, each with an electric charge, and a coin, which has no net charge. A Gaussian surface, shown in cross section, encloses three of the plastic objects and the coin.

CHECKPOINT 2: The figure shows three situations in which a Gaussian cube sits in an electric field. The arrows and values indicate the directions and magnitudes (in N·m²/C) of the flux through the six sides of each cube. (The lighter arrows are for the hidden faces.) In which situations does the cube enclose (a) a positive net charge, (b) a negative net charge, and (c) zero net charge?

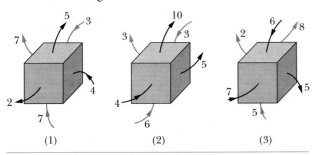

(1)   (2)   (3)

24-5 GAUSS' LAW AND COULOMB'S LAW

If Gauss' law and Coulomb's law are equivalent, we should be able to derive each from the other. Here we derive Coulomb's law from Gauss' law and some symmetry considerations.

Figure 24-8 shows a positive point charge q, around which we have drawn a concentric spherical Gaussian surface of radius r. Imagine dividing this surface into differential areas dA. By definition, the area vector $d\mathbf{A}$ at any point is perpendicular to the surface and directed outward from the interior. From the symmetry of the situation, we know that at any point the electric field $\mathbf{E}$ is also perpendicular to the surface and directed outward from the interior. Thus, since the angle θ between $\mathbf{E}$ and $d\mathbf{A}$ is zero, we can rewrite Eq. 24-7 for Gauss' law as

$$\epsilon_0 \oint \mathbf{E} \cdot d\mathbf{A} = \epsilon_0 \oint E \, dA = q_{\text{enc}}. \qquad (24\text{-}8)$$

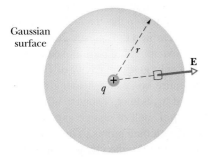

FIGURE 24-8 A spherical Gaussian surface centered on a point charge q.

Here $q_{\text{enc}} = q$. Although E varies radially with the distance from q, it has the same value everywhere on the spherical surface. Since the integral in Eq. 24-8 is taken over that surface, E is a constant in the integration and can be brought out in front of the integral sign. That gives us

$$\epsilon_0 E \oint dA = q. \qquad (24\text{-}9)$$

The integral is now merely the sum of all the differential areas dA on the sphere and thus is just the surface area, $4\pi r^2$. Substituting this, we have

$$\epsilon_0 E(4\pi r^2) = q$$

or

$$E = \frac{1}{4\pi\epsilon_0} \frac{q}{r^2}. \qquad (24\text{-}10)$$

This is exactly the electric field due to a point charge (Eq. 23-3), which we found using Coulomb's law. Thus Gauss' law is equivalent to Coulomb's law.

PROBLEM SOLVING TACTICS

TACTIC 1: *Choosing a Gaussian Surface*
The derivation of Eq. 24-10 using Gauss' law is a warm-up for derivations of electric fields produced by other charge configurations. So let us go back over the steps involved. We started with a given positive point charge q; we know that electric field lines extend radially outward from q in a spherically symmetric pattern.

To find the magnitude of the electric field E at a distance r by Gauss' law (Eq. 24-7), we had to place a hypothetical closed Gaussian surface around q, through a point that is a distance r from q. Then we had to sum via integration the values of $\mathbf{E} \cdot d\mathbf{A}$ over the full Gaussian surface. To make this integration as simple as possible, we chose a spherical Gaussian surface (to mimic the spherical symmetry of the electric field). That choice produced three simplifying features. (1) The dot product $\mathbf{E} \cdot d\mathbf{A}$ became simple, because at all points on the Gaussian surface the angle between $\mathbf{E}$ and $d\mathbf{A}$ is just zero, and so at all points we have $\mathbf{E} \cdot d\mathbf{A} = E \, dA$. (2) The electric field magnitude E is the same at all points on the spherical Gaussian surface, so E was a constant in the integration and could be brought out in front of the integral sign. (3) The result was a very simple integration—just a summation of the differential areas of the sphere, which we could immediately write as $4\pi r^2$.

Note that Gauss' law holds regardless of the shape of the Gaussian surface we choose to place around charge q_{enc}. However, if we had chosen, say, a cubical Gaussian surface, our three simplifying features would have disappeared and the integration of $\mathbf{E} \cdot d\mathbf{A}$ over the cubical surface would have been very difficult. The moral here is to choose the Gaussian surface that most simplifies the integration in Gauss' law.

C HECKPOINT 3: There is a certain net flux Φ_i through a Gaussian sphere of radius r enclosing an isolated charged particle. Suppose the enclosing Gaussian surface is changed to (a) a larger Gaussian sphere, (b) a Gaussian cube with edge length equal to r, and (c) a Gaussian cube with edge length equal to $2r$. In each case, is the net flux through the new Gaussian surface larger than, smaller than, or equal to Φ_i?

24-6 A CHARGED ISOLATED CONDUCTOR

Gauss' law permits us to prove an important theorem about isolated conductors:

> If an excess charge is placed on an isolated conductor, that amount of charge will move entirely to the surface of the conductor. None of the excess charge will be found within the body of the conductor.

This might seem reasonable, considering that charges with the same sign repel each other. You might imagine that, by moving to the surface, the added charges are getting as far away from each other as they can. We turn to Gauss' law for verification of this speculation.

Figure 24-9a shows, in cross section, an isolated lump of copper hanging from an insulating thread and having an excess charge q. We place a Gaussian surface just inside the actual surface of the conductor.

The electric field inside the conductor must be zero. If this were not so, the field would exert forces on the conduction (free) electrons, which are always present in the

conductor, and thus current would always exist within the conductor. (That is, charge would flow from place to place within the conductor.) Of course, there are no such perpetual currents in an isolated conductor, and so the internal electric field is zero.

(An internal electric field *does* appear as the conductor is being charged. However, the added charge quickly distributes itself in such a way that the net internal electric field—the vector sum of the electric fields due to all the charges—is zero. The movement of charge then ceases, and the net force on each charge is zero; the charges are then in *electrostatic equilibrium*.)

If $\mathbf{E}$ is zero everywhere inside the conductor, it must be zero for all points on the Gaussian surface because that surface, though close to the surface of the conductor, is definitely inside it. This means that the flux through the Gaussian surface must be zero. Gauss' law then tells us that the net charge inside the Gaussian surface must also be zero. If the excess charge is not inside the Gaussian surface, it must be outside that surface, which means that it must lie on the actual surface of the conductor.

An Isolated Conductor with a Cavity

Figure 24-9b shows the same hanging conductor, but now with a cavity that is totally within the conductor. It is perhaps reasonable to suppose that when we scoop out the electrically neutral material to form the cavity, we should not change the distribution of charge or the pattern of the electric field that exists in Fig. 24-9a. Again, we must turn to Gauss' law for a quantitative proof.

We draw a Gaussian surface surrounding the cavity, close to its surface but inside the conducting body. Because $\mathbf{E} = 0$ inside the conductor, there can be no flux through this new Gaussian surface. Therefore, from Gauss' law, that surface can enclose no net charge. We conclude that there is no net charge on the cavity walls; all the excess charge remains on the outer surface of the conductor, as in Fig. 24-9a.

The Conductor Removed

Suppose that, by some magic, the excess charges could be "frozen" into position on the conductor's surface, perhaps by embedding them in a thin plastic coating, and suppose that then the conductor could be removed completely. This is equivalent to enlarging the cavity of Fig. 24-9b until it consumes the entire conductor, leaving only the charges. The electric field would not change at all; it would remain zero inside the thin shell of charge and would remain unchanged for all external points. This shows us that the electric field is set up by the charges and not by the conductor. The conductor simply provides an initial pathway for the charges to take up their positions.

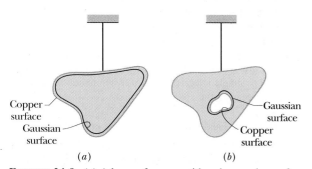

FIGURE 24-9 (a) A lump of copper with a charge q hangs from an insulating thread. A Gaussian surface is drawn within the metal, just inside the actual surface. (b) The lump of copper now has a cavity within it. A Gaussian surface lies within the metal, close to the cavity surface.

The External Electric Field

You have seen that the excess charge on an isolated conductor moves entirely to the conductor's surface. However, unless the conductor is spherical, the charge does not distribute itself uniformly. Put another way, the surface charge density σ (charge per unit area) varies over the surface of any nonspherical conductor. Generally, this variation makes the determination of the electric field set up by the surface charges very difficult.

However, the electric field just outside the surface of a conductor is easy to determine using Gauss' law. To do this, we consider a section of the surface that is small enough to permit us to neglect any curvature and take the section to be flat. We then imagine a tiny cylindrical Gaussian surface to be embedded in the section as in Fig. 24-10: one end cap is fully inside the conductor, the other is fully outside, and the cylinder is perpendicular to the conductor's surface.

The electric field **E** at and just outside the conductor's surface must also be perpendicular to that surface. If it were not, then it would have a component along the conductor's surface that would exert forces on the surface charges, causing them to move. But such motion would violate our implicit assumption that we are dealing with electrostatic equilibrium. So **E** is perpendicular to the conductor's surface.

We now sum the flux through the Gaussian surface. There is no flux through the internal end cap, because the electric field there is zero. There is no flux through the curved surface of the cylinder, because internally (in the conductor) there is no electric field and externally the electric field is parallel to the curved surface. The only flux through the Gaussian surface is that through the external end cap, where **E** is perpendicular to the plane of the cap. We assume that the cap area A is small enough that the field magnitude E is constant over the cap. Then the flux through the cap is EA, and that is the net flux Φ through the Gaussian surface.

The charge q_{enc} enclosed by the Gaussian surface lies on the conductor's surface in an area A. If σ is the charge per unit area, then q_{enc} is equal to σA. When we substitute σA for q_{enc} and EA for Φ, Gauss' law (Eq. 24-6) becomes

$$\epsilon_0 EA = \sigma A,$$

from which we find

$$E = \frac{\sigma}{\epsilon_0} \qquad \text{(conducting surface).} \quad (24\text{-}11)$$

Thus the magnitude of the electric field at a location just outside a conductor is proportional to the surface charge density at that location on the conductor. If the charge on the conductor is positive, the electric field points away from the conductor as in Fig. 24-10. It points toward the conductor if the charge is negative.

The field lines in Fig. 24-10 must terminate on negative charges somewhere in the environment. If we bring those charges near the conductor, the charge density at any given location changes and so does the magnitude of the electric field. However, the relation between σ and E is still given by Eq. 24-11.

SAMPLE PROBLEM 24-4

Figure 24-11a shows a cross section of a spherical metal shell of inner radius R. A point charge of $-5.0\ \mu C$ is located at a distance $R/2$ from the center of the shell. If the shell is electrically neutral, what are the (induced) charges on its inner and outer surfaces? Are those charges uniformly distributed? What is the field pattern inside and outside the shell?

SOLUTION: Figure 24-11b shows a cross section of a spherical Gaussian surface within the metal, just outside the inner wall of the shell. Since the electric field must be zero inside the metal (and thus on the Gaussian surface inside the metal), the electric flux through the Gaussian surface must also be zero. Gauss' law then tells us that the *net* charge enclosed by the Gaussian surface must be zero. With a point charge of $-5.0\ \mu C$ within the shell, a charge of $+5.0\ \mu C$ must lie on the inner wall of the shell.

If the point charge were centered, this positive charge would be uniformly distributed along the inner wall. However, since the point charge is off-center, the distribution of positive charge is skewed, as suggested by Fig. 24-11b, because the positive charge tends to collect on the section of the inner wall nearest the point charge.

Since the shell is electrically neutral, its inner wall can have a charge of $+5.0\ \mu C$ only if electrons, with a total charge of $-5.0\ \mu C$, leave the inner wall and move to the outer wall.

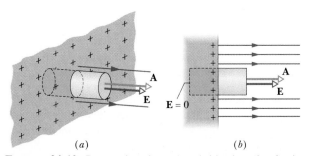

FIGURE 24-10 Perspective view (a) and side view (b) of a tiny portion of a large, isolated conductor with excess positive charge on its surface. A (closed) cylindrical Gaussian surface, embedded perpendicularly in the conductor, encloses some of the charge. Electric field lines pierce the external end cap of the cylinder, but not the internal end cap. The external end cap has area A and area vector **A**.

There they spread out uniformly, as is also suggested by Fig. 24-11*b*. This distribution of negative charge is uniform because the shell is spherical and because the skewed distribution of positive charge on the inner wall cannot produce an electric field in the shell to affect the distribution of charge on the outer wall.

The field lines inside and outside the shell are shown approximately in Fig. 24-11*b*. All the field lines intersect the shell and the point charge perpendicularly. Inside the shell the pattern of field lines is skewed owing to the skew of the positive charge distribution. Outside the shell the pattern is the same as if the point charge were centered and the shell were missing. In fact, this would be true no matter where inside the shell the point charge happened to be located.

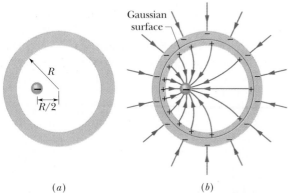

(*a*)   (*b*)

FIGURE 24-11   Sample Problem 24-4. (*a*) A negative point charge is located within a spherical metal shell that is electrically neutral. (*b*) As a result, positive charge is nonuniformly distributed on the inner wall of the shell, and an equal amount of negative charge is uniformly distributed on the outer wall. The electric field lines are shown.

CHECKPOINT **4:** A ball of charge $-50e$ lies at the center of a hollow spherical metal shell that has a net charge of $-100e$. What is the charge on (a) the shell's inner surface and (b) its outer surface?

24-7 APPLYING GAUSS' LAW: CYLINDRICAL SYMMETRY

Figure 24-12 shows a section of an infinitely long cylindrical plastic rod with a uniform (positive) linear charge density λ. Let us find an expression for the magnitude of the electric field **E** at a distance r from the axis of the rod.

Our Gaussian surface should match the symmetry of the problem, which is cylindrical. We choose a circular cylinder of radius r and length h, coaxial with the rod. The Gaussian surface must be closed, so we include two end caps as part of the surface.

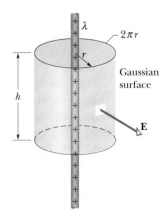

FIGURE 24-12   A Gaussian surface in the form of a closed cylinder surrounds a section of a very long, uniformly charged, cylindrical plastic rod.

Imagine now that, while you are not watching, someone rotates the plastic rod around its longitudinal axis or turns it end for end. When you look again at the rod, you will not be able to detect any change. We conclude from this symmetry that the only uniquely specified direction in this problem is along a radial line. Thus **E** must have a constant magnitude E and (for a positively charged rod) must be directed radially outward at every point on the cylindrical part of the Gaussian surface.

Since $2\pi r$ is the circumference of the cylinder and h is its height, the area of the cylindrical surface is $2\pi rh$. The flux of **E** through this cylindrical surface is then

$$\Phi = EA \cos \theta = E(2\pi rh).$$

There is no flux through the end caps because **E**, being radially directed, is parallel to the end caps at every point.

The charge enclosed by the surface is λh so that Gauss' law,

$$\epsilon_0 \Phi = q_{\text{enc}},$$

reduces to

$$\epsilon_0 E(2\pi rh) = \lambda h,$$

yielding

$$E = \frac{\lambda}{2\pi\epsilon_0 r} \qquad \text{(line of charge).} \qquad (24\text{-}12)$$

This is the electric field due to an infinitely long, straight line of charge, at a point that is a radial distance r from the line. The direction of **E** is radially outward if the charge is positive, and radially inward if it is negative.

SAMPLE PROBLEM 24-5

The visible portion of a lightning strike is preceded by an invisible stage in which a column of electrons is extended from a cloud to the ground. These electrons come from the cloud and from air molecules that are ionized within the column. The linear charge density λ along the column is typically -1×10^{-3} C/m. Once the column reaches the ground, electrons within it are rapidly dumped to the ground. During the dumping, collisions between the electrons and the air within the column result in a brilliant flash of light. If air molecules break down (ionize) in an electric field exceeding 3×10^6 N/C, what is the radius of the column?

SOLUTION: Although the column is not straight or infinitely long, we can approximate it as being a line of charge as in Fig. 24-12. (Since it contains a net negative charge, the electric field **E** points radially inward.) According to Eq. 24-12, the electric field E decreases with distance from the axis of the column of charge. The surface of the column of charge must be at a radius r where the magnitude of **E** is 3×10^6 N/C, because air molecules within that radius ionize while those farther out do not. Solving Eq. 24-12 for r and inserting the known data, we find the radius of the column to be

$$r = \frac{\lambda}{2\pi\epsilon_0 E}$$

$$= \frac{1 \times 10^{-3} \text{ C/m}}{(2\pi)(8.85 \times 10^{-12} \text{ C}^2/\text{N} \cdot \text{m}^2)(3 \times 10^6 \text{ N/C})}$$

$$= 6 \text{ m.} \qquad \text{(Answer)}$$

(The radius of the luminous portion of a lightning strike is smaller, perhaps only 0.5 m. You can get an idea of the width from Fig. 24-13.) Although the radius of the column may be

FIGURE 24-13 Lightning strikes a 20 m high sycamore. Because the tree was wet, most of the charge traveled through the water on it and the tree was unharmed.

FIGURE 24-14 Ground currents from a lightning strike have burned grass off this golf course, exposing the soil.

only 6 m, do not assume that you are safe if you are at a somewhat greater distance from the strike point, because the electrons dumped by the strike travel along the ground. Such *ground currents* are lethal. Figure 24-14 shows evidence of ground currents.

24-8 APPLYING GAUSS' LAW: PLANAR SYMMETRY

Nonconducting Sheet

Figure 24-15 shows a portion of a thin, infinite, nonconducting sheet with a uniform (positive) surface charge density σ. A sheet of thin plastic wrap, uniformly charged on one side, can serve as a simple model. Let us find the electric field **E** a distance r in front of the sheet.

A useful Gaussian surface is a closed cylinder with end caps of area A, arranged to pierce the sheet perpendicularly as shown. From symmetry, **E** must be perpendicular to the sheet, hence to the end caps. Furthermore, since the charge is positive, **E** must point *away* from the sheet, and thus the electric field lines pierce the two Gaussian end caps in an outward direction. Because the field lines do not pierce the cylinder walls, there is no flux through this portion of the Gaussian surface. Thus $\mathbf{E} \cdot d\mathbf{A}$ is simply $E\, dA$; then Gauss' law,

$$\epsilon_0 \oint \mathbf{E} \cdot d\mathbf{A} = q_{\text{enc}},$$

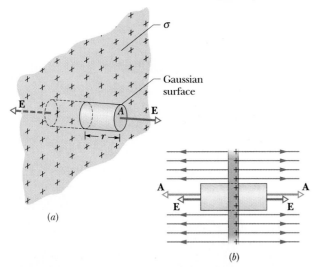

(a)

(b)

FIGURE 24-15   Perspective view (a) and side view (b) of a portion of a very large, thin plastic sheet, uniformly charged on one side to surface charge density σ. A closed cylindrical Gaussian surface passes through the sheet and is perpendicular to it.

becomes

$$\epsilon_0(EA + EA) = \sigma A,$$

where σA is the charge enclosed by the Gaussian surface. This gives

$$E = \frac{\sigma}{2\epsilon_0} \qquad \text{(sheet of charge).} \qquad (24\text{-}13)$$

Since we are considering an infinite sheet with uniform charge density, this result holds for any point at a finite distance from the sheet. Equation 24-13 agrees with Eq. 23-25, which we found by integration of the electric field components that are produced by individual charges. (Look back to that time-consuming and challenging integration, and note how much more easily we obtain the result with Gauss' law. That is one reason for devoting a whole chapter to that law: for certain symmetric arrangements of charge, it is very much easier to use than integration of field components.)

Two Conducting Plates

Figure 24-16a shows a cross section of a thin, infinite conducting plate with excess positive charge. From Section 24-6 we know that this excess charge lies on the surface of the plate. Since the plate is thin and very large, we can assume that essentially all the excess charge is on the two large faces of the plate.

If there is no external electric field to force the positive charge into some particular distribution, it will spread out on the two faces with a uniform surface charge density of

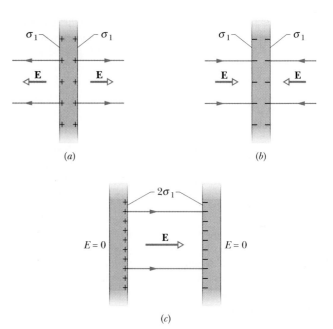

(a)

(b)

(c)

FIGURE 24-16   (a) A thin, very large conducting plate with excess positive charge. (b) An identical plate with excess negative charge. (c) The two plates arranged to be parallel and close.

magnitude σ_1. From Eq. 24-11 we know that just outside the plate this charge sets up an electric field of magnitude $E = \sigma_1/\epsilon_0$. Because the excess charge is positive, the field points away from the plate.

Figure 24-16b shows an identical plate with excess negative charge having the same magnitude of surface charge density σ_1. The only difference is that now the electric field points toward the plate.

Suppose we arrange for the plates of Figs. 24-16a and b to be close to each other and parallel (Fig. 24-16c). Since the plates are conductors, when we bring them into this arrangement, the excess charge on one plate attracts the excess charge on the other plate, and all the excess charge moves onto the inner faces of the plates as in Fig. 24-16c. With twice as much charge now on each inner face, the new surface charge density (call it σ) on each inner face is twice σ_1. Thus the electric field at any point between the plates has the magnitude

$$E = \frac{2\sigma_1}{\epsilon_0} = \frac{\sigma}{\epsilon_0}. \qquad (24\text{-}14)$$

This field points directly away from the positively charged plate and toward the negatively charged plate. Since no excess charge is left on the outer faces, the electric field to the left and right of the plates is zero.

Because the charges on the plates moved when we brought the plates close to each other, Fig. 24-16c is *not* the superposition of Figs. 24-16a and b; that is, the charge distribution of the two-plate system is not merely the sum of the charge distributions of the individual plates.

You may wonder why we discuss such seemingly unrealistic situations as the field set up by an infinite line of charge, an infinite sheet of charge, or a pair of infinite plates of charge. It is not enough to say that we do so because it is simple to analyze such situations with Gauss' law, although that is indeed true. The proper answer is that analyses for "infinite" situations yield good approximations to many real-world problems. Thus Eq. 24-13 holds well for a finite nonconducting sheet as long as you are close to the sheet and not too near its edges. And Eq. 24-14 holds well for a pair of finite conducting plates as long as you consider a point that is not too close to their edges.

The trouble with the edges of a sheet or a plate, and the reason we take care not to be near them, is that near an edge we can no longer use planar symmetry to find expressions for the fields. In fact, the field lines there are curved (said to be an *edge effect* or *fringing*), and the fields can be very difficult to express algebraically.

SAMPLE PROBLEM 24-6

Figure 24-17a shows portions of two large, parallel, nonconducting sheets, each with a fixed uniform charge on one side. The magnitudes of the surface charge densities are $\sigma_{(+)} = 6.8$ μC/m^2 for the positively charged sheet and $\sigma_{(-)} = 4.3$ μC/m^2 for the negatively charged sheet.

Find the electric field **E** (a) to the left of the sheets, (b) between the sheets, and (c) to the right of the sheets.

SOLUTION: Since the charges are fixed in place, we can find the electric field of the sheets in Fig. 24-17a by (1) finding the field of each sheet as if that sheet were isolated and (2) algebraically adding the fields of the isolated sheets via the superposition principle. (We can add the fields algebraically because they are parallel to each other.) From Eq. 24-13, the magnitude $E_{(+)}$ of the electric field due to the positive sheet at any point is

$$E_{(+)} = \frac{\sigma_{(+)}}{2\epsilon_0} = \frac{6.8 \times 10^{-6} \text{ C/m}^2}{(2)(8.85 \times 10^{-12} \text{ C}^2/\text{N} \cdot \text{m}^2)}$$

$$= 3.84 \times 10^5 \text{ N/C}.$$

Similarly, the magnitude $E_{(-)}$ of the electric field at any point due to the negative sheet is

$$E_{(-)} = \frac{\sigma_{(-)}}{2\epsilon_0} = \frac{4.3 \times 10^{-6} \text{ C/m}^2}{(2)(8.85 \times 10^{-12} \text{ C}^2/\text{N} \cdot \text{m}^2)}$$

$$= 2.43 \times 10^5 \text{ N/C}.$$

Figure 24-17b shows the fields set up by the sheets to the left of the sheets (*L*), between them (*B*), and to their right (*R*).

The resultant fields in these three regions follow from the superposition principle. To the left of the sheets, the field

magnitude is

$$E_L = E_{(+)} - E_{(-)}$$

$$= 3.84 \times 10^5 \text{ N/C} - 2.43 \times 10^5 \text{ N/C}$$

$$= 1.4 \times 10^5 \text{ N/C}. \qquad \text{(Answer)}$$

Because $E_{(+)}$ is larger than $E_{(-)}$, the net electric field $\mathbf{E}_L$ in this region points to the left, as Fig. 24-17c shows. To the right of the sheets, the electric field $\mathbf{E}_R$ has the same magnitude but points to the right, as Fig. 24-17c shows.

Between the sheets, the two fields add and we have

$$E_B = E_{(+)} + E_{(-)}$$

$$= 3.84 \times 10^5 \text{ N/C} + 2.43 \times 10^5 \text{ N/C}$$

$$= 6.3 \times 10^5 \text{ N/C}. \qquad \text{(Answer)}$$

The electric field $\mathbf{E}_B$ points to the right.

Note that outside the sheets, the electric field is the same as that from a single sheet whose surface charge density is $\sigma_{(+)} - \sigma_{(-)}$, or $+2.5 \times 10^{-6}$ C/m^2.

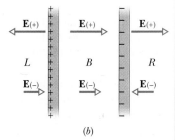

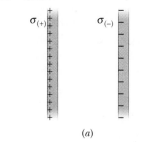

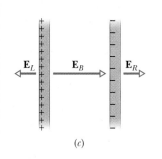

FIGURE 24-17 Sample Problem 24-6. (a) Two large parallel sheets, uniformly charged on one side. (b) The individual electric fields resulting from the two charged sheets. (c) The net field due to both charged sheets, found by superposition.

24-9 APPLYING GAUSS' LAW: SPHERICAL SYMMETRY

Here we use Gauss' law to prove the two shell theorems presented without proof in Section 22-4:

A shell of uniform charge attracts or repels a charged particle that is outside the shell as if all the shell's charge were concentrated at the center of the shell.

A shell of uniform charge exerts no electrostatic force on a charged particle that is located inside the shell.

Figure 24-18 shows a charged spherical shell of total charge q and radius R and two concentric spherical Gaussian surfaces, S_1 and S_2. Following the procedure of Section 24-5 and applying Gauss' law to surface S_2, for which $r \geq R$, we find

$$E = \frac{1}{4\pi\epsilon_0}\frac{q}{r^2} \qquad \text{(spherical shell, field at } r \geq R\text{).} \qquad (24\text{-}15)$$

This is the same field that would be set up by a point charge q at the center of the shell of charge. Thus the magnitude of the force exerted by the shell on a charged particle placed outside the shell is the same as if the shell were replaced with a point charge q at the center of the shell. This proves the first shell theorem.

Applying Gauss' law to surface S_1, for which $r < R$, leads directly to

$$E = 0 \qquad \text{(spherical shell, field at } r < R\text{),} \qquad (24\text{-}16)$$

because this Gaussian surface encloses no charge. Thus if a charged particle were enclosed by the shell, the shell would exert no net electrostatic force on it. This proves the second shell theorem.

Any spherically symmetric charge distribution, such as that of Fig. 24-19, can be constructed with a nest of concentric spherical shells. For purposes of applying the two shell theorems, the volume charge density ρ should have a uniform value for each shell but need not be the same from shell to shell. That is, for the charge distribution

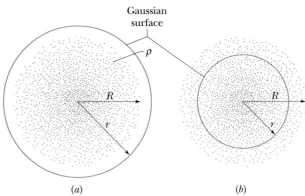

FIGURE 24-19 The dots represent a spherically symmetric distribution of charge of radius R, whose volume charge density ρ is a function only of distance from the center. The charged object is not a conductor, and the charge is assumed to be fixed in position. (a) A concentric spherical Gaussian surface with $r > R$ is included. (b) A similar Gaussian surface with $r < R$ is included.

as a whole, ρ can vary, but only with r, the radial distance from the center. We can then examine the effect of the charge distribution "shell by shell."

In Fig. 24-19a the entire charge lies within a Gaussian surface with $r > R$. The charge produces an electric field on the Gaussian surface as if the charge were a point charge located at the center, and Eq. 24-15 holds.

Figure 24-19b shows a Gaussian surface with $r < R$. To find the electric field at points on this Gaussian surface, we consider two sets of charged shells—one set inside the Gaussian surface and one set outside. Equation 24-16 says that the charge lying *outside* the Gaussian surface does not set up an electric field on the Gaussian surface. And Eq. 24-15 says that the charge *enclosed* by the surface sets up an electric field as if that enclosed charge were concentrated at the center. Letting q' represent that enclosed charge, we can then rewrite Eq. 24-15 as

$$E = \frac{1}{4\pi\epsilon_0}\frac{q'}{r^2} \qquad \text{(spherical distribution, field at } r \leq R\text{).} \qquad (24\text{-}17)$$

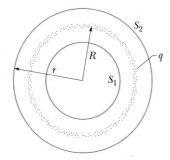

FIGURE 24-18 A thin, uniformly charged, spherical shell with total charge q, in cross section. Two Gaussian surfaces S_1 and S_2 are also shown in cross section. Surface S_2 encloses the shell, and S_1 encloses only the empty interior of the shell.

SAMPLE PROBLEM 24-7

The nucleus of an atom of gold has a radius $R = 6.2 \times 10^{-15}$ m and a positive charge $q = Ze$, where the atomic number Z of gold is 79. Plot the magnitude of the electric field from the center of the gold nucleus outward to a distance of about twice its radius. Assume that the nucleus is spherical with a uniform charge distribution.

SOLUTION: The total charge q on the nucleus is

$$q = Ze = (79)(1.60 \times 10^{-19} \text{ C}) = 1.264 \times 10^{-17} \text{ C.}$$

Outside the nucleus, the situation is represented by Fig. 24-19a and by Eq. 24-15. From this equation we have, for a point on the surface of the nucleus,

$$E = \frac{1}{4\pi\epsilon_0} \frac{q}{r^2}$$

$$= \frac{1.264 \times 10^{-17} \text{ C}}{(4\pi)(8.85 \times 10^{-12} \text{ C}^2/\text{N}\cdot\text{m}^2)(6.2 \times 10^{-15} \text{ m})^2}$$

$$= 3.0 \times 10^{21} \text{ N/C}.$$

Inside the nucleus, Fig. 24-19b and Eq. 24-17 apply. Let q' represent the charge enclosed by a Gaussian sphere of radius $r \leq R$. Since the charge is distributed uniformly throughout the volume of the nucleus, a charge enclosed by a sphere is proportional to the volume of that sphere. In particular,

$$\frac{q'}{q} = \frac{\frac{4}{3}\pi r^3}{\frac{4}{3}\pi R^3}, \tag{24-18}$$

so

$$q' = q\frac{r^3}{R^3}. \tag{}$$

If we substitute this result into Eq. 24-17, we find

$$E = \frac{1}{4\pi\epsilon_0} \frac{q'}{r^2} = \left(\frac{q}{4\pi\epsilon_0 R^3}\right)r. \tag{24-19}$$

The quantity in parentheses is a constant, so, within the nucleus, E is directly proportional to r and is zero at the nuclear center. (Comparison of Eqs. 24-19 and 24-15 shows that they give the same result, 3.0×10^{21} N/C, at $r = R$. This simply tells us that the "inside equation" and the "outside equation," Eqs. 24-17 and 24-15, are compatible where they both apply.) Figure 24-20 shows these results graphically. To ob-

tain it, we plot Eq. 24-19 for $0 \leq r \leq 6.2 \times 10^{-15}$ m, and Eq. 24-15 for $r \geq 6.2 \times 10^{-15}$ m.

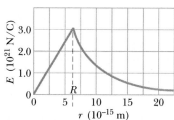

FIGURE 24-20 Sample Problem 24-7. The variation of electric field with distance from the center for the nucleus of a gold atom. The positive charge is assumed to be distributed uniformly throughout the volume of the nucleus.

CHECKPOINT 5: The figure shows two large, parallel, nonconducting sheets with identical (positive) uniform surface charge densities, and a sphere with a uniform (positive) volume charge density. Rank the four numbered points according to the magnitude of the net electric field there, greatest first.

REVIEW & SUMMARY

Gauss' Law

Gauss' law and *Coulomb's law*, although expressed in different forms, are equivalent ways of describing the relation between charge and electric field in static situations. Gauss' law is

$$\epsilon_0\Phi = q_{\text{enc}} \qquad \text{(Gauss' law)}, \tag{24-6}$$

in which q_{enc} is the net charge inside an imaginary closed surface (a **Gaussian surface**) and Φ is the net **flux** of the electric field through the surface:

$$\Phi = \oint \mathbf{E} \cdot d\mathbf{A} \qquad \begin{array}{l}\text{(electric flux through a}\\ \text{Gaussian surface).}\end{array} \tag{24-4}$$

Coulomb's law can readily be derived from Gauss' law.

Applications of Gauss' Law

Using Gauss' law and, in some cases, symmetry arguments, we can derive several important results in electrostatic situations. Among these are:

1. An excess charge on an *isolated conductor* is located entirely on the outer surface of the conductor.

2. The external electric field near the *surface of a charged conductor* is perpendicular to the surface and has magnitude

$$E = \frac{\sigma}{\epsilon_0} \qquad \text{(conducting surface).} \tag{24-11}$$

Inside the conductor, $E = 0$.

3. The electric field at a point due to an infinite *line of charge* with uniform linear charge density λ is in a direction perpendicular to the line of charge and has magnitude

$$E = \frac{\lambda}{2\pi\epsilon_0 r} \qquad \text{(line of charge),} \tag{24-12}$$

where r is the perpendicular distance from the line of charge to the point.

4. The electric field due to an *infinite nonconducting sheet* with uniform surface charge density σ is perpendicular to the plane of the sheet and has magnitude

$$E = \frac{\sigma}{2\epsilon_0} \qquad \text{(sheet of charge).} \qquad (24\text{-}13)$$

5. The electric field outside a *spherical shell of charge* with radius R and total charge q is directed radially and has magnitude

$$E = \frac{1}{4\pi\epsilon_0}\frac{q}{r^2} \qquad \text{(spherical shell, for } r \geq R). \quad (24\text{-}15)$$

Here r is the distance from the center of the shell to the point at

which E is measured. (The charge behaves, for external points, as if it were all at the center of the sphere.) The field *inside* a uniform spherical shell of charge is exactly zero:

$$E = 0 \qquad \text{(spherical shell, for } r < R). \quad (24\text{-}16)$$

6. The electric field *inside a uniform sphere of charge* is directed radially and has magnitude

$$E = \left(\frac{q}{4\pi\epsilon_0 R^3}\right) r. \qquad (24\text{-}19)$$

QUESTIONS

1. A surface has the area vector $\mathbf{A} = (2\mathbf{i} + 3\mathbf{j})$ m². What is the flux of an electric field through it if the field is (a) $\mathbf{E} = 4\mathbf{i}$ N/C and (b) $\mathbf{E} = 4\mathbf{k}$ N/C?

2. What is $\int dA$ for (a) a square of edge length a, (b) a circle of radius r, and (c) the curved surface of a cylinder of length h and radius r?

3. Figure 24-21 shows four Gaussian surfaces consisting of identical cylindrical midsections but different end caps. The surfaces are in a uniform electric field $\mathbf{E}$ that is directed parallel to the central axis of the cylindrical midsections. The end caps of surface S_1 are convex hemispheres; those of surface S_2 are concave hemispheres; those of surface S_3 are cones; and those of surface S_4 are flat disks. Rank the surfaces according to (a) the net electric flux through them and (b) the electric flux through the top end caps, greatest first.

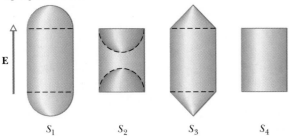

E

S_1 $\qquad$ S_2 $\qquad$ S_3 $\qquad$ S_4

FIGURE 24-21 Question 3.

4. In Fig. 24-22, a Gaussian surface encloses two of the four positively charged particles. (a) Which of the particles contribute to the electric field at point P on the surface? (b) Which net flux of electric field through the surface is greater (if either): that due to q_1 and q_2 or that due to all four charges?

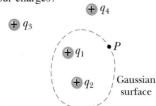

FIGURE 24-22 Question 4.

5. You are given a collection of eight particles with charges $+2q$, $+3q$, $+4q$, $+5q$, $-2q$, $-3q$, $-4q$, and $-5q$. You are also given

the goal of enclosing one or more of them with various Gaussian surfaces in turn, so that the net fluxes through the surfaces are 0, $+q/\epsilon_0$, $+2q/\epsilon_0$, $\cdots$, $+14q/\epsilon_0$. Which is impossible to produce?

6. There is a certain electric flux Φ_i through a spherical Gaussian surface of radius r when the surface encloses a proton. For the following situations, tell whether the net flux through that surface is greater than, less than, or equal to Φ_i. (a) The proton is outside the surface. (b) Two protons are inside the surface. (c) A proton is inside and another proton is outside. (d) A proton and an electron are inside.

7. Figure 24-23 shows, in cross section, a central metal ball, two spherical metal shells, and three spherical Gaussian surfaces of radii R, $2R$, and $3R$, all with the same center. The charges on the three objects are: ball, Q; smaller shell, $3Q$; larger shell, $5Q$. Rank the Gaussian surfaces according to the magnitude of the electric field at any point on the surface, greatest first.

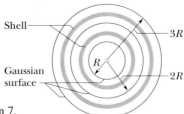

Shell

$3R$

R

Gaussian surface

$2R$

FIGURE 24-23 Question 7.

8. Figure 24-24 shows three Gaussian surfaces half-submerged in a large, thick metal plate with a uniform surface charge density. Surface S_1 is the tallest and has the smallest square end caps; surface S_3 is shortest and has the largest square end caps; and S_2 has intermediate values. Rank the surfaces according to (a) the charge they enclose, (b) the magnitude of the electric field at points on their top end cap, (c) the net electric flux through that top end cap, and (d) the net electric flux through their bottom end cap, greatest first.

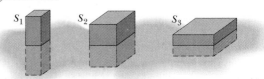

S_1 $\qquad$ S_2 $\qquad$ S_3

FIGURE 24-24 Question 8.

9. Figure 24-25 shows, in cross section, three cylinders, each of uniform charge Q. Concentric with each cylinder is a cylindrical Gaussian surface, all three with the same radius. Rank the Gaussian surfaces according to the electric field at any point on the surface, greatest first.

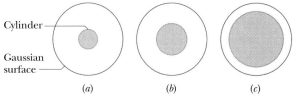

Cylinder

Gaussian surface

(a)   (b)   (c)

FIGURE 24-25 Question 9.

10. Figure 24-26 shows, in section, three long, uniformly charged cylinders centered on the same axis. Central cylinder A has a uniform charge of $q_A = +3q_0$. What uniform charges q_B and q_C should be on cylinders B and C so that (if possible) the net electric field is zero (a) at point 1, (b) at point 2, and (c) at point 3?

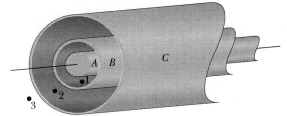

FIGURE 24-26 Question 10.

11. Three infinite nonconducting sheets, with uniform surface charge densities σ, 2σ, and 3σ, are arranged to be parallel like the two sheets in Fig. 24-17a. What is their order, from left to right, if the electric field **E** produced by the arrangement has magnitude $E = 0$ in one region and $E = 2\sigma/\epsilon_0$ in another region?

12. A small charged ball lies within the hollow of a metallic spherical shell of radius R. Here, for three situations, are the net charges on the ball and shell, respectively: (1) $+4q$, 0; (2) $-6q$, $+10q$; (3) $+16q$, $-12q$. Rank the situations according to the charge on (a) the inner surface of the shell and (b) the outer surface, most positive first.

13. Rank the situations of Question 12 according to the magnitude of the electric field (a) halfway through the shell and (b) at a point $2R$ from the center of the shell, greatest first.

14. In Checkpoint 4, what are the magnitude and direction of the electric field at a point that is a distance r from the center of the ball and spherical shell if the point is (a) between the ball and shell, (b) within the metal of the shell, (c) outside the shell?

15. A spherical nonconducting balloon has a uniform positive charge on its surface. If the balloon is expanded, does the magnitude of the electric field due to the charge increase, decrease, or remain the same at points that (a) are inside the balloon, (b) are on the balloon's surface, (c) were outside and are now inside, and (d) were and still are outside?

EXERCISES & PROBLEMS

SECTION 24-2 Flux

1E. Water in an irrigation ditch of width $w = 3.22$ m and depth $d = 1.04$ m flows with a speed of 0.207 m/s. The *mass flux* of the flowing water through an imaginary surface is the product of the water's density (1000 kg/m³) and its volume flux through that surface. Find the mass flux through the following imaginary surfaces: (a) a surface of area wd, entirely in the water, perpendicular to the flow; (b) a surface with area $3wd/2$, of which wd is in the water, perpendicular to the flow; (c) a surface of area $wd/2$, entirely in the water, perpendicular to the flow; (d) a surface of area wd, half in the water and half out, perpendicular to the flow; (e) a surface of area wd, entirely in the water, with its normal 34° from the direction of flow.

SECTION 24-3 Flux of An Electric Field

2E. The square surface shown in Fig. 24-27 measures 3.2 mm on each side. It is immersed in a uniform electric field with magnitude $E = 1800$ N/C. The field lines make an angle of 35° with a normal to the surface, as shown. Take the normal to be "outward," as though the surface were one face of a box. Calculate the electric flux through the surface.

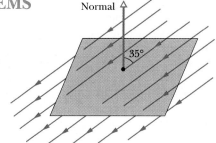

Normal

35°

FIGURE 24-27
Exercise 2.

3E. A cube with 1.40 m edges is oriented as shown in Fig. 24-28 in a region of uniform electric field. Find the electric flux through the right face if the electric field, in newtons per coulomb, is given by (a) 6.00**i**, (b) -2.00**j**, and (c) -3.00**i** $+ 4.00$**k**. (d) What is the total flux through the cube for each of these fields?

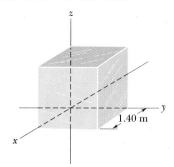

FIGURE 24-28
Exercise 3 and Problem 12.

4P. An electric field given by $\mathbf{E} = 4\mathbf{i} - 3(y^2 + 2)\mathbf{j}$ pierces the Gaussian cube of Fig. 24-5. (E is in newtons per coulomb and x is in meters.) What is the electric flux through (a) the top face, (b) the bottom face, (c) the left face, and (d) the back face? (e) What is the net electric flux through the cube?

SECTION 24-4 Gauss' Law

5E. You have four charges, $2q$, q, $-q$, and $-2q$. If possible, describe how you would place a closed surface that encloses at least the charge $2q$ and through which the net electric flux is (a) 0, (b) $+3q/\epsilon_0$, and (c) $-2q/\epsilon_0$.

6E. In Fig. 24-29, the charge on a neutral isolated conductor is separated by a nearby positively charged rod. What is the net flux through each of the five Gaussian surfaces shown in cross section? Assume that the charges enclosed by S_1, S_2, and S_3 are equal in magnitude.

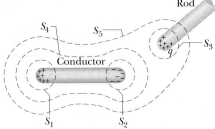

FIGURE 24-29 Exercise 6.

7E. A point charge of 1.8 μC is at the center of a cubical Gaussian surface 55 cm on edge. What is the net electric flux through the surface?

8E. The net electric flux through each face of a die (singular of dice) has a magnitude in units of 10^3 N·m²/C that is exactly equal to the number of spots N on the face (1 through 6). The flux is inward for N odd and outward for N even. What is the net charge inside the die?

9E. In Fig. 24-30, a point charge $+q$ is a distance $d/2$ directly above the center of a square of side d. What is the magnitude of the electric flux through the square? (*Hint:* Think of the square as one face of a cube with edge d.)

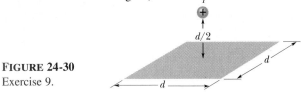

FIGURE 24-30
Exercise 9.

10E. In Fig. 24-31, a butterfly net is in a uniform electric field of magnitude E. The rim, a circle of radius a, is aligned perpendicular to the field. Find the electric flux through the netting.

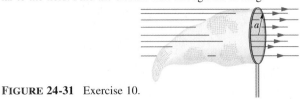

FIGURE 24-31 Exercise 10.

11E. Calculate Φ through (a) the flat base and (b) the curved surface of a hemisphere of radius R. The field $\mathbf{E}$ is uniform and perpendicular to the flat base of the hemisphere, and the field lines enter through the flat base.

12P. Find the net flux through the cube of Exercise 3 and Fig. 24-28 if the electric field is given by (a) $\mathbf{E} = 3.00y\mathbf{j}$ and (b) $\mathbf{E} = -4.00\mathbf{i} + (6.00 + 3.00y)\mathbf{j}$. E is in newtons per coulomb, and y is in meters. (c) In each case, how much charge is enclosed by the cube?

13P. What net charge is enclosed by the Gaussian cube of Problem 4 and Fig. 24-5?

14P. It is found experimentally that the electric field in a certain region of Earth's atmosphere is directed vertically down. At an altitude of 300 m the field has magnitude 60.0 N/C; at an altitude of 200 m, 100 N/C. Find the net amount of charge contained in a cube 100 m on edge, with horizontal faces at altitudes of 200 and 300 m. Neglect the curvature of Earth.

15P. A point charge q is placed at one corner of a cube of edge a. What is the flux through each of the cube faces? (*Hint:* Use Gauss' law and symmetry arguments.)

16P. "Gauss' law for gravitation" is

$$\frac{1}{4\pi G}\,\Phi_g = \frac{1}{4\pi G}\oint \mathbf{g}\cdot d\mathbf{A} = -m,$$

in which Φ_g is the net flux of the *gravitational field* $\mathbf{g}$ through a Gaussian surface that encloses a mass m. The field $\mathbf{g}$ is defined to be the acceleration of a test particle on which m exerts a gravitational force. Derive Newton's law of gravitation from this. What is the significance of the minus sign?

SECTION 24-6 A Charged Isolated Conductor

17E. The electric field just above the surface of the charged drum of a photocopying machine has a magnitude E of 2.3×10^5 N/C. What is the surface charge density on the drum, assuming that the drum is a conductor?

18E. A uniformly charged conducting sphere of 1.2 m diameter has a surface charge density of 8.1 μC/m². (a) Find the net charge on the sphere. (b) What is the total electric flux leaving the surface of the sphere?

19E. Space vehicles traveling through Earth's radiation belts can intercept a significant number of electrons. The resulting charge buildup can damage electronic components and disrupt operations. Suppose a spherical metallic satellite 1.3 m in diameter accumulates 2.4 μC of charge in one orbital revolution. (a) Find the resulting surface charge density. (b) Calculate the magnitude of the resulting electric field just outside the surface of the satellite due to the surface charge.

20E. A conducting sphere with positive charge Q is surrounded by a spherical conducting shell. (a) What is the net charge on the inner surface of the shell? (b) Another positive charge q is placed outside the shell. Now what is the net charge on the inner surface of the shell? (c) If q is moved to a position between the shell and the sphere, what then is the net charge on the inner surface of the shell? (d) Are your answers valid if the sphere and the shell are not concentric?

21P. An isolated conductor of arbitrary shape has a net charge of $+10 \times 10^{-6}$ C. Inside the conductor is a cavity within which is a point charge $q = +3.0 \times 10^{-6}$ C. What is the charge (a) on the cavity wall and (b) on the outer surface of the conductor?

SECTION 24-7 Applying Gauss' Law: Cylindrical Symmetry

22E. An infinite line of charge produces a field of 4.5×10^4 N/C at a distance of 2.0 m. Calculate the linear charge density.

23E. (a) The drum of the photocopying machine in Exercise 17 has a length of 42 cm and a diameter of 12 cm. What is the total charge on the drum? (b) The manufacturer wishes to produce a desktop version of the machine. This requires reducing the size of the drum to a length of 28 cm and a diameter of 8.0 cm. The electric field at the drum surface must remain unchanged. What must be the charge on this new drum?

24P. Figure 24-32 shows a section of a long, thin-walled metal tube of radius R, carrying a charge per unit length λ on its surface. Derive expressions for E in terms of distance r from the tube axis, considering both (a) $r > R$ and (b) $r < R$. Plot your results for the range $r = 0$ to $r = 5.0$ cm, assuming that $\lambda = 2.0 \times 10^{-8}$ C/m and $R = 3.0$ cm. (*Hint:* Use cylindrical Gaussian surfaces, coaxial with the metal tube.)

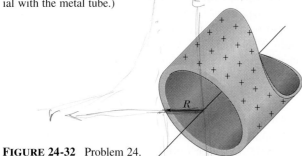

FIGURE 24-32 Problem 24.

25P. Figure 24-33 shows a section through two long thin concentric cylinders of radii a and b with $a < b$. The cylinders have equal and opposite charges per unit length λ. Using Gauss' law, prove (a) that $E = 0$ for $r < a$ and (b) that between the cylinders, where $a < r < b$,

$$E = \frac{1}{2\pi\epsilon_0} \frac{\lambda}{r}.$$

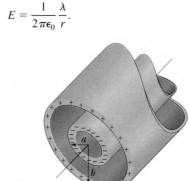

FIGURE 24-33 Problem 25.

26P. A long straight wire has fixed negative charge with a linear charge density of magnitude 3.6 nC/m. The wire is to be enclosed by a thin, nonconducting cylinder of outside radius 1.5 cm, coaxial with the wire. The cylinder is to have positive charge on its outside surface with a surface charge density σ such that the net external electric field is zero. Calculate the required σ.

27P. A very long conducting cylindrical rod of length L with a total charge $+q$ is surrounded by a conducting cylindrical shell (also of length L) with total charge $-2q$, as shown in the section in Fig. 24-34. Use Gauss' law to find (a) the electric field at points outside the conducting shell, (b) the distribution of charge on the conducting shell, and (c) the electric field in the region between the shell and rod.

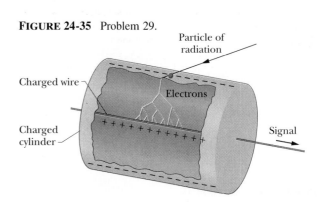

FIGURE 24-34 Problem 27.

28P. Two long, charged, concentric cylinders have radii of 3.0 and 6.0 cm. The charge per unit length is 5.0×10^{-6} C/m on the inner cylinder and -7.0×10^{-6} C/m on the outer cylinder. Find the electric field at (a) $r = 4.0$ cm and (b) $r = 8.0$ cm, where r is the radial distance from the common central axis.

29P. Figure 24-35 shows a Geiger counter, a device used to detect ionizing radiation (radiation that causes ionization of atoms). The counter consists of a thin, positively charged central wire surrounded by a concentric circular conducting cylinder with an equal negative charge. Thus a strong radial electric field is set up inside the cylinder. The cylinder contains a low-pressure inert gas. When a particle of radiation enters the device through the cylinder wall, it ionizes a few of the gas atoms. The resulting free electrons are drawn to the positive wire. However, the electric field is so intense that, between collisions with other gas atoms, the free electrons gain energy sufficient to ionize these atoms also. More free electrons are thereby created, and the process is repeated until the electrons reach the wire. The resulting

FIGURE 24-35 Problem 29.

"avalanche" of electrons is collected by the wire, generating a signal that is used to record the passage of the original particle of radiation. Suppose that the radius of the central wire is 25 μm, the radius of the cylinder 1.4 cm, and the length of the tube 16 cm. If the electric field at the cylinder's inner wall is 2.9×10^4 N/C, what is the total positive charge on the central wire?

30P. A positron, of charge 1.60×10^{-19} C, revolves in a circular path of radius r, between and concentric with the cylinders of Problem 25. What must be its kinetic energy K in electron-volts? Assume that $a = 2.0$ cm, $b = 3.0$ cm, and $\lambda = 30$ nC/m.

31P. Charge is distributed uniformly throughout the volume of an infinitely long cylinder of radius R. (a) Show that at a distance r from the cylinder axis (for $r < R$),

$$E = \frac{\rho r}{2\epsilon_0},$$

where ρ is the volume charge density. (b) Write an expression for E when $r > R$.

SECTION 24-8 Applying Gauss' Law: Planar Symmetry

32E. Figure 24-36 shows cross-sections through two large, parallel, nonconducting sheets with identical distributions of positive charge with surface charge density σ. What is **E** at points (a) above the sheets, (b) between them, and (c) below them?

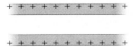

FIGURE 24-36 Exercise 32.

33E. A square metal plate of edge length 8.0 cm and negligible thickness has a total charge of 6.0×10^{-6} C. (a) Estimate the magnitude E of the electric field just off the center of the plate (at, say, a distance of 0.50 mm) by assuming that the charge is spread uniformly over the two faces of the plate. (b) Estimate E at a distance of 30 m (large relative to the plate size) by assuming that the plate is a point charge.

34E. A large, flat, nonconducting surface has a uniform charge density σ. A small circular hole of radius R has been cut in the middle of the surface, as shown in Fig. 24-37. Ignore fringing of the field lines around all edges, and calculate the electric field at point P, a distance z from the center of the hole along its axis. (*Hint:* See Eq. 23-24 and use superposition.)

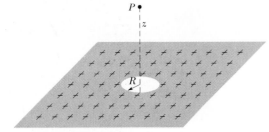

FIGURE 24-37 Exercise 34.

35P. In Fig. 24-38, a small, nonconducting ball of mass $m = 1.0$ mg and charge $q = 2.0 \times 10^{-8}$ C (distributed uniformly through its volume) hangs from an insulating thread that makes an angle $\theta = 30°$ with a vertical, uniformly charged nonconducting sheet (shown in cross section). Considering the weight of the ball and assuming that the sheet extends far vertically and into and out of the page, calculate the surface charge density σ of the sheet.

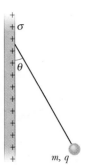

FIGURE 24-38 Problem 35.

36P. Two large thin metal plates are parallel and close to each other as in Fig. 24-16c, but with the negative plate on the left. On their inner faces, the plates have surface charge densities of opposite signs and of magnitude 7.0×10^{-22} C/m². What are the magnitude and direction of the electric field **E** (a) to the left of the plates, (b) to the right of the plates, and (c) between the plates?

37P. An electron is fired directly toward the center of a large metal plate that has excess negative charge with surface charge density 2.0×10^{-6} C/m². If the initial kinetic energy of the electron is 100 eV and if the electron is to stop (owing to electrostatic repulsion from the plate) just as it reaches the plate, how far from the plate must it be fired?

38P. Two large metal plates of area 1.0 m² face each other. They are 5.0 cm apart and have equal but opposite charges on their inner surfaces. If the magnitude E of the electric field between the plates is 55 N/C, what is the magnitude of the charge on each plate? Neglect edge effects.

39P. In a laboratory experiment, an electron's weight is just balanced by the force exerted on the electron by an electric field. If the electric field is due to charges on two large, parallel, nonconducting plates, oppositely charged and separated by 2.3 cm, (a) what is the magnitude of the surface charge density, assumed to be uniform, on the plates, and (b) in which direction does the field point?

40P*. A charge $+q$ placed a distance a from an infinite conducting plane induces negative charge on the plane with a surface charge density $\sigma = -qa/(2\pi r^3)$, where r is the distance from the charge $+q$ to a point P on the plane (Fig. 24-39). What are (a) the magnitude E of the electric field normal to the plane due to this induced charge and (b) the total negative charge induced on the plane? (c) What is the electrostatic force between charge $+q$ and the induced charge on the conducting plane? Is the force attractive or repulsive? (d) What charge, placed diametrically opposite charge $+q$ (on the other side of the plane, at the same distance from the plane) will give this same force?

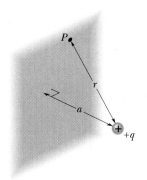

FIGURE 24-39
Problem 40.

41P*. A planar slab of thickness d has a uniform volume charge density ρ. Find the magnitude of the electric field at all points in space both (a) inside and (b) outside the slab, in terms of x, the distance measured from the central plane of the slab.

SECTION 24-9 Applying Gauss' Law: Spherical Symmetry

42E. A conducting sphere of radius 10 cm has an unknown charge. If the electric field 15 cm from the center of the sphere is 3.0×10^3 N/C and points radially inward, what is the net charge on the sphere?

43E. A point charge causes an electric flux of -750 N·m²/C to pass through a spherical Gaussian surface of 10.0 cm radius centered on the charge. (a) If the radius of the Gaussian surface were doubled, how much flux would pass through the surface? (b) What is the value of the point charge?

44E. A thin-walled metal sphere has a radius of 25 cm and a charge of 2.0×10^{-7} C. Find E for a point (a) inside the sphere, (b) just outside the sphere, and (c) 3.0 m from the center.

45E. A point charge $q = 1.0 \times 10^{-7}$ C is at the center of a spherical cavity of radius 3.0 cm in a chunk of metal. Use Gauss' law to find the electric field (a) at point P_1, halfway from the center to the surface of the cavity, and (b) at point P_2, within the metal wall.

46E. Two charged concentric spheres have radii of 10.0 and 15.0 cm. The charge on the inner sphere is 4.00×10^{-8} C and that on the outer sphere is 2.00×10^{-8} C. Find the electric field (a) at $r = 12.0$ cm and (b) at $r = 20.0$ cm.

47E. A thin, metallic, spherical shell of radius a has a charge q_a. Concentric with it is another thin, metallic, spherical shell of radius b (where $b > a$) and charge q_b. Find the electric field at radial points r where (a) $r < a$, (b) $a < r < b$, and (c) $r > b$. (d) Discuss the criterion one would use to determine how the charges are distributed on the inner and outer surfaces of the shells.

48E. In a 1911 paper, Ernest Rutherford said: ''In order to form some idea of the forces required to deflect an α particle through a large angle, consider an atom containing a point positive charge Ze at its centre and surrounded by a distribution of negative electricity, $-Ze$ uniformly distributed within a sphere of radius R.

The electric field E . . . at a distance r from the center for a point *inside* the atom [is]

$$E = \frac{Ze}{4\pi\epsilon_0} \left(\frac{1}{r^2} - \frac{r}{R^3} \right).\text{''}$$

Verify this equation.

49E. Equation 24-11 ($E = \sigma/\epsilon_0$) gives the electric field at points near a charged conducting surface. Apply this equation to a conducting sphere of radius r and charge q, and show that the electric field outside the sphere is the same as the field of a point charge located at the center of the sphere.

50P. A proton with speed $v = 3.00 \times 10^5$ m/s orbits just outside a charged sphere of radius $r = 1.00$ cm. What is the charge on the sphere?

51P. A point charge $+q$ is placed at the center of an electrically neutral, spherical conducting shell with inner radius a and outer radius b. What charge appears on (a) the inner surface of the shell and (b) the outer surface? Find expressions for the net electric field at a distance r from the center of the shell if (c) $r < a$, (d) $b > r > a$, and (e) $r > b$. Sketch field lines for those three regions. For $r > b$, what is the net electric field due to (f) the central point charge and inner surface charge and (g) the outer surface charge? A point charge $-q$ is now placed outside the shell. Does this point charge change the charge distribution on (h) the outer surface and (i) the inner surface? Sketch the field lines now. (j) Is there an electrostatic force on the second point charge? (k) Is there a net electrostatic force on the first point charge? (l) Does this situation violate Newton's third law?

52P. A solid nonconducting sphere of radius R has a nonuniform charge distribution of volume charge density $\rho = \rho_s r/R$, where ρ_s is a constant and r is the distance from the center of the sphere. Show that (a) the total charge on the sphere is $Q = \pi\rho_s R^3$ and (b) the electric field inside the sphere has a magnitude given by

$$E = \frac{1}{4\pi\epsilon_0} \frac{Q}{R^4} r^2.$$

53P. In Fig. 24-40 a sphere, of radius a and charge $+q$ uniformly distributed throughout its volume, is concentric with a spherical conducting shell of inner radius b and outer radius c. This shell has a net charge of $-q$. Find expressions for the electric field, as a function of the radius r, (a) within the sphere ($r < a$); (b) between the sphere and the shell ($a < r < b$); (c) inside the shell ($b < r < c$); and (d) outside the shell ($r > c$). (e) What are the charges on the inner and outer surfaces of the shell?

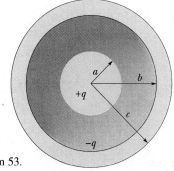

FIGURE 24-40 Problem 53.

54P. Figure 24-41a shows a spherical shell of charge of uniform volume charge density ρ. Plot E due to the shell for distances r from the center of the shell ranging from zero to 30 cm. Assume that $\rho = 1.0 \times 10^{-6}$ C/m^3, $a = 10$ cm, and $b = 20$ cm.

55P. In Fig. 24-41b, a nonconducting spherical shell, of inner radius a and outer radius b, has a volume charge density $\rho = A/r$ (within its thickness), where A is a constant and r is the distance from the center of the shell. In addition, a point charge q is located at the center. What value should A have if the electric field in the shell ($a \leq r \leq b$) is to be uniform? (*Hint:* The constant A depends on a but not on b.)

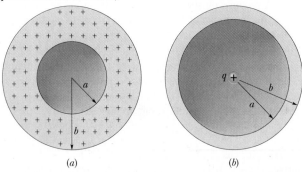

(a)   (b)

FIGURE 24-41   Problems 54 and 55.

56P. A hydrogen atom can be considered as having a central pointlike proton of positive charge $+e$ and an electron of negative charge $-e$ that is distributed about the proton according to the volume charge density $\rho = A \exp(-2r/a_0)$. Here A is a constant, $a_0 = 0.53 \times 10^{-10}$ m is the *Bohr radius,* and r is the distance from the center of the atom. (a) Using the fact that hydrogen is electrically neutral, find A. (b) Then find the electric field produced by the atom at the Bohr radius.

57P*. A nonconducting sphere has a uniform volume charge density ρ. Let **r** be the vector from the center of the sphere to a general point P within the sphere. (a) Show that the electric field at P is given by $\mathbf{E} = \rho\mathbf{r}/3\epsilon_0$. (Note that the result is independent of the radius of the sphere.) (b) A spherical cavity is hollowed out of the sphere, as shown in Fig. 24-42. Using superposition concepts, show that the electric field at all points within the cavity is $\mathbf{E} = \rho\mathbf{a}/3\epsilon_0$ (uniform field), where **a** is the position vector pointing from the center of the sphere to the center of the cavity. (Note that this result is independent of the radius of the sphere and also the radius of the cavity.)

FIGURE 24-42   Problem 57.

58P*. A spherically symmetrical but nonuniform distribution of charge produces an electric field of magnitude $E = Kr^4$, directed radially outward from the center of the sphere. Here r is the radial distance from that center. What is the volume density ρ of the charge distribution?

Electronic Computation

59. From Exercise 48, rewrite Rutherford's equation for the magnitude E of the electric field inside an atom by substituting $r = \alpha R$. Use the rewritten equation to plot E versus α for the range $0 < \alpha < R$. Also plot the magnitude E' of the electric field that would be produced by the nucleus alone. From the two plots determine the value of α, for which $E = 0.500E'$.

60. A computer can be used to demonstrate Gauss' law for a situation in which the electric field is not everywhere perpendicular to the Gaussian surface. Suppose a cube with edges that are 1.00 m long is centered at the origin of a coordinate system whose axes are parallel to the edges. A charge of 1.00 μC is on the y axis at a position given below. Divide each face of the cube into a large number of small squares, calculate the electric flux through each square, and sum the results to obtain the total flux through the face. Finally, sum the fluxes through all the faces to obtain the total flux through the cube surface. Compare the result with q/ϵ_0, where q is the charge inside the cube. The more squares you use for each calculation, the better your result will be; but you should obtain accuracy to three significant figures if each square has a side that is one-thirtieth of a cube edge. If y' is the coordinate of the charge, then the electric field at the point with coordinates x, y, and z has an x component that is given by $E_x = (q/4\pi\epsilon_0)x/r^3$, a y component that is given by $E_y = (q/4\pi\epsilon_0)(y - y')/r^3$, and a z component that is given by $E_z = (q/4\pi\epsilon_0)z/r^3$, in which $r = [x^2 + (y - y')^2 + z^2]^{1/2}$. (a) Take $y' = 0$ (the charge is at the center of the cube). (b) Take $y' = 0.200$ m (the charge is inside the cube). (c) Take $y' = 0.400$ m (the charge is inside the cube). (d) Take $y' = 0.600$ m (the charge is outside the cube).

25
Electric Potential

While enjoying the Sequoia National Park from a lookout platform, this woman found her hair rising from her head. Amused, her brother took her photograph. Five minutes after they left, lightning struck the platform, killing one person and injuring seven. What had caused the woman's hair to rise? From her look, it was not fear—but she certainly should have been fearful.

25-1 ELECTRIC POTENTIAL ENERGY

Newton's law for the gravitational force and Coulomb's law for the electrostatic force are mathematically identical. Thus, the general features we have discussed for the gravitational force should apply to the electrostatic force.

In particular, we can infer that the electrostatic force is a *conservative force*. Thus when that force acts between two or more charged particles within a system of particles, we can assign an **electric potential energy** U to the system. Moreover, if the system changes its configuration from an initial state i to a different final state f, the electrostatic force does work W on the particles. From Eq. 8-1, we then know that the resulting change ΔU in the potential energy of the system is

$$\Delta U = U_f - U_i = -W. \qquad (25\text{-}1)$$

As with other conservative forces, the work done by the electrostatic force is *path independent*: Suppose a charged particle within the system moves from point i to point f while an electrostatic force between it and the rest of the system acts on it. Provided the rest of the system does not change, the work W done by the force is the same for *any* path between points i and f.

For convenience, we usually take the *reference configuration* of a system of charged particles to be that in which the particles are all infinitely separated from each other. And we usually set the corresponding *reference potential energy* to be zero. Suppose that several charged particles come together from initially infinite separations (state i) to form a system of nearby particles (state f). Let the initial potential energy U_i be zero, and let W_∞ represent the work done by the electrostatic forces between the particles during the move in from infinity. Then from Eq. 25-1, the final potential energy U of the system is

$$U = -W_\infty. \qquad (25\text{-}2)$$

As is true of other kinds of potential energy, electric potential energy is considered to be a type of mechanical energy. Recall from Chapter 8 that if only conservative forces act within a (closed) system, the mechanical energy of the system is conserved. We shall use this fact extensively in the rest of this chapter.

PROBLEM SOLVING TACTICS

TACTIC 1: *Electric Potential Energy; Work Done by a Field*

An electric potential energy is associated with a system of particles as a whole. However, you will see statements (starting with Sample Problem 25-1) that associate it with only one

particle within a system. For example, you might read, "An electron in an electric field has a potential energy of 10^{-7} J." Such statements are often acceptable, but you should always keep in mind that the potential energy is actually associated with a system—here the electron plus the charged particles that set up the electric field. Also keep in mind that it makes sense to assign a particular potential energy value, such as 10^{-7} J here, to a particle or even a system *only* if the reference potential energy value is known.

When the potential energy is associated with only one particle within a system, you often will read that the work done on the particle is *by the electric field*. What is meant is that the work is done by the force on the particle due to the charges that set up the field.

SAMPLE PROBLEM 25-1

Electrons are continually being knocked out of air molecules in the atmosphere by cosmic-ray particles coming in from space. Once released, an electron experiences an electrostatic force **F** due to the electric field **E** that is produced in the atmosphere by charged particles already on Earth. Near Earth's surface the electric field has the magnitude $E = 150$ N/C and is directed downward. What is the change ΔU in the electric potential energy of a released electron when the electrostatic force causes it to move vertically upward through a distance $d = 520$ m (Fig. 25-1)?

SOLUTION: Equation 25-1 relates the change ΔU in the electric potential energy of the electron to the work W done on the electron by the electric field. From Chapter 7 we know that the work done by a constant force **F** on a particle undergoing a displacement **d** is

$$W = \mathbf{F} \cdot \mathbf{d}. \qquad (25\text{-}3)$$

From Eq. 23-28, we know that the electrostatic force and the electric field are related by $\mathbf{F} = q\mathbf{E}$. Recall that the sign of charge q is to be used in this vector equation—here q ($= -1.6 \times 10^{-19}$ C) is the charge of an electron. Substituting for **F** in Eq. 25-3 and taking the dot product yield

$$W = q\mathbf{E} \cdot \mathbf{d} = qEd \cos \theta, \qquad (25\text{-}4)$$

where θ is the angle between the directions of **E** and **d.** The field **E** is directed downward and the displacement **d** is di-

FIGURE 25-1 Sample Problem 25-1. An electron in the atmosphere is moved upward through displacement **d** by an electrostatic force **F** due to an electric field **E**.

rected upward. Thus, $\theta = 180°$. Substituting this and other data into Eq. 25-4, we find

$$W = (-1.6 \times 10^{-19} \text{ C})(150 \text{ N/C})(520 \text{ m}) \cos 180°$$
$$= 1.2 \times 10^{-14} \text{ J}.$$

Equation 25-1 then yields

$$\Delta U = -W = -1.2 \times 10^{-14} \text{ J}. \qquad \text{(Answer)}$$

This result tells us that during the 520 m ascent, the electric potential energy of the electron decreases by 1.2×10^{-14} J.

CHECKPOINT **1:** In the figure, a proton moves from point i to point f in a uniform electric field directed as shown. (a) Does the electric field do positive or negative work on the proton? (b) Does the electric potential energy of the proton increase or decrease?

25-2 ELECTRIC POTENTIAL

As you can infer from Sample Problem 25-1, the potential energy of a charged particle in an electric field depends on the magnitude of the charge. However, the potential energy *per unit charge* has a unique value at any point in an electric field.

For example, suppose we place a test particle of positive charge 1.60×10^{-19} C at a point in an electric field where the particle has an electric potential energy of 2.40×10^{-17} J. Then the potential energy per unit charge is

$$\frac{2.40 \times 10^{-17} \text{ J}}{1.60 \times 10^{-19} \text{ C}} = 150 \text{ J/C}.$$

Next, suppose we replace that test particle with one having twice as much positive charge, 3.20×10^{-19} C. We would find that the second particle has an electric potential energy of 4.80×10^{-17} J, twice that of the first particle. However, the potential energy per unit charge would be the same, still 150 J/C.

Thus the potential energy per unit charge, which can be symbolized as U/q, is independent of the charge q of the particle we happen to use and is *characteristic only of the electric field* we are investigating. The potential energy per unit charge at a point in an electric field is called the **electric potential** V (or simply the **potential**) at that point. Thus,

$$V = \frac{U}{q}. \qquad (25\text{-}5)$$

Note that electric potential is a scalar, not a vector.

The *electric potential difference* ΔV between any two points i and f in an electric field is equal to the difference in potential energy per unit charge between the two points:

$$\Delta V = V_f - V_i = \frac{U_f}{q} - \frac{U_i}{q} = \frac{\Delta U}{q}. \qquad (25\text{-}6)$$

Using Eq. 25-1 to substitute $-W$ for ΔU in Eq. 25-6, we can define the potential difference between points i and f as

$$\Delta V = V_f - V_i = -\frac{W}{q} \qquad \begin{array}{l}\text{(potential}\\ \text{difference defined).}\end{array} \qquad (25\text{-}7)$$

The potential difference between two points is thus the negative of the work done by the electrostatic force per unit charge that moves from one point to the other. Potential difference can be positive, negative, or zero, depending on the signs and magnitudes of q and W.

If we set $U_i = 0$ at infinity as our reference potential energy, then by Eq. 25-5, the electric potential must also be zero there. Then from Eq. 25-7, we can define the electric potential V at any point f in an electric field to be

$$V = -\frac{W_\infty}{q} \qquad \text{(potential defined),} \qquad (25\text{-}8)$$

where W_∞ is the work done by the electric field on a charged particle as that particle moves in from infinity to point f. A potential V can be positive, negative, or zero, depending on the signs and magnitudes of q and W_∞.

The SI unit for potential that follows from Eq. 25-8 is the joule per coulomb. This combination occurs so often that a special unit, the *volt* (abbreviated V) is used to represent it. That is,

$$1 \text{ volt} = 1 \text{ joule per coulomb.} \qquad (25\text{-}9)$$

This new unit allows us to adopt a more conventional unit for the electric field **E**, which we have measured up to now in newtons per coulomb. With two unit conversions, we obtain

$$1 \text{ N/C} = \left(1 \frac{\text{N}}{\text{C}}\right)\left(\frac{1 \text{ V} \cdot \text{C}}{1 \text{ J}}\right)\left(\frac{1 \text{ J}}{1 \text{ N} \cdot \text{m}}\right)$$
$$= 1 \text{ V/m}. \qquad (25\text{-}10)$$

The conversion factor in the second set of parentheses comes from Eq. 25-9; that in the third set of parentheses is derived from the definition of the joule. From now on, we shall express values of the electric field in volts per meter rather than in newtons per coulomb.

Finally, we are now in a position to define the electron-volt, the energy unit that was introduced in Section 7-1 as a convenient one for energy measurements in the atomic and subatomic domain. One *electron-volt* (eV) is the energy equal to the work required to move a single elementary charge e, such as that of the electron or the proton, through a potential difference of exactly one volt. Equation 25-7 tells us that the magnitude of this work is $q\,\Delta V$, so

$$1\ \text{eV} = e(1\ \text{V})$$
$$= (1.60 \times 10^{-19}\ \text{C})(1\ \text{J/C}) = 1.60 \times 10^{-19}\ \text{J}.$$

PROBLEM SOLVING TACTICS

TACTIC 2: *Electric Potential and Electric Potential Energy*

Electric potential V and electric potential energy U are quite different quantities and should not be confused.

Electric potential is a property of an electric field, regardless of whether a charged object has been placed in that field; it is measured in joules per coulomb, or volts.

Electric potential energy is an energy of a charged object in an external electric field (or more precisely, an energy of the system consisting of the object and the external electric field); it is measured in joules.

Work Done by an Applied Force

Suppose we move a particle of charge q from point i to point f in an electric field by applying a force to it. During the move, our applied force does work W_{app} on the charge while the electric field does work W on it. By the work–kinetic energy theorem of Eq. 7-15, the change ΔK in the kinetic energy of the particle is

$$\Delta K = K_f - K_i = W_{\text{app}} + W. \qquad (25\text{-}11)$$

Now suppose the particle is stationary before and after the move. Then K_f and K_i are both zero, and Eq. 25-11 reduces to

$$W_{\text{app}} = -W. \qquad (25\text{-}12)$$

In words, the work W_{app} done by our applied force during the move is equal to the negative of the work W done by the electric field.

By substituting Eq. 25-12 into Eq. 25-1, we can relate the work done by our applied force to the change in the potential energy of the particle during the move. We find

$$\Delta U = U_f - U_i = W_{\text{app}}. \qquad (25\text{-}13)$$

By similarly substituting Eq. 25-12 into Eq. 25-7, we can relate our work W_{app} to the electric potential difference ΔV

between the initial and final points of the particle. We find

$$W_{\text{app}} = q\,\Delta V. \qquad (25\text{-}14)$$

W_{app} can be positive, negative, or zero depending on the signs and magnitudes of q and ΔV. It is the work we must do to move a particle of charge q through a potential difference ΔV with no change in the particle's kinetic energy.

$\mathbf{C}$HECKPOINT **2:** In the figure of Checkpoint 1, we move a proton from point i to point f in a uniform electric field directed as shown. (a) Does our force do positive or negative work? (b) Does the proton move to higher or lower potential?

25-3 EQUIPOTENTIAL SURFACES

Adjacent points that have the same electric potential form an **equipotential surface,** which can be either an imaginary surface or a real, physical surface. No net work W is done on a charged particle by an electric field when the particle moves between two points i and f on the same equipotential surface. This follows from Eq. 25-7, which tells us that W must be zero if $V_f = V_i$. Because of the path independence of work (and thus of potential energy and potential), $W = 0$ for *any* path connecting points i and f, regardless of whether that path lies entirely on the equipotential surface.

Figure 25-2 shows a *family* of equipotential surfaces, associated with the electric field due to some distribution of charges. The work done by the electric field on a charged particle as the particle moves from one end to the other of paths I and II is zero because each of these paths begins and ends on the same equipotential surface. The work done as the charged particle moves from one end to the other of paths III and IV is not zero but has the same

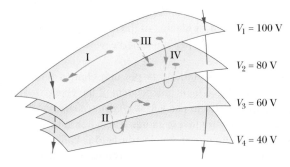

FIGURE 25-2 Portions of four equipotential surfaces. Four paths along which a test charge may move are also shown. Two electric field lines are indicated.

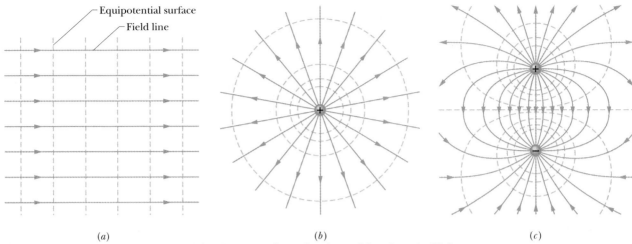

(a) (b) (c)

FIGURE 25-3 Electric field lines (purple) and cross sections of equipotential surfaces (gold) for (a) a uniform field, (b) the field of a point charge, and (c) the field of an electric dipole.

value for both these paths because the initial and final potentials are identical for the two paths. That is, paths III and IV connect the same pair of equipotential surfaces.

From symmetry, the equipotential surfaces produced by a point charge or a spherically symmetrical charge distribution are a family of concentric spheres. For a uniform field, the surfaces are a family of planes perpendicular to the field lines. In fact, equipotential surfaces are always perpendicular to electric field lines and thus to **E**, which is always tangent to these lines. If **E** were *not* perpendicular to an equipotential surface, it would have a component lying along that surface. This component would then do work on a charged particle as it moved along the surface. But by Eq. 25-7 work cannot be done if the surface is truly an equipotential surface; the only possible conclusion is that **E** must be everywhere perpendicular to the surface. Figure 25-3 shows electric field lines and cross sections of the equipotential surfaces for a uniform electric field and for the field associated with a point charge and with an electric dipole.

We now return to the woman in the opening photograph for this chapter. Because she was standing on a platform that was connected to the mountainside, she was at about the same potential as the mountainside. A highly charged cloud system created a strong electric field around her and the mountainside, with **E** pointing outward from her and the mountain. Electrostatic forces due to this field drove some of the conduction electrons in the woman downward through her body, leaving the strands of hair positively charged. The magnitude of **E** was apparently large, but less than the value of about 3×10^6 V/m that would have caused electrical breakdown of the air mole-

cules. (That value was exceeded shortly later when the lightning struck the platform.)

The equipotential surfaces surrounding the woman on the mountainside platform can be inferred from her hair: the strands are extended along the direction of **E** and thus are perpendicular to the equipotential surfaces, as drawn in Fig. 25-4. The magnitude of **E** was apparently greatest (the

FIGURE 25-4 This enhancement of the chapter's opening photograph shows the result of an overhead cloud system creating a strong electric field **E** near a woman's head. Many of the hair strands extended along the field, which was perpendicular to the equipotential surfaces and greatest where those surfaces were closest, near the top of her head.

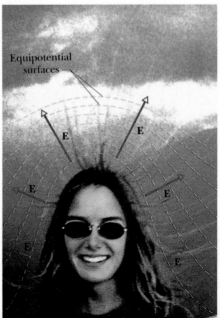

equipotential surfaces were most closely spaced) just above her head, because the hair there was extended farther than the hair around the side.

The lesson here is simple. If an electric field causes the hairs on your head to stand up, you had better run for shelter—not pose for a snapshot.

25-4 CALCULATING THE POTENTIAL FROM THE FIELD

We can calculate the potential difference between any two points i and f in an electric field if we know the field vector **E** at all positions along any path connecting those points. To make the calculation, we find the work done on a positive test charge by the field as the charge moves from i to f, and then use Eq. 25-7.

Consider an arbitrary electric field, represented by the field lines in Fig. 25-5, and a positive test charge q_0 that moves along the path shown from point i to point f. At any point on the path, an electrostatic force $q_0\mathbf{E}$ acts on the charge as it moves through a differential displacement $d\mathbf{s}$. From Chapter 7, we know that the differential work dW done on a particle by a force **F** during a displacement $d\mathbf{s}$ is

$$dW = \mathbf{F} \cdot d\mathbf{s}. \qquad (25\text{-}15)$$

For the situation of Fig. 25-5, $\mathbf{F} = q_0\mathbf{E}$ and Eq. 25-15 becomes

$$dW = q_0\mathbf{E} \cdot d\mathbf{s}. \qquad (25\text{-}16)$$

To find the total work W done on the particle by the field as the particle moves from point i to point f, we sum—via integration—the differential work done on the charge for all the differential displacements $d\mathbf{s}$ along the path:

$$W = q_0 \int_i^f \mathbf{E} \cdot d\mathbf{s}. \qquad (25\text{-}17)$$

If we substitute the total work W from Eq. 25-17 into Eq.

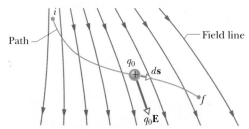

FIGURE 25-5   A test charge q_0 moves from point i to point f along the path shown in a nonuniform electric field. During a displacement $d\mathbf{s}$, an electrostatic force $q_0\mathbf{E}$ acts on the test charge. This force points in the direction of the field line at the location of the test charge.

25-7 we find

$$V_f - V_i = -\int_i^f \mathbf{E} \cdot d\mathbf{s}. \qquad (25\text{-}18)$$

Thus the potential difference $V_f - V_i$ between any two points i and f in an electric field is equal to the negative of the *line integral* (meaning the integral along the path) of $\mathbf{E} \cdot d\mathbf{s}$ from i to f. Note that this result is independent of the value of q_0 that we used to obtain it.

If the electric field is known throughout a certain region, Eq. 25-18 allows us to calculate the difference in potential between any two points in the field. Because the electrostatic force is conservative, all paths (whether easy or difficult to use) yield the same result.

If we choose the potential V_i at point i to be zero, then Eq. 25-18 becomes

$$V = -\int_i^f \mathbf{E} \cdot d\mathbf{s}, \qquad (25\text{-}19)$$

in which we have dropped the subscript f on V_f. Equation 25-19 gives us the potential V at any point f in the electric field *relative to the zero potential* at point i. If we let point i be at infinity, then Eq. 25-19 gives us the potential V at any point f relative to the zero potential at infinity.

SAMPLE PROBLEM 25-2

(a) Figure 25-6a shows two points i and f in a uniform electric field **E**. The points lie on the same electric field line (not shown) and are separated by a distance d. Find the potential difference $V_f - V_i$ by moving a positive test charge q_0 from i to f along a path that is parallel to the field direction.

SOLUTION: As the test charge moves from i to f in Fig. 25-6a, its differential displacement $d\mathbf{s}$, which is always in the direction of motion, points in the same direction as the electric field **E**. The angle θ between these two vectors is then zero, and Eq. 25-18 becomes

$$V_f - V_i = -\int_i^f \mathbf{E} \cdot d\mathbf{s} = -\int_i^f E(\cos 0)\, ds$$

$$= -\int_i^f E\, ds.$$

Since the field is uniform, E is constant over the path and can be moved outside the integral, giving us

$$V_f - V_i = -E \int_i^f ds = -Ed, \qquad \text{(Answer)}$$

in which the integral is simply the length d of the path. The minus sign in the result shows that the potential at point f in Fig. 25-6a is lower than the potential at point i. This is a general result: the potential always decreases along a path that extends in the direction of the electric field lines.

(b) Now find the potential difference $V_f - V_i$ by moving the positive test charge q_0 from i to f along the path icf shown in Fig. 25-6b.

SOLUTION: At all points along line ic, $\mathbf{E}$ and $d\mathbf{s}$ are perpendicular to each other. Thus $\mathbf{E} \cdot d\mathbf{s} = 0$ everywhere along this part of the path. Equation 25-18 then tells us that points i and c are at the same potential. In other words, i and c lie on the same equipotential surface.

For line cf we have $\theta = 45°$ and, from Eq. 25-18,

$$V_f - V_i = -\int_c^f \mathbf{E} \cdot d\mathbf{s} = -\int_c^f E(\cos 45°)\, ds$$

$$= -\frac{E}{\sqrt{2}} \int_c^f ds.$$

The integral in this equation is the length of line cf, which is $d/\sin 45° = \sqrt{2}d$. Thus

$$V_f - V_i = -\frac{E}{\sqrt{2}}\sqrt{2}d = -Ed. \qquad \text{(Answer)}$$

This is the same result we obtained in (a), as it must be: the potential difference between two points does not depend on the path connecting them. Moral: When you want to find the potential difference between two points by moving a test charge between them, you can save time and work by choosing a path that simplifies the use of Eq. 25-18.

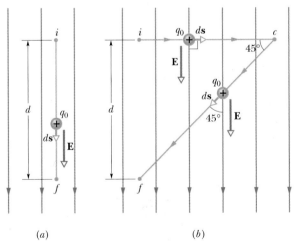

(a)   (b)

FIGURE 25-6 Sample Problem 25-2. (a) A test charge q_0 moves in a straight line from point i to point f, along the direction of a uniform electric field. (b) Charge q_0 moves along path icf in the same electric field.

CHECKPOINT 3: The figure shows a family of parallel equipotential surfaces (in cross section) and five paths along which we shall move an electron from one surface to another. (a) What is the direction of the electric field associated with the surfaces? (b) For each path, is the work we do positive, negative, or zero? (c) Rank the paths according to the work we do, greatest first.

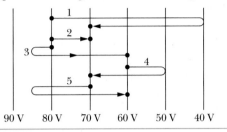

90 V   80 V   70 V   60 V   50 V   40 V

25-5 POTENTIAL DUE TO A POINT CHARGE

We are now going to use Eq. 25-19 to derive an expression for the electric potential V in the space around a point charge, relative to the zero potential at infinity.

Consider a point P at a distance r from a fixed point charge of magnitude q (Fig. 25-7). To use Eq. 25-19, we imagine that a positive test charge moves from infinity to point P. Because the path followed by the test charge does not matter, we make the simplest choice: a line that extends from infinity to P along a radius from the point charge q.

We next must evaluate the dot product $\mathbf{E} \cdot d\mathbf{s}$ in Eq. 25-19 along the path taken by the test charge. In Fig. 25-7 the test particle is at some intermediate point, at distance r' from the point charge. The electric field $\mathbf{E}$ at the location of the test particle is directed radially outward. The differential displacement $d\mathbf{s}$ of the test charge as it moves toward point P is radially inward. Thus the angle between $\mathbf{E}$ and $d\mathbf{s}$

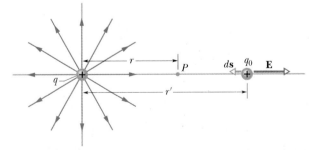

FIGURE 25-7 The positive point charge q produces an electric field $\mathbf{E}$ and an electric potential V at point P. We find the potential by moving a test charge q_0 to P from infinity. The test charge is shown at distance r' from the point charge, undergoing differential displacement $d\mathbf{s}$.

is 180°. Using this angle and writing the magnitude of the displacement ds as dr', we can write the dot product of Eq. 25-19 as

$$\mathbf{E} \cdot d\mathbf{s} = |E| (dr')(\cos 180°) = -|E| dr', \quad (25\text{-}20)$$

where $|E|$ is the absolute value of $\mathbf{E}$. (We use the absolute value sign to show explicitly that we want only the magnitude of $\mathbf{E}$ here.) Substituting Eq. 25-20 and the limits of our integration into Eq. 25-19 gives us

$$V = -\int_i^f \mathbf{E} \cdot d\mathbf{s} = \int_\infty^r |E| \, dr'. \quad (25\text{-}21)$$

The magnitude of the electric field at the site of the test charge is given by Eq. 23-3 as

$$E = \frac{1}{4\pi\epsilon_0} \frac{q}{r'^2}. \quad (25\text{-}22)$$

Substituting this result into Eq. 25-21 and integrating lead to

$$V = \frac{q}{4\pi\epsilon_0} \int_\infty^r \left| \frac{1}{r'^2} \right| dr' = \frac{q}{4\pi\epsilon_0} \left[\frac{1}{r'} \right]_\infty^r, \quad (25\text{-}23)$$

or

$$V = \frac{1}{4\pi\epsilon_0} \frac{q}{r} \quad \text{(point charge } +q). \quad (25\text{-}24)$$

Thus the potential V at any point around a positive point charge is positive, relative to the zero potential at infinity.

If the point charge is negative (but still of magnitude q), the electric field at P points toward q and the angle between $d\mathbf{s}$ and $\mathbf{E}$ at r' is zero. The integration in Eq. 25-23 now yields

$$V = -\frac{1}{4\pi\epsilon_0} \frac{q}{r} \quad \text{(point charge } -q). \quad (25\text{-}25)$$

Thus the potential V at any point around a negative point charge is negative, relative to the zero potential at infinity.

If we allow the symbol q to be either positive or negative instead of representing just the magnitude of the charge, we can generalize Eqs. 25-24 and 25-25 as

$$V = \frac{1}{4\pi\epsilon_0} \frac{q}{r} \quad \begin{array}{l} \text{(positive or negative} \\ \text{point charge } q). \end{array} \quad (25\text{-}26)$$

Now the sign of V is the same as the sign of q. Figure 25-8 shows a computer-generated plot of Eq. 25-26 for a positive point charge. Note also that the magnitude of V increases as $r \rightarrow 0$. In fact, according to Eq. 25-26, V for a point charge is infinite at $r = 0$, although Fig. 25-8 shows a finite, smoothed-off value there.

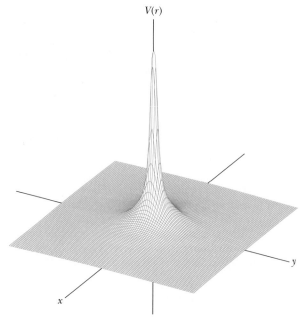

FIGURE 25-8 A computer-generated plot of electric potential $V(r)$ due to a positive point charge, which is located at the origin of an xy plane. The potentials at points in that plane are plotted vertically. (Curved lines have been added to help you visualize the plot.) The infinite value of V predicted by Eq. 25-26 for $r = 0$ is not plotted.

Equation 25-26 also gives the electric potential *outside or on the external surface of* a spherically symmetric charge distribution. We can prove this by using one of the shell theorems of Sections 22-4 and 24-9 to replace the actual charge with an equal charge concentrated at the center of the spherical distribution. Then the derivation leading to Eq. 25-26 follows, provided we do not consider a point within the actual distribution.

PROBLEM SOLVING TACTICS

TACTIC 3: *Finding a Potential Difference*
To find the potential difference ΔV between any two points in the field of an isolated point charge, we can evaluate Eq. 25-26 at each point and then subtract the results. The value of ΔV will be the same for any choice of reference potential energy because that choice is eliminated by the subtraction.

SAMPLE PROBLEM 25-3

(a) What is the electric potential V at a distance $r = 2.12 \times 10^{-10}$ m from the nucleus of a hydrogen atom (the nucleus consists of a single proton)?

SOLUTION: Substituting the given distance and the charge of a proton into Eq. 25-26 yields

$$V = \frac{1}{4\pi\epsilon_0}\frac{e}{r}$$
$$= \frac{(8.99 \times 10^9 \text{ N}\cdot\text{m}^2/\text{C}^2)(1.60 \times 10^{-19}\text{ C})}{2.12 \times 10^{-10}\text{ m}}$$
$$= 6.78 \text{ V}. \qquad \text{(Answer)}$$

(b) What is the electric potential energy U in electron-volts of an electron at the given distance from the nucleus? (The potential energy is actually that of the electron–proton system—the hydrogen atom.)

SOLUTION: Substituting $V = 6.78$ V and the charge of an electron into Eq. 25-5 yields

$$U = qV = (-1.60 \times 10^{-19}\text{ C})(6.78 \text{ V})$$
$$= -1.09 \times 10^{-18}\text{ J} = -6.78 \text{ eV}. \qquad \text{(Answer)}$$

(c) If the electron moves closer to the proton, does the electric potential energy increase or decrease?

SOLUTION: The electric potential V due to the proton at the electron's position increases. Thus the value of V in (b) increases. Because the electron is negatively charged, this means that the value of U becomes more negative. Hence, the potential energy U of the electron (that is, of the system or atom) decreases.

25-6 POTENTIAL DUE TO A GROUP OF POINT CHARGES

We can find the net potential at a point due to a group of point charges with the help of the superposition principle. We calculate the potential resulting from each charge at the given point separately, using Eq. 25-26 with the sign of the charge included. Then we sum the potentials. For n charges, the net potential is

$$V = \sum_{i=1}^{n} V_i = \frac{1}{4\pi\epsilon_0}\sum_{i=1}^{n}\frac{q_i}{r_i} \qquad \begin{array}{l}(n\text{ point}\\ \text{charges}).\end{array} \quad (25\text{-}27)$$

Here q_i is the value of the ith charge, and r_i is the radial distance of the given point from the ith charge. The sum in Eq. 25-27 is an *algebraic sum,* not a vector sum like the sum that would be used to calculate the electric field resulting from a group of point charges. Herein lies an important computational advantage of potential over electric field: it is a lot easier to sum several scalar quantities than to sum several vector quantities whose directions and components must be considered.

CHECKPOINT **4:** The figure shows three arrangements of two protons. Rank the arrangements according to the net electric potential produced at point P by the protons, greatest first.

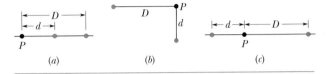

(a)   (b)   (c)

SAMPLE PROBLEM 25-4

What is the potential at point P, located at the center of the square of point charges shown in Fig. 25-9a? Assume that $d = 1.3$ m and that the charges are

$$q_1 = +12 \text{ nC}, \qquad q_3 = +31 \text{ nC},$$
$$q_2 = -24 \text{ nC}, \qquad q_4 = +17 \text{ nC}.$$

SOLUTION: Since each charge is the same distance r from P, Eq. 25-27 gives us

$$V = \sum_{i=1}^{4} V_i = \frac{1}{4\pi\epsilon_0}\frac{q_1 + q_2 + q_3 + q_4}{r}.$$

The distance r is $d/\sqrt{2}$, which is 0.919 m, and the sum of the charges is

$$q_1 + q_2 + q_3 + q_4 = (12 - 24 + 31 + 17) \times 10^{-9}\text{ C}$$
$$= 36 \times 10^{-9}\text{ C}.$$

So, $$V = \frac{(8.99 \times 10^9 \text{ N}\cdot\text{m}^2/\text{C}^2)(36 \times 10^{-9}\text{ C})}{0.919 \text{ m}}$$
$$\approx 350 \text{ V}. \qquad \text{(Answer)}$$

Close to the three positive charges in Fig. 25-9a, the potential has very large positive values. Close to the single negative charge, the potential has very large negative values. Thus there must be points within the square that have the same

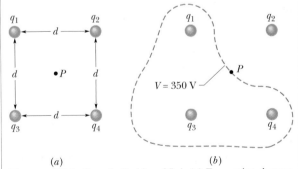

(a)   (b)

FIGURE 25-9   Sample Problem 25-4. (*a*) Four point charges are held fixed at the corners of a square. (*b*) The closed curve is a cross section, in the plane of the figure, of the equipotential surface that contains point P.

intermediate potential as that at point P. The curve in Fig. 25-9b shows the intersection of the plane of the figure with the equipotential surface that contains point P. Any point along that curve has the same potential as point P.

SAMPLE PROBLEM 25-5

(a) In Fig. 25-10a, 12 electrons (of charge $-e$) are equally spaced and fixed around a circle of radius R. Relative to $V = 0$ at infinity, what are the electric potential and electric field at the center C of the circle due to these electrons?

SOLUTION: Since electrons all have the same negative charge, and since all the electrons here are the same distance R from C, the potential at C must be, from Eq. 25-27,

$$V = -12 \frac{1}{4\pi\epsilon_0} \frac{e}{R}. \qquad \text{(Answer)} \qquad (25\text{-}28)$$

Since electric potential is a scalar, the orientation of the charges with respect to C is irrelevant to the potential V. However, since electric field is a vector, that orientation *is* important to $\mathbf{E}$. In fact, here, because of the symmetry of the arrangement, the electric field vector at C due to any given electron is canceled by the field vector due to the electron that is diametrically opposite it. Thus at C,

$$\mathbf{E} = 0. \qquad \text{(Answer)}$$

(b) If the electrons are moved along the circle until they are nonuniformly spaced over a 120° arc (Fig. 25-10b), what then is the potential at C? How does the electric field at C change (if at all)?

SOLUTION: The potential is still given by Eq. 25-28, because the distance between C and each electron is unchanged and orientation is irrelevant. The electric field is no longer zero, because the arrangement is no longer symmetric. There is now a net field that points toward the charge distribution.

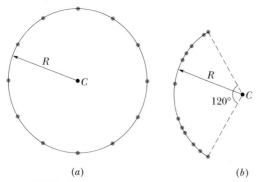

(a)   (b)

FIGURE 25-10   Sample Problem 25-5. (a) Twelve electrons uniformly spaced around a circle. (b) Those electrons are now nonuniformly spaced along an arc of the original circle.

25-7 POTENTIAL DUE TO AN ELECTRIC DIPOLE

Now let us apply Eq. 25-27 to an electric dipole to find the potential at an arbitrary point P in Fig. 25-11a. At P, the positive point charge (at distance $r_{(+)}$) sets up potential $V_{(+)}$ and the negative point charge (at distance $r_{(-)}$) sets up potential $V_{(-)}$, both potentials as given by Eq. 25-26. So the net potential at P is given by Eq. 25-27 as

$$V = \sum_{i=1}^{2} V_i = V_{(+)} + V_{(-)} = \frac{1}{4\pi\epsilon_0} \left(\frac{q}{r_{(+)}} + \frac{-q}{r_{(-)}} \right)$$

$$= \frac{q}{4\pi\epsilon_0} \frac{r_{(-)} - r_{(+)}}{r_{(-)}r_{(+)}}. \qquad (25\text{-}29)$$

Because naturally occurring dipoles—such as those possessed by many molecules—are small, we are usually interested only in points far from the dipole, such that $r \gg d$, where d is the distance between the charges. Under these conditions, the approximations that follow from Fig. 25-11b are

$$r_{(-)} - r_{(+)} \approx d \cos \theta \quad \text{and} \quad r_{(-)}r_{(+)} \approx r^2.$$

If we substitute these quantities into Eq. 25-29, we can approximate V to be

$$V = \frac{q}{4\pi\epsilon_0} \frac{d \cos \theta}{r^2},$$

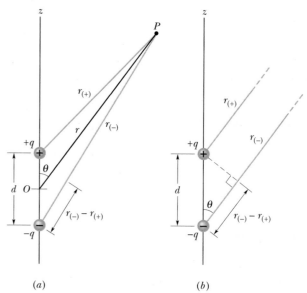

(a)   (b)

FIGURE 25-11   (a) Point P is a distance r from the midpoint O of a dipole. The line OP makes an angle θ with the dipole axis. (b) If P is far from the dipole, the lines of lengths $r_{(+)}$ and $r_{(-)}$ are approximately parallel to the line of length r, and the dashed black line is approximately perpendicular to the line of length $r_{(-)}$.

where θ is measured from the dipole axis as shown in Fig. 25-11a. We can now write V as

$$V = \frac{1}{4\pi\epsilon_0} \frac{p \cos\theta}{r^2} \quad \text{(electric dipole),} \quad (25\text{-}30)$$

in which p ($= qd$) is the magnitude of the electric dipole moment **p** defined in Section 23-5. The vector **p** is along the dipole axis, pointing from the negative to the positive charge. (So, θ is measured from the direction of **p**.)

CHECKPOINT 5: Suppose that three points are set at equal (large) distances r from the center of the dipole in Fig. 25-11: point a is on the dipole axis above the positive charge, point b is on the axis below the negative charge, and point c is on a perpendicular bisector through the line connecting the two charges. Rank the points according to the electric potential of the dipole there, greatest (most positive) first.

Induced Dipole Moment

Many molecules such as water have *permanent* electric dipole moments. In other molecules (*nonpolar molecules*) and in every atom, the centers of the positive and negative charges coincide (Fig. 25-12a) and thus no dipole moment is set up. However, if we place an atom or a nonpolar molecule in an external electric field, the field distorts the electron orbits and separates the centers of positive and negative charge (Fig. 25-12b). Because the electrons are negatively charged, they tend to be shifted in a direction opposite the field. This shift sets up a dipole moment **p** that points in the direction of the field. This dipole moment is said to be *induced* by the field, and the atom or molecule is then said to be *polarized* by the field (it has a positive side and a negative side). When the field is removed, the induced dipole moment and the polarization disappear.

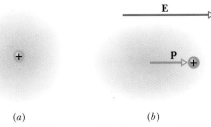

(a)   (b)

FIGURE 25-12 (a) An atom, showing the positively charged nucleus (green) and the negatively charged electrons (gold shading). The centers of positive and negative charge coincide. (b) If the atom is placed in an external electric field, the electron orbits are distorted so that the centers of positive and negative charge no longer coincide. An induced dipole moment appears. The distortion is greatly exaggerated here.

25-8 POTENTIAL DUE TO A CONTINUOUS CHARGE DISTRIBUTION

When a charge distribution q is continuous (as on a uniformly charged thin rod or disk), we cannot use the summation of Eq. 25-27 to find the potential V at a point P. Instead, we must choose a differential element of charge dq, determine the potential dV at P due to dq, and then integrate over the entire charge distribution.

Let us again take the zero of potential to be at infinity. If we treat the element of charge dq as a point charge, then we can use Eq. 25-26 to express the potential dV at point P due to dq:

$$dV = \frac{1}{4\pi\epsilon_0} \frac{dq}{r} \quad \text{(positive or negative } dq\text{).} \quad (25\text{-}31)$$

Here r is the distance between P and dq. To find the total potential V at P, we integrate to sum the potentials due to all the charge elements:

$$V = \int dV = \frac{1}{4\pi\epsilon_0} \int \frac{dq}{r}. \quad (25\text{-}32)$$

The integral is to be taken over the entire charge distribution. Note that because the electric potential is a scalar, there are *no vector components* to consider in Eq. 25-32.

We now examine two continuous charge distributions, a line of charge and a charged disk.

Line of Charge

In Fig. 25-13a, a thin nonconducting rod of length L has a positive charge of uniform linear density λ. Let us determine the electric potential V due to the rod at point P, a perpendicular distance d from the left end of the rod.

We consider a differential element dx of the rod as shown in Fig. 25-13b. This (or any other) element of the rod has a differential charge of

$$dq = \lambda \, dx. \quad (25\text{-}33)$$

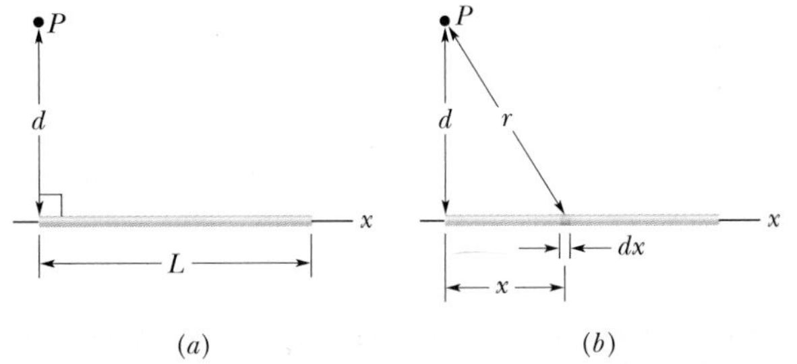

FIGURE 25-13 (a) A thin, uniformly charged rod produces an electric potential V at point P. (b) An element of charge produces a differential potential dV at P.

This element produces a potential dV at point P, which is a distance $r = (x^2 + d^2)^{1/2}$ from the element. Treating the element as a point charge, we can use Eq. 25-31 to write the potential dV as

$$dV = \frac{1}{4\pi\epsilon_0} \frac{dq}{r} = \frac{1}{4\pi\epsilon_0} \frac{\lambda \, dx}{(x^2 + d^2)^{1/2}}. \quad (25\text{-}34)$$

Since the charge on the rod is positive and we have taken $V = 0$ at infinity, we know from Section 25-5 that dV in Eq. 25-34 must be positive.

We now find the total potential V produced by the rod at point P by integrating Eq. 25-34 along the length of the rod, from $x = 0$ to $x = L$, using integral 17 in Appendix E. We find

$$V = \int dV = \int_0^L \frac{1}{4\pi\epsilon_0} \frac{\lambda}{(x^2 + d^2)^{1/2}} \, dx$$

$$= \frac{\lambda}{4\pi\epsilon_0} \int_0^L \frac{dx}{(x^2 + d^2)^{1/2}}$$

$$= \frac{\lambda}{4\pi\epsilon_0} \left[\ln \left(x + (x^2 + d^2)^{1/2} \right) \right]_0^L$$

$$= \frac{\lambda}{4\pi\epsilon_0} \left[\ln[L + (L^2 + d^2)^{1/2}] - \ln d \right].$$

We can simplify this result by using the general relation $\ln A - \ln B = \ln (A/B)$. We then find

$$V = \frac{\lambda}{4\pi\epsilon_0} \ln \left[\frac{L + (L^2 + d^2)^{1/2}}{d} \right]. \quad (25\text{-}35)$$

Because V is the sum of positive values of dV, it should be positive. But does Eq. 25-35 give a positive V? Since the argument of the logarithm is greater than one, the logarithm is a positive number and V is indeed positive.

infinity, positive charge gives a positive potential and negative charge gives a negative potential.)

If you happen to reverse the limits on the integral used to calculate a potential, you will obtain a negative value for V. The magnitude will be correct, but discard the minus sign. Then determine the proper sign for V from the sign of the charge. As an example, we would have obtained a minus sign in Eq. 25-35 if we had reversed the limits in the integral above that equation. We would then have discarded that minus sign and noted that the potential is positive because the charge producing it is positive.

Charged Disk

In Section 23-7, we calculated the magnitude of the electric field at points on the central axis of a plastic disk of radius R that has a uniform charge density σ on one surface. Here we derive an expression for $V(z)$, the potential at any point on the central axis.

In Fig. 25-14, consider a differential element consisting of a flat ring of radius R' and radial width dR'. Its charge has magnitude

$$dq = \sigma(2\pi R')(dR'),$$

in which $(2\pi R')(dR')$ is the upper surface area of the ring. All parts of this charged element are the same distance r from point P on the disk's axis. With the aid of Fig. 25-14, we can now use Eq. 25-31 to write the contribution of this ring to the electric field at P as

$$dV = \frac{1}{4\pi\epsilon_0} \frac{dq}{r} = \frac{1}{4\pi\epsilon_0} \frac{\sigma(2\pi R')(dR')}{\sqrt{z^2 + R'^2}}. \quad (25\text{-}36)$$

We find the net potential at P by adding (via integration) the contributions of all the strips from $R' = 0$ to $R' = R$:

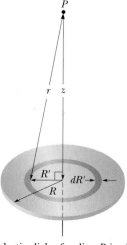

FIGURE 25-14   A plastic disk of radius R is charged on its top surface to a uniform surface charge density σ. We wish to find the potential V at point P on the central axis of the disk.

$$V = \int dV = \frac{\sigma}{2\epsilon_0} \int_0^R \frac{R'\, dR'}{\sqrt{z^2 + R'^2}}$$

$$= \frac{\sigma}{2\epsilon_0} (\sqrt{z^2 + R^2} - z). \qquad (25\text{-}37)$$

Note that the variable in the second integral of Eq. 25-37 is R' and not z, which remains constant while the integration over the surface of the disk is carried out. (Note also that, in evaluating the integral, we have assumed that $z \geq 0$.)

SAMPLE PROBLEM 25-6

The potential at the center of a uniformly charged circular disk of radius $R = 3.5$ cm is $V_0 = 550$ V.

(a) What is the total charge q on the disk?

SOLUTION: At the center of the disk, z in Eq. 25-37 is zero, so that equation reduces to

$$V_0 = \frac{\sigma R}{2\epsilon_0},$$

from which

$$\sigma = \frac{2\epsilon_0 V_0}{R}. \qquad (25\text{-}38)$$

Since σ is the surface charge density, the total charge q on the disk is $\sigma(\pi R^2)$. Using Eq. 25-38, we can now write

$$q = \sigma(\pi R^2) = 2\pi\epsilon_0 R V_0$$
$$= (2\pi)(8.85 \times 10^{-12} \text{ C}^2/\text{N} \cdot \text{m}^2)(0.035 \text{ m})(550 \text{ V})$$
$$= 1.1 \times 10^{-9} \text{ C} = 1.1 \text{ nC}, \qquad \text{(Answer)}$$

in which we use Eq. 25-9 to write $1 \text{ V} = 1 \text{ J/C} = 1 \text{ N} \cdot \text{m/C}$.

(b) What is the potential at a point on the axis of the disk a distance $z = 5.0R$ from the center of the disk?

SOLUTION: From Eq. 25-37 we find

$$V = \frac{\sigma}{2\epsilon_0} [\sqrt{(5.0R)^2 + R^2} - 5.0R].$$

Substituting σ from Eq. 25-38 then yields

$$V = \frac{V_0}{R} (\sqrt{26R^2} - 5.0R) = V_0(\sqrt{26} - 5.0)$$
$$= (550 \text{ V})(0.099) = 54 \text{ V}. \qquad \text{(Answer)}$$

25-9 CALCULATING THE FIELD FROM THE POTENTIAL

In Section 25-4, you saw how to find the potential at a point f if you know the electric field along a path from a reference point to point f. In this section, we propose to go

the other way, that is, to find the electric field when we know the potential. As Fig. 25-3 shows, solving this problem graphically is easy: if we know the potential V at all points near an assembly of charges, we can draw in a family of equipotential surfaces. The electric field lines, sketched perpendicular to those surfaces, reveal the variation of **E**. What we are seeking here is the mathematical equivalent of this graphical procedure.

Figure 25-15 shows cross sections of a family of closely spaced equipotential surfaces, the potential difference between each pair of adjacent surfaces being dV. As the figure suggests, the field **E** at any point P is perpendicular to the equipotential surface through P.

Suppose that a positive test charge q_0 moves through a displacement $d\mathbf{s}$ from one equipotential surface to the adjacent surface. From Eq. 25-7, we see that the work that the electric field does on the test charge during the move is $-q_0\, dV$. From Eq. 25-16 and Fig. 25-15, we see that the work done by the electric field may also be written as $(q_0\mathbf{E}) \cdot d\mathbf{s}$, or $q_0 E(\cos\theta)\, ds$. Equating these two expressions for the work yields

$$-q_0\, dV = q_0 E(\cos\theta)\, ds,$$

or

$$E \cos\theta = -\frac{dV}{ds}. \qquad (25\text{-}39)$$

Since $E \cos\theta$ is the component of **E** in the direction of $d\mathbf{s}$, Eq. 25-39 becomes

$$E_s = -\frac{\partial V}{\partial s}. \qquad (25\text{-}40)$$

We have added a subscript to E and switched to the partial derivative symbols to emphasize that Eq. 25-40 involves only the variation of V along a specified axis (here called the s axis) and only the component of **E** along that axis. In

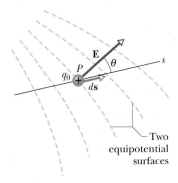

FIGURE 25-15 A test charge q_0 moves a distance $d\mathbf{s}$ from one equipotential surface to another. (The separation between the surfaces has been exaggerated for clarity.) The displacement $d\mathbf{s}$ makes an angle θ with the direction of the electric field **E**.

words, Eq. 25-40 (which is essentially the inverse of Eq. 25-18) states:

The component of **E** in any direction is the negative of the rate of change of the electric potential with distance in that direction.

If we take the s axis to be, in turn, the x, y, and z axes, we find that the x, y, and z components of **E** at any point are

$$E_x = -\frac{\partial V}{\partial x}; \; E_y = -\frac{\partial V}{\partial y}; \; E_z = -\frac{\partial V}{\partial z}. \quad (25\text{-}41)$$

Thus if we know V for all points in the region around a charge distribution, that is, if we know the function V(x, y, z), we can find the components of **E**—and thus **E** itself—at any point by taking partial derivatives.

In the simple situation where the electric field **E** is uniform, Eq. 25-40 becomes

$$E = -\frac{\Delta V}{\Delta s}, \quad (25\text{-}42)$$

where s is perpendicular to an equipotential surface. The electric field is zero in any direction tangent to an equipotential surface.

CHECKPOINT **6:** The figure shows three pairs of parallel plates with the same separation, and the electric potential of each plate. The electric field between the plates is uniform and perpendicular to the plates. (a) Rank the pairs according to the magnitude of the electric field between the plates, greatest first. (b) For which pair is the electric field pointing rightward? (c) If an electron is released midway between the third pair of plates, does it remain there, move rightward at constant speed, move leftward at constant speed, accelerate rightward, or accelerate leftward?

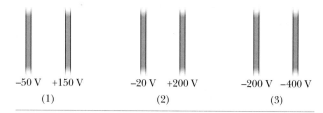

−50 V +150 V   −20 V +200 V   −200 V −400 V
   (1)   (2)   (3)

SAMPLE PROBLEM 25-7

The potential at any point on the axis of a charged disk is given by Eq. 25-37, which we can write as

$$V = \frac{\sigma}{2\epsilon_0} (\sqrt{z^2 + R^2} - z).$$

Starting with this expression, derive an expression for the electric field at any point on the axis of the disk.

SOLUTION: From symmetry, **E** must lie along the axis of the disk. If we choose the s axis to coincide with the z axis, then Eq. 25-40 gives us

$$E_z = -\frac{\partial V}{\partial z} = -\frac{\sigma}{2\epsilon_0} \frac{d}{dz} (\sqrt{z^2 + R^2} - z)$$

$$= \frac{\sigma}{2\epsilon_0} \left(1 - \frac{z}{\sqrt{z^2 + R^2}} \right). \quad \text{(Answer)}$$

This is the same expression that we derived in Section 23-7 by integration, using Coulomb's law.

25-10 ELECTRIC POTENTIAL ENERGY OF A SYSTEM OF POINT CHARGES

In Section 25-1, we discussed the electric potential energy of a test charge as a function of its position in an external electric field. In that section, we assumed that the charges that produced the field were fixed in place, so that the field could not be influenced by the presence of the test charge. In this section we can take a broader view, to find the electric potential energy of a *system* of charges due to the electric field produced *by* those charges.

For a simple example, if you push together two bodies that have charges of the same electrical sign, the work that you must do is stored as electric potential energy in the two-charge system (provided the kinetic energy of the bodies does not change). If you later release the charges, you can recover this stored energy, in whole or in part, as kinetic energy of the charged bodies as they rush away from each other.

We define the electric potential energy *of a system of point charges,* held in fixed positions by forces not specified, as follows:

The electric potential energy of a system of fixed point charges is equal to the work that must be done by an external agent to assemble the system, bringing each charge in from an infinite distance.

We assume that the charges are stationary both in their initial infinitely distant positions and in their final assembled configuration.

Figure 25-16 shows two point charges q_1 and q_2, separated by a distance r. To find the electric potential energy

FIGURE 25-16 Two charges held a fixed distance r apart. What is the electric potential energy of the configuration?

of this two-charge system, we mentally build the system, starting with both charges infinitely far away and at rest. When we bring q_1 in from infinity and put it in place, we do no work, because no electrostatic force acts on q_1. But when we next bring q_2 in from infinity and put it in place, we must do work, because q_1 exerts an electrostatic force on q_2 during the move.

We can calculate that work with Eq. 25-8 by dropping the minus sign (so that the equation gives the work *we* do rather than the field's work) and substituting q_2 for the general charge q. Our work is then equal to $q_2 V$, where V is the potential that has been set up by q_1 at the point where we put q_2. From Eq. 25-26, that potential is

$$V = \frac{1}{4\pi\epsilon_0} \frac{q_1}{r}.$$

Thus, from our definition, the electric potential energy of the pair of point charges of Fig. 25-16 is

$$U = W = \frac{1}{4\pi\epsilon_0} \frac{q_1 q_2}{r}. \qquad (25\text{-}43)$$

If the charges have the same sign, we have to do positive work to push them together against their mutual repulsion. Hence, as Eq. 25-43 shows, the potential energy of the system is then positive. If the charges have opposite signs, we have to do negative work against their mutual attraction to bring them together, so that they are stationary. The potential energy of the system is then negative. Sample Problem 25-8 shows how to extend this process to more than two charges.

SAMPLE PROBLEM 25-8

Figure 25-17 shows three charges held in fixed positions by forces that are not shown. What is the electric potential energy of this system of charges? Assume that $d = 12$ cm and that

$$q_1 = +q, \quad q_2 = -4q, \quad \text{and} \quad q_3 = +2q,$$

in which $q = 150$ nC.

SOLUTION: To answer, we mentally build the system of Fig. 25-17, starting with one of the charges, say q_1, in place and the others at infinity. Then we bring another one, say q_2, in from infinity and put it in place. From Eq. 25-43, with d substituted for r, the potential energy U_{12} associated with the

pair of charges q_1 and q_2 is

$$U_{12} = \frac{1}{4\pi\epsilon_0} \frac{q_1 q_2}{d}.$$

We then bring the last charge q_3 in from infinity and put it in place. The work that we must do in this last step is equal to the sum of the work we must do to bring q_3 near q_1 and the work we must do to bring it near q_2. From Eq. 25-43, with d substituted for r, that sum is

$$W_{13} + W_{23} = U_{13} + U_{23} = \frac{1}{4\pi\epsilon_0} \frac{q_1 q_3}{d} + \frac{1}{4\pi\epsilon_0} \frac{q_2 q_3}{d}.$$

The total potential energy U of the three-charge system is the sum of the potential energies associated with the three pairs of charges. This sum (which is actually independent of the order in which the charges are brought together) is

$$U = U_{12} + U_{13} + U_{23}$$
$$= \frac{1}{4\pi\epsilon_0}$$
$$\times \left(\frac{(+q)(-4q)}{d} + \frac{(+q)(+2q)}{d} + \frac{(-4q)(+2q)}{d} \right)$$
$$= -\frac{10q^2}{4\pi\epsilon_0 d}$$
$$= -\frac{(8.99 \times 10^9 \text{ N} \cdot \text{m}^2/\text{C}^2)(10)(150 \times 10^{-9} \text{ C})^2}{0.12 \text{ m}}$$
$$= -1.7 \times 10^{-2} \text{ J} = -17 \text{ mJ}. \qquad \text{(Answer)}$$

The negative potential energy means that negative work would have to be done to assemble this structure, starting with the three charges infinitely separated and at rest. Put another way, an external agent would have to do 17 mJ of work to disassemble the structure completely, ending with the three charges infinitely far apart.

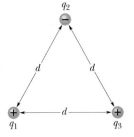

FIGURE 25-17 Sample Problem 25-8. Three charges are fixed at the vertices of an equilateral triangle. What is the electric potential energy of the configuration?

SAMPLE PROBLEM 25-9

An alpha particle (which consists of two protons and two neutrons) passes through the region of electron orbits in a gold atom, moving directly toward the gold nucleus, which has 79 protons and 118 neutrons. The alpha particle slows and then comes to a momentary stop, at a center-to-center separation r of 9.23 fm, before it begins to move back along its original path (Fig. 25-18). (Because the gold nucleus is much more massive than the alpha particle, we can assume the gold nucleus does not move.) What was the kinetic energy K of the alpha particle when it was initially far away (hence external to the gold atom)? Neglect the effect of the nuclear strong force.

SOLUTION: During the entire process, the mechanical energy of the *alpha particle + gold atom* system is conserved. When the alpha particle is outside the atom, the electric potential energy of the system is zero, because the atom has an equal number of electrons and protons, is thus electrically neutral, and so does not produce an external electric field. However, once the alpha particle has passed through the region of electron orbits on its way toward the nucleus, it is acted on by a repulsive electrostatic force due to its protons and those in the nucleus. (The neutrons, being electrically neutral, do not participate in producing this force. The electrons, now being outside the location of the alpha particle, act like a uniformly charged spherical shell, which produces no internal force.)

As the alpha particle slows because of the repulsive force, its kinetic energy is transferred to electric potential energy of the system. The transfer is complete when the alpha particle momentarily stops. Using the principle of conservation of mechanical energy, we can equate the initial kinetic energy K of the alpha particle to the electric potential energy U of the system at the instant the alpha particle stops:

$$K = U. \qquad (25\text{-}44)$$

By substituting Eq. 25-43 with $q_1 = 2e$, $q_2 = 79e$ (in which e is the elementary charge, 1.60×10^{-19} C), and $r = 9.23$ fm, we can rewrite Eq. 25-44 as

$$
\begin{aligned}
K &= \frac{1}{4\pi\epsilon_0} \frac{(2e)(79e)}{9.23 \text{ fm}} \\
&= \frac{(8.99 \times 10^9 \text{ N}\cdot\text{m}^2/\text{C}^2)(158)(1.60 \times 10^{-19} \text{ C})^2}{9.23 \times 10^{-15} \text{ m}} \\
&= 3.94 \times 10^{-12} \text{ J} = 24.6 \text{ MeV.} \qquad \text{(Answer)}
\end{aligned}
$$

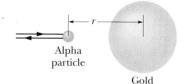

Alpha
particle

Gold
nucleus

FIGURE 25-18 Sample Problem 25-9. An alpha particle, traveling head-on toward the center of a gold nucleus, has come to a momentary stop, at which time all its kinetic energy has been transferred to electric potential energy.

CHECKPOINT **7:** If the alpha particle of Sample Problem 25-9 is replaced by a single proton of the same kinetic energy, will the proton momentarily stop at the same distance of 9.23 fm from the gold nucleus, farther from it, or closer to it?

25-11 POTENTIAL OF A CHARGED ISOLATED CONDUCTOR

In Section 24-6, we concluded that $\mathbf{E} = 0$ for all points inside an isolated conductor. We then used Gauss' law to prove that an excess charge placed on an isolated conductor lies entirely on its surface. (This is true even if the conductor has an empty internal cavity.) Here we use the fact that $\mathbf{E} = 0$ for all points inside an isolated conductor to prove another fact about such conductors:

> An excess charge placed on an isolated conductor will distribute itself on the surface of that conductor so that all points of the conductor—whether on the surface or inside—come to the same potential. This is true regardless of whether the conductor has an internal cavity.

Our proof follows directly from Eq. 25-18, which is

$$V_f - V_i = -\int_i^f \mathbf{E}\cdot d\mathbf{s}.$$

Since $\mathbf{E} = 0$ for all points within a conductor, it follows directly that $V_f = V_i$ for all possible pairs of points i and f in the conductor.

Figure 25-19a is a plot of potential against radial distance r from the center for an isolated spherical conducting shell of 1.0 m radius, having a charge of 1.0 μC. For points outside the shell, we can calculate $V(r)$ from Eq. 25-26 because the charge q behaves for such external points as if it were concentrated at the center of the shell. That equation holds right up to the surface of the shell. Now let us push a small test charge through the shell—assuming a small hole exists—to its center. No extra work is needed to do this because no net electric force acts on the test charge once it is inside the shell. Thus the potential at all points inside the shell has the same value as that on the surface, as Fig. 25-19a shows.

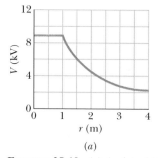
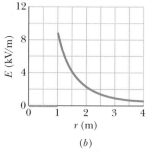

(a)   (b)

FIGURE 25-19 (a) A plot of $V(r)$ for a charged spherical shell. (b) A plot of $E(r)$ for the same shell.

FIGURE 25-20 A large spark jumps to the car's body and then exits by moving across the insulated left front tire (note the flash there), leaving the person inside unharmed.

Figure 25-19*b* shows the variation of electric field with radial distance for the same shell. Note that $E = 0$ everywhere inside the shell. The curves of Fig. 25-19*b* can be derived from the curve of Fig. 25-19*a* by differentiating with respect to *r*, using Eq. 25-40 (the derivative of a constant, recall, is zero). The curve of Fig. 25-19*a* can be derived from the curves of Fig. 25-19*b* by integrating with respect to *r*, using Eq. 25-19.

On nonspherical conductors, a surface charge does not distribute itself uniformly over the surface of the conductor. At sharp points or edges, the surface charge density—and thus the external electric field, which is proportional to it—may reach very high values. The air around such sharp points may become ionized, producing the corona discharge that golfers and mountaineers see on the tips of bushes, golf clubs, and rock hammers when thunderstorms

threaten. Such corona discharges, like hair that stands on end, are often the precursors of lightning strikes. In such circumstances, it is wise to enclose oneself in a cavity inside a conducting shell, where the electric field is guaranteed to be zero. A car (unless it is a convertible) is almost ideal (Fig. 25-20).

If an isolated conductor is placed in an *external electric field*, as in Fig. 25-21, all points of the conductor still come to a single potential regardless of whether the conductor has an excess charge. The free conduction electrons distribute themselves on the surface in such a way that the electric field they produce at interior points cancels the external electric field that would otherwise be there. Furthermore, the electron distribution causes the net electric field at all points on the surface to be perpendicular to the surface. If the conductor in Fig. 25-21 could be somehow removed, leaving the surface charges frozen in place, the pattern of the electric field would remain absolutely unchanged, for both exterior and interior points.

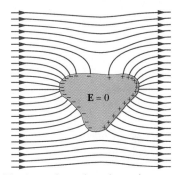

FIGURE 25-21 An uncharged conductor is suspended in an external electric field. The free electrons in the conductor distribute themselves on the surface as shown, reducing the net electric field inside the conductor to zero and making the net field at the surface perpendicular to the surface.

REVIEW & SUMMARY

Electric Potential Energy

The change ΔU in the electric potential energy U of a point charge as the charge moves from an initial point *i* to a final point *f* in an electric field is

$$\Delta U = U_f - U_i = -W, \qquad (25\text{-}1)$$

where *W* is the work done by the electric field on the point charge during the move from *i* to *f*. If the potential energy is defined to be zero at infinity, the **electric potential energy** U of the point charge at a particular point is

$$U = -W_\infty. \qquad (25\text{-}2)$$

Here W_∞ is the work done by the electric field on the point charge as the charge moves from infinity to the particular point.

Electric Potential Difference and Electric Potential

We define the **potential difference** ΔV between two points in an electric field as

$$\Delta V = V_f - V_i = -\frac{W}{q}, \qquad (25\text{-}7)$$

where *q* is the charge of a test particle on which work is done by the field. The **potential** at a point is

$$V = -\frac{W_\infty}{q}. \qquad (25\text{-}8)$$

The SI unit of potential is the *volt*: 1 volt = 1 joule per coulomb.

Potential and potential difference can also be written in terms of the electric potential energy U of a particle of charge q in an electric field:

$$V = \frac{U}{q}, \qquad (25\text{-}5)$$

$$\Delta V = V_f - V_i = \frac{U_f}{q} - \frac{U_i}{q} = \frac{\Delta U}{q}. \qquad (25\text{-}6)$$

Equipotential Surfaces

The points on an **equipotential surface** all have the same potential. The work done on a test charge in moving it from one such surface to another is independent of the locations of the initial and final points on these surfaces and of the path that joins the points. The electric field **E** is always directed perpendicularly to equipotential surfaces.

Finding V from E

The electric potential difference between any two points is

$$V_f - V_i = -\int_i^f \mathbf{E} \cdot d\mathbf{s}, \qquad (25\text{-}18)$$

where the integral is taken over any path connecting the points. If we choose $V_i = 0$ we have, for the potential at a particular point,

$$V = -\int_i^f \mathbf{E} \cdot d\mathbf{s}. \qquad (25\text{-}19)$$

Potential Due to Point Charges

The electric potential due to a single point charge at a distance r from that point charge is

$$V = \frac{1}{4\pi\epsilon_0} \frac{q}{r}. \qquad (25\text{-}26)$$

V has the same sign as q. The potential due to a collection of point charges is

$$V = \sum_{i=1}^n V_i = \frac{1}{4\pi\epsilon_0} \sum_{i=1}^n \frac{q_i}{r_i}. \qquad (25\text{-}27)$$

Potential Due to an Electric Dipole

At a distance r from an electric dipole with dipole moment $p = qd$, the electric potential of the dipole is

$$V = \frac{1}{4\pi\epsilon_0} \frac{p \cos\theta}{r^2} \qquad (25\text{-}30)$$

for $r \gg d$; the angle θ is defined in Fig. 25-11.

Potential Due to a Continuous Charge Distribution

For a continuous distribution of charge, Eq. 25-27 becomes

$$V = \frac{1}{4\pi\epsilon_0} \int \frac{dq}{r}, \qquad (25\text{-}32)$$

in which the integral is taken over the entire distribution.

Calculating E from V

The component of **E** in any direction is the negative of the rate of change of the potential with distance in that direction:

$$E_s = -\frac{\partial V}{\partial s}. \qquad (25\text{-}40)$$

The x, y, and z components of **E** may be found from

$$E_x = -\frac{\partial V}{\partial x}; \quad E_y = -\frac{\partial V}{\partial y}; \quad E_z = -\frac{\partial V}{\partial z}. \qquad (25\text{-}41)$$

When **E** is uniform, Eq. 25-40 reduces to

$$E = -\frac{\Delta V}{\Delta s}, \qquad (25\text{-}42)$$

where s is perpendicular to an equipotential surface. The electric field is zero in any direction parallel to an equipotential surface.

Electric Potential Energy of a System of Point Charges

The electric potential energy of a system of point charges is equal to the work needed to assemble the system with the charges initially at rest and infinitely distant from each other. For two charges at separation r,

$$U = W = \frac{1}{4\pi\epsilon_0} \frac{q_1 q_2}{r}. \qquad (25\text{-}43)$$

Potential of a Charged Conductor

An excess charge placed on a conductor will, in the equilibrium state, be located entirely on the outer surface of the conductor. The charge distributes itself so that the entire conductor, including interior points, is at a uniform potential.

QUESTIONS

1. Figure 25-22 shows three paths along which we can move positively charged sphere A closer to positively charged sphere B, which is fixed in place. (a) Would sphere A be moved to a higher or lower electric potential? Is the work done (b) by our force and (c) by the electric field (due to the second sphere) positive, negative, or zero? (d) Rank the paths according to the work our force does, greatest first.

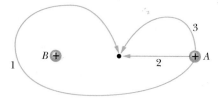

FIGURE 25-22 Question 1.

2. (a) In Fig. 25-3a, does the electric potential increase toward the right or toward the left? (b) If the adjacent equipotential surfaces differ by 10 V and the right-most one is at an electric potential of -100 V, what is the electric potential of the left-most one? If we move an electron toward the right, is the work done on the electron by (c) our force and (d) the electric field positive or negative?

3. Figure 25-23 shows four pairs of charged particles. Let $V = 0$ at infinity. For which pairs is there another point of zero net electric potential *on the axis* (a) between the particles and (b) to their right? (c) Where such a zero potential point exists, is the net electric field **E** due to the particles equal to zero? (d) For each pair, are there points off the axis (other than at infinity, of course) where $V = 0$?

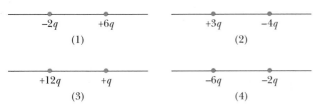

$-2q$ $+6q$ $+3q$ $-4q$
      (1)   (2)

$+12q$ $+q$ $-6q$ $-2q$
      (3)   (4)

FIGURE 25-23   Questions 3 and 14.

4. Figure 25-24 shows a square array of charged particles, with distance d between adjacent particles. What is the electric potential at point P at the center of the square if the electric potential is zero at infinity?

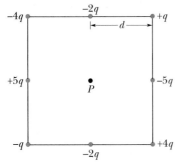

FIGURE 25-24   Question 4.

5. Figure 25-25 shows four arrangements of charged particles, all the same distance from the origin. Rank the situations according to the net electric potential at the origin, most positive first. Take the potential to be zero at infinity.

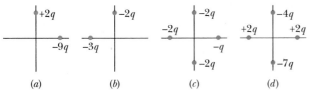

      (a)   (b)   (c)   (d)
FIGURE 25-25   Question 5.

6. Figure 25-26 shows a proton at the origin, three choices for point a at distance r, and three choices for point b at distance $2r$. There are nine different ways to pair the choices of a with the choices of b. Rank those nine ways according to the potential difference $V_a - V_b$ between points a and b, greatest first.

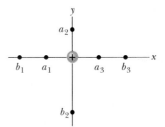

FIGURE 25-26
Question 6.

7. Figure 25-27 shows two situations in which we move an electron in from an infinite distance to a point midway between two charged particles (either proton or electron) fixed in place. In each situation, is the work done on the incoming electron *by the electric field* positive, negative, or zero?

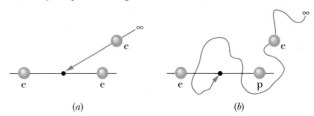

      (a)   (b)
FIGURE 25-27   Question 7.

8. (a) In Fig. 25-28a, what is the potential at point P due to charge Q at distance R from P? Set $V = 0$ at infinity. (b) In Fig. 25-28b, the same charge Q has been spread uniformly over a circular arc of radius R and central angle 40°. What is the potential at point P, the center of curvature of the arc? (c) In Fig. 25-28c, the same charge Q has been spread uniformly over a circle of radius R. What is the potential at point P, the center of the circle? (d) Rank the three situations according to the magnitude of the electric field that is set up at P, greatest first.

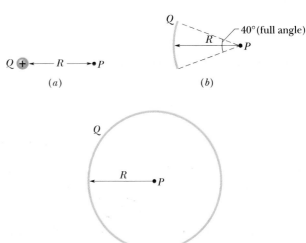

FIGURE 25-28
Question 8.   (c)

9. Figure 25-29 shows three sets of cross sections of equipotential surfaces; all three cover the same size region of space. (a) Rank the arrangements according to the magnitude of the electric field present in the region, greatest first. (b) In which is the electric field directed down the page?

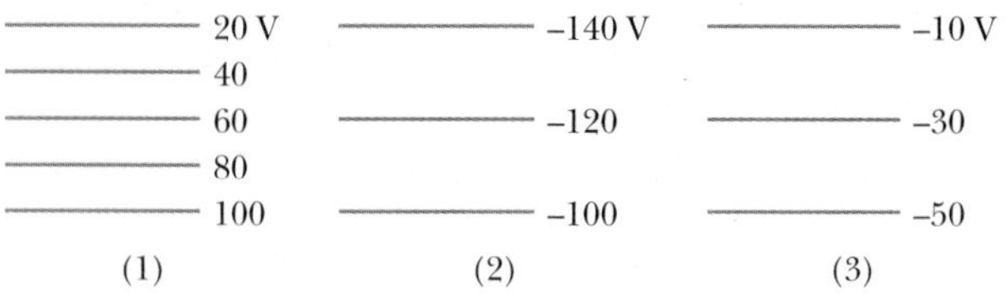

20 V	−140 V	−10 V
40		
60	−120	−30
80		
100	−100	−50
(1)	(2)	(3)

FIGURE 25-29 Question 9.

10. The electric potential at the coordinates (2 m, 0.5 m, 0.2 m) is given by $V = 2x - 3y + 4z$. Rank the magnitudes of the electric field components E_x, E_y, and E_z there, greatest first.

11. In Fig. 25-2, is the magnitude E of the electric field greater at the left or at the right?

12. Figure 25-30 gives the electric potential V as a function of distance through five regions on an x axis. (a) Rank the regions according to the magnitude of the x component of the electric field within them, greatest first. What is the direction of that component in (b) region 2 and (c) region 4?

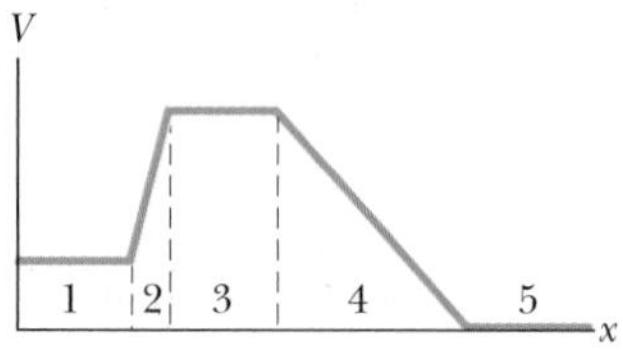

FIGURE 25-30
Question 12.

13. Rank the arrangements of Checkpoint 4 according to the electric potential energy of the system, greatest first.

14. Figure 25-23 shows four pairs of charged particles with identical separations. (a) Rank the pairs according to their electric potential energy, greatest (most positive) first. (b) For each pair, if the separation between the particles is increased, does the potential energy of the pair increase or decrease?

15. Figure 25-31 shows three systems of particles of charge $+q$ or $-q$, forming one equilateral and two isosceles triangles with edge lengths of either d, $2d$, or $d/2$. (a) Rank the systems ac-

cording to their electric potential energy, most positive first. (b) How much work must we do to make the system of Fig. 25-31b if the particles are initially infinitely far apart?

16. Figure 25-32 shows a system of three charged particles. If you move the particle of charge $+q$ from point A to point D, are the following positive, negative, or zero: (a) the change in the electric potential energy of the three-particle system, (b) the work done by the net electrostatic force on the particle you moved, and (c) the work done by your force? (d) What are the answers to (a) through (c) if, instead, the move is from point B to point C?

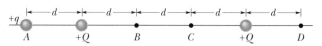

FIGURE 25-32 Questions 16 and 17.

17. Consider again the situation in Question 16. Is the work done by your force positive, negative, or zero if the move is (a) from A to B, (b) from A to C, and (c) from B to D? (d) Rank those moves according to the magnitude of the work done by your force, greatest first.

18. If the alpha particle of Sample Problem 25-9 had a smaller initial kinetic energy than the calculated 24.6 MeV, would it momentarily stop farther from, closer to, or at the same distance of 9.23 fm from the gold nucleus? (Again, neglect the strong force due to the gold nucleus.)

19. (a) If the surface of a charged conductor is an equipotential surface, does that mean that the charge is spread uniformly over the surface? (b) If the electric field is constant in magnitude over the surface of a charged conductor, does *that* mean that the charge is spread uniformly?

20. We have seen that, inside a hollow conductor, you are shielded from the fields of outside charges. If you are *outside* a hollow conductor that contains charges, are you shielded from the fields of these charges?

21. Can two equipotential surfaces of different values intersect? Explain your answer.

22. Three isolated, empty spherical shells of the same radius have the following charges: shell A, $+q$; shell B, $+2q$; shell C, $+3q$. Set the electric potential to be zero at an infinite distance from the shells. Then rank the shells, greatest first, according to (a) the electric potential at the surface of the shell, (b) the electric potential at the center of the shell, (c) the electric field magnitude at the surface of the shell, and (d) the electric field magnitude at the center of the shell.

23. Repeat Question 22 but with the electric potential now set to zero at the center of each shell.

24. Particle A of charge $+q$ and mass m and particle B of charge $-q$ and mass m are initially separated by distance d. In situation 1, we release both particles. In situation 2, we release only particle A. In which situation does particle A have greater kinetic energy when the separation between the particles has decreased to $d/2$, or does it have the same kinetic energy in the two situations?

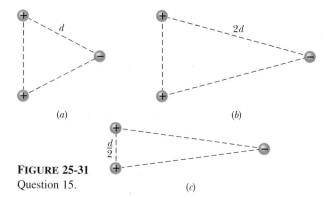

FIGURE 25-31
Question 15.

EXERCISES & PROBLEMS

SECTION 25-2 Electric Potential

1E. The electric potential difference between the ground and a cloud in a particular thunderstorm is 1.2×10^9 V. What is the magnitude of the change in the electric potential energy (in multiples of the electron-volt) of an electron that moves between the ground and the cloud?

2E. A particular 12 V car battery can send a total charge of 84 A·h (ampere-hours) through a circuit, from one terminal to the other. (a) How many coulombs of charge does this represent? (b) If this entire charge undergoes a potential difference of 12 V, how much energy is involved?

3P. In a given lightning flash, the potential difference between a cloud and the ground is 1.0×10^9 V and the quantity of charge transferred is 30 C. (a) What is the change in energy of that transferred charge? (b) If all the energy released by the transfer could be used to accelerate a 1000 kg automobile from rest, what would be the automobile's final speed? (c) If the energy could be used to melt ice, how much ice would it melt at 0°C? The heat of fusion of ice is 3.33×10^5 J/kg.

SECTION 25-4 Calculating the Potential from the Field

4E. Two infinite lines of charge are parallel to and in the same plane with the z axis. One, of charge per unit length $+\lambda$, is a distance a to the right of this axis. The other, of charge per unit length $-\lambda$, is a distance a to the left of this axis. Sketch some of the equipotential surfaces due to this arrangement.

5E. In Fig. 25-33, three long parallel lines of charge, with the relative linear charge densities shown, extend perpendicular to the page in both directions. Sketch some electric field lines; also sketch the cross sections in the plane of the figure of some equipotential surfaces.

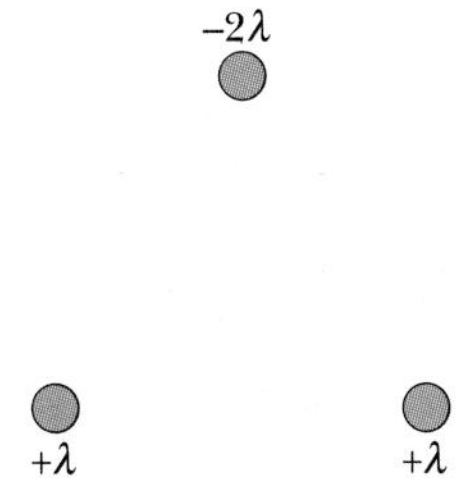

FIGURE 25-33 Exercise 5.

-2λ

$+\lambda$ $+\lambda$

6E. When an electron moves from A to B along an electric field line in Fig. 25-34, the electric field does 3.94×10^{-19} J of work on it. What are the electric potential differences (a) $V_B - V_A$, (b) $V_C - V_A$, and (c) $V_C - V_B$?

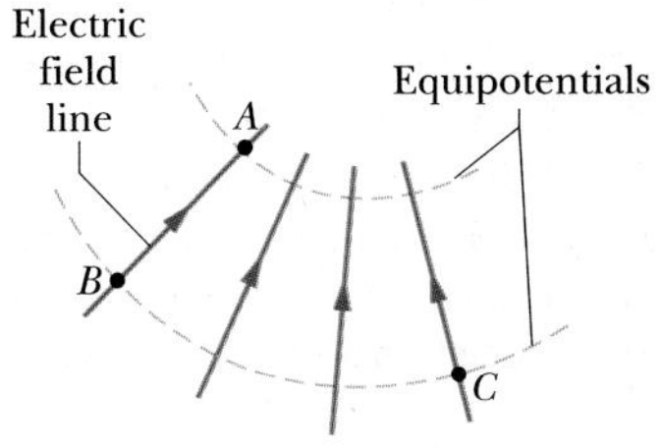

FIGURE 25-34 Exercise 6.

7E. In the Millikan oil-drop experiment (see Section 23-8), a uniform electric field of 1.92×10^5 N/C is maintained in the region between two plates separated by 1.50 cm. Find the potential difference between the plates.

8E. Two large parallel conducting plates are 12 cm apart and have charges of equal magnitude and opposite sign on their facing surfaces. An electrostatic force of 3.9×10^{-15} N acts on an electron placed anywhere between the two plates. (Neglect fringing.) (a) Find the electric field at the position of the electron. (b) What is the potential difference between the plates?

9E. An infinite nonconducting sheet has a surface charge density $\sigma = 0.10$ μC/m^2 on one side. How far apart are equipotential surfaces whose potentials differ by 50 V?

10P. Figure 25-35 shows, edge-on, an infinite nonconducting sheet with positive surface charge density σ on one side. (a) How much work is done by the electric field of the sheet as a small positive test charge q_0 is moved from an initial position on the sheet to a final position located a perpendicular distance z from the sheet? (b) Use Eq. 25-18 and the result from (a) to show that the electric potential of an infinite sheet of charge can be written $V = V_0 - (\sigma/2\epsilon_0)z$, where V_0 is the electric potential at the surface of the sheet.

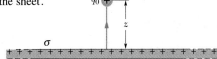

FIGURE 25-35 Problem 10.

11P. A Geiger counter has a metal cylinder 2.00 cm in diameter along whose axis is stretched a wire 1.30×10^{-4} cm in diameter. If the potential difference between them is 850 V, what is the electric field at the surface of (a) the wire and (b) the cylinder? (*Hint:* Use the result of Problem 29 of Chapter 24.)

12P. The electric field inside a nonconducting sphere of radius R, with charge spread uniformly throughout its volume, is radially directed and has magnitude

$$E(r) = \frac{qr}{4\pi\epsilon_0 R^3}.$$

Here q (positive or negative) is the total charge in the sphere, and r is the distance from the sphere center. (a) Taking $V = 0$ at the center of the sphere, find the potential $V(r)$ inside the sphere. (b) What is the difference in electric potential between a point on the surface and the sphere's center? (c) If q is positive, which of those two points is at the higher potential?

13P*. A charge q is distributed uniformly throughout a spherical volume of radius R. (a) Setting $V = 0$ at infinity, show that the potential at a distance r from the center, where $r < R$, is given by

$$V = \frac{q(3R^2 - r^2)}{8\pi\epsilon_0 R^3}.$$

(*Hint:* See Sample Problem 24-7.) (b) Why does this result differ from that in (a) of Problem 12? (c) What is the potential difference between a point on the surface and the sphere's center? (d) Why doesn't this result differ from that of (b) of Problem 12?

14P*. A thick spherical shell of charge Q and uniform volume charge density ρ is bounded by radii r_1 and r_2, where $r_2 > r_1$. With $V = 0$ at infinity, find the electric potential V as a function of the distance r from the center of the distribution, considering the regions (a) $r > r_2$, (b) $r_2 > r > r_1$, and (c) $r < r_1$. (d) Do these solutions agree at $r = r_2$ and $r = r_1$? (*Hint:* See Sample Problem 24-7.)

SECTION 25-6 Potential Due to a Group of Point Charges

15E. Consider a point charge $q = 1.0$ μC, point A at distance $d_1 = 2.0$ m from q, and point B at distance $d_2 = 1.0$ m. (a) If these points are diametrically opposite each other, as in Fig. 25-36a, what is the electric potential difference $V_A - V_B$? (b) What is that electric potential difference if points A and B are located as in Fig. 25-36b?

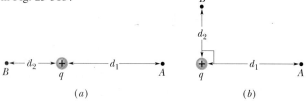

FIGURE 25-36 Exercise 15.

16E. Consider a point charge $q = 1.5 \times 10^{-8}$ C, and take $V = 0$ at infinity. (a) What are the shape and dimensions of an equipotential surface having a potential of 30 V due to q alone? (b) Are surfaces whose potentials differ by a constant amount (1.0 V, say) evenly spaced?

17E. A charge of 1.50×10^{-8} C lies on an isolated metal sphere of radius 16.0 cm. With $V = 0$ at infinity, what is the electric potential at points on the sphere's surface?

18E. As a space shuttle moves through the dilute ionized gas of Earth's ionosphere, its potential is typically changed by -1.0 V during one revolution. By assuming that the shuttle is a sphere of radius 10 m, estimate the amount of charge it collects.

19E. Much of the material making up Saturn's rings is in the form of tiny dust grains having radii on the order of 10^{-6} m. These grains are located in a region containing a dilute ionized gas, and they pick up excess electrons. As an approximation, suppose each grain is spherical, with radius $R = 1.0 \times 10^{-6}$ m. How many electrons would one grain have to pick up to have a potential of -400 V on its surface (taking $V = 0$ at infinity)?

20E. Figure 25-37 shows two charged particles on an axis. Sketch the electric field lines and the equipotential surfaces in the plane of the page for (a) $q_1 = +q$ and $q_2 = +2q$ and (b) $q_1 = +q$ and $q_2 = -3q$.

FIGURE 25-37
Exercises 20 through 23.

21E. In Fig. 25-37, set $V = 0$ at infinity and let the particles have charges $q_1 = +q$ and $q_2 = -3q$. Then locate (in terms of the separation distance d) any point on the x axis (other than at infinity) at which the net potential due to the two particles is zero.

22E. Let the separation d between the particles in Fig. 25-37 be 1.0 m; let their charges be $q_1 = +q$ and $q_2 = +2q$; and let $V = 0$ at infinity. Then locate any point on the x axis (other than at infinity) at which (a) the net electric potential due to the two particles is zero and (b) the net electric field due to them is zero.

23E. Two particles of charges q_1 and q_2 are separated by distance d in Fig. 25-37. The net electric field of the particles is zero at $x = d/4$. With $V = 0$ at infinity, locate (in terms of d) any point on the x axis (other than at infinity) at which the electric potential due to the two particles is zero.

24E. (a) If an isolated conducting sphere 10 cm in radius has a charge of 4.0 μC, and $V = 0$ at infinity, what is the potential on the surface of the sphere? (b) Can this situation actually occur, given that the air around the sphere undergoes electrical breakdown when the field exceeds 3.0 MV/m?

25P. What are (a) the charge and (b) the charge density on the surface of a conducting sphere of radius 0.15 m whose potential is 200 V (with $V = 0$ at infinity)?

26P. A spherical drop of water carrying a charge of 30 pC has a potential of 500 V at its surface (with $V = 0$ at infinity). (a) What is the radius of the drop? (b) If two such drops of the same charge and radius combine to form a single spherical drop, what is the potential at the surface of the new drop?

27P. An electric field of approximately 100 V/m is often observed near the surface of Earth. If this were the field over the entire surface, what would be the electric potential of a point on the surface? (Set $V = 0$ at infinity.)

28P. In Fig. 25-38, what is the net potential at point P due to the four point charges, if $V = 0$ at infinity?

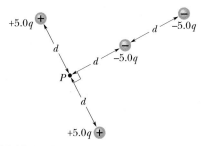

FIGURE 25-38 Problem 28.

29P. Suppose that the negative charge in a copper one-cent coin were removed to a very large distance from Earth—perhaps to a distant galaxy—and that the positive charge were distributed uniformly over Earth's surface. By how much would the electric potential at the surface change? (See Sample Problem 22-4.)

30P. In Fig. 25-39, point P is at the center of the rectangle. With $V = 0$ at infinity, what is the net electric potential at P due to the six charged particles?

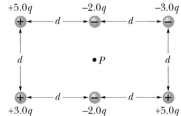

FIGURE 25-39
Problem 30.

31P. A point charge $q_1 = +6.0e$ is fixed at the origin of a rectangular coordinate system, and a second point charge $q_2 = -10e$ is fixed at $x = 8.6$ nm, $y = 0$. The locus of all points in the xy plane with $V = 0$ (other than at infinity) is a circle centered on the x axis, as shown in Fig. 25-40. Find (a) the location x_c of the center of the circle and (b) the radius R of the circle. (c) Is the xy cross section of the 5 V equipotential surface also a circle?

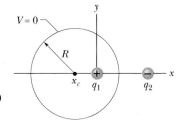

FIGURE 25-40
Problem 31.

32P. A solid copper sphere whose radius is 1.0 cm has a very thin surface coating of nickel. Some of the nickel atoms are radioactive, each atom emitting an electron as it decays. Half of these electrons enter the copper sphere, each depositing 100 keV of energy there. The other half of the electrons escape, each carrying away a charge of $-e$. The nickel coating has an activity of 10 mCi ($= 10$ millicuries $= 3.70 \times 10^8$ radioactive decays per second). The sphere is hung from a long, nonconducting string and isolated from its surroundings. (a) How long will it take for the potential of the sphere to increase by 1000 V? (b) How long will it take for the temperature of the sphere to increase by 5.0 K due to the energy deposited by the electrons? The heat capacity of the sphere is 14.3 J/K.

SECTION 25-7 Potential Due to an Electric Dipole

33E. The ammonia molecule NH_3 has a permanent electric dipole moment equal to 1.47 D, where 1 D = 1 debye unit = 3.34×10^{-30} C·m. Calculate the electric potential due to an ammonia molecule at a point 52.0 nm away along the axis of the dipole. (Set $V = 0$ at infinity.)

34P. For the charge configuration of Fig. 25-41, show that $V(r)$ for points such as P on the axis, assuming $r \gg d$, is given by

$$V = \frac{1}{4\pi\epsilon_0} \frac{q}{r} \left(1 + \frac{2d}{r}\right).$$

(*Hint:* The charge configuration can be viewed as the sum of an isolated charge and a dipole.)

FIGURE 25-41 Problem 34.

SECTION 25-8 Potential Due to a Continuous Charge Distribution

35E. (a) Figure 25-42a shows a positively charged plastic rod of length L and uniform linear charge density λ. Setting $V = 0$ at infinity and considering Fig. 25-13 and Eq. 25-35, find the electric potential at point P without written calculation. (b) Figure 25-42b shows an identical rod, except that it is split in half and the right half is negatively charged; the left and right halves have the same magnitude λ of uniform linear charge density. With V still zero at infinity, what is the electric potential at point P in Fig. 25-42b?

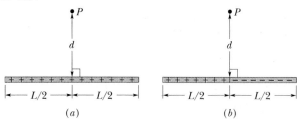

FIGURE 25-42 Exercise 35.

36E. In Fig. 25-43, a plastic rod having a uniformly distributed charge $-Q$ has been bent into a circular arc of radius R and central angle 120°. With $V = 0$ at infinity, what is the electric potential at P, the center of curvature of the rod?

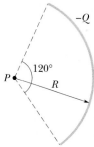

FIGURE 25-43 Exercise 36.

37E. A circular plastic rod of radius R has a positive charge $+Q$ uniformly distributed along one-quarter of its circumference and a negative charge of $-6Q$ uniformly distributed along the rest of the circumference (Fig. 25-44). With $V = 0$ at infinity, what is the electric potential (a) at the center C of the circle and (b) at point P, which is on the central axis of the circle at a distance z from the center?

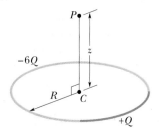

FIGURE 25-44 Exercise 37.

38E. A plastic disk is charged on one side with a uniform surface charge density λ, and then three quadrants of the disk are removed. The remaining quadrant is shown in Fig. 25-45. With $V = 0$ at infinity, what is the potential due to the remaining quadrant at point P, which is on the central axis of the original disk at a distance z from the original center?

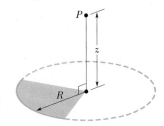

FIGURE 25-45
Exercise 38.

39P. Figure 25-46 shows a ring of outer radius R and inner radius $r = 0.200R$; the ring has a uniform surface charge density σ. With $V = 0$ at infinity, find an expression for the electric potential at point P on the central axis of the ring, at a distance $z = 2.00R$ from the center of the ring.

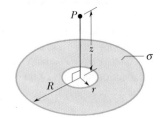

FIGURE 25-46
Problem 39.

40P. A disk like that of Fig. 25-14 has radius $R = 2.20$ cm. Its surface charge density is 1.50×10^{-6} C/m² from $r = 0$ to $R/2$ and 8.00×10^{-7} C/m² from $r = R/2$ to R. (a) What is the total charge on the disk? (b) With $V = 0$ at infinity, what is the electric potential at a point on the central axis of the disk, at a distance $z = R/2$ from the center of the disk?

41P. Figure 25-47 shows a plastic rod of length L and uniform positive charge Q lying on an x axis. With $V = 0$ at infinity, find the electric potential at point P_1 on the axis, at distance d from one end of the rod.

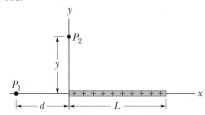

FIGURE 25-47 Problems 41, 42, 50, and 51.

42P. The plastic rod shown in Fig. 25-47 has length L and a nonuniform linear charge density $\lambda = cx$, where c is a positive constant. With $V = 0$ at infinity, find the electric potential at point P_1 on the axis, at distance d from one end.

SECTION 25-9 Calculating the Field from the Potential

43E. Two large parallel metal plates are 1.5 cm apart and have equal but opposite charges on their facing surfaces. Take the potential of the negative plate to be zero. If the potential halfway between the plates is then $+5.0$ V, what is the electric field in the region between the plates?

44E. In a certain situation, the electric potential varies along the x axis as shown in the graph of Fig. 25-48. For each of the intervals ab, bc, cd, de, ef, fg, and gh, determine the x component of the electric field, and then plot E_x versus x. (Ignore behavior at the interval end points.)

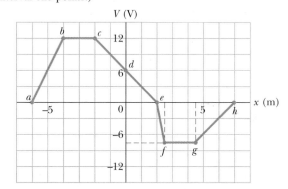

FIGURE 25-48   Exercise 44.

45E. Starting from Eq. 25-30, find the electric field due to a dipole at a point on the dipole axis.

46E. The electric potential at points in an xy plane is given by $V = (2.0 \text{ V/m}^2)x^2 - (3.0 \text{ V/m}^2)y^2$. What are the magnitude and direction of the electric field at point $(3.0 \text{ m}, 2.0 \text{ m})$?

47E. The electric potential V in the space between two flat parallel plates is given by $V = 1500x^2$, where V is in volts if x, the distance from one of the plates, is in meters. Calculate the magnitude and direction of the electric field at $x = 1.3$ cm.

48E. Exercise 48 in Chapter 24 deals with Rutherford's calculation of the electric field at a distance r from the center of an atom and inside the atom. He also gave the electric potential as

$$V = \frac{Ze}{4\pi\epsilon_0}\left(\frac{1}{r} - \frac{3}{2R} + \frac{r^2}{2R^3}\right).$$

(a) Show how the expression for the electric field given in Exercise 48 of Chapter 24 follows from the above expression for V. (b) Why does this expression for V not go to zero as $r \to \infty$?

49P. (a) Using Eq. 25-32, show that the electric potential at a point at distance z on the central axis of a thin ring of charge of radius R is

$$V = \frac{1}{4\pi\epsilon_0}\frac{q}{\sqrt{z^2 + R^2}}.$$

(b) From this result, derive an expression for E at points on the ring's axis; compare your result with the calculation of E in Section 23-6.

50P. (a) Use the result of Problem 41 to find the electric field component E_x at point P_1 in Fig. 25-47. (*Hint:* First substitute the variable x for the distance d in the result.) (b) Use symmetry to determine the electric field component E_y at P_1.

51P. The plastic rod of length L in Fig. 25-47 has the nonuniform linear charge density $\lambda = cx$, where c is a positive constant. (a) With $V = 0$ at infinity, find the electric potential at point P_2 on the y axis, a distance y from one end. (b) From that result, find the electric field component E_y at P_2. (c) Why cannot the field component E_x at P_2 be found using the result of (a)?

SECTION 25-10 Electric Potential Energy of a System of Point Charges

52E. (a) What is the electric potential energy of two electrons separated by 2.00 nm? (b) If the separation increases, does the potential energy increase or decrease?

53E. Two charges $q = +2.0\ \mu C$ are fixed in space a distance $d = 2.0$ cm apart, as shown in Fig. 25-49. (a) With $V = 0$ at infinity, what is the electric potential at point C? (b) You bring a third charge $q = +2.0\ \mu C$ from infinity to C. How much work must you do? (c) What is the potential energy U of the three-charge configuration when the third charge is in place?

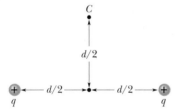

FIGURE 25-49 Exercise 53.

54E. The charges and coordinates of two point charges located in the xy plane are: $q_1 = +3.0 \times 10^{-6}$ C, $x = +3.5$ cm, $y = +0.50$ cm; and $q_2 = -4.0 \times 10^{-6}$ C, $x = -2.0$ cm, $y = +1.5$ cm. How much work must be done to locate these charges at their given positions, starting from infinite separation?

55E. A decade before Einstein published his theory of relativity, J. J. Thomson proposed that the electron might consist of small parts and attributed its mass to the electrical interaction of the parts. Furthermore, he suggested that the energy equals mc^2. Make a rough estimate of the electron mass in the following way: assume that the electron is composed of three identical parts that are brought in from infinity and placed at the vertices of an equilateral triangle having sides equal to the *classical radius* of the electron, 2.82×10^{-15} m. (a) Find the total electric potential energy of this arrangement. (b) Divide by c^2 and compare your result to the accepted electron mass (9.11×10^{-31} kg). (The result improves if more parts are assumed.)

56E. Derive an expression for the work required to set up the four-charge configuration of Fig. 25-50, assuming the charges are initially infinitely far apart.

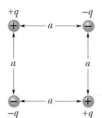

FIGURE 25-50
Exercise 56.

57E. In the quark model of fundamental particles, a proton is composed of three quarks: two "up" quarks, each having charge $+2e/3$, and one "down" quark, having charge $-e/3$. Suppose that the three quarks are equidistant from one another. Take the distance to be 1.32×10^{-15} m and calculate (a) the potential energy of the subsystem of two "up" quarks and (b) the total electric potential energy of the three-particle system.

58E. What is the electric potential energy of the charge configuration of Fig. 25-9a? Use the numerical values provided in Sample Problem 25-4.

59P. Three $+0.12$ C charges form an equilateral triangle, 1.7 m on a side. Using energy that is supplied at the rate of 0.83 kW, how many days would be required to move one of the charges to the midpoint of the line joining the other two charges?

60P. In the rectangle of Fig. 25-51, the sides have lengths 5.0 cm and 15 cm, $q_1 = -5.0\ \mu C$, and $q_2 = +2.0\ \mu C$. With $V = 0$ at infinity, what are the electric potentials (a) at corner A and (b) at corner B? (c) How much work is required to move a third charge $q_3 = +3.0\ \mu C$ from B to A along a diagonal of the rectangle? (d) Does this work increase or decrease the electric energy of the three-charge system? Is more, less, or the same work required if q_3 is moved along paths that are (e) inside the rectangle but not on a diagonal and (f) outside the rectangle?

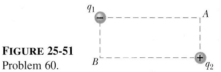

FIGURE 25-51
Problem 60.

61P. In Fig. 25-52, how much work is required to bring the charge of $+5q$ in from infinity along the dashed line and place it as shown near the two fixed charges $+4q$ and $-2q$? Take $d = 1.40$ cm and $q = 1.6 \times 10^{-19}$ C.

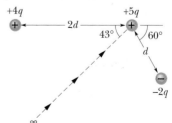

FIGURE 25-52 Problem 61.

62P. A particle of positive charge Q is fixed at point P. A second particle of mass m and negative charge $-q$ moves at constant speed in a circle of radius r_1, centered at P. Derive an expression for the work W that must be done by an external agent on the second particle to increase the radius of the circle of motion to r_2.

63P. Calculate (a) the electric potential established by the nucleus of a hydrogen atom at the average distance of the circulating electron ($r = 5.29 \times 10^{-11}$ m), (b) the electric potential energy of the atom when the electron is at this radius, and (c) the kinetic energy of the electron, assuming it to be moving in a circular orbit of this radius centered on the nucleus. (d) How much energy is required to ionize the hydrogen atom (that is, to remove the electron from the nucleus so that the separation is effectively infinite)? Express all energies in electron-volts.

64P. A particle of charge q is kept in a fixed position at a point P, and a second particle of mass m and the same charge q is initially held a distance r_1 from P. The second particle is then released. Determine its speed when it is a distance r_2 from P. Let $q = 3.1$ μC, $m = 20$ mg, $r_1 = 0.90$ mm, and $r_2 = 2.5$ mm.

65P. A charge of -9.0 nC is uniformly distributed around a ring of radius 1.5 m that lies in the yz plane with its center at the origin. A point charge of -6.0 pC is located on the x axis at $x = 3.0$ m. Calculate the work done in moving the point charge to the origin.

66P. Two tiny metal spheres A and B of mass $m_A = 5.00$ g and $m_B = 10.0$ g have equal positive charges $q = 5.00$ μC. The spheres are connected by a massless nonconducting string of length $d = 1.00$ m, a distance that is much greater than the radii of the spheres. (a) What is the electric potential energy of the system? (b) Suppose you cut the string. At that instant, what is the acceleration of each sphere? (c) A long time after you cut the string, what is the speed of each sphere?

67P. Two charged, parallel, flat conducting surfaces are spaced $d = 1.00$ cm apart and produce a potential difference $\Delta V = 625$ V between them. An electron is projected from one surface directly toward the second. What is the initial speed of the electron if it comes to rest just at the second surface?

68P. (a) A proton of kinetic energy 4.80 MeV travels head-on toward a lead nucleus. Assuming that the proton does not penetrate the nucleus and considering only electrostatic interactions, calculate the smallest center-to-center separation that occurs between the proton and the nucleus when the proton momentarily stops. (b) If the proton is replaced with an alpha particle of the same initial kinetic energy, how would the smallest center-to-center separation compare with that in (a)?

69P. A particle of mass m, positive charge q, and initial kinetic energy K is projected (from a large distance) toward a heavy nucleus of charge Q that is fixed in place. Assuming that the particle approaches head-on, how close to the center of the nucleus is the particle when it comes momentarily to rest?

70P. A thin, conducting, spherical shell of radius R is mounted on an isolating support and charged to a potential of $-V$. An electron is then fired from point P at a distance r from the center of the shell ($r \gg R$) with an initial speed v_0, directed radially inward. What value of v_0 is needed for the electron to just reach the shell before reversing direction?

71P. Two electrons are fixed 2.0 cm apart. Another electron is shot from infinity and comes to rest midway between the two. What was its initial speed?

72P. Consider an electron on the surface of a uniformly charged sphere of radius 1.0 cm and total charge 1.6×10^{-15} C. What is the *escape speed* for this electron? That is, what initial speed must it have to reach an infinite distance from the sphere and there have zero kinetic energy? (This escape speed is defined similarly to that in Chapter 14 for escaping the gravitational force, but here neglect that force.)

73P. An electron is projected with an initial speed of 3.2×10^5 m/s directly toward a proton that is fixed in place. If the electron is initially a great distance from the proton, at what distance from the proton is the speed of the electron instantaneously equal to twice the initial value?

SECTION 25-11 Potential of a Charged Isolated Conductor

74E. An empty hollow metal sphere has a potential of $+400$ V with respect to ground (defined to be at $V = 0$) and has a charge of 5.0×10^{-9} C. Find the electric potential at the center of the sphere.

75E. A thin, conducting, spherical shell of outer radius 20 cm has a charge of $+3.0$ μC. Sketch graphs of (a) the magnitude of the electric field **E** and (b) the potential V, both versus the distance r from the center of the shell. (Set $V = 0$ at infinity.)

76E. What is the excess charge on a conducting sphere of radius $r = 0.15$ m if the potential of the sphere is 1500 V and $V = 0$ at infinity?

77E. Consider two widely separated conducting spheres, 1 and 2, the second having twice the diameter of the first. The smaller sphere initially has a positive charge q, and the larger one is initially uncharged. You now connect the spheres with a long thin wire. (a) How are the final potentials V_1 and V_2 of the spheres related? (b) Find the final charges q_1 and q_2 on the spheres in terms of q. (c) What is the ratio of the final surface charge density of sphere 1 to that of sphere 2?

78P. The metal object shown in cross section in Fig. 25-53 is a figure of revolution about a horizontal axis. Suppose that it is charged negatively, and sketch a few equipotential surfaces and electric field lines. Use physical reasoning rather than mathematical analysis.

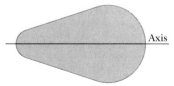

FIGURE 25-53 Problem 78.

79P. (a) If Earth had a net surface charge density of 1.0 electron per square meter (a very artificial assumption), what would its potential be? (Set $V = 0$ at infinity.) (b) What would be the electric field due to Earth just outside its surface?

80P. Two metal spheres, each of radius 3.0 cm, have a center-to-center separation of 2.0 m. One has a charge of $+1.0 \times 10^{-8}$ C; the other has a charge of -3.0×10^{-8} C. Assume that the separation is large enough relative to the size of the spheres to permit us to consider the charge on each to be uniformly distributed (the spheres are electrically isolated from each other). With $V = 0$ at infinity, calculate (a) the potential at the point halfway between their centers and (b) the potential of each sphere.

81P. A charged metal sphere of radius 15 cm has a net charge of 3.0×10^{-8} C. (a) What is the electric field at the sphere's surface? (b) If $V = 0$ at infinity, what is the electric potential at the sphere's surface? (c) At what distance from the sphere's surface has the electric potential decreased by 500 V?

82P. Two thin, isolated, concentric conducting spheres of radii R_1 and R_2 carry charges q_1 and q_2. With $V = 0$ at infinity, derive expressions for $E(r)$ and $V(r)$, where r is distance from the center of the spheres. Plot $E(r)$ and $V(r)$ from $r = 0$ to $r = 4.0$ m for $R_1 = 0.50$ m, $R_2 = 1.0$ m, $q_1 = +2.0$ μC, and $q_2 = +1.0$ μC.

Electronic Computation

83. Charge $q_1 = -1.2 \times 10^{-9}$ C is at the origin, and charge $q_2 = 2.5 \times 10^{-9}$ C is on the y axis at $y = 0.50$ m. Take the electric potential to be zero far from both charges. (a) Plot the inter-section of the $V = 5.0$ V equipotential surface with the xy plane. It encloses one of the charges. (b) There are two equipotential surfaces corresponding to $V = 3.0$ V. One encloses one of the charges and the other encloses both charges. Plot their intersections with the xy plane. (c) Find the value of the potential for which the pattern of the electric potential switches from one to two equipotential surfaces.

84. Suppose that N electrons are to be placed in a ring of radius R and that they can be placed in either of two configurations. In the first configuration they are all placed on the circumference and are uniformly distributed so that the distance between adjacent electrons is the same everywhere; in the second configuration $N - 1$ of the electrons are placed on the circumference as before and one electron is placed in the center of the ring. (a) For which configuration is the electrostatic potential energy less? Answer for N equal to all integer values from 2 to 15. (b) What is the smallest value of N for which the second configuration is less energetic than the first? (c) For the value of N found in (b), how many rim electrons are closer to any given electron there than is the electron at the center?

26
Capacitance

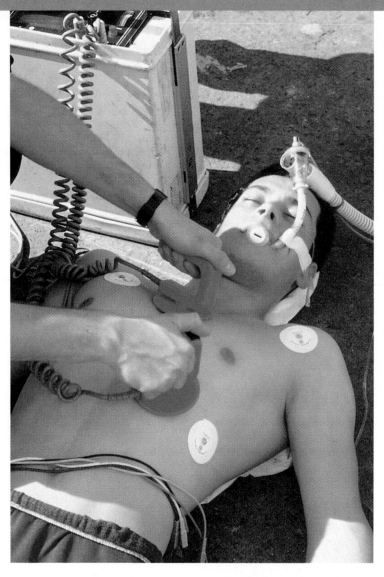

During ventricular fibrillation, a common type of heart attack, the chambers of the heart fail to pump blood because their muscle fibers randomly contract and relax. To save a victim of ventricular fibrillation, the heart muscle must be shocked to reestablish its normal rhythm. For that, 20 A of current must be sent through the chest cavity to transfer 200 J of electrical energy in about 2.0 ms. This requires about 100 kW of electrical power. The requirement may easily be met in a hospital, but what could produce that much power on, say, a remote road? Certainly not the electrical system of a car or ambulance, even if such a vehicle is available?

26-1 THE USES OF CAPACITORS

You can store energy as potential energy by pulling a bowstring, stretching a spring, compressing a gas, or lifting a book. You can also store energy as potential energy in an electric field, and a **capacitor** is a device you can use to do exactly that.

There is a capacitor in a portable battery-operated photoflash unit, for example. It accumulates charge relatively slowly during the charging process, building up an electric field as it does so. It holds this field and its energy until the energy is rapidly released during the flash.

Capacitors have many uses in our electronic and microelectronic age beyond serving as storehouses for potential energy. For one example, they are vital elements in the circuits with which we tune radio and television transmitters and receivers. For another example, microscopic capacitors form the memory banks of computers. These tiny devices are important—not so much for their stored energy as for the ON–OFF information that the presence or absence of their electric fields provides.

26-2 CAPACITANCE

Figure 26-1 shows some of the many sizes and shapes of capacitors. Figure 26-2 shows the basic elements of *any* capacitor—two isolated conductors of arbitrary shape. No matter what their geometry, flat or not, we call these conductors *plates*.

Figure 26-3*a* shows a less general but more conventional arrangement, called a *parallel-plate capacitor*, consisting of two parallel conducting plates of area A separated by a distance d. The symbol that we use to represent a

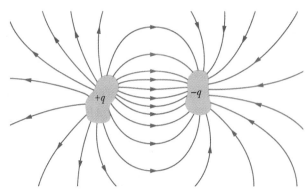

FIGURE 26-2 Two conductors, isolated electrically from each other and from their surroundings, form a *capacitor*. When the capacitor is charged, the conductors, or *plates* as they are called, carry equal but opposite charges of magnitude q.

capacitor (⊣⊢) is based on the structure of a parallel-plate capacitor but is used for capacitors of all geometries. We assume for the time being that no material medium (such as glass or plastic) is present in the region between the plates. In Section 26-6, we shall remove this restriction.

When a capacitor is *charged*, its plates have equal but opposite charges of $+q$ and $-q$. However, we refer to the

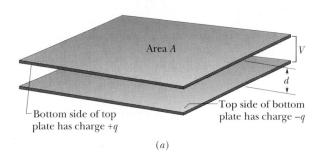

(*a*)

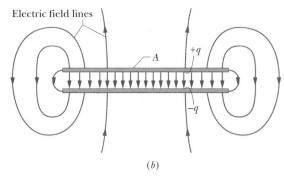

(*b*)

FIGURE 26-3 (*a*) A parallel-plate capacitor, made up of two plates of area A separated by a distance d. The plates have equal and opposite charges of magnitude q on their facing surfaces. (*b*) As the field lines show, the electric field is uniform in the central region between the plates. The field is not uniform at the edges of the plates, as indicated by the "fringing" of the field lines there.

FIGURE 26-1 An assortment of capacitors.

charge of a capacitor as being q, the absolute value of these charges on the plates. (Note that q is not the net charge on the capacitor, which is zero.)

Because the plates are conductors, they are equipotential surfaces: all points on a plate are at the same electric potential. Moreover, there is a potential difference between the two plates. For historical reasons, we represent the absolute value of this potential difference with V rather than with ΔV as we would with previous notation.

The charge q and the potential difference V for a capacitor are proportional to each other. That is,

$$q = CV. \qquad (26\text{-}1)$$

The proportionality constant C is called the **capacitance** of the capacitor. Its value depends only on the geometry of the plates and *not* on their charge or potential difference. The capacitance is a measure of how much charge must be put on the plates to produce a certain potential difference between them: the *greater the capacitance, the more charge is required.*

The SI unit of capacitance that follows from Eq. 26-1 is the coulomb per volt. This unit occurs so often that it is given a special name, the *farad* (F):

$$1 \text{ farad} = 1 \text{ F} = 1 \text{ coulomb per volt}$$
$$= 1 \text{ C/V}. \qquad (26\text{-}2)$$

As you will see, the farad is a very large unit. Submultiples of the farad, such as the microfarad ($1 \ \mu\text{F} = 10^{-6} \text{ F}$) and the picofarad ($1 \ \text{pF} = 10^{-12} \text{ F}$), are more convenient units in practice.

Charging a Capacitor

One way to charge a capacitor is to place it in an electric circuit with a battery. An *electric circuit* is a path through which charge can flow. A *battery* is a device that maintains a certain potential difference between its *terminals* (points at which charge can enter or leave the battery) by means of internal electrochemical reactions.

In Fig. 26-4a, a battery B, a switch S, an uncharged capacitor C, and interconnecting wires form a circuit. The same circuit is shown in the *schematic diagram* of Fig. 26-4b, in which the symbols for a battery, a switch, and a capacitor represent those devices. The battery maintains potential difference V between its terminals. The terminal of higher potential is labeled $+$ and is often called the positive terminal; the terminal of lower potential is labeled $-$ and is often called the negative terminal.

The circuit shown in Figs. 26-4a and b is said to be incomplete because switch S is *open*; that is, it does not electrically connect the wires attached to it. When the switch is *closed*, electrically connecting those wires, the

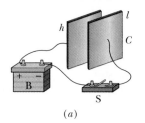

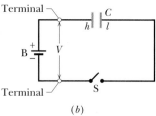

FIGURE 26-4 (*a*) Battery B, switch S, and plates h and l of capacitor C, connected in a circuit. (*b*) A schematic diagram with the *circuit elements* represented by their symbols.

circuit is complete and charge can then flow through the switch and the wires. As we discussed in Chapter 22, the charge that can flow through a conductor, such as a wire, is that of electrons. When the circuit of Fig. 26-4 is completed, electrons are driven through the wires by an electric field that the battery sets up in the wires. The field drives electrons from capacitor plate h to the positive terminal of the battery; thus plate h, losing electrons, becomes positively charged. The field drives just as many electrons from the negative terminal of the battery to capacitor plate l; thus plate l, gaining electrons, becomes negatively charged *just as much* as plate h becomes positively charged.

The potential difference between the initially uncharged plates is zero. As the plates become oppositely charged, that potential difference increases until it equals the potential difference V between the terminals of the battery. Then plate h and the positive terminal of the battery are at the same potential, and there is no longer an electric field in the wire between them. Similarly, plate l and the negative terminal reach the same potential and there is then no electric field in the wire between them. Thus, with the field zero, there is no further drive of electrons. The capacitor is then said to be *fully charged,* with a potential difference V and a charge q, which are related by Eq. 26-1.

PROBLEM SOLVING TACTICS

TACTIC 1: *The Symbol V and Potential Difference*

In previous chapters, the symbol V represents an electric potential at a point or along an equipotential surface. However, in matters concerning electrical devices, V often represents a *potential difference* between two points or two equipotential surfaces. Equation 26-1 is an example of this second use of the symbol. In Section 26-3, you will see a mixture of the two meanings of *V*. There and in later chapters, you need to be alert as to the intent of this symbol.

You will also be seeing, in this book and elsewhere, a variety of phrases regarding potential difference. A potential difference or a "potential" or a "voltage" may be *applied* to a device, or it may be *across* a device. A capacitor can be charged to a potential difference, as in "a capacitor is charged to 12 V." And a battery can be characterized by the potential difference across it, as in "a 12 V battery." Always keep in mind what is meant by such phrases: there is a potential difference between two points, such as two points in a circuit or at the terminals of a device such as a battery.

CHECKPOINT **1**: Does the capacitance *C* of a capacitor increase, decrease, or remain the same (a) when the charge *q* on it is doubled and (b) when the potential difference *V* across it is tripled?

26-3 CALCULATING THE CAPACITANCE

Our task here is to calculate the capacitance of a capacitor once we know its geometry. Because we are going to consider a number of different geometries, it seems wise to develop a general plan to simplify the work. In brief our plan is as follows: (1) assume a charge *q* on the plates; (2) calculate the electric field **E** between the plates in terms of this charge, using Gauss' law; (3) knowing **E**, calculate the potential difference *V* between the plates from Eq. 25-18; (4) calculate *C* from Eq. 26-1.

Before we start, we can simplify the calculation of both the electric field and the potential difference by making certain assumptions. We discuss each in turn.

Calculating the Electric Field
The electric field **E** between the plates of a capacitor is related to the charge *q* on a plate by Gauss' law:

$$\epsilon_0 \oint \mathbf{E} \cdot d\mathbf{A} = q. \qquad (26\text{-}3)$$

Here *q* is the charge enclosed by a Gaussian surface, and $\oint \mathbf{E} \cdot d\mathbf{A}$ is the net electric flux through that surface. In all cases that we shall consider, the Gaussian surface will be such that whenever electric flux passes through it, **E** will have a magnitude *E* and the vectors **E** and *d***A** will be parallel. Equation 26-3 then reduces to

$$q = \epsilon_0 E A \qquad \text{(special case of Eq. 26-3)}, \qquad (26\text{-}4)$$

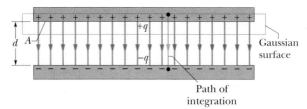

FIGURE 26-5 A charged parallel-plate capacitor. A Gaussian surface encloses the charge on the positive plate. The integration of Eq. 26-6 is taken along a path extending directly from the positive plate to the negative plate.

in which *A* is the area of that part of the Gaussian surface through which flux passes. For convenience, we shall always draw the Gaussian surface in such a way that it completely encloses the charge on the positive plate; see Fig. 26-5 for an example.

Calculating the Potential Difference
In the notation of Chapter 25 (Eq. 25-18), the potential difference between the plates is related to the electric field **E** by

$$V_f - V_i = -\int_i^f \mathbf{E} \cdot d\mathbf{s}, \qquad (26\text{-}5)$$

in which the integral is to be evaluated along any path that starts on one plate and ends on the other. We shall always choose a path that follows an electric field line from the positive plate to the negative plate. For this path, the vectors **E** and *d***s** will always point in the same direction, so the dot product **E** · *d***s** will be equal to the positive quantity *E ds*. Equation 26-5 then tells us that the quantity $V_f - V_i$ will always be negative. Since we are looking for *V*, the *absolute value* of the potential difference between the plates, we can set $V_f - V_i = -V$. Thus we can recast Eq. 26-5 as

$$V = \int_+^- E \, ds \qquad \text{(special case of Eq. 26-5)}, \qquad (26\text{-}6)$$

in which the + and − remind us that our path of integration starts on the positive plate and ends on the negative plate.

We are now ready to apply Eqs. 26-4 and 26-6 to some particular cases.

A Parallel-Plate Capacitor
We assume, as Fig. 26-5 suggests, that the plates of our parallel-plate capacitor are so large and so close together that we can neglect the fringing of the electric field at the edges of the plates, taking **E** to be constant throughout the volume between the plates.

We draw a Gaussian surface that encloses just the charge q on the positive plate, as in Fig. 26-5. From Eq. 26-4 we can then write

$$q = \epsilon_0 EA, \qquad (26\text{-}7)$$

where A is the area of the plate.

Equation 26-6 yields

$$V = \int_+^- E\, ds = E \int_0^d ds = Ed. \qquad (26\text{-}8)$$

In Eq. 26-8, E can be placed outside the integral because it is a constant; the second integral then is simply the plate separation d.

If we substitute q from Eq. 26-7 and V from Eq. 26-8 into the relation $q = CV$ (Eq. 26-1), we find

$$C = \frac{\epsilon_0 A}{d} \qquad \text{(parallel-plate capacitor).} \qquad (26\text{-}9)$$

So the capacitance does indeed depend only on geometrical factors, namely, the plate area A and the plate separation d. Note that C increases as we increase area A or decrease separation d.

As an aside we point out that Eq. 26-9 suggests one of our reasons for writing the electrostatic constant in Coulomb's law in the form $1/4\pi\epsilon_0$. If we had not done so, Eq. 26-9—which is used more often in engineering practice than Coulomb's law—would have been less simple in form. We note further that Eq. 26-9 permits us to express the permittivity constant ϵ_0 in a unit more appropriate for use in problems involving capacitors, namely,

$$\epsilon_0 = 8.85 \times 10^{-12}\ \text{F/m} = 8.85\ \text{pF/m}. \qquad (26\text{-}10)$$

We have previously expressed this constant as

$$\epsilon_0 = 8.85 \times 10^{-12}\ \text{C}^2/\text{N}\cdot\text{m}^2, \qquad (26\text{-}11)$$

using units that are useful for problems involving Coulomb's law (see Section 22-4).

A Cylindrical Capacitor

Figure 26-6 shows, in cross section, a cylindrical capacitor of length L formed by two coaxial cylinders of radii a and b. We assume that $L \gg b$ so that we can neglect the fringing of the electric field that occurs at the ends of the cylinders. Each plate contains a charge of magnitude q.

As a Gaussian surface, we choose a cylinder of length L and radius r, closed by end caps and placed as is shown in Fig. 26-6. Equation 26-4 then yields

$$q = \epsilon_0 EA = \epsilon_0 E(2\pi rL),$$

in which $2\pi rL$ is the area of the curved part of the Gaussian surface. There is no flux through the end caps. Solving

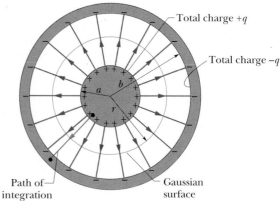

FIGURE 26-6 A cross section of a long cylindrical capacitor, showing a cylindrical Gaussian surface of radius r and the radial path of integration along which Eq. 26-6 is to be applied. This figure also serves to illustrate a spherical capacitor in a cross section through its center.

for E yields

$$E = \frac{q}{2\pi\epsilon_0 Lr}. \qquad (26\text{-}12)$$

Substitution of this result into Eq. 26-6 yields

$$V = \int_+^- E\, ds = \frac{q}{2\pi\epsilon_0 L} \int_a^b \frac{dr}{r}$$

$$= \frac{q}{2\pi\epsilon_0 L} \ln\!\left(\frac{b}{a}\right), \qquad (26\text{-}13)$$

where we have used the fact that here $ds = dr$. From the relation $C = q/V$, we then have

$$C = 2\pi\epsilon_0 \frac{L}{\ln(b/a)} \qquad \begin{array}{l}\text{(cylindrical}\\ \text{capacitor).}\end{array} \qquad (26\text{-}14)$$

We see that the capacitance of a cylindrical capacitor, like that of a parallel-plate capacitor, depends only on geometrical factors, in this case L, b, and a.

A Spherical Capacitor

Figure 26-6 can also serve as a central cross section of a capacitor that consists of two concentric spherical shells, of radii a and b. As a Gaussian surface we draw a sphere of radius r concentric with the two shells. Applying Eq. 26-4 to this surface yields

$$q = \epsilon_0 EA = \epsilon_0 E(4\pi r^2),$$

in which $4\pi r^2$ is the area of the spherical Gaussian surface. We solve this equation for E, obtaining

$$E = \frac{1}{4\pi\epsilon_0} \frac{q}{r^2}, \qquad (26\text{-}15)$$

which we recognize as the expression for the electric field due to a uniform spherical charge distribution (Eq. 24-15).

If we substitute this expression into Eq. 26-6, we find

$$V = \int_+^- E \, ds = \frac{q}{4\pi\epsilon_0} \int_a^b \frac{dr}{r^2} = \frac{q}{4\pi\epsilon_0} \left(\frac{1}{a} - \frac{1}{b} \right)$$

$$= \frac{q}{4\pi\epsilon_0} \frac{b-a}{ab}. \tag{26-16}$$

If we substitute Eq. 26-16 into Eq. 26-1 and solve for C, we find

$$C = 4\pi\epsilon_0 \frac{ab}{b-a} \quad \begin{array}{l}\text{(spherical}\\\text{capacitor).}\end{array} \tag{26-17}$$

An Isolated Sphere

We can assign a capacitance to a *single* isolated spherical conductor of radius R by assuming that the "missing plate" is a conducting sphere of infinite radius. After all, the field lines that leave the surface of a charged isolated conductor must end somewhere; the walls of the room in which the conductor is housed can serve effectively as our sphere of infinite radius.

To find the capacitance of the isolated conductor, we first rewrite Eq. 26-17 as

$$C = 4\pi\epsilon_0 \frac{a}{1 - a/b}.$$

If we then let $b \to \infty$ and substitute R for a, we find

$$C = 4\pi\epsilon_0 R \quad \text{(isolated sphere).} \tag{26-18}$$

Note that this formula and the others we have derived for capacitance (Eqs. 26-9, 26-14 and 26-17) involve the constant ϵ_0 multiplied by a quantity that has the dimensions of a length.

CHECKPOINT **2:** For capacitors charged by the same battery, does the charge stored by the capacitor increase, decrease, or remain the same in each of the following situations? (a) The plate separation of a parallel-plate capacitor is increased. (b) The radius of the inner cylinder of a cylindrical capacitor is increased. (c) The radius of the outer spherical shell of a spherical capacitor is increased.

SAMPLE PROBLEM 26-1

The plates of a parallel-plate capacitor are separated by a distance $d = 1.0$ mm. What must be the plate area if the capacitance is to be 1.0 F?

SOLUTION: From Eq. 26-9 we have

$$A = \frac{Cd}{\epsilon_0} = \frac{(1.0 \text{ F})(1.0 \times 10^{-3} \text{ m})}{8.85 \times 10^{-12} \text{ F/m}}$$

$$= 1.1 \times 10^8 \text{ m}^2. \tag{Answer}$$

This is the area of a square more than 10 km on edge. The farad is indeed a large unit. Modern technology, however, has permitted the construction of 1 F capacitors of very modest size. These "supercaps" are used as backup voltage sources for computers; they can maintain the computer memory for up to 30 days in case of power failure.

SAMPLE PROBLEM 26-2

A storage capacitor on a random access memory (RAM) chip has a capacitance of 55 fF. If the capacitor is charged to 5.3 V, how many excess electrons are on its negative plate?

SOLUTION: The number n of excess electrons is given by q/e, where e is the fundamental charge. Then, using Eq. 26-1, we have

$$n = \frac{q}{e} = \frac{CV}{e} = \frac{(55 \times 10^{-15} \text{ F})(5.3 \text{ V})}{1.60 \times 10^{-19} \text{ C}}$$

$$= 1.8 \times 10^6 \text{ electrons.} \tag{Answer}$$

For electrons, this is a very small number. A speck of household dust, so tiny that it essentially never settles, contains about 10^{17} electrons (and the same number of protons).

26-4 CAPACITORS IN PARALLEL AND IN SERIES

When there is a combination of capacitors in a circuit, we can sometimes replace that combination with an **equivalent capacitor,** that is, a single capacitor that has the same capacitance as the actual combination of capacitors. With such a replacement, we can simplify the circuit, affording easier solutions for unknown quantities of the circuit. Here we discuss two basic combinations of capacitors that allow such a replacement.

Capacitors in Parallel

Figure 26-7a shows three capacitors connected *in parallel* to a battery B. They are connected "in parallel" because the terminals of the battery are effectively wired directly to the plates of each of the three capacitors. Because the battery maintains a potential difference V between its terminals, it applies the same potential difference V across each capacitor.

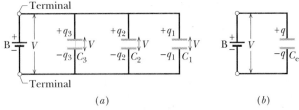

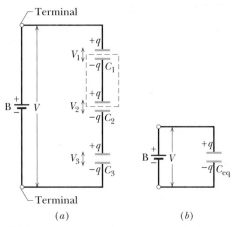

FIGURE 26-7 (*a*) Three capacitors connected in parallel to battery B. The battery maintains potential difference V across its terminals and thus across *each* capacitor. (*b*) The equivalent capacitance C_{eq} replaces the parallel combination. The charge q on C_{eq} is equal to the sum of the charges q_1, q_2, and q_3 on the capacitors of (*a*).

> Connected capacitors are said to be in parallel when a potential difference that is applied across their combination results in that same potential difference across each capacitor.

We seek the single capacitance C_{eq} that is equivalent to this parallel combination and thus can replace the combination (as in Fig. 26-7*b*). By *equivalent*, we mean that when the same potential difference V is applied across it, a capacitor with capacitance C_{eq} will store the same total charge q as is stored in the combination being replaced.

For the three capacitors we can write, from Eq. 26-1,

$$q_1 = C_1V, \quad q_2 = C_2V, \quad \text{and} \quad q_3 = C_3V.$$

The total charge on the parallel combination is then

$$q = q_1 + q_2 + q_3 = (C_1 + C_2 + C_3)V.$$

The equivalent capacitance, with the same total charge q and applied potential difference V as the combination, is then

$$C_{eq} = \frac{q}{V} = C_1 + C_2 + C_3,$$

a result that we can easily extend to any number n of capacitors, as

$$C_{eq} = \sum_{j=1}^{n} C_j \qquad \text{(\textit{n} capacitors in parallel).} \quad (26\text{-}19)$$

Thus, to find the equivalent capacitance of a parallel combination, we simply add the individual capacitances.

Capacitors in Series

Figure 26-8*a* shows three capacitors connected *in series* to battery B, which maintains a potential difference V across the left and right terminals of the series combination. This arrangement produces potential differences V_1, V_2, and V_3

FIGURE 26-8 (*a*) Three capacitors connected in series to battery B. The battery maintains potential difference V between the left and right sides of the series combination. (*b*) The equivalent capacitance C_{eq} replaces the series combination. The potential difference V across C_{eq} is equal to the sum of the potential differences V_1, V_2, and V_3 across the capacitors of (*a*).

across capacitors C_1, C_2, and C_3, respectively, such that $V_1 + V_2 + V_3 = V$.

> Connected capacitors are said to be in series when a potential difference that is applied across their combination is the sum of the resulting potential differences across each capacitor.

We seek the single capacitance C_{eq} that is equivalent to this series combination and thus can replace the combination (as in Fig. 26-8*b*). Again, *equivalent* means that the same total charge q is stored with the same applied potential difference V.

When the battery is connected, each capacitor in Fig. 26-8*a* must have the same charge q. This is true even though the three capacitors may be of different types and may have different capacitances. To understand this fact, note that the part of the circuit enclosed by the dashed lines in Fig. 26-8*a* is electrically isolated from the rest of the circuit. Thus it cannot gain or lose charge. However, the battery can *induce* charge on the isolated part, that is, redistribute the charge that is already there: When the battery produces a charge of $+q$ on the top plate of C_1, that charge attracts electrons within the isolated part, redistributing some of them. The redistribution leaves the bottom plate of C_1 with charge $-q$ and the top plate of C_2 with charge $+q$. At the same time, the battery produces a charge of $-q$ on the bottom plate of C_3, causing redistribution of charges on the connected plates of C_2 and C_3.

The net effect is that all three capacitors in the series combination have the same charge q. However, the total charge produced by the battery on the series combination is *not* the sum of these three charges. The battery has produced only charge q, using the top-most and bottom-most plates in the series. The other plates are charged only because of the redistribution of electrons already there.

Application of Eq. 26-1 to each capacitor yields

$$V_1 = \frac{q}{C_1}, \quad V_2 = \frac{q}{C_2}, \quad \text{and} \quad V_3 = \frac{q}{C_3}.$$

The potential difference across the entire series combination is then

$$V = V_1 + V_2 + V_3$$
$$= q \left(\frac{1}{C_1} + \frac{1}{C_2} + \frac{1}{C_3} \right).$$

The equivalent capacitance is then

$$C_{eq} = \frac{q}{V} = \frac{1}{1/C_1 + 1/C_2 + 1/C_3},$$

or

$$\frac{1}{C_{eq}} = \frac{1}{C_1} + \frac{1}{C_2} + \frac{1}{C_3}.$$

We can easily extend this to any number n of capacitors as

$$\frac{1}{C_{eq}} = \sum_{j=1}^{n} \frac{1}{C_j} \quad \text{(}n\text{ capacitors in series).} \quad \text{(26-20)}$$

From Eq. 26-20 you can deduce that the equivalent series capacitance is always less than the least capacitance in the series of capacitors.

As you will see in Sample Problem 26-3, some complicated combinations of capacitors can be subdivided into parallel and series combinations, which can then be replaced with equivalent capacitances. This simplifies the original combination and the analysis of circuits.

CHECKPOINT **3:** A battery of potential V stores charge q on a combination of two identical capacitors. What are the potential difference across and the charge on either capacitor if the capacitors are (a) in parallel and (b) in series?

SAMPLE PROBLEM 26-3

(a) Find the equivalent capacitance of the combination shown in Fig. 26-9a. Assume

$$C_1 = 12.0 \ \mu\text{F}, \quad C_2 = 5.30 \ \mu\text{F}, \quad \text{and} \quad C_3 = 4.50 \ \mu\text{F}.$$

SOLUTION: Capacitors C_1 and C_2 are in parallel. From Eq.

26-19, their equivalent capacitance is

$$C_{12} = C_1 + C_2 = 12.0 \ \mu\text{F} + 5.30 \ \mu\text{F} = 17.3 \ \mu\text{F}.$$

As Fig. 26-9b shows, C_{12} and C_3 now form a series combination. From Eq. 26-20, their equivalent capacitance (shown in Fig. 26-9c) is given by

$$\frac{1}{C_{123}} = \frac{1}{C_{12}} + \frac{1}{C_3} = \frac{1}{17.3 \ \mu\text{F}} + \frac{1}{4.50 \ \mu\text{F}} = 0.280 \ \mu\text{F}^{-1},$$

from which

$$C_{123} = \frac{1}{0.280 \ \mu\text{F}^{-1}} = 3.57 \ \mu\text{F}. \quad \text{(Answer)}$$

(b) A potential difference $V = 12.5$ V is applied to the input terminals in Fig. 26-9a. What is the charge on C_1?

SOLUTION: We treat the equivalent capacitors C_{12} and C_{123} exactly as we would real capacitors of the same capacitances. For the charge on C_{123} in Fig. 26-9c we then have

$$q_{123} = C_{123}V = (3.57 \ \mu\text{F})(12.5 \text{ V}) = 44.6 \ \mu\text{C}.$$

This same charge exists on each capacitor in the series combination of Fig. 26-9b. Let $q_{12} \ (= q_{123})$ represent the charge on C_{12} in that figure. The potential difference across C_{12} is then

$$V_{12} = \frac{q_{12}}{C_{12}} = \frac{44.6 \ \mu\text{C}}{17.3 \ \mu\text{F}} = 2.58 \text{ V}.$$

This same potential difference appears across both C_1 and C_2 in Fig. 26-9a. Let $V_1 \ (= V_{12})$ represent the potential difference across C_1. We then have

$$q_1 = C_1 V_1 = (12.0 \ \mu\text{F})(2.58 \text{ V})$$
$$= 31.0 \ \mu\text{C}. \quad \text{(Answer)}$$

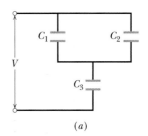

(a)

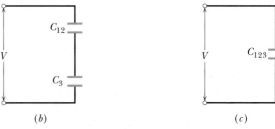

(b)　　　　　　　　　(c)

FIGURE 26-9　Sample Problem 26-3. (a) Three capacitors. (b) C_1 and C_2, a parallel combination, are replaced by C_{12}. (c) C_{12} and C_3, a series combination, are replaced by the equivalent capacitance C_{123}.

SAMPLE PROBLEM 26-4

A 3.55 μF capacitor C_1 is charged to a potential difference $V_0 = 6.30$ V, using a 6.30 V battery. The battery is then removed and the capacitor is connected as in Fig. 26-10 to an uncharged 8.95 μF capacitor C_2. When switch S is closed, charge flows from C_1 to C_2 until the capacitors have the same potential difference V. What is this common potential difference they reach?

SOLUTION: The original charge q_0 is now shared by two capacitors, so

$$q_0 = q_1 + q_2.$$

Applying the relation $q = CV$ to each term of this equation yields

$$C_1 V_0 = C_1 V + C_2 V,$$

from which

$$V = V_0 \frac{C_1}{C_1 + C_2} = \frac{(6.30 \text{ V})(3.55 \ \mu\text{F})}{3.55 \ \mu\text{F} + 8.95 \ \mu\text{F}}$$

$$= 1.79 \text{ V}. \qquad \text{(Answer)}$$

When the capacitors reach this value of electric potential difference, the charge flow stops.

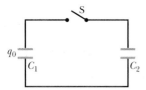

FIGURE 26-10 Sample Problems 26-4 and 26-5. A potential difference V_0 is applied to C_1 and the charging battery is removed. Switch S is then closed so that the charge on C_1 is shared with C_2.

PROBLEM SOLVING TACTICS

TACTIC 2: *Multiple Capacitor Circuits*

Let us review the procedure used in the solution of Sample Problem 26-3, in which several capacitors are connected to a battery. To find a single equivalent capacitance, we simplify the given arrangement of capacitances by replacing them, in steps, with equivalent capacitances, using Eq. 26-19 when we find capacitances in parallel and Eq. 26-20 when we find capacitances in series. Then, to find the charge stored by that single equivalent capacitance, we use Eq. 26-1 and the potential difference V imposed by the battery.

That result tells us the net charge stored on the actual arrangement of capacitors. However, to find the charge on, or the potential difference across, any particular capacitor in the actual arrangement, we need to reverse our steps of simplification. With each reversed step, we use these two rules: when capacitances are in parallel, they have the same potential difference their equivalent capacitance has, and we use Eq. 26-1 to find the charge on each capacitance; when they are in series,

they have the same charge their equivalent capacitance has, and we use Eq. 26-1 to find the potential difference across each capacitance.

TACTIC 3: *Batteries and Capacitors*

A battery maintains a certain potential difference across its terminals. So when capacitor C_1 of Sample Problem 26-4 is connected to the 6.30 V battery, charge flows between the capacitor and the battery until the capacitor has the same potential difference as the battery.

A capacitor differs from a battery in that a capacitor lacks internal electrochemical reactions to release charged particles (electrons) from internal atoms and molecules. So when the charged capacitor C_1 of Sample Problem 26-4 is disconnected from the battery and then connected to the uncharged capacitor C_2 with switch S closed, the potential difference across C_1 is not maintained. The quantity that *is* maintained is the total charge q_0 of the two-capacitor system; that is, charge obeys a conservation law, *not* electric potential.

Here is what happens to that charge. When switch S is open as shown in Fig. 26-10, charge q_0 is entirely on C_1. Charge cannot be transferred between the capacitors until there is a complete circuit, or loop, through which the charge can flow. When switch S is closed, there is such a complete circuit, and a portion of q_0 flows from C_1 to C_2, increasing the potential difference of C_2 and decreasing that of C_1, until the two capacitors have the same potential difference V. The top plates of the capacitors are then at the same electric potential, and so are the bottom plates; thus the capacitors are in equilibrium and there is no further charge flow.

CHECKPOINT **4:** In Sample Problem 26-4 and Fig. 26-10, suppose capacitor C_2 is replaced by a series combination of capacitors C_3 and C_4. (a) After the switch is closed and charge has stopped flowing, what is the relation between the initial charge q_0, the charge q_1 then on C_1, and the charge q_{34} then on the equivalent capacitance C_{34}? (b) If $C_3 > C_4$, is the charge q_3 on C_3 more than, less than, or equal to the charge q_4 on C_4?

26-5 STORING ENERGY IN AN ELECTRIC FIELD

Work must be done by an external agent to charge a capacitor. Starting with an uncharged capacitor, for example, imagine that—using "magic tweezers"—you remove electrons from one plate and transfer them one at a time to the other plate. The electric field that builds up in the space between the plates has a direction that tends to oppose further transfer. Thus, as charge accumulates on the capacitor plates, you have to do increasingly larger amounts of work to transfer additional electrons. In practice, this work is done not by "magic tweezers" but by a battery, at the expense of its store of chemical energy.

We visualize the work required to charge a capacitor as being stored in the form of **electric potential energy** U in the electric field between the plates. You can recover this energy at will, by discharging the capacitor in a circuit, just as you can recover the potential energy stored in a stretched bow by releasing the bowstring to transfer the energy to the kinetic energy of an arrow.

Suppose that, at a given instant, a charge q' has been transferred from one plate to the other. The potential difference V' between the plates at that instant will be q'/C. If an extra increment of charge dq' is then transferred, the increment of work required will be, from Eq. 25-7,

$$dW = V' \, dq' = \frac{q'}{C} \, dq'.$$

The work required to bring the total capacitor charge up to a final value q is

$$W = \int dW = \frac{1}{C} \int_0^q q' \, dq' = \frac{q^2}{2C}.$$

This work is stored as potential energy U in the capacitor, so that

$$U = \frac{q^2}{2C} \qquad \text{(potential energy).} \qquad (26\text{-}21)$$

From Eq. 26-1, we can also write this as

$$U = \tfrac{1}{2}CV^2 \qquad \text{(potential energy).} \qquad (26\text{-}22)$$

Equations 26-21 and 26-22 hold no matter what the geometry of the capacitor is.

To gain some physical insight into energy storage, consider two parallel-plate capacitors C_1 and C_2 that are identical except that C_1 has twice the plate separation of C_2. Then C_1 has twice the volume between its plates and also, from Eq. 26-9, half the capacitance of C_2. Equation 26-4 tells us that if both capacitors have the same charge q, the electric fields between their plates are identical. And Eq. 26-21 tells us that C_1 has twice the stored potential energy of C_2. Thus, of two otherwise identical capacitors with the same charge and same electric field, the one with twice the volume between its plates has twice the stored potential energy. Arguments like this tend to verify our earlier assumption:

The potential energy of a charged capacitor may be viewed as stored in the electric field between its plates.

The Medical Defibrillator

The ability of a capacitor to store potential energy is the basis of *defibrillator* devices, which are used by emergency medical teams to stop the fibrillation of heart attack victims. In the portable version, a battery charges a capacitor to a high potential difference, storing a large amount of energy in less than a minute. The battery maintains only a modest potential difference; an electronic circuit repeatedly uses that potential difference to greatly increase the potential difference of the capacitor. The power, or rate of energy transfer, during this process is also modest.

Conducting leads ("paddles") are placed on the victim's chest. When a control switch is closed, the capacitor sends a portion of its stored energy from paddle to paddle through the victim. As an example, when a 70 μF capacitor in a defibrillator is charged to 5000 V, Eq. 26-22 gives the energy stored in the capacitor as

$$U = \tfrac{1}{2}CV^2 = \tfrac{1}{2}(70 \times 10^{-6} \text{ F})(5000 \text{ V})^2 = 875 \text{ J}.$$

About 200 J of this energy is sent through the victim during a pulse of about 2.0 ms. The power of the pulse is

$$P = \frac{U}{t} = \frac{200 \text{ J}}{2.0 \times 10^{-3} \text{ s}} = 100 \text{ kW},$$

which is much greater than the power of the battery itself.

Energy Density

In a parallel-plate capacitor, neglecting fringing, the electric field has the same value for all points between the plates. Thus the **energy density** u, that is, the potential energy per unit volume between the plates, should also be uniform. We can find u by dividing the total potential energy by the volume Ad of the space between the plates. Using Eq. 26-22, we obtain

$$u = \frac{U}{Ad} = \frac{CV^2}{2Ad}.$$

With Eq. 26-9 ($C = \epsilon_0 A/d$), this result becomes

$$u = \tfrac{1}{2}\epsilon_0 \left(\frac{V}{d}\right)^2.$$

But, from Eq. 25-42, V/d equals the electric field magnitude E, so

$$u = \tfrac{1}{2}\epsilon_0 E^2 \qquad \text{(energy density).} \qquad (26\text{-}23)$$

Although we derived this result for the special case of a parallel-plate capacitor, it holds generally, whatever may be the source of the electric field. If an electric field **E** exists at any point in space, we can think of that point as the site of potential energy whose amount per unit volume is given by Eq. 26-23.

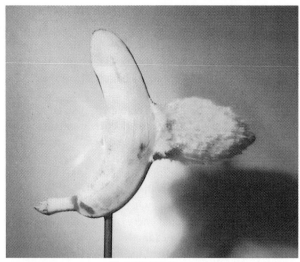

To photograph a bullet blowing apart a banana, Harold Edgerton, the inventor of the stroboscope, used a capacitor to dump electrical energy into one of his stroboscopic lamps, which then brightly illuminated the banana for only 0.3 μs.

SAMPLE PROBLEM 26-5

What is the potential energy of the two-capacitor system in Sample Problem 26-4, before and after switch S in Fig. 26-10 is closed?

SOLUTION: Initially, only capacitor C_1 is charged and has a potential energy; its potential difference is $V_0 = 6.30$ V. So from Eq. 26-22, the initial potential energy is

$$U_i = \tfrac{1}{2}C_1V_0^2 = (\tfrac{1}{2})(3.55 \times 10^{-6} \text{ F})(6.30 \text{ V})^2$$
$$= 7.04 \times 10^{-5} \text{ J} = 70.4 \ \mu\text{J}. \qquad \text{(Answer)}$$

After the switch has been closed, the capacitors come to the same final potential difference $V = 1.79$ V. The final potential energy is then

$$U_f = \tfrac{1}{2}C_1V^2 + \tfrac{1}{2}C_2V^2 = \tfrac{1}{2}(C_1 + C_2)V^2$$
$$= (\tfrac{1}{2})(3.55 \times 10^{-6} \text{ F} + 8.95 \times 10^{-6} \text{ F})(1.79 \text{ V})^2$$
$$= 2.00 \times 10^{-5} \text{ J} = 20.0 \ \mu\text{J}. \qquad \text{(Answer)}$$

Thus $U_f < U_i$, by about 72%.

This is not a violation of the principle of energy conservation. The "missing" energy appears as thermal energy in the connecting wires (as we shall discuss in Chapter 27) and as radiated energy.

SAMPLE PROBLEM 26-6

An isolated conducting sphere whose radius R is 6.85 cm has a charge $q = 1.25$ nC.

(a) How much potential energy is stored in the electric field of this charged conductor?

SOLUTION: From Eqs. 26-21 and 26-18 we have

$$U = \frac{q^2}{2C} = \frac{q^2}{8\pi\epsilon_0 R}$$
$$= \frac{(1.25 \times 10^{-9} \text{ C})^2}{(8\pi)(8.85 \times 10^{-12} \text{ F/m})(0.0685 \text{ m})}$$
$$= 1.03 \times 10^{-7} \text{ J} = 103 \text{ nJ}. \qquad \text{(Answer)}$$

(b) What is the energy density at the surface of the sphere?

SOLUTION: From Eq. 26-23,

$$u = \tfrac{1}{2}\epsilon_0 E^2,$$

so we must first find E at the surface of the sphere. This is given by Eq. 24-15:

$$E = \frac{1}{4\pi\epsilon_0}\frac{q}{R^2}.$$

The energy density is then

$$u = \tfrac{1}{2}\epsilon_0 E^2 = \frac{q^2}{32\pi^2\epsilon_0 R^4} \qquad (26\text{-}24)$$
$$= \frac{(1.25 \times 10^{-9} \text{ C})^2}{(32\pi^2)(8.85 \times 10^{-12} \text{ C}^2/\text{N} \cdot \text{m}^2)(0.0685 \text{ m})^4}$$
$$= 2.54 \times 10^{-5} \text{ J/m}^3 = 25.4 \ \mu\text{J/m}^3. \qquad \text{(Answer)}$$

(c) What is the radius R_0 of an imaginary spherical surface such that half of the stored potential energy lies within it?

SOLUTION: This situation requires that

$$\int_R^{R_0} dU = \frac{1}{2}\int_R^\infty dU. \qquad (26\text{-}25)$$

The lower limit on the integrals is R rather than 0 because no electric field, and thus no stored potential energy, lies within the conducting sphere of radius R.

The energy dU that lies in a spherical shell between inner and outer radii r and $r + dr$ is

$$dU = (u)(4\pi r^2)(dr), \qquad (26\text{-}26)$$

where u is again the energy density and $(4\pi r^2)(dr)$ is the volume of the spherical shell. Substituting Eq. 26-24 into Eq. 26-26 with r replacing R, we have

$$dU = \frac{q^2}{8\pi\epsilon_0}\frac{dr}{r^2}. \qquad (26\text{-}27)$$

Substituting Eq. 26-26 into both sides of Eq. 26-25 and simplifying give us

$$\int_R^{R_0}\frac{dr}{r^2} = \frac{1}{2}\int_R^\infty\frac{dr}{r^2},$$

which, after integration, becomes

$$\frac{1}{R} - \frac{1}{R_0} = \frac{1}{2R}.$$

Solving for R_0 yields

$$R_0 = 2R = (2)(6.85 \text{ cm}) = 13.7 \text{ cm}. \quad \text{(Answer)}$$

Thus half the stored energy is contained within a spherical surface whose radius is twice the radius of the conducting sphere.

26-6 CAPACITOR WITH A DIELECTRIC

If you fill the space between the plates of a capacitor with a *dielectric,* which is an insulating material such as mineral oil or plastic, what happens to the capacitance? Michael Faraday—to whom the whole concept of capacitance is largely due and for whom the SI unit of capacitance is named—first looked into this matter in 1837. Using simple equipment much like that shown in Fig. 26-11, he found that the capacitance *increased* by a numerical factor κ, which he called the **dielectric constant** of the introduced material. Table 26-1 shows some dielectric materials and their dielectric constants. The dielectric constant of a vacuum is unity by definition. Because air is mostly empty space, its measured dielectric constant is only slightly greater than unity.

Another effect of the introduction of a dielectric is to limit the potential difference that can be applied between the plates to a certain value $V_{\max}$, called the *breakdown potential.* If this value is substantially exceeded, the dielectric material will break down and form a conducting path between the plates. Every dielectric material has a

FIGURE 26-11 The simple electrostatic apparatus used by Faraday. An assembled apparatus (second from left) forms a spherical capacitor consisting of a central brass ball and a concentric brass shell. Faraday placed dielectric materials in the space between the sphere and shell.

TABLE 26-1 **SOME PROPERTIES OF DIELECTRICS**[a]

MATERIAL	DIELECTRIC CONSTANT κ	DIELECTRIC STRENGTH (kV/mm)
Air (1 atm)	1.00054	3
Polystyrene	2.6	24
Paper	3.5	16
Transformer oil	4.5	
Pyrex	4.7	14
Ruby mica	5.4	
Porcelain	6.5	
Silicon	12	
Germanium	16	
Ethanol	25	
Water (20°C)	80.4	
Water (25°C)	78.5	
Titania ceramic	130	
Strontium titanate	310	8

For a vacuum, $\kappa =$ unity.

[a]Measured at room temperature, except for the water.

characteristic *dielectric strength,* which is the maximum value of the electric field that it can tolerate without breakdown. A few such values are listed in Table 26-1.

As we discussed in connection with Eq. 26-18, the capacitance of any capacitor can be written in the form

$$C = \epsilon_0 \mathscr{L}, \qquad (26\text{-}28)$$

in which $\mathscr{L}$ has the dimensions of a length. For example, $\mathscr{L} = A/d$ for a parallel-plate capacitor. Faraday's discovery was that, with a dielectric *completely* filling the space between the plates, Eq. 26-28 becomes

$$C = \kappa \epsilon_0 \mathscr{L} = \kappa C_{\text{air}}, \qquad (26\text{-}29)$$

where C_{air} is the value of the capacitance with only air between the plates.

Figure 26-12 provides some insight into Faraday's experiments. In Fig. 26-12a the battery ensures that the potential difference V between the plates will remain constant. When a dielectric slab is inserted between the plates, the charge q increases by a factor of κ, the additional charge being delivered to the capacitor plates by the battery. In Fig. 26-12b there is no battery and therefore the charge q must remain constant as the dielectric slab is inserted; then the potential difference V between the plates decreases by a factor of κ. Both these observations are consistent (through the relation $q = CV$) with the increase in capacitance caused by the dielectric.

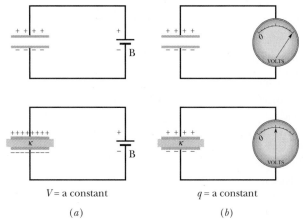

V = a constant $\qquad$ q = a constant

$\qquad$ (a) $\qquad\qquad\qquad$ (b)

FIGURE 26-12 (a) If the potential difference between the plates of a capacitor is maintained, as by battery B, the effect of a dielectric is to increase the charge on the plates. (b) If the charge on the capacitor plates is maintained, as in this case, the effect of a dielectric is to reduce the potential difference between the plates. The scale shown is that of a *potentiometer*, a device used to measure potential difference (here, between the plates). A capacitor cannot discharge through a potentiometer.

Comparison of Eqs. 26-28 and 26-29 suggests that the effect of a dielectric can be summed up in more general terms:

> In a region completely filled by a dielectric material of dielectric constant κ, all electrostatic equations containing the permittivity constant ϵ_0 are to be modified by replacing ϵ_0 with $\kappa\epsilon_0$.

Thus a point charge inside a dielectric produces an electric field that, by Coulomb's law, has magnitude

$$E = \frac{1}{4\pi\kappa\epsilon_0}\frac{q}{r^2}. \qquad (26\text{-}30)$$

Also, the expression for the electric field just outside an isolated conductor immersed in a dielectric (see Eq. 24-11) becomes

$$E = \frac{\sigma}{\kappa\epsilon_0}. \qquad (26\text{-}31)$$

Both these expressions show that *for a fixed distribution of charges, the effect of a dielectric is to weaken the electric field* that would otherwise be present.

SAMPLE PROBLEM 26-7

A parallel-plate capacitor whose capacitance C is 13.5 pF is charged to a potential difference $V = 12.5$ V between its plates. The charging battery is now disconnected and a porcelain slab ($\kappa = 6.50$) is slipped between the plates. What is the potential energy of the device, both before and after the slab is introduced?

SOLUTION: The initial potential energy is given by Eq. 26-22 as

$$U_i = \tfrac{1}{2}CV^2 = (\tfrac{1}{2})(13.5 \times 10^{-12}\text{ F})(12.5\text{ V})^2$$
$$= 1.055 \times 10^{-9}\text{ J} = 1055\text{ pJ} \approx 1100\text{ pJ}. \quad \text{(Answer)}$$

We can also write the initial potential energy, from Eq. 26-21, in the form

$$U_i = \frac{q^2}{2C}.$$

We choose to do so because, from the conditions of the problem statement, q (but not V) remains constant as the slab is introduced. After the slab is in place, C increases to κC so that

$$U_f = \frac{q^2}{2\kappa C} = \frac{U_i}{\kappa} = \frac{1055\text{ pJ}}{6.50}$$
$$= 162\text{ pJ} \approx 160\text{ pJ}. \quad \text{(Answer)}$$

When the slab is introduced, the energy decreases by a factor of $1/\kappa$.

The "missing" energy, in principle, would be apparent to the person who introduced the slab. The capacitor would exert a tiny tug on the slab and would do work on it, in amount

$$W = U_i - U_f = (1055 - 162)\text{ pJ} = 893\text{ pJ}.$$

If the slab were allowed to slide between the plates with no restraint and if there were no friction, the slab would oscillate back and forth between the plates with a (constant) mechanical energy of 893 pJ, and this system energy would transfer back and forth between kinetic energy of the moving slab and potential energy stored in the electric field.

$\mathbb{C}$HECKPOINT **5:** If the battery in Sample Problem 26-7 remains connected, do the following increase, decrease, or remain the same when the slab is introduced: (a) the potential difference between the capacitor plates, (b) the capacitance, (c) the charge on the capacitor, (d) the potential energy of the device, (e) the electric field between the plates? (*Hint:* For (e), note that the charge is not fixed.)

26-7 DIELECTRICS: AN ATOMIC VIEW

What happens, in atomic and molecular terms, when we put a dielectric in an electric field? There are two possibilities, depending on the nature of the molecules:

1. *Polar dielectrics.* The molecules of some dielectrics, like water, have permanent electric dipole moments. In

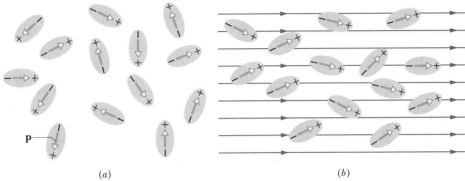

(a) (b)

FIGURE 26-13 (a) Molecules with a permanent electric dipole moment, showing their random orientation in the absence of an external electric field. (b) An electric field is applied, producing partial alignment of the dipoles. Thermal agitation prevents complete alignment.

such materials (called *polar dielectrics*), the electric dipoles tend to line up with an external electric field as in Fig. 26-13. Because the molecules are continuously jostling each other as a result of their random thermal motion, this alignment is not complete, but it becomes more complete as the magnitude of the applied field is increased (or as the temperature, and thus the jostling, is decreased). The alignment of the electric dipoles produces an electric field that is opposite the applied field and smaller in magnitude than that field.

2. Nonpolar dielectrics. Regardless of whether they have permanent electric dipole moments, molecules acquire dipole moments by induction when placed in an external electric field. In Section 25-7 (see Fig. 25-12), we saw that this external field tends to "stretch" the molecule, separating slightly the centers of negative and positive charge.

Figure 26-14a shows a nonpolar dielectric slab with no external electric field applied. In Fig. 26-14b, an electric field E_0 is applied via a capacitor, whose plates are charged as shown. The effect of field E_0 is a slight separation of the centers of the positive and negative charge distributions within the slab, producing positive charge on one face of the slab (due to the positive ends of dipoles there) and negative charge on the opposite face (due to the negative ends of dipoles there). The slab as a whole remains electrically neutral and—within the slab—there is no excess charge in any volume element.

Figure 26-14c shows that the induced surface charges on the faces produce an electric field E' in the direction opposite that of the applied electric field E_0. The resultant field E inside the dielectric (the vector sum of E_0 and E') has the direction of E_0 but is smaller in magnitude.

Both the field E' produced by the surface charges and the electric field produced by the permanent electric dipoles in Fig. 26-13 act in the same way: they oppose the applied field E. Thus, the effect of both polar and nonpolar dielectrics is to weaken any applied field within them—as between the plates of a capacitor.

We can now see why the dielectric porcelain slab in

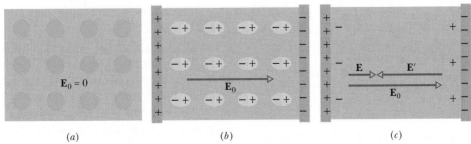

(a) (b) (c)

FIGURE 26-14 (a) A dielectric slab. The circles represent the electrically neutral atoms within the slab. (b) An electric field is applied via charged capacitor plates; the field slightly stretches the atoms, separating the centers of positive and negative charge. (c) The separation produces surface charges on the slab faces. These charges set up a field E', which opposes the applied field E_0. The resultant field E inside the dielectric (the vector sum of E_0 and E') has the same direction as E_0 but less magnitude.

Sample Problem 26-7 is pulled into the capacitor: as it enters the space between the plates, the surface charge that appears on each slab face has the opposite sign as the charge on the adjacent capacitor plate. Thus, slab and plate attract each other.

26-8 DIELECTRICS AND GAUSS' LAW

In our discussion of Gauss' law in Chapter 24, we assumed that the charges existed in a vacuum. Here we shall see how to modify and generalize that law if dielectric materials, such as those listed in Table 26-1, are present. Figure 26-15 shows a parallel-plate capacitor of plate area A, both with and without a dielectric. We assume that the charge q on the plates is the same in both situations. Note that the field between the plates induces charges on the faces of the dielectric by one of the methods of Section 26-7.

For the situation of Fig. 26-15a, without a dielectric, we can find the electric field $\mathbf{E}_0$ between the plates as we did in Fig. 26-5: we enclose the charge $+q$ on the top plate with a Gaussian surface and then apply Gauss' law. Letting E_0 represent the magnitude of the field, we find

$$\epsilon_0 \oint \mathbf{E} \cdot d\mathbf{A} = \epsilon_0 E_0 A = q, \qquad (26\text{-}32)$$

or

$$E_0 = \frac{q}{\epsilon_0 A}. \qquad (26\text{-}33)$$

In Fig. 26-15b, with the dielectric in place, we can find the electric field between the plates (and within the dielectric) by using the same Gaussian surface. However, now the surface encloses two types of charge: it still encloses charge $+q$ on the top plate, but it now also encloses the induced charge $-q'$ on the top face of the dielectric. The charge on the conducting plate is said to be *free charge* because it can move if we can change the electric potential of the plate; the induced charge on the surface of the dielectric is not free charge because it cannot move from that surface.

The net charge enclosed by the Gaussian surface in Fig. 26-15b is $q - q'$. So, Gauss' law now gives

$$\epsilon_0 \oint \mathbf{E} \cdot d\mathbf{A} = \epsilon_0 E A = q - q', \qquad (26\text{-}34)$$

or

$$E = \frac{q - q'}{\epsilon_0 A}. \qquad (26\text{-}35)$$

The effect of the dielectric is to weaken the original field E_0 by a factor κ. So we may write

$$E = \frac{E_0}{\kappa} = \frac{q}{\kappa \epsilon_0 A}. \qquad (26\text{-}36)$$

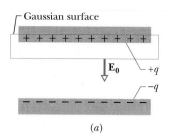

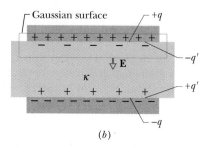

FIGURE 26-15
Parallel-plate capacitor (*a*) without and (*b*) with a dielectric slab inserted. The charge q on the plates is assumed to be the same in both cases.

Comparison of Eqs. 26-35 and 26-36 shows that

$$q - q' = \frac{q}{\kappa}. \qquad (26\text{-}37)$$

Equation 26-37 shows correctly that the magnitude q' of the induced surface charge is less than that of the free charge q and is zero if no dielectric is present, that is, if $\kappa = 1$ in Eq. 26-37.

By substituting for $q - q'$ from Eq. 26-37 in Eq. 26-34, we can write Gauss' law in the form

$$\epsilon_0 \oint \kappa \mathbf{E} \cdot d\mathbf{A} = q \qquad \text{(Gauss' law with dielectric).} \qquad (26\text{-}38)$$

This important equation, although derived for a parallel-plate capacitor, is true generally and is the most general form in which Gauss' law can be written. Note the following:

1. The flux integral now deals with $\kappa \mathbf{E}$, not with $\mathbf{E}$. (The vector $\epsilon_0 \kappa \mathbf{E}$ is sometimes called the *electric displacement* $\mathbf{D}$, so that Eq. 26-38 can be written in the simplified form $\oint \mathbf{D} \cdot d\mathbf{A} = q$.)

2. The charge q enclosed by the Gaussian surface is now taken to be the *free charge only*. The induced surface charge is deliberately ignored on the right side of this equation, having been taken fully into account by introducing the dielectric constant κ on the left side.

3. Equation 26-38 differs from Eq. 24-7, our original statement of Gauss' law, only in that ϵ_0 in the latter equation has been replaced by $\kappa\epsilon_0$. We take κ inside the integral to allow for cases in which κ is not constant over the entire Gaussian surface.

SAMPLE PROBLEM 26-8

Figure 26-16 shows a parallel-plate capacitor of plate area A and plate separation d. A potential difference V_0 is applied between the plates. The battery is then disconnected, and a dielectric slab of thickness b and dielectric constant κ is placed between the plates as shown. Assume

$$A = 115 \text{ cm}^2, \qquad d = 1.24 \text{ cm}, \qquad V_0 = 85.5 \text{ V},$$

$$b = 0.780 \text{ cm}, \qquad \kappa = 2.61.$$

(a) What is the capacitance C_0 before the dielectric slab is inserted?

SOLUTION: From Eq. 26-9 we have

$$C_0 = \frac{\epsilon_0 A}{d} = \frac{(8.85 \times 10^{-12} \text{ F/m})(115 \times 10^{-4} \text{ m}^2)}{1.24 \times 10^{-2} \text{ m}}$$

$$= 8.21 \times 10^{-12} \text{ F} = 8.21 \text{ pF.} \qquad \text{(Answer)}$$

(b) What free charge appears on the plates?

SOLUTION: From Eq. 26-1,

$$q = C_0 V_0 = (8.21 \times 10^{-12} \text{ F})(85.5 \text{ V})$$

$$= 7.02 \times 10^{-10} \text{ C} = 702 \text{ pC.} \qquad \text{(Answer)}$$

Because the charging battery was disconnected before the slab was introduced, the free charge remains unchanged as the slab is put into place.

(c) What is the electric field E_0 in the gaps between the plates and the dielectric slab?

SOLUTION: Let us apply Gauss' law in the form given in Eq. 26-38 to Gaussian surface I in Fig. 26-16, which encloses only the free charge on the upper capacitor plate. Because the area vector $d\mathbf{A}$ and the field vector $\mathbf{E}_0$ both point downward, we have

$$\epsilon_0 \oint \kappa \mathbf{E} \cdot d\mathbf{A} = \epsilon_0 (1) E_0 A = q,$$

or

$$E_0 = \frac{q}{\epsilon_0 A} = \frac{7.02 \times 10^{-10} \text{ C}}{(8.85 \times 10^{-12} \text{ F/m})(115 \times 10^{-4} \text{ m}^2)}$$

$$= 6900 \text{ V/m} = 6.90 \text{ kV/m.} \qquad \text{(Answer)}$$

Note that we put $\kappa = 1$ in this equation because the Gaussian surface over which Gauss' law was integrated does not pass through any dielectric. Note too that the value of E_0 does not change when the slab is introduced because the amount of charge enclosed by Gaussian surface I in Fig. 26-16 does not change.

(d) What is the electric field E_1 in the dielectric slab?

SOLUTION: We now apply Eq. 26-38 to Gaussian surface II in Fig. 26-16. That surface encloses free charge $-q$ and induced charge $+q'$, but we do not consider the latter when we use Eq. 26-38. We find

$$\epsilon_0 \oint \kappa \mathbf{E}_1 \cdot d\mathbf{A} = -\epsilon_0 \kappa E_1 A = -q. \qquad (26\text{-}39)$$

(The first minus sign in Eq. 26-39 comes from the dot product $\mathbf{E}_1 \cdot d\mathbf{A}$, because now the field vector $\mathbf{E}_1$ points downward and the area vector $d\mathbf{A}$ points upward.) Equation 26-39 gives us

$$E_1 = \frac{q}{\kappa \epsilon_0 A} = \frac{E_0}{\kappa} = \frac{6.90 \text{ kV/m}}{2.61}$$

$$= 2.64 \text{ kV/m.} \qquad \text{(Answer)}$$

(e) What is the potential difference V between the plates after the slab has been introduced?

SOLUTION: We answer by applying Eq. 26-6, integrating along a straight-line path extending directly from the top plate to the bottom plate. Within the dielectric, the path length is b and the electric field is E_1. Within the two gaps above and below the dielectric, the total path length is $d - b$ and the electric field is E_0. Equation 26-6 then yields

$$V = \int_+^- E \, ds = E_0(d - b) + E_1 b$$

$$= (6900 \text{ V/m})(0.0124 \text{ m} - 0.00780 \text{ m})$$

$$+ (2640 \text{ V/m})(0.00780 \text{ m})$$

$$= 52.3 \text{ V.} \qquad \text{(Answer)}$$

This contrasts with the original potential difference of 85.5 V.

(f) What is the capacitance with the slab in place?

SOLUTION: From Eq. 26-1,

$$C = \frac{q}{V} = \frac{7.02 \times 10^{-10} \text{ C}}{52.3 \text{ V}}$$

$$= 1.34 \times 10^{-11} \text{ F} = 13.4 \text{ pF.} \qquad \text{(Answer)}$$

This is more than the original capacitance of 8.21 pF.

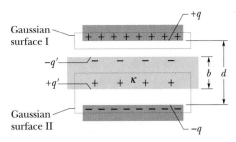

FIGURE 26-16 Sample Problem 26-8. A parallel-plate capacitor containing a dielectric slab that only partially fills the space between the plates.

CHECKPOINT 6: In Sample Problem 26-8, if the thickness b of the slab increases, do the following increase, decrease, or remain the same: (a) the electric field E_1, (b) the potential difference between the plates, and (c) the capacitance of the capacitor?

REVIEW & SUMMARY

Capacitor; Capacitance

A **capacitor** consists of two isolated conductors (the *plates*) with equal and opposite charges $+q$ and $-q$. Its **capacitance** C is defined from

$$q = CV, \qquad (26\text{-}1)$$

where V is the potential difference between the plates. The SI unit of capacitance is the farad (1 farad = 1 F = 1 coulomb per volt).

Determining Capacitance

We generally determine the capacitance of a particular capacitor configuration by (1) assuming a charge q to have been placed on the plates, (2) finding the electric field **E** due to this charge, (3) evaluating the potential difference V, and (4) calculating C from Eq. 26-1. Some specific results are the following:

A *parallel-plate capacitor* with flat parallel plates of area A and spacing d has capacitance

$$C = \frac{\epsilon_0 A}{d}. \qquad (26\text{-}9)$$

A *cylindrical capacitor* (two long coaxial cylinders) of length L and inner and outer radii a and b has capacitance

$$C = 2\pi\epsilon_0 \frac{L}{\ln(b/a)}. \qquad (26\text{-}14)$$

A *spherical capacitor* with concentric spherical plates of inner and outer radii a and b has capacitance

$$C = 4\pi\epsilon_0 \frac{ab}{b-a}. \qquad (26\text{-}17)$$

If we let $b \to \infty$ and $a = R$ in Eq. 26-17, we obtain the capacitance of an *isolated sphere* of radius R:

$$C = 4\pi\epsilon_0 R. \qquad (26\text{-}18)$$

Capacitors in Parallel and in Series

The **equivalent capacitances** C_{eq} of combinations of individual capacitors connected in **parallel** and in **series** are

$$C_{eq} = \sum_{j=1}^{n} C_j \qquad (n \text{ capacitors in parallel}) \qquad (26\text{-}19)$$

and

$$\frac{1}{C_{eq}} = \sum_{j=1}^{n} \frac{1}{C_j} \qquad (n \text{ capacitors in series}). \qquad (26\text{-}20)$$

These equivalent capacitances can be used to calculate the capacitances of more complicated series-parallel combinations.

Potential Energy and Energy Density

The **electric potential energy** U of a charged capacitor, given by

$$U = \frac{q^2}{2C} = \tfrac{1}{2}CV^2, \qquad (26\text{-}21, 26\text{-}22)$$

is the work required to charge it. This energy can be associated with the capacitor's electric field **E**. By extension we can associate stored energy with an electric field. In vacuum, the **energy density** u, or potential energy per unit volume, is given by

$$u = \tfrac{1}{2}\epsilon_0 E^2. \qquad (26\text{-}23)$$

Capacitance with a Dielectric

If the space between the plates of a capacitor is completely filled with a dielectric material, the capacitance C is increased by a factor κ, called the **dielectric constant,** which is characteristic of the material. In a region that is completely filled by a dielectric, all electrostatic equations containing ϵ_0 must be modified by replacing ϵ_0 with $\kappa\epsilon_0$.

The effects of adding a dielectric can be understood physically in terms of the action of an electric field on the permanent or induced electric dipoles in the dielectric slab. The result is the formation of induced charges on the surfaces of the dielectric, which results in a weakening of the field within the dielectric.

Gauss' Law with a Dielectric

When a dielectric is present, Gauss' law may be generalized to

$$\epsilon_0 \oint \kappa \mathbf{E} \cdot d\mathbf{A} = q. \qquad (26\text{-}38)$$

Here q is the free charge; the induced surface charge is accounted for by including the dielectric constant κ inside the integral.

QUESTIONS

1. Figure 26-17 shows plots of charge versus potential difference for three parallel-plate capacitors, which have the plate areas and separations given in the table. Which of the plots goes with which of the capacitors?

CAPACITOR	AREA	SEPARATION
1	A	d
2	$2A$	d
3	A	$2d$

FIGURE 26-17 Question 1.

2. Figure 26-18 shows, in cross section, an isolated solid metal sphere A of radius R and two spherical capacitors B and C with inner and outer radii R and $2R$. The inner spherical "plate" of capacitor B is a spherical shell; that of capacitor C is a solid sphere. Rank objects A, B, and C according to their capacitance, greatest first.

A   *B*   *C*

FIGURE 26-18 Question 2.

3. When a flat sheet of aluminum foil of negligible thickness is placed midway between the parallel plates of a capacitor, does the capacitance increase, decrease, or remain the same if the sheet is (a) electrically connected to one of the plates and (b) electrically isolated? (*Hint:* For (b), consider the equivalent capacitance.)

4. For each circuit in Fig. 26-19, are the capacitors connected in series, in parallel, or in neither mode?

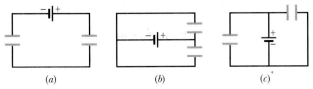

(*a*)   (*b*)   (*c*)

FIGURE 26-19 Question 4.

5. Two capacitors are wired to a battery. (a) In which arrangement, parallel or series, is the potential difference across each capacitor the same and the same as that across the equivalent capacitance? (b) In which is the charge on each capacitor the same and the same as that on the equivalent capacitance?

6. (a) In Fig. 26-20a, are capacitors C_1 and C_3 in series? (b) In the same figure, are capacitors C_1 and C_2 in parallel? (c) Rank the equivalent capacitances of the four circuits shown in Fig. 26-20, greatest first.

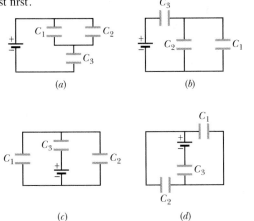

(*a*)   (*b*)

(*c*)   (*d*)

FIGURE 26-20 Question 6.

7. What is the equivalent capacitance of three capacitors, each of capacitance C, if they are connected to a battery (a) in series with one another and (b) in parallel? (c) In which arrangement is there more charge on the equivalent capacitance?

8. You are to connect capacitors C_1 and C_2, with $C_1 > C_2$, to a battery, first individually, then in series, and then in parallel. Rank those arrangements according to the amount of charge stored, greatest first.

9. (a) In Sample Problem 26-3, is the potential difference across capacitor C_2 more than, less than, or equal to that across capacitor C_1? (b) Is the charge on capacitor C_2 more than, less than, or equal to that on capacitor C_1?

10. Initially, a single capacitor C_1 is wired to a battery. Then capacitor C_2 is added in parallel. Are (a) the potential difference across C_1 and (b) the charge q_1 on C_1 now more than, less than, or the same as previously? (c) Is the equivalent capacitance C_{12} of C_1 and C_2 more than, less than, or equal to C_1? (d) Is the total charge stored on C_1 and C_2 together more than, less than, or equal to the charge stored previously on C_1?

11. Repeat Question 10 for C_2 added in series, not in parallel.

12. In Sample Problem 26-4, if we increase the capacitance of C_2, do the following increase, decrease, or remain the same: (a) the final potential difference across each capacitor, and the share of q_0 received by (b) C_1 and (c) C_2?

13. Figure 26-21 shows three circuits, each consisting of a switch and two capacitors, initially charged as indicated. After the switches have been closed, in which circuit (if any) will the charge on the left-hand capacitor (a) increase, (b) decrease, and (c) remain the same?

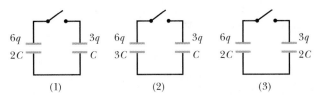

(1)   (2)   (3)

FIGURE 26-21 Question 13.

14. Two isolated metal spheres A and B have radii R and $2R$, respectively, and the same charge q. (a) Is the capacitance of A more than, less than, or equal to that of B? (b) Is the energy density just outside the surface of A more than, less than, or equal to that of B? (c) Is the energy density at radius $3R$ from the center of A more than, less than, or equal to that at the same radius from the center of B? (d) Is the total energy of the electric field due to A more than, less than, or equal to that of B?

15. An oil-filled parallel-plate capacitor was designed to have a capacitance C and to operate safely at or below a certain potential difference V_m without undergoing breakdown. However, the design is flawed and the capacitor occasionally breaks down. What can be done to redesign the capacitor, keeping C and V_m unchanged and using the same dielectric (the oil)?

16. When a dielectric slab is inserted between the plates of one of the two identical capacitors in Fig. 26-22, do the following properties of that capacitor increase, decrease, or remain the same: (a) capacitance, (b) charge, (c) potential difference, and (d) potential energy? (e) How about the same properties of the other capacitor?

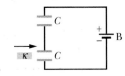

FIGURE 26-22 Question 16.

EXERCISES & PROBLEMS

SECTION 26-2 Capacitance

1E. An electrometer is a device used to measure static charge: an unknown charge is placed on the plates of the meter's capacitor, and the potential difference is measured. What minimum charge can be measured by an electrometer with a capacitance of 50 pF and a voltage sensitivity of 0.15 V?

2E. The two metal objects in Fig. 26-23 have net charges of +70 pC and −70 pC, which result in a 20 V potential difference between them. (a) What is the capacitance of the system? (b) If the charges are changed to +200 pC and −200 pC, what does the capacitance become? (c) What does the potential difference become?

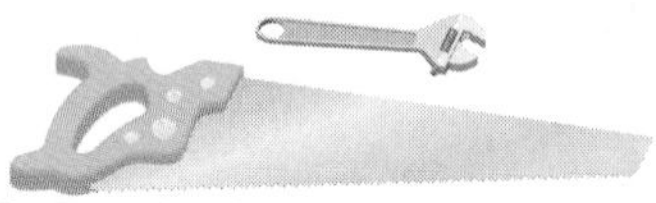

FIGURE 26-23 Exercise 2.

3E. The capacitor in Fig. 26-24 has a capacitance of 25 μF and is initially uncharged. The battery provides a potential difference of 120 V. After switch S is closed, how much charge will pass through it?

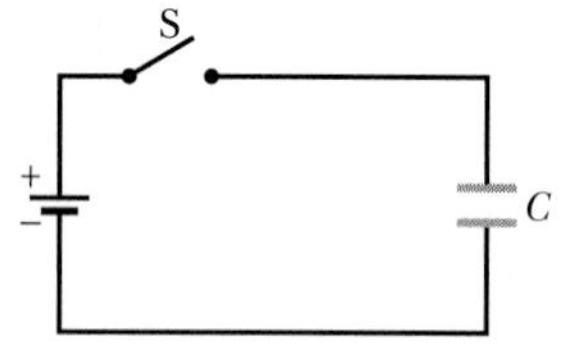

FIGURE 26-24 Exercise 3.

SECTION 26-3 Calculating the Capacitance

4E. If we solve Eq. 26-9 for ϵ_0, we see that its SI unit is the farad per meter. Show that this unit is equivalent to that obtained earlier for ϵ_0, namely, the coulomb squared per newton-meter squared.

5E. A parallel-plate capacitor has circular plates of 8.2 cm radius and 1.3 mm separation. (a) Calculate the capacitance. (b) What charge will appear on the plates if a potential difference of 120 V is applied?

6E. You have two flat metal plates, each of area 1.00 m², with which to construct a parallel-plate capacitor. If the capacitance of the device is to be 1.00 F, what must be the separation between the plates? Could this capacitor actually be constructed?

7E. The plates of a spherical capacitor have radii 38.0 and 40.0 mm. (a) Calculate the capacitance. (b) What must be the plate area of a parallel-plate capacitor with the same plate separation and capacitance?

8E. After you walk over a carpet on a dry day, your hand comes close to a metal doorknob and a 5 mm spark results. Such a spark means that there must have been a potential difference of possibly 15 kV between you and the doorknob. Assuming this potential difference, how much charge did you accumulate in walking over the carpet? For this extremely rough calculation, assume that your body can be represented by a uniformly charged conducting sphere 25 cm in radius and electrically isolated from its surroundings.

9E. Two sheets of aluminum foil have the same area, a separation of 1.0 mm, and a capacitance of 10 pF, and are charged to 12 V. (a) Calculate the area of each sheet. The separation is now decreased by 0.10 mm with the charge held constant. (b) What is the new capacitance? (c) By how much does the potential difference change? Explain how a microphone might be constructed using this principle.

10E. A spherical drop of mercury of radius R has a capacitance given by $C = 4\pi\epsilon_0 R$. If two such drops combine to form a single larger drop, what is its capacitance?

11P. Using the approximation that $\ln(1 + x) \approx x$ when $x \ll 1$ (see Appendix E), show that the capacitance of a cylindrical capacitor approaches that of a parallel-plate capacitor when the spacing between the two cylinders is small.

12P. Suppose that the two spherical shells of a spherical capacitor have approximately equal radii. Under these conditions the device approximates a parallel-plate capacitor with $b - a = d$. Show that Eq. 26-17 does indeed reduce to Eq. 26-9 in this case.

13P. A capacitor is to be designed to operate with constant capacitance in an environment of fluctuating temperature. As shown in Fig. 26-25, the capacitor is a parallel-plate type with plastic spacers to keep the plates aligned. (a) Show that the rate of change of the capacitance C with temperature T is given by

$$\frac{dC}{dT} = C\left(\frac{1}{A}\frac{dA}{dT} - \frac{1}{x}\frac{dx}{dT}\right),$$

where A is the plate area and x the plate separation. (b) If the plates are aluminum, what should be the coefficient of thermal expansion of the spacers to ensure that the capacitance does not vary with temperature? (Ignore the effect of the spacers on the capacitance.)

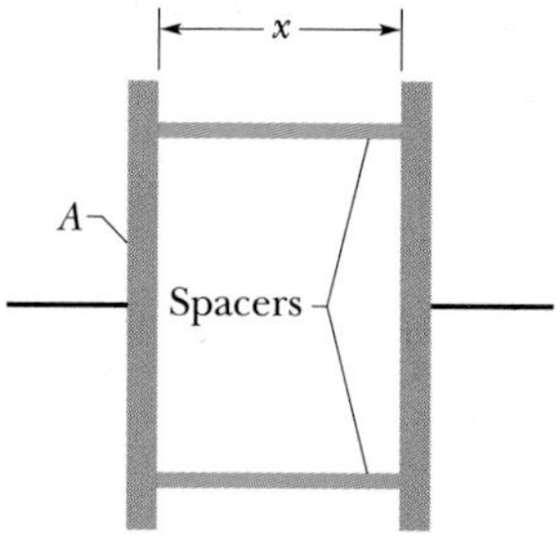

FIGURE 26-25
Problem 13.

SECTION 26-4 Capacitors in Parallel and in Series

14E. How many 1.00 μF capacitors must be connected in parallel to store a charge of 1.00 C with a potential of 110 V across the capacitors?

15E. In Fig. 26-26 find the equivalent capacitance of the combination. Assume that $C_1 = 10.0$ μF, $C_2 = 5.00$ μF, and $C_3 = 4.00$ μF.

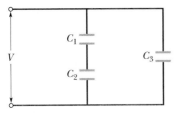

FIGURE 26-26 Exercise 15 and Problem 47.

16E. In Fig. 26-27 find the equivalent capacitance of the combination. Assume that $C_1 = 10.0 \ \mu F$, $C_2 = 5.00 \ \mu F$, and $C_3 = 4.00 \ \mu F$.

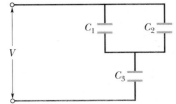

FIGURE 26-27 Exercise 16 and Problems 24 and 45.

17E. Each of the uncharged capacitors in Fig. 26-28 has a capacitance of 25.0 μF. A potential difference of 4200 V is established when the switch is closed. How many coulombs of charge then pass through meter A?

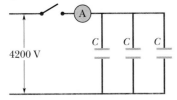

FIGURE 26-28 Exercise 17.

18E. A capacitance $C_1 = 6.00 \ \mu F$ is connected in series with a capacitance $C_2 = 4.00 \ \mu F$, and a potential difference of 200 V is applied across the pair. (a) Calculate the equivalent capacitance. (b) What is the charge on each capacitor? (c) What is the potential difference across each capacitor?

19E. Repeat Exercise 18 for the same two capacitors but with them now connected in parallel.

20P. Figure 26-29 shows two capacitors in series; the center section of length b is movable vertically. Show that the equivalent capacitance of this series combination is independent of the position of the center section and is given by $C = \epsilon_0 A/(a - b)$.

FIGURE 26-29 Problem 20.

21P. (a) Three capacitors are connected in parallel. Each has plate area A and plate spacing d. What must be the spacing of a single capacitor of plate area A if its capacitance equals that of the parallel combination? (b) What must be the spacing if the three capacitors are connected in series?

22P. (a) A potential difference of 300 V is applied to a series connection of two capacitors, of capacitance $C_1 = 2.0 \ \mu F$ and capacitance $C_2 = 8.0 \ \mu F$. What are the charge on and the potential difference across each capacitor? (b) The charged capacitors are disconnected from each other and from the battery. They are then reconnected, positive plate to positive plate and negative plate to negative plate, with no external voltage being applied. What are the charge and the potential difference for each now? (c) Suppose the charged capacitors in (a) were reconnected with plates of *opposite* sign together. What then would be the steady-state charge and potential difference for each?

23P. Figure 26-30 shows a variable "air gap" capacitor of the type used in manually tuned radios. Alternate plates are connected together; one group is fixed in position and the other group is capable of rotation. Consider a pile of n plates of alternate polarity, each having an area A and separated from adjacent plates by a distance d. Show that this capacitor has a maximum capacitance of

$$C = \frac{(n - 1)\epsilon_0 A}{d}.$$

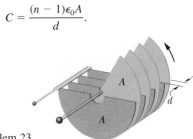

FIGURE 26-30 Problem 23.

24P. In Fig. 26-27 suppose that capacitor C_3 breaks down electrically, becoming equivalent to a conducting path. What *changes* in (a) the charge and (b) the potential difference occur for capacitor C_1? Assume that $V = 100$ V.

25P. You have several 2.0 μF capacitors, each capable of withstanding 200 V without electrical breakdown (in which they conduct charge instead of storing it). How would you assemble a combination having an equivalent capacitance of (a) 0.40 μF or (b) 1.2 μF, each capable of withstanding 1000 V?

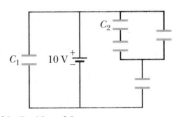

FIGURE 26-31 Problem 26.

26P. In Fig. 26-31, the battery has a potential difference of 10 V and the five capacitors each have a capacitance of 10 μF. What is the charge on (a) capacitor C_1 and (b) capacitor C_2?

27P. A 100 pF capacitor is charged to a potential difference of 50 V, and the charging battery is disconnected. The capacitor is then connected in parallel with a second (initially uncharged) capacitor. If the measured potential difference drops to 35 V, what is the capacitance of this second capacitor?

28P. In Fig. 26-32, the battery has a potential difference of 20 V. Find (a) the equivalent capacitance of all the capacitors and (b) the charge stored on that equivalent capacitance. Give the potential across and charge on (c) capacitor C_1, (d) capacitor C_2, and (e) capacitor C_3.

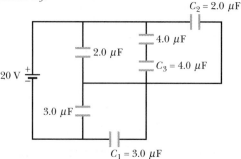

FIGURE 26-32 Problem 28.

29P. In Fig. 26-33, capacitors $C_1 = 1.0 \ \mu F$ and $C_2 = 3.0 \ \mu F$ are each charged to a potential difference of $V = 100$ V but with opposite polarity as shown. Switches S_1 and S_2 are now closed. (a) What is now the potential difference between points a and b? What are now the charges on (b) C_1 and (c) C_2?

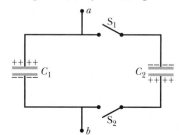

FIGURE 26-33
Problem 29.

30P. When switch S is thrown to the left in Fig. 26-34, the plates of capacitor C_1 acquire a potential difference V_0. Capacitors C_2 and C_3 are initially uncharged. The switch is now thrown to the right. What are the final charges q_1, q_2, and q_3 on the corresponding capacitors?

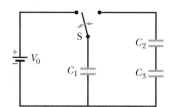

FIGURE 26-34 Problem 30.

31P. In Fig. 26-35, battery B supplies 12 V. (a) Find the charge on each capacitor first when only switch S_1 is closed and (b) later when switch S_2 is also closed. Take $C_1 = 1.0 \ \mu F$, $C_2 = 2.0 \ \mu F$, $C_3 = 3.0 \ \mu F$, and $C_4 = 4.0 \ \mu F$.

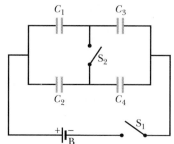

FIGURE 26-35 Problem 31.

32P. Figure 26-36 shows two identical capacitors C in a circuit with two (ideal) diodes D. (An ideal diode has the property that positive charge flows through it only in the direction of the arrow and negative charge flows through it only in the opposite direction.) A 100 V battery is connected across the input terminals, first with terminal a connected to the positive battery terminal and later with terminal b connected there. In each case, what is the potential difference across the output terminals?

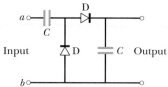

FIGURE 26-36 Problem 32.

SECTION 26-5 Storing Energy in an Electric Field

33E. How much energy is stored in one cubic meter of air due to the "fair weather" electric field of magnitude 150 V/m?

34E. Attempts to build a controlled thermonuclear fusion reactor, which, if successful, could provide the world with a vast supply of energy from heavy hydrogen in seawater, usually involve huge electric currents for short periods of time in magnetic field windings. For example, ZT-40 at Los Alamos Scientific Laboratory has rooms full of capacitors. One of the capacitor banks provides 61.0 mF at 10.0 kV. Calculate the stored energy (a) in joules and (b) in kilowatt-hours.

35E. What capacitance is required to store an energy of 10 kW·h at a potential difference of 1000 V?

36E. A parallel-plate air-filled capacitor has a capacitance of 130 pF. (a) What is the stored energy if the applied potential difference is 56.0 V? (b) Can you calculate the energy density for points between the plates? Explain.

37E. A certain capacitor is charged to a potential difference V. If you wish to increase its stored energy by 10%, by what percentage should you increase V?

38E. A parallel-plate air-filled capacitor having area 40 cm² and plate spacing 1.0 mm is charged to a potential difference of 600 V. Find (a) the capacitance, (b) the magnitude of the charge on each plate, (c) the stored energy, (d) the electric field between the plates, and (e) the energy density between the plates.

39E. Two capacitors, of 2.0 and 4.0 μF capacitance, are connected in parallel across a 300 V potential difference. Calculate the total energy stored in the capacitors.

40E. (a) Calculate the energy density of the electric field at distance r from the center of an electron at rest. (b) If the electron is assumed to be an infinitesimal point, what does this calculation yield for the energy density in the limit of $r \rightarrow 0$?

41P. A charged isolated metal sphere of diameter 10 cm has a potential of 8000 V relative to $V = 0$ at infinity. Calculate the energy density in the electric field near the surface of the sphere.

42P. A parallel-connected bank of 5.00 μF capacitors is used to store electric energy. What does it cost to charge the 2000 capacitors of the bank to 50,000 V, assuming a unit cost of 3.0¢/kW·h?

43P. One capacitor is charged until its stored energy is 4.0 J. A second uncharged capacitor is then connected to it in parallel. (a) If the charge distributes equally, what is now the total energy stored in the electric fields? (b) Where did the excess energy go?

44P. Compute the energy stored for the three different connections of the capacitors of Problem 22. Compare these stored energies and explain any differences.

45P. In Fig. 26-27 find (a) the charge, (b) the potential difference, and (c) the stored energy for each capacitor. Assume the numerical values of Exercise 16, with $V = 100$ V.

46P. A parallel-plate capacitor has plates of area A and separation d and is charged to a potential difference V. The charging battery is then disconnected, and the plates are pulled apart until their separation is $2d$. Derive expressions in terms of A, d, and V for (a) the new potential difference, (b) the initial and final stored energy, and (c) the work required to separate the plates.

47P. In Fig. 26-26 find (a) the charge, (b) the potential difference, and (c) the stored energy for each capacitor. Assume the numerical values of Exercise 15, with $V = 100$ V.

48P. A cylindrical capacitor has radii a and b as in Fig. 26-6. Show that half the stored electric potential energy lies within a cylinder whose radius is $r = \sqrt{ab}$.

49P. Show that the plates of a parallel-plate capacitor attract each other with a force given by $F = q^2/2\epsilon_0 A$. Do so by calculating the work necessary to increase the plate separation from x to $x + dx$, with the charge q remaining constant.

50P. Using the result of Problem 49, show that the force per unit area (the *electrostatic stress*) acting on either capacitor plate is given by $\frac{1}{2}\epsilon_0 E^2$. (Actually, this result is true in general, for a conductor of *any* shape with an electric field $\mathbf{E}$ at its surface.)

51P*. A soap bubble of radius R_0 is slowly given a charge q. Because of mutual repulsion of the surface charges, the radius increases slightly to R. Because of the expansion, the air pressure inside the bubble drops to $p(V_0/V)$, where p is the atmospheric pressure, V_0 is the initial volume, and V is the final volume. Show that these quantities are related by

$$q^2 = 32\pi^2\epsilon_0 pR(R^3 - R_0^3).$$

(*Hint:* Consider the forces acting on a small area of the charged bubble. These forces are due to gas pressure, atmospheric pressure, and electrostatic stress; see Problem 50.)

SECTION 26-6 Capacitor with a Dielectric

52E. An air-filled parallel-plate capacitor has a capacitance of 1.3 pF. The separation of the plates is doubled and wax is inserted between them. The new capacitance is 2.6 pF. Find the dielectric constant of the wax.

53E. Given a 7.4 pF air-filled capacitor, you are asked to convert it to a capacitor that can store up to 7.4 μJ with a maximum potential difference of 652 V. What dielectric in Table 26-1 should you use to fill the gap in the air capacitor if you do not allow for a margin of error?

54E. For making a parallel-plate capacitor, you have available two plates of copper, a sheet of mica (thickness = 0.10 mm, $\kappa = 5.4$), a sheet of glass (thickness = 2.0 mm, $\kappa = 7.0$), and a slab of paraffin (thickness = 1.0 cm, $\kappa = 2.0$). To obtain the largest capacitance, which sheet should you place between the copper plates?

55E. A parallel-plate air-filled capacitor has a capacitance of 50 pF. (a) If each of its plates has an area of 0.35 m², what is the separation? (b) If the region between the plates is now filled with material having $\kappa = 5.6$, what is the capacitance?

56E. A coaxial cable used in a transmission line has an inner radius of 0.10 mm and an outer radius of 0.60 mm. Calculate the capacitance per meter for the cable. Assume that the space between the conductors is filled with polystyrene.

57P. A certain substance has a dielectric constant of 2.8 and a dielectric strength of 18 MV/m. If it is used as the dielectric material in a parallel-plate capacitor, what minimum area should the plates of the capacitor have to obtain a capacitance of 7.0×10^{-2} μF and to ensure that the capacitor will be able to withstand a potential difference of 4.0 kV?

58P. You are asked to construct a capacitor having a capacitance near 1 nF and a breakdown potential in excess of 10,000 V. You think of using the sides of a tall Pyrex drinking glass as a dielectric, lining the inside and outside curved surfaces with aluminum foil. The glass is 15 cm tall with an inner radius of 3.6 cm and an outer radius of 3.8 cm. What are the (a) capacitance and (b) breakdown potential?

59P. You have been assigned to design a transportable capacitor that can store 250 kJ of energy. You decide on a parallel-plate type with dielectric. (a) What is the minimum capacitor volume possible if you use a dielectric whose dielectric strength is listed in Table 26-1? (b) Modern high-performance capacitors that can store 250 kJ have volumes of 0.0870 m³. Assuming that the dielectric used has the same dielectric strength as in (a), what must be its dielectric constant?

60P. Two parallel-plate capacitors have the same plate area A and separation d, but the dielectric constants of the materials between their plates are $\kappa + \Delta\kappa$ in one and $\kappa - \Delta\kappa$ in the other. (a) Find the equivalent capacitance when they are connected in parallel. (b) If the total charge on the parallel combination is Q, what is the charge on the capacitor with the larger capacitance?

61P. A slab of copper of thickness b is thrust into a parallel-plate capacitor of plate area A, as shown in Fig. 26-37; it is exactly halfway between the plates. (a) What is the capacitance after the slab is introduced? (b) If a charge q is maintained on the plates, what is the ratio of the stored energy before to that after the slab is inserted? (c) How much work is done on the slab as it is inserted? Is the slab sucked in or must it be pushed in?

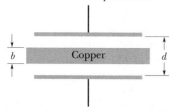

FIGURE 26-37   Problem 61.

62P. Repeat Problem 61, assuming that the potential difference rather than the charge is held constant.

63P. A parallel-plate capacitor of plate area A is filled with two dielectrics as in Fig. 26-38a. Show that the capacitance is

$$C = \frac{\epsilon_0 A}{d} \frac{\kappa_1 + \kappa_2}{2}.$$

Check this formula for limiting cases. (*Hint:* Can you justify this arrangement as being two capacitors in parallel?)

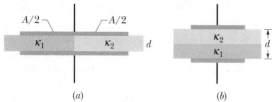

FIGURE 26-38   Problems 63 and 64.

64P. A parallel-plate capacitor of plate area A is filled with two dielectrics as in Fig. 26-38b. Show that the capacitance is

$$C = \frac{2\epsilon_0 A}{d} \frac{\kappa_1 \kappa_2}{\kappa_1 + \kappa_2}.$$

Check this formula for limiting cases. (*Hint:* Can you justify this arrangement as being two capacitors in series?)

65P. What is the capacitance of the capacitor, of plate area A, shown in Fig. 26-39? (*Hint:* See Problems 63 and 64.)

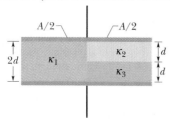

FIGURE 26-39   Problem 65.

SECTION 26-8   Dielectrics and Gauss' Law

66E. A parallel-plate capacitor has a capacitance of 100 pF, a plate area of 100 cm², and a mica dielectric ($\kappa = 5.4$). At 50 V potential difference, calculate (a) E in the mica, (b) the magnitude of the free charge on the plates, and (c) the magnitude of the induced surface charge on the mica.

67E. In Sample Problem 26-8, suppose that the battery remains connected while the dielectric slab is being introduced. Calculate (a) the capacitance, (b) the charge on the capacitor plates, (c) the electric field in the gap, and (d) the electric field in the slab, after the slab is in place.

68P. The space between two concentric conducting spherical shells of radii b and a (where $b > a$) is filled with a substance of dielectric constant κ. A potential difference V exists between the inner and outer shells. Determine (a) the capacitance of the device, (b) the free charge q on the inner shell, and (c) the charge q' induced along the surface of the inner shell.

69P. Two parallel plates of area 100 cm² are given charges of equal magnitude 8.9×10^{-7} C but opposite signs. The electric field within the dielectric material filling the space between the plates is 1.4×10^6 V/m. (a) Calculate the dielectric constant of the material. (b) Determine the magnitude of the charge induced on each dielectric surface.

70P. A parallel-plate capacitor has plates of area 0.12 m² and a separation of 1.2 cm. A battery charges the plates to a potential difference of 120 V and is then disconnected. A dielectric slab of thickness 4.0 mm and dielectric constant 4.8 is then placed symmetrically between the plates. (a) What is the capacitance before the slab is inserted? (b) What is the capacitance with the slab in place? (c) What is the free charge q before and after the slab is inserted? (d) What is the electric field in the space between the plates and dielectric? (e) What is the electric field in the dielectric? (f) With the slab in place, what is the potential difference across the plates? (g) How much external work is involved in the process of inserting the slab?

71P. In the capacitor of Sample Problem 26-8 (Fig. 26-16), (a) what fraction of the energy is stored in the air gaps? (b) What fraction is stored in the slab?

72P. A dielectric slab of thickness b is inserted between the plates of a parallel-plate capacitor of plate separation d. Show that the capacitance is then given by

$$C = \frac{\kappa \epsilon_0 A}{\kappa d - b(\kappa - 1)}.$$

(*Hint:* Derive the formula following the pattern of Sample Problem 26-8.) Does this formula predict the correct numerical result of Sample Problem 26-8? Verify that the formula gives reasonable results for the special cases of $b = 0$, $\kappa = 1$, and $b = d$.

27
Current and Resistance

The pride of Germany and a wonder of its time, the zeppelin <u>Hindenburg</u> was almost the length of three football fields—the largest flying machine that had ever been built. Although it was kept aloft by 16 cells of highly flammable hydrogen gas, it made many trans-Atlantic trips without incident. In fact, German zeppelins, which all depended on hydrogen, had never suffered an accident due to the hydrogen. But shortly after 7:21 p.m. on May 6, 1937, as the <u>Hindenburg</u> was ready to land at the U.S. Naval Air Station at Lakehurst, New Jersey, the ship burst into flames. Its crew had been waiting for a rainstorm to diminish, and handling ropes had just been let down to a navy ground crew, when ripples were sighted on the outer fabric of the ship about one-third of the way forward from the stern. Seconds later a flame erupted from that region, and a red glow illuminated the interior of the ship. Within 32 seconds the burning ship fell to the ground. Why, after so many successful flights of hydrogen-floated zeppelins, did this zeppelin burst into flames?

27-1 MOVING CHARGES AND ELECTRIC CURRENTS

Chapters 22 through 26 deal largely with *electrostatics*, that is, with charges at rest. With this chapter we begin to focus on **electric currents**, that is, charges in motion.

Examples of electric currents abound, ranging from the large currents that constitute lightning strokes to the tiny nerve currents that regulate our muscular activity. The currents in household wiring, in lightbulbs, and in electrical appliances are familiar to all. A beam of electrons—a current—moves through an evacuated space in the picture tube of a television set. Charged particles of *both* signs flow in the ionized gases of fluorescent lamps, in the batteries of transistor radios, and in car batteries. Electric currents can also be found in the semiconductors in pocket calculators and in the chips that control microwave ovens and electric dishwashers.

On a global scale, charged particles trapped in the Van Allen radiation belts surge back and forth above the atmosphere between Earth's north and south magnetic poles. On the scale of the solar system, enormous currents of protons, electrons, and ions fly radially outward from the Sun as the *solar wind*. On the galactic scale, cosmic rays, which are largely energetic protons, stream through our Milky Way galaxy, some reaching Earth.

Although an electric current is a stream of moving charges, not all moving charges constitute an electric current. If we are to say that an electric current passes through a given surface, there must be a net flow of charge through that surface. Two examples clarify our meaning.

1. The free electrons, conduction electrons, in an isolated length of copper wire are in random motion at speeds of the order of 10^6 m/s. If you pass a hypothetical plane through such a wire, conduction electrons pass through it *in both directions* at the rate of many billions per second. Hence, there is no *net* transport of charge and thus no current through the wire. However, if you connect the ends of the wire to a battery, you slightly bias the flow in one direction, with the result that there now is a net transport of charge and thus an electric current through the wire.

2. The flow of water through a garden hose represents the directed flow of positive charge (the protons in the water molecules) at a rate of perhaps several million coulombs per second. There is no net transport of charge, however, because there is a parallel flow of negative charge (the electrons in the water molecules) of exactly the same amount moving in exactly the same direction.

In this chapter we restrict ourselves largely to the study—within the framework of classical physics—of *steady* currents of *conduction electrons* moving through *metallic conductors* such as copper wires.

27-2 ELECTRIC CURRENT

As Fig. 27-1*a* reminds us, an isolated conducting loop—regardless of whether it has an excess charge—is all at the same potential. No electric field can exist within it or parallel to its surface. Although conduction electrons are available, no net electric force acts on them and thus there is no current.

If, as in Fig. 27-1*b*, we insert a battery in the loop, the conducting loop is no longer at a single potential. Electric fields act inside the material making up the loop, exerting forces on the conduction electrons, causing them to move, and thus establishing a current. After a very short time, the electron flow reaches a final, constant value and the current is in its *steady state* (it is not a function of time).

Figure 27-2 shows a section of a conductor, part of a conducting loop in which current has been established. If charge dq passes through a hypothetical plane (such as aa') in time dt, then the current through that plane is defined as

$$i = \frac{dq}{dt} \qquad \text{(definition of current).} \qquad (27\text{-}1)$$

We can find the charge that passes through the plane in a time interval extending from 0 to t by integration:

$$q = \int dq = \int_0^t i \, dt, \qquad (27\text{-}2)$$

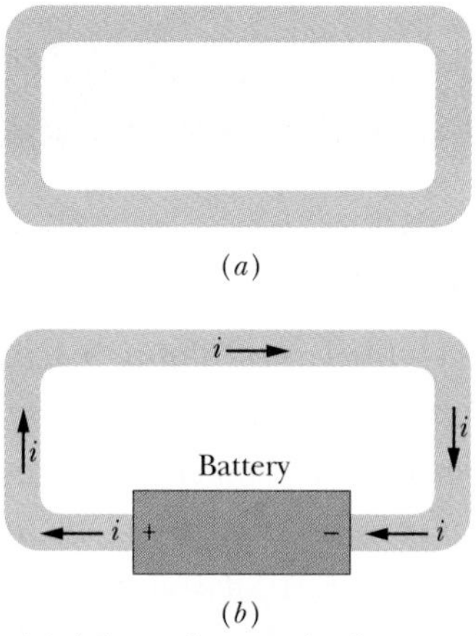

(a)

(b)

FIGURE 27-1 (*a*) A loop of copper in electrostatic equilibrium. The entire loop is at a single potential, and the electric field is zero at all points inside the copper. (*b*) Adding a battery imposes an electric potential difference between the ends of the loop that are connected to the terminals of the battery. The battery thus produces an electric field within the loop, from terminal to terminal, and the field causes charges to move around the loop. This movement of charges is a current i.

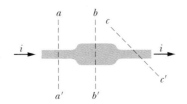

FIGURE 27-2 The current i through the conductor has the same value at planes aa', bb', and cc'.

in which the current i may be a function of time.

Under steady-state conditions, the current is the same for planes bb' and cc' and indeed for all planes that pass completely through the conductor, no matter what their location or orientation. This follows from the fact that charge is conserved. Under the steady-state conditions assumed here, an electron must enter the conductor at one end for every electron that leaves at the other. In the same way, if we have a steady flow of water through a garden hose, a drop of water must leave the nozzle for every drop that enters the hose at the other end. The amount of water in the hose is a conserved quantity.

The SI unit for current is the coulomb per second, also called the *ampere* (A):

1 ampere = 1 A = 1 coulomb per second = 1 C/s.

The ampere is an SI base unit; the coulomb is defined in terms of the ampere, as we discussed in Chapter 22. The formal definition of the ampere is presented in Chapter 30.

Current, as defined by Eq. 27-1, is a scalar because both charge and time in that equation are scalars. Yet, as in Fig. 27-1b, we often represent a current with an arrow to indicate the direction in which the charge is moving. Such arrows are not vectors, however, and they do not require vector addition. Figure 27-3a shows a conductor splitting at a junction into two branches. Because charge is conserved, the magnitudes of the currents in the branches must add to yield the magnitude of the current in the original

conductor, so that

$$i_0 = i_1 + i_2. \qquad (27\text{-}3)$$

As Fig. 27-3b suggests, bending or reorienting the wires in space does not change the validity of Eq. 27-3. Current arrows show only a direction (or sense) of flow along a conductor, not a direction in space.

The Directions of Currents

In Fig. 27-1b we drew the current arrows in the direction in which positively charged particles would be forced to move through the loop by the electric field. Such positive *charge carriers,* as they are often called, would move away from the positive battery terminal and toward the negative terminal. Actually, the charge carriers in the copper loop of Fig. 27-1b are electrons and thus are negatively charged. The electric field forces them to move in a direction opposite the current arrows, from the negative terminal to the positive terminal. Still, for historical reasons, we use the following convention:

> A current arrow is drawn in the direction in which positive charge carriers would move, even if the actual charge carriers are negative and move in the opposite direction.

We can use this convention because in *most* situations, the assumed motion of positive charge carriers in one direction has the same effect as the actual motion of negative charge carriers in the opposite direction. (When the effect is not the same, we shall, of course, drop the convention and describe the actual motion.)

CHECKPOINT **1:** The figure shows a portion of a circuit. What are the magnitude and direction of the current i in the lower right-hand wire?

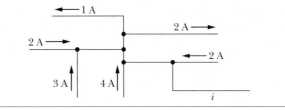

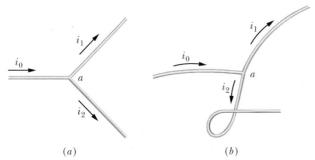

(a) (b)

FIGURE 27-3 The relation $i_0 = i_1 + i_2$ is true at junction a no matter what the orientation in space of the three wires. Currents are scalars, not vectors.

SAMPLE PROBLEM 27-1

Water flows through a garden hose at a rate R of 450 cm³/s. To what current of negative charge does this correspond?

SOLUTION: The current of negative charge carried by the water is the rate at which water molecules pass through any plane that cuts across the hose times the amount of negative charge carried by each molecule. If ρ is the density of water and M is its molar mass, then the rate (in moles per second) at which water is flowing through the plane is $R\rho/M$. If N is the number of water molecules and N_A is Avogadro's number, the rate dN/dt at which molecules pass through the plane is

$$\frac{dN}{dt} = \frac{R\rho N_A}{M}$$

$$= (450 \times 10^{-6} \text{ m}^3/\text{s})(1000 \text{ kg/m}^3)$$

$$\times \frac{(6.02 \times 10^{23} \text{ molecules/mol})}{0.018 \text{ kg/mol}}$$

$$= 1.51 \times 10^{25} \text{ molecules/s}.$$

Each water molecule contains 10 electrons: 8 in the oxygen atom and 1 in each of the two hydrogens. Each electron has a charge of $-e$, so the current corresponding to this movement of negative charge is

$$i = \frac{dq}{dt} = 10e\frac{dN}{dt} = (10 \text{ electrons/molecule})$$

$$\times (1.60 \times 10^{-19} \text{ C/electron})$$

$$\times (1.51 \times 10^{25} \text{ molecules/s})$$

$$= 2.42 \times 10^7 \text{ C/s} = 2.42 \times 10^7 \text{ A}$$

$$= 24.2 \text{ MA}. \quad \text{(Answer)}$$

This current of negative charge is exactly compensated by a current of positive charge associated with the nuclei of the three atoms that make up the water molecule. Thus there is no net flow of charge through the hose.

27-3 CURRENT DENSITY

Sometimes we are interested in the current i in a particular conductor. At other times we take a localized view and study the flow of charge at a particular point within a conductor. A (positive) charge carrier at a given point will flow in the direction of the electric field $\mathbf{E}$ at that point. To describe this flow, we can use the **current density J**. This vector quantity has the same direction as the electric field through a surface and has a magnitude J equal to the current per unit area through an element of that surface. We can write the amount of current through the element as $\mathbf{J} \cdot d\mathbf{A}$, where $d\mathbf{A}$ is the area vector of that element, perpendicular to the element. The total current through the surface is then

$$i = \int \mathbf{J} \cdot d\mathbf{A}. \quad (27\text{-}4)$$

If the current is uniform across the surface and parallel to $d\mathbf{A}$, then $\mathbf{J}$ is also uniform and parallel to $d\mathbf{A}$. Then Eq. 27-4 becomes

$$i = \int J \, dA = J \int dA = JA,$$

or

$$J = \frac{i}{A}, \quad (27\text{-}5)$$

where A is the total area of the surface. From Eq. 27-4 or 27-5 we see that the SI unit for current density is the ampere per square meter (A/m^2).

In Chapter 23 we saw that we can represent an array of electric field vectors with electric field lines. Figure 27-4 shows how an array of current density vectors can be represented with a similar set of lines, which we can call *streamlines*. The current, which is toward the right in Fig. 27-4, makes a transition from the wider conductor at the left to the narrower conductor at the right. Because charge is conserved during the transition, the amount of charge and thus the amount of current cannot change. However, the current density does change—it is greater in the narrower conductor. The spacing of the streamlines suggests this increase in current density: streamlines that are closer together imply greater current density.

Drift Speed

When a conductor does not have a current through it, its conduction electrons move randomly, with no net motion in any direction. When the conductor does have a current through it, these electrons actually still move randomly, but now they tend to *drift* with a **drift speed** v_d in the direction opposite that of the applied electric field that causes the current. The drift speed is tiny compared to the speeds in the random motion. For example, in the copper conductors of household wiring, electron drift speeds are perhaps 10^{-5} or 10^{-4} m/s, while the random motion speeds are around 10^6 m/s.

We can use Fig. 27-5 to relate the drift speed v_d of the conduction electrons in a current through a wire to the magnitude J of the current density in the wire. For convenience, Fig. 27-5 shows the equivalent drift of *positive*

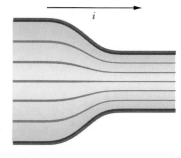

FIGURE 27-4 Streamlines representing the current density vectors in the flow of charge through a constricted conductor.

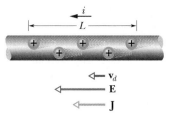

FIGURE 27-5 Positive charge carriers drift at speed v_d in the direction of the applied electric field **E**. By convention, the direction of the current density **J** and the sense of the current arrow are drawn in that same direction.

charge carriers in the direction of the applied electric field **E**. Let us assume that these charge carriers all move with the same drift speed v_d and that the current density J is uniform across the wire's cross-sectional area A. The number of charge carriers in a length L of the wire is nAL, where n is the number of carriers per unit volume. The total charge of the carriers in the length L, each with charge e, is then

$$q = (nAL)e.$$

Because the carriers all move along the wire with speed v_d, this total charge moves through any cross section of the wire in the time interval

$$t = \frac{L}{v_d}.$$

Equation 27-1 tells us that the current i is the time rate of transfer of charge across a cross section. So here we have

$$i = \frac{q}{t} = \frac{nALe}{L/v_d} = nAev_d. \qquad (27\text{-}6)$$

Solving for v_d and recalling Eq. 27-5 ($J = i/A$), we obtain

$$v_d = \frac{i}{nAe} = \frac{J}{ne}$$

or, extended to vector form,

$$\mathbf{J} = (ne)\mathbf{v}_d. \qquad (27\text{-}7)$$

Here the product ne, whose SI unit is the coulomb per cubic meter (C/m³), is the *carrier charge density*. For positive carriers, which we always assume, ne is positive and Eq. 27-7 predicts that **J** and $\mathbf{v}_d$ point in the same direction.

$\mathbb{C}$HECKPOINT **2:** The figure shows conduction electrons moving leftward through a wire. Are the following leftward or rightward: (a) the current i, (b) the current density **J**, (c) the electric field **E** in the wire?

SAMPLE PROBLEM 27-2

(a) The current density in a cylindrical wire of radius $R = 2.0$ mm is uniform across a cross section of the wire and is given by $J = 2.0 \times 10^5$ A/m². What is the current through the outer portion of the wire between radial distances $R/2$ and R (Fig. 27-6a)?

SOLUTION: Because the current density is uniform across a cross section, we can use Eq. 27-5 ($J = i/A$) to find the current. However, we want only the current through a reduced cross-sectional area A' of the wire (rather than the entire area), where

$$A' = \pi R^2 - \pi \left(\frac{R}{2}\right)^2 = \pi \left(\frac{3R^2}{4}\right)$$

$$= \frac{\pi 3}{4} (0.002 \text{ m})^2 = 9.424 \times 10^{-6} \text{ m}^2.$$

We now rewrite Eq. 27-5 as

$$i = JA'$$

and then substitute the data to find

$$i = (2.0 \times 10^5 \text{ A/m}^2)(9.424 \times 10^{-6} \text{ m}^2)$$

$$= 1.9 \text{ A}. \qquad \text{(Answer)}$$

(b) Suppose, instead, that the current density through a cross section varies with radial distance r as $J = ar^2$, in which $a = 3.0 \times 10^{11}$ A/m⁴ and r is in meters. What now is the current through the same outer portion of the wire?

SOLUTION: Because the current density is not uniform across a cross section of the wire, we must use Eq. 27-4 ($i = \int \mathbf{J} \cdot d\mathbf{A}$) and integrate the current density over the portion of the wire from $r = R/2$ to $r = R$. The current density vector **J** (along the wire's length) and the differential area vector $d\mathbf{A}$ (perpendicular to a cross section of the wire) have the same direction. So,

$$\mathbf{J} \cdot d\mathbf{A} = J \, dA \cos 0 = J \, dA.$$

We need to replace the differential area dA with something we can actually integrate between the limits $r = R/2$ and $r = R$. The simplest replacement (because J is given as a

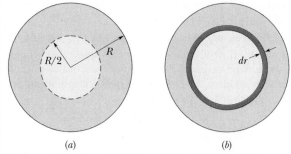

(a)   (b)

FIGURE 27-6 Sample Problem 27-2. (a) Cross section of a wire of radius R. (b) A thin ring has width dr and circumference $2\pi r$, and thus a differential area $dA = 2\pi r \, dr$.

function of r) is the area $2\pi r\,dr$ of a thin ring of circumference $2\pi r$ and width dr (Fig. 27-6b). We can then integrate with r as the variable of integration. Equation 27-4 then gives us

$$i = \int \mathbf{J}\cdot d\mathbf{A} = \int J\,dA$$

$$= \int_{R/2}^{R} ar^2\,2\pi r\,dr = 2\pi a \int_{R/2}^{R} r^3\,dr$$

$$= 2\pi a \left[\frac{r^4}{4}\right]_{R/2}^{R} = \frac{\pi a}{2}\left[R^4 - \frac{R^4}{16}\right] = \frac{15}{32}\,\pi a R^4$$

$$= \frac{15}{32}\,\pi(3.0 \times 10^{11}\ \text{A/m}^4)(0.002\ \text{m})^4 = 7.1\ \text{A}. \quad \text{(Answer)}$$

SAMPLE PROBLEM 27-3

One end of an aluminum wire whose diameter is 2.5 mm is welded to one end of a copper wire whose diameter is 1.8 mm. The composite wire carries a steady current i of 17 mA.

(a) What is the current density in each wire?

SOLUTION: We may take the current density as constant within each wire (except near the junction, where the diameter changes). The cross-sectional area A of the aluminum wire is

$$A_{\text{Al}} = \pi\left(\frac{d}{2}\right)^2 = \frac{\pi}{4}\,(2.5 \times 10^{-3}\ \text{M})^2$$

$$= 4.91 \times 10^{-6}\ \text{m}^2,$$

and the current density is given by Eq. 27-5:

$$J_{\text{Al}} = \frac{i}{A_{\text{Al}}} = \frac{17 \times 10^{-3}\ \text{A}}{4.91 \times 10^{-6}\ \text{m}^2}$$

$$= 3.5 \times 10^3\ \text{A/m}^2. \quad \text{(Answer)}$$

As you can verify, the cross-sectional area of the copper wire is $2.54 \times 10^{-6}\ \text{m}^2$, so

$$J_{\text{Cu}} = \frac{i}{A_{\text{Cu}}} = \frac{17 \times 10^{-3}\ \text{A}}{2.54 \times 10^{-6}\ \text{m}^2}$$

$$= 6.7 \times 10^3\ \text{A/m}^2. \quad \text{(Answer)}$$

(b) What is the drift speed of the conduction electrons in the copper wire? Assume that, on the average, each copper atom contributes one conduction electron.

SOLUTION: We can find the drift speed from Eq. 27-7 ($\mathbf{J} = ne v_d$) if we first find n, the number of electrons per unit volume. With the given assumption of about one conduction electron per atom, n is the same as the number of atoms per unit volume and can be found from

$$\frac{n}{N_A} = \frac{\rho}{M} \quad \text{or} \quad \frac{\text{atoms/m}^3}{\text{atoms/mol}} = \frac{\text{mass/m}^3}{\text{mass/mol}}$$

where ρ is the density of copper, N_A is Avogadro's number, and M is the molar mass of copper. Thus

$$n = \frac{N_A\rho}{M}$$

$$= \frac{(6.02 \times 10^{23}\ \text{mol}^{-1})(9.0 \times 10^3\ \text{kg/m}^3)}{64 \times 10^{-3}\ \text{kg/mol}}$$

$$= 8.47 \times 10^{28}\ \text{electrons/m}^3.$$

We then have from Eq. 27-7,

$$v_d = \frac{6.7 \times 10^3\ \text{A/m}^2}{\left(8.47 \times 10^{28}\ \dfrac{\text{electrons}}{\text{m}^3}\right)\left(1.6 \times 10^{-19}\ \dfrac{\text{C}}{\text{electron}}\right)}$$

$$= 4.9 \times 10^{-7}\ \text{m/s} = 1.8\ \text{mm/h}. \quad \text{(Answer)}$$

You may well ask: "If the electrons drift so slowly, why do the room lights turn on so quickly when I throw the switch?" Confusion on this point results from not distinguishing between the drift speed of the electrons and the speed at which *changes* in the electric field configuration travel along wires. This latter speed is nearly that of light; electrons everywhere in the wire begin drifting almost at once, including into the lightbulbs. Similarly, when you open the valve on your garden hose, with the hose full of water, a pressure wave travels along the hose at the speed of sound in water. The speed at which the water itself moves through the hose—measured perhaps with a dye marker—is much lower.

SAMPLE PROBLEM 27-4

Consider a strip of silicon that has a rectangular cross section with width $w = 3.2$ mm and height $h = 250$ μm, and through which there is a uniform current i of 5.2 mA. The silicon is an *n-type semiconductor*, having been "doped" with a controlled phosphorus impurity. As we shall discuss in Section 27-8, the doping has the effect of greatly increasing n, the number of charge carriers per unit volume, as compared with the value for pure silicon. In this case, $n = 1.5 \times 10^{23}\ \text{m}^{-3}$.

(a) What is the current density in the strip?

SOLUTION: From Eq. 27-5,

$$J = \frac{i}{wh} = \frac{5.2 \times 10^{-3}\ \text{A}}{(3.2 \times 10^{-3}\ \text{m})(250 \times 10^{-6}\ \text{m})}$$

$$= 6500\ \text{A/m}^2. \quad \text{(Answer)}$$

(b) What is the drift speed?

SOLUTION: From Eq. 27-7,

$$v_d = \frac{J}{ne} = \frac{6500\ \text{A/m}^2}{(1.5 \times 10^{23}\ \text{m}^{-3})(1.60 \times 10^{-19}\ \text{C})}$$

$$= 0.27\ \text{m/s} = 27\ \text{cm/s}. \quad \text{(Answer)}$$

Note that the current density (6500 A/m²) for this semiconductor turns out to be comparable to the current density (6700 A/m²) for the copper conductor in Sample Problem

27-3. That is, the rate at which charge flows through a unit area is about the same for the two devices. Yet, the drift speed (0.27 m/s) in the semiconductor is *much* greater than the drift speed (4.9×10^{-7} m/s) in the copper conductor.

If you recheck the calculations, you will see that this large difference in drift speed occurs because the number n of charge carriers per unit volume is much smaller in the semiconductor. Thus, if the current densities are to be comparable, the fewer conduction electrons in the semiconductor must move much faster than the electrons in the copper conductor.

27-4 RESISTANCE AND RESISTIVITY

If we apply the same potential difference between the ends of geometrically similar rods of copper and of glass, very different currents result. The characteristic of the conductor that enters here is its **resistance.** We determine the resistance between any two points of a conductor by applying a potential difference V between those points and measuring the current i that results. The resistance R is then

$$R = \frac{V}{i} \qquad \text{(definition of } R \text{)}. \qquad (27\text{-}8)$$

The SI unit for resistance that follows from Eq. 27-8 is the volt per ampere. This combination occurs so often that we give it a special name, the **ohm** (symbol Ω). That is,

$$1 \text{ ohm} = 1 \ \Omega = 1 \text{ volt per ampere}$$
$$= 1 \text{ V/A}. \qquad (27\text{-}9)$$

A conductor whose function in a circuit is to provide a specified resistance is called a **resistor** (see Fig. 27-7). We represent a resistor in a circuit diagram with the symbol -\/\/\/-. If we write Eq. 27-8 as

$$i = \frac{V}{R},$$

we see that "resistance" is aptly named. For a given potential difference, the greater the resistance (to current), the smaller the current.

The resistance of a conductor depends on the manner in which the potential difference is applied to it. Figure 27-8, for example, shows a given potential difference applied in two different ways to the same conductor. As the current density streamlines suggest, the currents in the two cases—hence the measured resistances—will be different. Unless otherwise stated, we shall assume that any given potential difference is applied as in Fig. 27-8*b*.

As we have done several times in other connections, we often wish to take a general view and deal not with particular objects but with materials. Here we do so by focusing not on the potential difference V across a particular resistor but on the electric field $\mathbf{E}$ at a point in a resistive material. Instead of dealing with the current i through the resistor, we deal with the current density $\mathbf{J}$ at the point in question. Instead of the resistance R of an object, we deal with the **resistivity** ρ of the *material*, defined as

$$\rho = \frac{E}{J} \qquad \text{(definition of } \rho \text{)}. \qquad (27\text{-}10)$$

(Compare this equation with Eq. 27-8.)

If we combine the SI units of E and J according to Eq. 27-10, we get, for the unit of ρ, the ohm-meter ($\Omega \cdot$ m):

$$\frac{\text{unit } (E)}{\text{unit } (J)} = \frac{\text{V/m}}{\text{A/m}^2} = \frac{\text{V}}{\text{A}} \text{ m} = \Omega \cdot \text{m}.$$

(Do not confuse the *ohm-meter,* the unit of resistivity, with the *ohmmeter,* which is an instrument that measures resistance.) Table 27-1 lists the resistivities of some materials.

We can write Eq. 27-10 in vector form as

$$\mathbf{E} = \rho \mathbf{J}. \qquad (27\text{-}11)$$

Equations 27-10 and 27-11 hold only for *isotropic* materials—materials whose electrical properties are the same in all directions.

We often speak of the **conductivity** σ of a material. This is simply the reciprocal of its resistivity, so

$$\sigma = \frac{1}{\rho} \qquad \text{(definition of } \sigma \text{)}. \qquad (27\text{-}12)$$

FIGURE 27-7 An assortment of resistors. The circular bands are color coding marks that identify the value of the resistance.

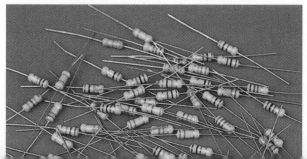

FIGURE 27-8 Two ways of applying a potential difference to a conducting rod. The heavy gray connectors are assumed to have negligible resistance. When they are arranged as in (*a*), the measured resistance is larger than when they are arranged as in (*b*).

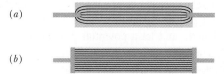

TABLE 27-1 RESISTIVITIES OF SOME MATERIALS
AT ROOM TEMPERATURE (20°C)

MATERIAL	RESISTIVITY, ρ ($\Omega \cdot$ m)	TEMPERATURE COEFFICIENT OF RESISTIVITY, α (K^{-1})
Typical Metals		
Silver	1.62×10^{-8}	4.1×10^{-3}
Copper	1.69×10^{-8}	4.3×10^{-3}
Aluminum	2.75×10^{-8}	4.4×10^{-3}
Tungsten	5.25×10^{-8}	4.5×10^{-3}
Iron	9.68×10^{-8}	6.5×10^{-3}
Platinum	10.6×10^{-8}	3.9×10^{-3}
Manganin[a]	48.2×10^{-8}	0.002×10^{-3}
Typical Semiconductors		
Silicon, pure	2.5×10^3	-70×10^{-3}
Silicon, n-type[b]	8.7×10^{-4}	
Silicon, p-type[c]	2.8×10^{-3}	
Typical Insulators		
Glass	$10^{10} - 10^{14}$	
Fused quartz	$\sim 10^{16}$	

[a]An alloy specifically designed to have a small value of α.

[b] Pure silicon doped with phosphorus impurities to a charge carrier density of 10^{23} m^{-3}.

[c] Pure silicon doped with aluminum impurities to a charge carrier density of 10^{23} m^{-3}.

The SI unit of conductivity is the reciprocal ohm-meter, $(\Omega \cdot \text{m})^{-1}$. The unit name mhos per meter is sometimes used (mho is ohm backward). The definition of σ allows us to write Eq. 27-11 in the alternative form

$$\mathbf{J} = \sigma\mathbf{E}. \qquad (27\text{-}13)$$

Calculating Resistance from Resistivity

We have just made an important distinction:

> Resistance is a property of an object. Resistivity is a property of a material.

If we know the resistivity of a substance such as copper, we can calculate the resistance of a length of wire made of that substance. Let A be the cross-sectional area of the wire, let L be its length, and let a potential difference V exist between its ends (Fig. 27-9). If the streamlines representing the current density are uniform throughout the wire, the electric field and the current density will be con-

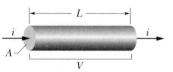

FIGURE 27-9 A potential difference V is applied between the ends of a wire of length L and cross section A, establishing a current i.

stant for all points within the wire and, from Eqs. 25-42 and 27-5, will have the values

$$E = V/L \quad \text{and} \quad J = i/A. \qquad (27\text{-}14)$$

We can then combine Eqs. 27-10 and 27-14 to write

$$\rho = \frac{E}{J} = \frac{V/L}{i/A}. \qquad (27\text{-}15)$$

But V/i is the resistance R, which allows us to recast Eq. 27-15 as

$$R = \rho \frac{L}{A}. \qquad (27\text{-}16)$$

Equation 27-16 can be applied only to a homogeneous isotropic conductor of uniform cross section, with the potential difference applied as in Fig. 27-8b.

The macroscopic quantities V, i, and R are of greatest interest when we are making electrical measurements on specific conductors. They are the quantities that we read directly on meters. We turn to the microscopic quantities E, J, and ρ when we are interested in the fundamental electrical properties of materials.

CHECKPOINT 3: The figure shows three cylindrical copper conductors along with their face areas and lengths. Rank them according to the current through them, greatest first, when the same potential difference V is placed across their lengths.

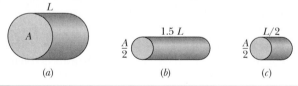

Variation with Temperature

The values of most physical properties vary with temperature, and resistivity is no exception. Figure 27-10, for example, shows the variation of this property for copper over a wide temperature range. The relation between temperature and resistivity for copper —and for metals in general —is fairly linear over a rather broad temperature range. For

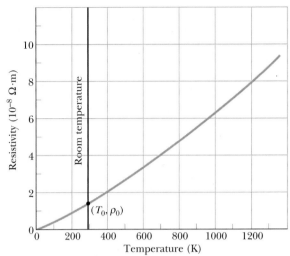

FIGURE 27-10 The resistivity of copper as a function of temperature. The dot on the curve marks a convenient reference point ($T_0 = 293$ K, $\rho_0 = 1.69 \times 10^{-8}$ $\Omega \cdot$m).

such linear relations we can write, as an empirical approximation that is good enough for most engineering purposes,

$$\rho - \rho_0 = \rho_0 \alpha (T - T_0). \qquad (27\text{-}17)$$

Here T_0 is a selected reference temperature and ρ_0 is the resistivity at that temperature. Usually $T_0 = 293$ K (room temperature), for which $\rho_0 = 1.69 \times 10^{-8}$ $\Omega \cdot$m.

Because temperature enters Eq. 27-17 only as a difference, it does not matter whether you use the Celsius or Kelvin scale in that equation because the sizes of the degree on these scales are identical. The quantity α in Eq. 27-17, called the *temperature coefficient of resistivity,* is chosen so that the equation gives the best agreement with experiment for temperatures in the chosen range. Some values of α for metals are listed in Table 27-1.

The *Hindenburg*

When the zeppelin *Hindenburg* was preparing to land, the handling ropes were let down to the ground crew. Exposed to the rain, the ropes became wet (and thus able to conduct a current). In this condition, the ropes "grounded" the metal framework of the zeppelin to which they were attached. That is, the wet ropes formed a conducting path between the framework and the ground, making the electric potential of the framework the same as that of the ground. This should have also grounded the outer fabric of the zeppelin, too. The *Hindenburg*, however, had been the first zeppelin to have its outer fabric painted with a sealant of large electrical resistivity. Thus the fabric remained at the electric potential of the atmosphere at the zeppelin's

altitude of about 43 m. Owing to the rainstorm, that potential was large relative to the potential at ground level.

The handling of the ropes apparently ruptured one of the hydrogen cells and released hydrogen between that cell and the zeppelin's outer fabric, causing the reported rippling of the fabric. There was then a dangerous situation: the fabric was wet with conducting rainwater and was at a potential much different from that of the framework of the zeppelin. Apparently, charge flowed along the wet fabric and then sparked through the released hydrogen to reach the metal framework of the zeppelin, igniting the hydrogen in the process. The burning rapidly ignited the cells of hydrogen in the zeppelin and brought the ship down. If the sealant on the outer fabric of the *Hindenburg* had been of less resistivity (like that of earlier and later zeppelins), the *Hindenburg* disaster probably would not have occurred.

SAMPLE PROBLEM 27-5

(a) What is the magnitude of the electric field applied to the copper conductor of Sample Problem 27-3?

SOLUTION: In Sample Problem 27-3(a) we found the current density J to be 6.7×10^3 A/m^2; from Table 27-1 we see that the resistivity ρ for copper is 1.69×10^{-8} $\Omega \cdot$m. Thus from Eq. 27-11

$$E = \rho J = (1.69 \times 10^{-8} \ \Omega \cdot \text{m})(6.7 \times 10^3 \ \text{A/m}^2)$$
$$= 1.1 \times 10^{-4} \ \text{V/m (copper)}. \qquad \text{(Answer)}$$

(b) What is the magnitude of the electric field in the *n*-type silicon semiconductor of Sample Problem 27-4?

SOLUTION: In that sample problem we found that $J = 6500$ A/m^2, and from Table 27-1 we see that $\rho = 8.7 \times 10^{-4}$ $\Omega \cdot$m. Thus from Eq. 27-11

$$E = \rho J = (8.7 \times 10^{-4} \ \Omega \cdot \text{m})(6500 \ \text{A/m}^2)$$
$$= 5.7 \ \text{V/m (}n\text{-type silicon)}. \qquad \text{(Answer)}$$

Note that the applied electric field in the semiconductor is much greater than that in the copper conductor. If you recheck the calculations, you will find that this difference is required by the large difference in the resistivities of the two devices. The need for a much greater electric field in the semiconductor is consistent with the need for a much greater drift speed that we found in Sample Problem 27-4: if the current densities are to be comparable in the two devices, the electric field applied to the semiconductor must be greater, to make possible the acceleration of the electrons to a greater drift speed.

SAMPLE PROBLEM 27-6

A rectangular block of iron has dimensions 1.2 cm $\times$ 1.2 cm $\times$ 15 cm.

(a) What is the resistance of the block measured between the two square ends?

SOLUTION: The resistivity of iron at room temperature is 9.68×10^{-8} $\Omega \cdot$m (Table 27-1). The area of a square end is $(1.2 \times 10^{-2}$ m$)^2$, or 1.44×10^{-4} m^2. From Eq. 27-16,

$$R = \frac{\rho L}{A} = \frac{(9.68 \times 10^{-8} \ \Omega \cdot \text{m})(0.15 \ \text{m})}{1.44 \times 10^{-4} \ \text{m}^2}$$

$$= 1.0 \times 10^{-4} \ \Omega = 100 \ \mu\Omega. \qquad \text{(Answer)}$$

(b) What is the resistance between two opposite rectangular faces?

SOLUTION: The area of a rectangular face is $(1.2 \times 10^{-2}$ m$)(0.15$ m$)$, or 1.80×10^{-3} m^2. From Eq. 27-16,

$$R = \frac{\rho L}{A} = \frac{(9.68 \times 10^{-8} \ \Omega \cdot \text{m})(1.2 \times 10^{-2} \ \text{m})}{1.80 \times 10^{-3} \ \text{m}^2}$$

$$= 6.5 \times 10^{-7} \ \Omega = 0.65 \ \mu\Omega. \qquad \text{(Answer)}$$

This result is much smaller than the previous result, because the distance L is smaller and the area A is larger. We assume in each part that the potential difference is applied to the block in such a way that the surfaces between which the resistance is desired are equipotential surfaces (as in Fig. 27-8b). Otherwise, Eq. 27-16 would not be valid.

27-5 OHM'S LAW

As we just discussed in Section 27-4, a resistor is a conductor with a specified resistance. It has that same resistance no matter what the magnitude and direction (*polarity*) of the applied potential difference. Other conducting devices, however, might have resistances that change with the applied potential difference.

Figure 27-11a shows how to distinguish such devices. A potential difference V is applied across the device being

tested, and the resulting current i through the device is measured as V is varied in both magnitude and polarity. The polarity of V is arbitrarily taken to be positive when the left terminal of the device is at a higher potential than the right terminal. The direction of the resulting current (from left to right) is arbitrarily assigned a plus sign. The reverse polarity of V (with the right terminal at a higher potential) is then negative; the current it causes is assigned a minus sign.

Figure 27-11b is a plot of i versus V for one device. This plot is a straight line passing through the origin, so the ratio i/V (which is the slope of the straight line) is the same for all values of V. This means that the resistance $R = V/i$ of the device is independent of the magnitude and polarity of the applied potential difference V.

Figure 27-11c is a plot for another conducting device. Current can exist in this device only when the polarity of V is positive and the applied potential difference is more than about 1.5 V. And when current does exist, the relation between i and V is not linear; it depends on the value of the applied potential difference V.

We distinguish between the two types of device by saying that one obeys Ohm's law and the other does not.

Ohm's law is an assertion that the current through a device is *always* directly proportional to the potential difference applied to the device.

(This assertion is correct only in certain situations; still, for historical reasons, the term "law" is used.) The device of Fig. 27-11b—which turns out to be a 1000 Ω resistor—obeys Ohm's law. The device of Fig. 27-11c—which turns out to be a so-called *pn* junction diode—does not.

A conducting device obeys Ohm's law when the resistance of the device is independent of the magnitude and polarity of the applied potential difference.

FIGURE 27-11 (*a*) A device to whose terminals a potential difference V is applied, establishing a current i. (*b*) A plot of current i versus applied potential difference V when the device is a 1000 Ω resistor. (*c*) A plot when the device is a semiconducting *pn* junction diode.

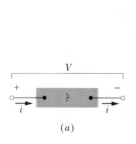

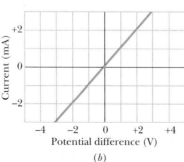

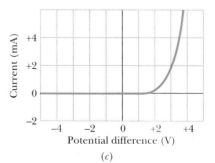

(a)   (b)   (c)

Modern microelectronics—and therefore much of the character of our present technological civilization—depends almost totally on devices that do *not* obey Ohm's law. Your calculator, for example, is full of them.

It is often contended that $V = iR$ is a statement of Ohm's law. That is not true! This equation is the defining equation for resistance, and it applies to all conducting devices, whether they obey Ohm's law or not. If we measure the potential difference V across, and the current i through, any device, even a *pn* junction diode, we can find its resistance *at that value of V* as $R = V/i$. The essence of Ohm's law, however, is that a plot of i versus V is linear; that is, the value of R is independent of the value of V.

We can express Ohm's law in a more general way if we focus on conducting *materials* rather than on conducting *devices*. The relevant relation is then Eq. 27-11 ($\mathbf{E} = \rho\mathbf{J}$), which is the analog of $V = iR$.

> A conducting material obeys Ohm's law when the resistivity of the material is independent of the magnitude and direction of the applied electric field.

All homogeneous materials, whether they are conductors like copper or semiconductors like silicon (doped or pure), obey Ohm's law within some range of values of the electric field. If the field is too strong, however, there are departures from Ohm's law in all cases.

CHECKPOINT **4:** The following table gives the current i (in amperes) through two devices for several values of potential difference V (in volts). From these data, determine which device does not obey Ohm's law.

DEVICE 1		DEVICE 2	
V	i	V	i
2.00	4.50	2.00	1.50
3.00	6.75	3.00	2.20
4.00	9.00	4.00	2.80

27-6 A MICROSCOPIC VIEW OF OHM'S LAW

To find out *why* particular materials obey Ohm's law, we must look into the details of the conduction process at the atomic level. Here we consider only conduction in metals, such as copper. We base our analysis on the *free-electron model*, in which we assume that the conduction electrons

in the metal are free to move throughout the volume of the sample, like the molecules of a gas in a closed container. We also assume that the electrons collide not with one another but only with the atoms of the metal.

According to classical physics, the electrons should have a Maxwellian speed distribution somewhat like that of the molecules in a gas. In such a distribution (see Section 20-7), the average electron speed would be proportional to the square root of the absolute temperature. The motions of the electrons, however, are governed not by the laws of classical physics but by those of quantum physics. As it turns out, an assumption that is much closer to the quantum reality is that the electrons move with a single effective speed v_{eff}, and this motion is essentially independent of the temperature. For copper, $v_{\text{eff}} \approx 1.6 \times 10^6$ m/s.

When we apply an electric field to a metal sample, the electrons modify their random motions slightly and drift very slowly—in a direction opposite that of the field—with an average drift speed v_d. As we saw in Sample Problem 27-3(b), the drift speed in a typical metallic conductor is about 4×10^{-7} m/s, less than the effective speed (1.6×10^6 m/s) by many orders of magnitude. Figure 27-12 suggests the relation between these two speeds. The gray lines show a possible random path for an electron in the absence of an applied field; the electron proceeds from A to B, making six collisions along the way. The green lines show how the same events *might* occur when an electric field $\mathbf{E}$ is applied. We see that the electron drifts steadily to the right, ending at B' rather than at B. Figure 27-12 was drawn with the assumption that $v_d \approx 0.02v_{\text{eff}}$. Since, however, the actual value is more like $v_d \approx (10^{-13})v_{\text{eff}}$, the drift displayed in the figure is greatly exaggerated.

The motion of the electrons in an electric field $\mathbf{E}$ is thus a combination of the motion due to random collisions

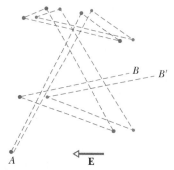

FIGURE 27-12 The gray lines show an electron moving from A to B, making six collisions en route. The green lines show what its path might be in the presence of an applied electric field $\mathbf{E}$. Note the steady drift in the direction of $-\mathbf{E}$. (Actually, the green lines should be slightly curved, to represent the parabolic paths followed by the electrons between collisions, under the influence of an electric field.)

and that due to **E**. When we consider all the free electrons, their random motions average to zero and make no contribution to the drift speed. Thus the drift speed is due only to the effect of the electric field on the electrons.

If an electron of mass m is placed in an electric field of magnitude E, the electron will experience an acceleration given by Newton's second law:

$$a = \frac{F}{m} = \frac{eE}{m}. \qquad (27\text{-}18)$$

The nature of the collisions experienced by electrons is such that, after a typical collision, the electron will—so to speak—completely lose its memory of its previous drift velocity. Each electron will then start off fresh after every encounter, moving off in a random direction. In the average time τ between collisions, the average electron will acquire a drift speed of $v_d = a\tau$. Moreover, if we measure the drift speeds of all the electrons at any instant, we will find that their average drift speed is also $a\tau$. Thus, at any instant, on average, the electrons will have drift speed $v_d = a\tau$. Then Eq. 27-18 gives us

$$v_d = a\tau = \frac{eE\tau}{m}. \qquad (27\text{-}19)$$

Combining this result with Eq. 27-7 ($J = nev_d$) yields

$$v_d = \frac{J}{ne} = \frac{eE\tau}{m},$$

which we can write as

$$E = \left(\frac{m}{e^2 n\tau}\right) J.$$

Comparing this with Eq. 27-11 ($E = \rho J$) leads to

$$\rho = \frac{m}{e^2 n\tau}. \qquad (27\text{-}20)$$

Equation 27-20 may be taken as a statement that metals obey Ohm's law if we can show that, for metals, ρ is a constant, independent of the strength of the applied electric field E. Because n, m, and e are constant, this reduces to convincing ourselves that τ, the average time (or *mean free time*) between collisions, is a constant, independent of the strength of the applied electric field. Indeed τ can be considered to be a constant because the drift speed v_d caused by the field is about a billion times smaller than the effective speed v_{eff}.

SAMPLE PROBLEM 27-7

(a) What is the mean free time τ between collisions for the conduction electrons in copper?

SOLUTION: From Eq. 27-20 we have

$$\tau = \frac{m}{ne^2\rho}.$$

We take the value of n, the number of conduction electrons per unit volume in copper, from Sample Problem 27-3(b). We take the value of ρ from Table 27-1. The denominator then becomes

$$(8.47 \times 10^{28} \text{ m}^{-3})(1.6 \times 10^{-19} \text{ C})^2(1.69 \times 10^{-8} \ \Omega\cdot\text{m})$$
$$= 3.66 \times 10^{-17} \text{ C}^2\cdot\Omega/\text{m}^2 = 3.66 \times 10^{-17} \text{ kg/s},$$

where we converted units as

$$\frac{\text{C}^2\cdot\Omega}{\text{m}^2} = \frac{\text{C}^2\cdot\text{V}}{\text{m}^2\cdot\text{A}} = \frac{\text{C}^2\cdot\text{J/C}}{\text{m}^2\cdot\text{C/s}} = \frac{\text{kg}\cdot\text{m}^2/\text{s}^2}{\text{m}^2/\text{s}} = \frac{\text{kg}}{\text{s}}.$$

For the mean free time we then have

$$\tau = \frac{9.1 \times 10^{-31} \text{ kg}}{3.66 \times 10^{-17} \text{ kg/s}} = 2.5 \times 10^{-14} \text{ s}. \quad \text{(Answer)}$$

(b) What is the mean free path λ for these collisions? Assume an effective speed v_{eff} of 1.6×10^6 m/s.

SOLUTION: In Section 20-6, we defined the mean free path as being the average distance traversed by a particle between collisions. Here the time between collisions of a free electron is τ and the speed of the electron is v_{eff}. So,

$$\lambda = \tau v_{\text{eff}} = (2.5 \times 10^{-14} \text{ s})(1.6 \times 10^6 \text{ m/s})$$
$$= 4.0 \times 10^{-8} \text{ m} = 40 \text{ nm}. \qquad \text{(Answer)}$$

This is about 150 times the distance between nearest-neighbor atoms in a copper lattice.

27-7 POWER IN ELECTRIC CIRCUITS

Figure 27-13 shows a circuit consisting of a battery B that is connected by wires, which we assume to have negligible resistance, to an unspecified conducting device. The device might be a resistor, a storage battery (a rechargeable battery), a motor, or some other electrical device. The battery

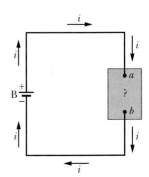

FIGURE 27-13 A battery B sets up a current i in a circuit containing an unspecified conducting device.

maintains a potential difference of magnitude V across its own terminals, and thus (because of the wires) across the terminals of the unspecified device, with a greater potential at terminal a of the device than at terminal b.

Since there is an external conducting path between the two terminals of the battery and since the potential differences set up by the battery are maintained, a steady current i is produced in the circuit, directed from terminal a to terminal b. The amount of charge dq that moves between those terminals in time interval dt is equal to $i\,dt$. This charge dq moves through a decrease in potential of magnitude V, and thus its electric potential energy decreases in magnitude by the amount

$$dU = dq\,V = i\,dt\,V.$$

The principle of conservation of energy tells us that the decrease in electric potential energy from a to b is accompanied by a transfer of energy to some other form. The power P associated with that transfer is the rate of transfer dU/dt, which is

$$P = iV \qquad \text{(rate of electrical energy transfer).} \qquad (27\text{-}21)$$

Moreover, this power P is the rate of energy transfer from the battery to the unspecified device. If that device is a

The wire coils within a toaster have appreciable resistance. When current is set up through them, electrical energy is transferred to thermal energy of the coils, increasing their temperature. The coils then emit infrared radiation and visible light that will toast (or burn) bread.

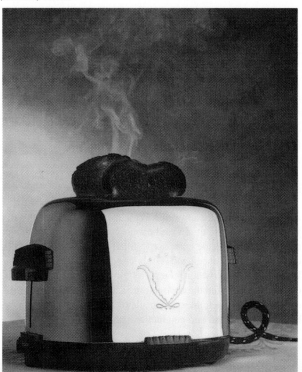

motor connected to a mechanical load, the energy is transferred as work on the load. If the device is a storage battery that is being charged, the energy is transferred to stored chemical energy in the storage battery. If the device is a resistor, the energy is transferred to internal thermal energy, tending to increase the resistor's temperature.

The unit of power that follows from Eq. 27-21 is the volt-ampere (V·A). We can write it as

$$1\text{ V}\cdot\text{A} = \left(1\,\frac{\text{J}}{\text{C}}\right)\left(1\,\frac{\text{C}}{\text{s}}\right) = 1\,\frac{\text{J}}{\text{s}} = 1\text{ W}.$$

The course of an electron moving through a resistor at constant drift speed is much like that of a stone falling through water at constant terminal speed. The average kinetic energy of the electron remains constant, and its lost electric potential energy appears as thermal energy in the resistor and the surroundings. On a microscopic scale this energy transfer is due to collisions between the electron and the molecules of the resistor, which leads to an increase in the temperature of the lattice. The mechanical energy thus transferred to thermal energy is *dissipated* (lost), because the transfer cannot be reversed.

For a resistor we can combine Eqs. 27-8 ($R = V/i$) and 27-21 to obtain, for the rate of electrical energy dissipation in a resistor, either

$$P = i^2R \qquad \text{(resistive dissipation)} \qquad (27\text{-}22)$$

or

$$P = \frac{V^2}{R} \qquad \text{(resistive dissipation).} \qquad (27\text{-}23)$$

However, we must be careful to distinguish these two new equations from Eq. 27-21: $P = iV$ applies to electrical energy transfers of all kinds; $P = i^2R$ and $P = V^2/R$ apply only to the transfer of electric potential energy to thermal energy in a resistance.

CHECKPOINT **5:** A potential difference V is connected across a resistance R, causing current i through the resistance. Rank the following variations according to the change in the rate at which electrical energy is converted to thermal energy in the resistance, greatest change first: (a) V is doubled with R unchanged, (b) i is doubled with R unchanged. (c) R is doubled with V unchanged, (d) R is doubled with i unchanged.

SAMPLE PROBLEM 27-8

You are given a length of uniform heating wire made of a nickel–chromium–iron alloy called Nichrome; it has a resistance R of 72 Ω. At what rate is energy dissipated in each of the following situations? (1) A potential difference of 120 V is applied across the full length of the wire. (2) The wire is cut in half, and a potential difference of 120 V is applied across the length of each half.

SOLUTION: From Eq. 27-23, the rate of energy dissipation in situation 1 is

$$P = \frac{V^2}{R} = \frac{(120 \text{ V})^2}{72 \ \Omega} = 200 \text{ W}. \quad \text{(Answer)}$$

In situation 2, the resistance of each half of the wire is (72 Ω)/2, or 36 Ω. Thus the dissipation rate for each half is

$$P' = \frac{(120 \text{ V})^2}{36 \ \Omega} = 400 \text{ W}. \quad \text{(Answer)}$$

The total power of the two halves is 800 W, or four times that for the full length of wire. This would seem to suggest that you could buy a heating coil, cut it in half, and reconnect it to obtain four times the heat output. Why is this unwise? (What would happen to the amount of current in the coil?)

SAMPLE PROBLEM 27-9

A wire of length $L = 2.35$ m and diameter $d = 1.63$ mm carries a current i of 1.24 A. The wire dissipates electrical energy at the rate P of 48.5 mW. Of what is the wire made?

SOLUTION: We can identify the material by its resistivity. From Eqs. 27-16 and 27-22 we have

$$P = i^2 R = \frac{i^2 \rho L}{A} = \frac{4 i^2 \rho L}{\pi d^2},$$

in which A $(= \frac{1}{4}\pi d^2)$ is the cross-sectional area of the wire. Solving for ρ, the resistivity of the material of which the wire is made, yields

$$\rho = \frac{\pi P d^2}{4 i^2 L} = \frac{(\pi)(48.5 \times 10^{-3} \text{ W})(1.63 \times 10^{-3} \text{ m})^2}{(4)(1.24 \text{ A})^2(2.35 \text{ m})}$$

$$= 2.80 \times 10^{-8} \ \Omega \cdot \text{m}. \quad \text{(Answer)}$$

Inspection of Table 27-1 tells us that the material is aluminum.

27-8 SEMICONDUCTORS

Semiconducting devices are at the heart of the microelectronic revolution that has so influenced our lives. Table 27-2 compares the properties of silicon—a typical semiconductor—and copper—a typical metallic conductor. We see that silicon has many fewer charge carriers, a much higher resistivity, and a temperature coefficient of resistivity that is both large and negative. That is, although the resistivity of copper increases with temperature, that of pure silicon decreases.

TABLE 27-2 SOME ELECTRICAL PROPERTIES OF COPPER AND SILICON[a]

PROPERTY	COPPER	SILICON
Type of material	Metal	Semiconductor
Charge carrier density, m^{-3}	9×10^{28}	1×10^{16}
Resistivity, $\Omega \cdot$m	2×10^{-8}	3×10^{3}
Temperature coefficient of resistivity, K^{-1}	$+4 \times 10^{-3}$	-70×10^{-3}

[a]Rounded to one significant figure for easy comparison.

The resistivity of pure silicon is so high that it is virtually an insulator and is thus not of much direct use in microelectronic circuits. The property that makes it useful is that—as Table 27-1 shows—its resistivity can be reduced in a controlled way by adding minute amounts of specific foreign "impurity" atoms, a process called *doping*.

We can explain the difference in resistivity (hence conductivity) between semiconductors and metallic conductors in terms of the energy levels of their electrons. We saw in Section 8-9 that the energies of electrons in isolated atoms are *quantized*; that is, they are restricted to certain values, *or levels*, as shown in Fig. 8-17. An electron can *occupy* (that is, have the energy of) any one of the energy levels but cannot have an intermediate energy.

Electrons in solids also occupy quantized levels, but the proximity of the atoms to each other tends to "squeeze" the many levels into a few *bands* (Fig. 27-14). An electron can occupy an energy level within a band but cannot have an energy value within the *gaps* separating the

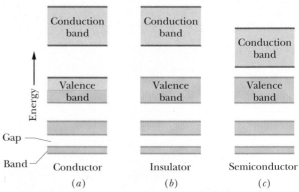

FIGURE 27-14 The allowed energy levels for the electrons in a solid form a pattern of allowed bands and forbidden gaps. Green denotes a partially or completely filled band. (*a*) In a metallic conductor, the valence band is only partially filled. (*b*) In an insulator, the valence band is completely filled and the gap between the valence band and the conduction band is relatively large. (*c*) A semiconductor resembles an insulator except that the gap between valence band and conduction band is relatively small.

bands. Moreover, the number of electrons that can occupy an energy level is limited by quantum physics. So, an electron can become more energetic *only* if it receives enough energy to reach a *vacant* higher energy level, either in the same band or in a higher band.

In a metallic conductor such as copper (Fig. 27-14*a*), the highest band that is occupied by electrons—called the *valence band*—has electrons only in its lower levels. Thus, those electrons can reach abundant vacant levels higher in the band if they receive even a modest amount of energy. We can supply such energy via an electric field applied across the conductor: the field propels some of the valence band electrons along the wire, giving them kinetic energy and thus elevating them to a higher energy level. Hence, these electrons are the conduction electrons that comprise the current through the conductor. Electrons in lower bands cannot participate in the current because the applied electric field cannot provide enough energy for them to reach vacant levels.

In an insulator (Fig. 27-14*b*), the valence band is completely filled. The next higher available vacant levels lie in an empty band (called the *conduction band*) separated from the valence band by a considerable energy gap. No current can occur when an electric field is applied because the field cannot provide enough energy for electrons to jump up to a vacant level.

A semiconductor (Fig. 27-14*c*) is like an insulator except that the energy gap between the conduction band and the valence band is small enough that the probability that electrons might ''jump the gap'' by thermal agitation is not vanishingly small. More important, controlled impurities —deliberately added—can contribute charge carriers to the conduction band. Most semiconducting devices, such as transistors and junction diodes, are fabricated by the selective doping of different regions of the silicon with impurity atoms of different kinds.

Let us now look again at Eq. 27-20, the expression for the resistivity of a conductor, with the band-gap picture in mind:

$$\rho = \frac{m}{e^2 n \tau}, \tag{27-24}$$

where n is the number of charge carriers per unit volume and τ is the mean time between collisions of the charge carriers. (We derived this equation for conductors, but it also applies to semiconductors.) Let us consider how the variables n and τ change as the temperature is increased.

In a conductor, n is large but very nearly constant; that is, its value does not change appreciably with temperature. The increase of resistivity with temperature for metals (Fig. 27-10) is caused by an increase in the collision rate of the charge carriers, which shows up in Eq. 27-24 as a decrease in τ, the mean time between collisions.

In a semiconductor, n is small but increases very rapidly with temperature as the increased thermal agitation makes more charge carriers available. This causes the *decrease* of resistivity with increasing temperature, as indicated by the negative temperature coefficient of resistivity for silicon in Table 27-2. The same increase in collision rate that we noted for metals also occurs for semiconductors, but its effect is swamped by the rapid increase in the number of charge carriers.

27-9 SUPERCONDUCTORS

In 1911 Dutch physicist Kamerlingh Onnes discovered that the resistivity of mercury absolutely disappears at temperatures below about 4 K (Fig. 27-15). This phenomenon of **superconductivity** is of vast potential importance in technology because it would be very useful to be able to cause charge to flow through a superconducting conductor without thermal energy losses. Currents created in a superconducting ring, for example, have persisted for several years without diminution; the electrons making up the current require a force and a source of energy at start-up time, but not thereafter.

Prior to 1986, the technological development of superconductivity was throttled by the cost of producing the extremely low temperatures that were required to achieve the effect. In 1986, however, new ceramic materials were discovered that become superconducting at considerably higher (and thus cheaper to produce) temperatures. Practical application of superconducting devices at room temperature may eventually become feasible.

Superconductivity is much different from conductivity. In fact, the best of the normal conductors, such as silver and copper, cannot become superconducting at any temperature, and the new ceramic superconductors are actually insulators when they are not at low enough temperatures to be in a superconducting state.

One explanation for superconductivity is that the electrons that make up the current move in coordinated pairs. One of the electrons in a pair may electrically distort the molecular structure of the superconducting material as it moves through, creating nearby a short-lived concentration

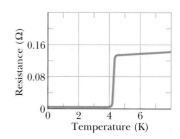

FIGURE 27-15 The resistance of mercury drops to zero at a temperature of about 4 K.

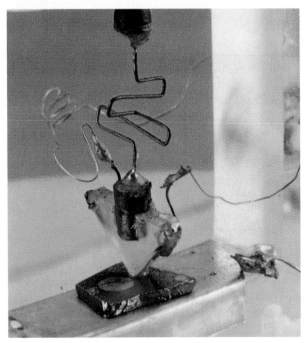

A model of the first transistor, an electronic device using semiconductor materials. Today, many thousands of these devices can be placed on a thin wafer a few millimeters wide.

A disk-shaped magnet is levitated above a superconducting material that has been cooled by liquid nitrogen. The goldfish is along for the ride.

of positive charge. The other electron in the pair may then be attracted toward this positive charge. According to the theory, such coordination between electrons would prevent them from colliding with the molecules and thus would eliminate electrical resistance. The theory worked well to explain the pre-1986, lower temperature superconductors, but new theories appear to be needed for the newer, higher temperature superconductors.

REVIEW & SUMMARY

Current

An **electric current** i in a conductor is defined by

$$i = \frac{dq}{dt}. \qquad (27\text{-}1)$$

Here dq is the amount of (positive) charge that passes in time dt through a hypothetical surface that cuts across the conductor. By convention, the direction of electric current is taken as the direction in which positive charge carriers would move. The SI unit of electric current is the **ampere** (A): 1 A = 1 C/s.

Current Density

Current (a scalar) is related to **current density J** (a vector) by

$$i = \int \mathbf{J} \cdot d\mathbf{A}, \qquad (27\text{-}4)$$

where $d\mathbf{A}$ is a vector perpendicular to a surface element of area dA, and the integral is taken over any surface cutting across the conductor. The direction of **J** is that of the electric field causing the current.

Drift Speed of the Charge Carriers

When an electric field **E** is established in a conductor, the charge carriers (assumed positive) acquire a **drift speed** v_d in the direction of **E**; the velocity $\mathbf{v}_d$ is related to the current density by

$$\mathbf{J} = (ne)\mathbf{v}_d, \qquad (27\text{-}7)$$

where ne is the carrier charge density.

Resistance of a Conductor

The **resistance** R of a conductor is defined as

$$R = \frac{V}{i} \qquad \text{(definition of } R\text{)}, \qquad (27\text{-}8)$$

where V is the potential difference across the conductor and i is the current. The SI unit of resistance is the **ohm** (Ω): 1 Ω = 1 V/A. Similar equations define the **resistivity** ρ and **conductivity** σ of a material:

$$\rho = \frac{1}{\sigma} = \frac{E}{J} \qquad \begin{array}{c}\text{(definitions} \\ \text{of } \rho \text{ and } \sigma\text{)},\end{array} \qquad (27\text{-}12, 27\text{-}10)$$

where E is the applied electric field. The SI unit of resistivity is the ohm-meter ($\Omega \cdot$m). Equation 27-10 corresponds to the vector

equation

$$\mathbf{E} = \rho\mathbf{J}. \qquad (27\text{-}11)$$

The resistance R of a conducting wire of length L and uniform cross section is

$$R = \frac{\rho L}{A}, \qquad (27\text{-}16)$$

where A is the cross-sectional area.

Change of ρ with Temperature

The resistivity ρ for most materials changes with temperature. For many materials, including metals, the relation between ρ and temperature T is approximated by the equation is

$$\rho - \rho_0 = \rho_0\alpha(T - T_0). \qquad (27\text{-}17)$$

Here T_0 is a reference temperature, ρ_0 is the resistivity at T_0, and α is a mean temperature coefficient of resistivity.

Ohm's Law

A given *conductor* obeys *Ohm's law* if its resistance R, defined by Eq. 27-8 as V/i, is independent of the applied potential difference V. A given *material* obeys Ohm's law if its resistivity, defined by Eq. 27-10, is independent of the magnitude and direction of the applied electric field $\mathbf{E}$.

Resistivity of a Metal

By assuming that the conduction electrons in a metal are free to move like the molecules of a gas, it is possible to derive an expression for the resistivity of a metal:

$$\rho = \frac{m}{e^2 n \tau}. \qquad (27\text{-}20)$$

Here n is the number of electrons per unit volume and τ is the mean time between the collisions of an electron with the atoms of the metal. We can explain why metals obey Ohm's law by pointing out that τ is essentially independent of E.

Power

The power P, or rate of energy transfer, in an electrical device across which a potential difference V is maintained is

$$P = iV \qquad \begin{array}{l}\text{(rate of electrical}\\ \text{energy transfer).}\end{array} \qquad (27\text{-}21)$$

Resistive Dissipation

If the device is a resistor, we can write Eq. 27-21 as

$$P = i^2 R = \frac{V^2}{R} \qquad \begin{array}{l}\text{(resistive}\\ \text{dissipation).}\end{array} \qquad (27\text{-}22,\ 27\text{-}23)$$

In a resistor, electric potential energy is converted to internal thermal energy via collisions between charge carriers and atoms.

Semiconductors

Semiconductors are materials with few conduction electrons but with available conduction-level states that are close, in energy, to their valence bands. These materials become conductors when they are *doped* with other atoms that contribute electrons to the conduction band.

Superconductors

Superconductors are materials that lose all electrical resistance at low temperatures. Recent research has discovered materials that are superconducting at surprisingly high temperatures.

QUESTIONS

1. Figure 27-16 shows plots of the current i through the cross section of a wire over four different time periods. Rank the periods according to the net charge that passes through the cross section during each, greatest first.

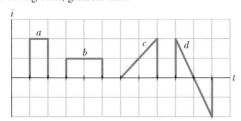

FIGURE 27-16   Question 1.

2. Figure 27-17 shows four situations in which positive and negative charges move horizontally through a region and gives the rate at which each charge moves. Rank the situations according to the effective current through the regions, greatest first.

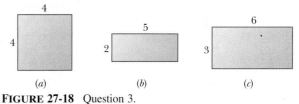

FIGURE 27-17   Question 2.

3. Figure 27-18 shows cross sections through three wires of equal length and of the same material. The figure also gives the length of each side in millimeters. Rank the wires according to their resistances (measured end to end along each wire's length), greatest first.

4

4

5

2

6

3

(a)

(b)

(c)

FIGURE 27-18   Question 3.

4. If you stretch a cylindrical wire and it remains cylindrical, does the resistance of the wire (measured end to end along its length) increase, decrease, or remain the same?

5. Figure 27-19 shows cross sections through three long square conductors of the same length and material, and with cross-sectional edge lengths as shown. Conductor B will fit snugly within conductor A, and conductor C will fit snugly within conductor B. Rank the following according to their end-to-end resistances, greatest first: the individual conductors and the combinations of $A + B$, $B + C$, and $A + B + C$.

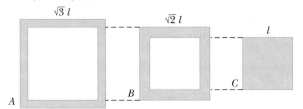

FIGURE 27-19 Question 5.

6. Figure 27-20 shows a rectangular solid conductor of edge lengths L, $2L$, and $3L$. A certain potential difference V is to be applied between pairs of opposite faces of the conductor as in Fig. 27-8b: left–right, top–bottom, and front–back. Rank those pairs according to (a) the magnitude of the electric field within the conductor, (b) the current density within the conductor, (c) the current through the conductor, and (d) the drift speed of the electrons through the conductor, greatest first.

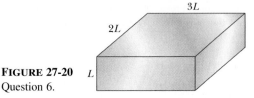

FIGURE 27-20 Question 6.

7. The following table gives the lengths of three copper rods, their diameters, and the potential differences between their ends. Rank the rods according to (a) the magnitude of the electric field within them, (b) the current density within them, and (c) the drift speed of electrons through them, greatest first.

ROD	LENGTH	DIAMETER	POTENTIAL DIFFERENCE
1	L	$3d$	V
2	$2L$	d	$2V$
3	$3L$	$2d$	$2V$

8. The following table gives the conductivity and the density of electrons for materials A, B, C, and D. Rank the materials according to the average time between collisions of the conduction electrons in the materials, greatest first.

	A	B	C	D
Conductivity	σ	2σ	2σ	σ
Electrons/m³	n	$2n$	n	$2n$

9. Three wires, of the same diameter, are connected in turn between two points maintained at a constant potential difference. Their resistivities and lengths are ρ and L (wire A), 1.2ρ and $1.2L$ (wire B), and 0.9ρ and L (wire C). Rank the wires according to the rate at which energy is transferred to thermal energy within them, greatest first.

10. In Fig. 27-21a, battery B_1 is recharging battery B_2. The current through B_2 and the potential across B_2 may be (a) 3 A and 4 V, (b) 2 A and 5 V, or (c) 6 A and 2 V. Rank these pairs of values according to the rate at which electrical energy is transferred from B_1 to B_2, greatest first.

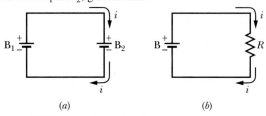

FIGURE 27-21 Questions 10 and 11.

11. In three situations, a battery B and a resistor of resistance R are connected as in Fig. 27-21b. The values of R and the current through the resistor in the three situations are (a) 4 Ω and 2 A, (b) 3 Ω and 3 A, and (c) 3 Ω and 2 A. Rank the situations according to the rate at which electrical energy is transferred to thermal energy in the resistor, greatest first.

12. Is the filament resistance lower or higher in a 500 W light-bulb than in a 100 W bulb? (The same potential difference is applied to them.)

13. Figure 27-22 gives the resistivities of four materials as a function of temperature. (a) Which materials are conductors and which semiconductors? In which materials does an increase in temperature result in (b) an increase in the number of conduction electrons per unit volume and (c) an increase in the collision rate of conduction electrons?

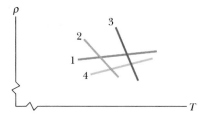

FIGURE 27-22 Question 13.

EXERCISES & PROBLEMS

SECTION 27-2 Electric Current

1E. The current in the electron beam producing a picture on a typical video display terminal is 200 μA. How many electrons strike the screen each second?

2E. A current of 5.0 A exists in a 10 Ω resistor for 4.0 min. How many (a) coulombs and (b) electrons pass through any cross section of the resistor in this time?

3P. A charged belt, 50 cm wide, travels at 30 m/s between a source of charge and a sphere. The belt carries charge into the sphere at a rate corresponding to 100 μA. Compute the surface charge density on the belt.

4P. An isolated conducting sphere has a 10 cm radius. One wire carries a current of 1.000 002 0 A into it. Another wire carries a current of 1.000 000 0 A out of it. How long would it take for the sphere to increase in potential by 1000 V?

SECTION 27-3 Current Density

5E. The (United States) National Electric Code, which sets maximum safe currents for rubber-insulated copper wires of various diameters, is given (in part) below. Plot the safe current density as a function of diameter. Which wire gauge has the maximum safe current density?

Gauge[a]	4	6	8	10	12	14	16	18
Diameter, mils[b]	204	162	129	102	81	64	51	40
Safe current, A	70	50	35	25	20	15	6	3

[a]A way of identifying the wire diameter.

[b]1 mil = 10^{-3} in.

6E. A beam contains 2.0×10^8 doubly charged positive ions per cubic centimeter, all of which are moving north with a speed of 1.0×10^5 m/s. (a) What are the magnitude and direction of the current density **J**? (b) Can you calculate the total current i in this ion beam? If not, what additional information is needed?

7E. A small but measurable current of 1.2×10^{-10} A exists in a copper wire whose diameter is 2.5 mm. Assuming the current is uniform, calculate (a) the current density and (b) the electron drift speed. (See Sample Problem 27-3.)

8E. A fuse in an electric circuit is a wire that is designed to melt, and thereby open the circuit, if the current exceeds a predetermined value. Suppose that the material to be used in a fuse melts when the current density rises to 440 A/cm². What diameter of cylindrical wire should be used to limit the current to 0.50 A?

9E. A current is established in a gas discharge tube when a sufficiently high potential difference is applied across the two electrodes in the tube. The gas ionizes; electrons move toward the positive terminal and singly charged positive ions toward the negative terminal. What are the magnitude and direction of the current in a hydrogen discharge tube in which 3.1×10^{18} electrons and 1.1×10^{18} protons move past a cross-sectional area of the tube each second?

10E. A *pn* junction is formed from two different semiconducting materials in the form of identical cylinders with radius 0.165 mm, as depicted in Fig. 27-23. In one application 3.50×10^{15} electrons per second flow across the junction from the *n* to the *p* side while 2.25×10^{15} holes per second flow from the *p* to the *n* side. (A hole acts like a particle with charge $+1.60 \times 10^{-19}$ C.) What are (a) the total current and (b) the current density?

FIGURE 27-23
Exercise 10.

11P. Near Earth, the density of protons in the solar wind is 8.70 cm⁻³ and their speed is 470 km/s. (a) Find the current density of these protons. (b) If Earth's magnetic field did not deflect them, the protons would strike the planet. What total current would Earth then receive?

12P. A steady beam of alpha particles ($q = +2e$) traveling with constant kinetic energy 20 MeV carries a current of 0.25 μA. (a) If the beam is directed perpendicular to a plane surface, how many alpha particles strike the surface in 3.0 s? (b) At any instant, how many alpha particles are there in a given 20 cm length of the beam? (c) Through what potential difference was it necessary to accelerate each alpha particle from rest to bring it to an energy of 20 MeV?

13P. How long does it take electrons to get from a car battery to the starting motor? Assume the current is 300 A and the electrons travel through a copper wire with cross-sectional area 0.21 cm² and length 0.85 m. (See Sample Problem 27-3.)

14P. In a hypothetical fusion research lab, high-temperature helium gas is completely ionized, each helium atom being separated into two free electrons and the remaining positively charged nucleus (alpha particle). An applied electric field causes the alpha particles to drift to the east at 25 m/s while the electrons drift to the west at 88 m/s. The alpha particle density is 2.8×10^{15} cm⁻³. Calculate the net current density; specify the current direction.

15P. (a) The current density across a cylindrical conductor of radius R varies according to the equation

$$J = J_0\left(1 - \frac{r}{R}\right),$$

where r is the distance from the central axis. Thus the current density is a maximum J_0 at the axis ($r = 0$) and decreases linearly to zero at the surface ($r = R$). Calculate the current in terms of J_0 and the conductor's cross-sectional area $A = \pi R^2$. (b) Suppose that, instead, the current density is a maximum J_0 at the cylinder's surface and decreases linearly to zero at the axis: $J = J_0 r/R$. Calculate the current. Why is the result different from that in (a)?

SECTION 27-4 Resistance and Resistivity

16E. A steel trolley-car rail has a cross-sectional area of 56.0 cm². What is the resistance of 10.0 km of rail? The resistivity of the steel is 3.00×10^{-7} $\Omega \cdot$m.

17E. A conducting wire has a 1.0 mm diameter, a 2.0 m length, and a 50 mΩ resistance. What is the resistivity of the material?

18E. A wire of Nichrome (a nickel–chromium–iron alloy commonly used in heating elements) is 1.0 m long and 1.0 mm² in cross-sectional area. It carries a current of 4.0 A when a 2.0 V potential difference is applied between its ends. Calculate the conductivity σ of Nichrome.

19E. A human being can be electrocuted if a current as small as 50 mA passes near the heart. An electrician working with sweaty hands makes good contact with the two conductors he is holding. If his resistance is 2000 Ω, what might the fatal voltage be?

20E. A coil is formed by winding 250 turns of insulated 16-gauge copper wire (diameter = 1.3 mm) in a single layer on a cylindrical form of radius 12 cm. What is the resistance of the coil? Neglect the thickness of the insulation. (Use Table 27-1.)

21E. A wire 4.00 m long and 6.00 mm in diameter has a resistance of 15.0 mΩ. A potential difference of 23.0 V is applied between the ends. (a) What is the current in the wire? (b) What is the current density? (c) Calculate the resistivity of the wire material. Identify the material. (Use Table 27-1.)

22E. The copper windings of a motor have a resistance of 50 Ω at 20°C when the motor is idle. After the motor has run for several hours, the resistance rises to 58 Ω. What is the temperature of the windings now? Ignore changes in the dimensions of the windings. (Use Table 27-1.)

23E. (a) At what temperature would the resistance of a copper conductor be double its resistance at 20.0°C? (Use 20.0°C as the reference point in Eq. 27-17; compare your answer with Fig. 27-10.) (b) Does this same "doubling temperature" hold for all copper conductors, regardless of shape or size?

24E. Using data taken from Fig. 27-11c, plot the resistance of the pn junction diode as a function of applied potential difference.

25E. A 4.0 cm long caterpillar crawls in the direction of electron drift along a 5.2 mm diameter bare copper wire that carries a current of 12 A. (a) What is the potential difference between the two ends of the caterpillar? (b) Is its tail positive or negative compared to its head? (c) How much time would the caterpillar take to crawl 1.0 cm if it crawls at the drift speed of the electrons in the wire?

26E. A cylindrical copper rod of length L and cross-sectional area A is re-formed to twice its original length with no change in volume. (a) Find the new cross-sectional area. (b) The resistance between its ends was R; what is it now?

27E. A wire with a resistance of 6.0 Ω is drawn out through a die so that its new length is three times its original length. Find the resistance of the longer wire, assuming that the resistivity and density of the material are unchanged.

28E. A certain wire has a resistance R. What is the resistance of a second wire, made of the same material, that is half as long and has half the diameter?

29P. Two conductors are made of the same material and have the same length. Conductor A is a solid wire of diameter 1.0 mm. Conductor B is a hollow tube of outside diameter 2.0 mm and inside diameter 1.0 mm. What is the resistance ratio R_A/R_B, measured between their ends?

30P. A copper wire and an iron wire of the same length have the same potential difference applied to them. (a) What must be the ratio of their radii if the currents in the two wires are to be the same? (b) Can the current densities be made the same by suitable choices of the radii?

31P. An aluminum rod with a square cross section is 1.3 m long and 5.2 mm on edge. (a) What is the resistance between its ends? (b) What must be the diameter of a cylindrical copper rod of length 1.3 m if its resistance is to be the same as that of the aluminum rod?

32P. A cylindrical metal rod is 1.60 m long and 5.50 mm in diameter. The resistance between its two ends (at 20°C) is 1.09×10^{-3} Ω. (a) What is the material? (b) A round disk, 2.00 cm in diameter and 1.00 mm thick, is formed of the same material. What is the resistance between the round faces, assuming that each face is an equipotential surface?

33P. An electrical cable consists of 125 strands of fine wire, each having 2.65 $\mu\Omega$ resistance. The same potential difference is applied between the ends of all the strands and results in a total current of 0.750 A. (a) What is the current in each strand? (b) What is the applied potential difference? (c) What is the resistance of the cable?

34P. When 115 V is applied across a wire that is 10 m long and has a 0.30 mm radius, the current density is 1.4×10^4 A/m². Find the resistivity of the wire.

35P. A common flashlight bulb is rated at 0.30 A and 2.9 V (the values of the current and voltage under operating conditions). If the resistance of the bulb filament at room temperature (20°C) is 1.1 Ω, what is the temperature of the filament when the bulb is on? The filament is made of tungsten.

36P. A block in the shape of a rectangular solid has a cross-sectional area of 3.50 cm² across its width, a front-to-rear length of 15.8 cm, and a resistance of 935 Ω. The material of which the block is made has 5.33×10^{22} conduction electrons/m³. A potential difference of 35.8 V is maintained between its front and rear. (a) What is the current in the block? (b) If the current density is uniform, what is its value? (c) What is the drift velocity of the conduction electrons? (d) What is the magnitude of the electric field in the block?

37P. Copper and aluminum are being considered for a high-voltage transmission line that must carry a current of 60.0 A. The resistance per unit length is to be 0.150 Ω/km. Compute for each choice of cable material (a) the current density and (b) the mass per meter of the cable. The densities of copper and aluminum are 8960 and 2700 kg/m³, respectively.

38P. In Earth's lower atmosphere there are negative and positive ions, created by radioactive elements in the soil and cosmic rays from space. In a certain region, the atmospheric electric field strength is 120 V/m, directed vertically down. This field causes singly charged positive ions, 620 per cm³, to drift downward and singly charged negative ions, 550 per cm³, to drift upward (Fig. 27-24). The measured conductivity is $2.70 \times 10^{-14}/\Omega \cdot m$. Calculate (a) the ion drift speed, assumed to be the same for positive and negative ions, and (b) the current density.

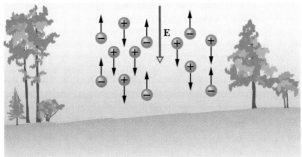

FIGURE 27-24 Problem 38.

39P. If the gauge number of a wire is increased by 6, the diameter is halved; if a gauge number is increased by 1, the diameter decreases by the factor $2^{1/6}$ (see the table in Exercise 5). Knowing this, and knowing that 1000 ft of 10-gauge copper wire has a resistance of approximately 1.00 Ω, estimate the resistance of 25 ft of 22-gauge copper wire.

40P. When a metal rod is heated, not only its resistance but also its length and its cross-sectional area change. The relation $R = \rho L/A$ suggests that all three factors should be taken into account in measuring ρ at various temperatures. (a) If the temperature changes by 1.0 C°, what percentage changes in R, L, and A occur for a copper conductor? The coefficient of linear expansion is $1.7 \times 10^{-5}/K$. (b) What conclusion do you draw?

41P. A resistor has the shape of a truncated right-circular cone (Fig. 27-25). The end radii are a and b, and the altitude is L. If the taper is small, we may assume that the current density is uniform across any cross section. (a) Calculate the resistance of this object. (b) Show that your answer reduces to $\rho(L/A)$ for the special case of zero taper (that is, for $a = b$).

FIGURE 27-25 Problem 41.

SECTION 27-6 A Microscopic View of Ohm's Law

42P. Show that according to the free-electron model of electrical conduction in metals and classical physics, the resistivity of metals should be proportional to $\sqrt{T}$, where T is the temperature in kelvins. (See Eq. 20-27.)

SECTION 27-7 Power in Electric Circuits

43E. A student kept his 9.0 V, 7.0 W radio turned on at full volume from 9:00 p.m. until 2:00 a.m. How much charge went through it?

44E. A certain x-ray tube operates at a current of 7.0 mA and a potential difference of 80 kV. What is its power in watts?

45E. Thermal energy is developed in a resistor at a rate of 100 W when the current is 3.00 A. What is the resistance?

46E. The headlights of a moving car draw about 10 A from the 12 V alternator, which is driven by the engine. Assume the alternator is 80% efficient (its output electrical power is 80% of its input mechanical power), and calculate the horsepower the engine must supply to run the lights.

47E. A 120 V potential difference is applied to a space heater whose resistance is 14 Ω when hot. (a) At what rate is electrical energy transferred to heat? (b) At 5.0¢/kW·h, what does it cost to operate the device for 5.0 h?

48E. A 120 V potential difference is applied to a space heater that dissipates 500 W during operation. (a) What is its resistance during operation? (b) At what rate do electrons flow through any cross section of the heater element?

49E. An unknown resistor is connected between the terminals of a 3.00 V battery. Energy is dissipated in the resistor at the rate of 0.540 W. The same resistor is then connected between the terminals of a 1.50 V battery. At what rate is energy now dissipated?

50E. The National Board of Fire Underwriters has fixed safe current-carrying capacities for various sizes and types of wire. For 10-gauge rubber-coated copper wire (diameter = 0.10 in.), the maximum safe current is 25 A. At this current, find (a) the current density, (b) the electric field, (c) the potential difference across 1000 ft of wire, and (d) the rate at which thermal energy is developed in 1000 ft of wire.

51E. A potential difference of 1.20 V will be applied to a 33.0 m length of 18-gauge copper wire (diameter = 0.0400 in.). Calculate (a) the current, (b) the current density, (c) the electric field, and (d) the rate at which thermal energy will appear in the wire.

52P. A potential difference V is applied to a wire of cross section A, length L, and resistivity ρ. You want to change the applied potential difference and stretch the wire so that the energy dissipation rate is multiplied by 30 and the current is multiplied by 4. What should be the new values of L and A?

53P. A cylindrical resistor of radius 5.0 mm and length 2.0 cm is made of material that has a resistivity of 3.5×10^{-5} Ω·m. What are (a) the current density and (b) the potential difference when the energy dissipation rate in the resistor is 1.0 W?

54P. A heating element is made by maintaining a potential difference of 75.0 V along the length of a Nichrome wire with a 2.60×10^{-6} m² cross section and a resistivity of 5.00×10^{-7} Ω·m. (a) If the element dissipates 5000 W, what is its length? (b) If a potential difference of 100 V is used to obtain the same dissipation rate, what should the length be?

55P. A 100 W lightbulb is plugged into a standard 120 V outlet. (a) How much does it cost per month to leave the light turned on continuously? Assume electrical energy costs 6¢/kW·h. (b) What is the resistance of the bulb? (c) What is the current in the bulb? (d) Is the resistance different when the bulb is turned off?

56P. A 1250 W radiant heater is constructed to operate at 115 V. (a) What will be the current in the heater? (b) What is the resistance of the heating coil? (c) How much thermal energy is generated in 1.0 h by the heater?

57P. A Nichrome heater dissipates 500 W when the applied potential difference is 110 V and the wire temperature is 800°C. What would be the dissipation rate if the wire temperature were held at 200°C by immersing the wire in a bath of cooling oil? The applied potential difference remains the same, and α for Nichrome at 800°C is 4.0×10^{-4}/K.

58P. A beam of 16 MeV deuterons from a cyclotron falls on a copper block. The beam is equivalent to a current of 15 μA. (a) At what rate do deuterons strike the block? (b) At what rate is thermal energy produced in the block?

59P. A linear accelerator produces a pulsed beam of electrons. The pulse current is 0.50 A, and the pulse duration is 0.10 μs. (a) How many electrons are accelerated per pulse? (b) What is the average current for a machine operating at 500 pulses/s? (c) If the electrons are accelerated to an energy of 50 MeV, what are the average and peak powers of the accelerator?

60P. A coil of current-carrying Nichrome wire is immersed in a liquid contained in a calorimeter. When the potential difference across the coil is 12 V and the current through the coil is 5.2 A, the liquid boils at a steady rate, evaporating at the rate of 21 mg/s. Calculate the heat of vaporization of the liquid, in joules per kilogram (see Section 19-7).

61P. In Fig. 27-26, a resistance coil, wired to an external battery, is placed inside a thermally insulated cylinder fitted with a fric-tionless piston and containing an ideal gas. A current $i = 240$ mA exists in the coil, which has a resistance $R = 550$ Ω. At what speed v must the piston, of mass $m = 12$ kg, move upward to keep the temperature of the gas unchanged?

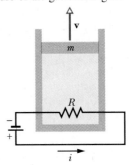

FIGURE 27-26   Problem 61.

62P. A 500 W heating unit is designed to operate with an applied potential difference of 115 V. (a) By what percentage will its heat output drop if the applied potential difference drops to 110 V? Assume no change in resistance. (b) If you took the variation of resistance with temperature into account, would the actual drop in heat output be larger or smaller than that calculated in (a)?

Electronic Computation

63. The resistance of a resistor is measured at several temperatures, as shown below. Enter the data in your graphing calculator and perform a linear regression fit of R versus T. Have your calculator graph the results of the linear regression fit; using the TRACE capability of the calculator (and perhaps the parameters of the fit), find the value of the resistance (a) at 20°C and (b) at 0°C. (c) Find the temperature coefficient *of resistance* (instead of resistivity) with a reference temperature of 20°C. (d) Find the temperature coefficient of resistance with a reference temperature of 0°C. (e) Find the resistance of the resistor at 265°C.

T, °C	50	100	150	200	250	300
R, Ω	139	171	203	234	266	298

28
Circuits

The electric eel (<u>Electrophorus</u>) lurks in rivers of South America, killing the fish on which it preys with pulses of current. It does so by producing a potential difference of several hundred volts along its length; the resulting current in the surrounding water, from near the eel's head to the tail region, can be as much as one ampere. If you were to brush up against this eel while swimming, you might wonder (after recovering from the very painful stun): How can the creature manage to produce a current that large without shocking itself?

28-1 "PUMPING" CHARGES

If you want to make charge carriers flow through a resistor, you must establish a potential difference between the ends of the device. One way to do this is to connect each end of the resistor to a separate conducting sphere, with one sphere charged negatively and the other positively. The trouble with this scheme is that the flow of charge acts to discharge the spheres, bringing them quickly to the same potential. When that happens, the flow of charge stops.

To produce a steady flow of charge, you need a "charge pump," a device that—by doing work on the charge carriers—maintains a potential difference between a pair of terminals. We call such a device an **emf device,** and the device is said to provide an **emf** $\mathcal{E}$, which means that it does work on charge carriers. An emf device is sometimes called a *seat of emf*. The term *emf* comes from the outdated phrase *electromotive force,* which was adopted before scientists clearly understood the function of an emf device.

A common emf device is the *battery,* used to power devices from wristwatches to submarines. The emf device that most influences our daily lives, however, is the *electric generator,* which by means of electrical lines from a generating plant, creates a potential difference in our homes and workplaces. The emf devices known as *solar cells,* long familiar as the winglike panels on spacecraft, also dot the countryside for domestic applications. Less familiar emf devices are the *fuel cells* that power the space shuttles and the *thermopiles* that provide onboard electrical power for some spacecraft and for remote stations in Antarctica and elsewhere. An emf device does not have to be an instrument: living systems, ranging from electric eels and human beings to plants, have physiological emf devices.

Although the devices we have listed differ widely in their modes of operation, they all perform the same basic function: they do work on charge carriers and thus maintain a potential difference between their terminals.

28-2 WORK, ENERGY, AND EMF

Figure 28-1 shows an emf device (consider it to be a battery) that is part of a simple circuit. The device keeps one

terminal (called the positive terminal and often labeled +) at a higher electric potential than the other terminal (called the negative terminal and labeled −). We can represent the emf of the device with an arrow that points from the negative terminal toward the positive terminal as in Fig. 28-1. This is the direction in which the device causes positive charge carriers (which make up a current) to move through itself. The device also produces a current around the circuit in the same direction (clockwise in Fig. 28-1). A small circle on the emf arrow distinguishes it from the arrows that indicate current direction.

Within the emf device, positive charge carriers move from a region of low electric potential and thus low electric potential energy (at the negative terminal) to a region of higher electric potential and higher electric potential energy (at the positive terminal). This motion is just the opposite of what the electric field between the terminals (which points from the positive terminal toward the negative terminal) would cause the charge carriers to do.

So there must be some source of energy within the device, enabling it to do work on the charges and thus

The world's largest battery, housed in Chino, California, has a power capability of 10 MW, which is put to use during peak power demands on the electric system served by Southern California Edison. Because the battery does work on charge carriers, it is an emf device.

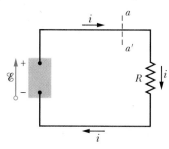

FIGURE 28-1 A simple electric circuit, in which a device of emf $\mathcal{E}$ does work on the charge carriers and maintains a steady current i through the resistor.

forcing them to move as they do. The energy source may be chemical, as in a battery or a fuel cell. It may involve mechanical forces, as in an electric generator. Temperature differences may supply the energy, as in a thermopile; or the Sun may supply it, as in a solar cell.

Let us now analyze the circuit of Fig. 28-1 from the point of view of work and energy transfers. In any time interval dt, a charge dq passes through any cross section of this circuit, such as aa'. This same amount of charge must enter the emf device at its low-potential end and leave at its high-potential end. The device must do an amount of work dW on the charge dq to force it to move in this way. We define the emf of the emf device in terms of this work:

$$\mathscr{E} = \frac{dW}{dq} \qquad \text{(definition of } \mathscr{E}\text{).} \qquad (28\text{-}1)$$

In words, the emf of an emf device is the work per unit charge that the device does in moving charge from its low-potential terminal to its high-potential terminal. The SI unit for emf is the joule per coulomb; in Chapter 25 we defined that unit as the *volt*.

An **ideal emf device** is one that lacks any internal resistance to the internal movement of charge from terminal to terminal. The potential difference between the terminals of an ideal emf device is equal to the emf of the device. For example, an ideal battery with an emf of 12.0 V always has a potential difference of 12.0 V between its terminals.

A **real emf device,** such as any real battery, has internal resistance to the internal movement of charge. When a real emf device is not connected to a circuit, and thus does not have current through it, the potential difference between its terminals is equal to its emf. But when that device has current through it, the potential difference between its terminals differs from its emf. We will discuss such real batteries in Section 28-4.

When an emf device is connected to a circuit, the device transfers energy to the charge carriers passing through it. This energy can then be transferred from the charge carriers to other devices in the circuit, for example, to light a bulb. Figure 28-2a shows a circuit containing two ideal rechargeable (*storage*) batteries A and B, a resistor R, and an electric motor M that can lift an object by using energy it obtains from charge carriers in the circuit. Note that the batteries are connected so that they tend to send charges around the circuit in opposite directions. The actual direction of the current in the circuit is determined by the battery with the larger emf, which happens to be battery B. So the chemical energy within battery B is decreasing as energy is transferred to the charge carriers passing through it. But the chemical energy within battery A is increasing

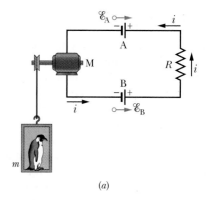

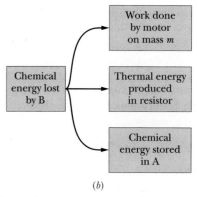

FIGURE 28-2 (*a*) In the circuit, $\mathscr{E}_B > \mathscr{E}_A$; so battery B determines the direction of the current. (*b*) The energy transfers in the circuit, assuming that no dissipation occurs in the motor.

because the current in it is directed from the positive terminal to the negative terminal. Thus battery B is charging battery A. Battery B is also providing energy to motor M and energy that is being dissipated in resistor R. Figure 28-2b shows all three energy transfers from battery B; each decreases that battery's chemical energy.

28-3 CALCULATING THE CURRENT IN A SINGLE-LOOP CIRCUIT

We discuss here two equivalent ways to calculate the current in the simple *single-loop* circuit of Fig. 28-3; one method is based on energy conservation considerations and

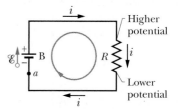

FIGURE 28-3 A single-loop circuit in which a resistance R is connected across an ideal battery B with emf $\mathscr{E}$. The resulting current i is the same throughout the circuit.

the other on the concept of potential. The circuit consists of an ideal battery B with emf $\mathcal{E}$, a resistor of resistance R, and two connecting wires. (Unless otherwise indicated, we assume that wires in circuits have negligible resistance. Their function, then, is merely to provide pathways along which charge carriers can move.)

Energy Method

Equation 27-22, $P = i^2R$, tells us that in a time interval dt an amount of energy given by $i^2R\, dt$ will appear in the resistor of Fig. 28-3 as thermal energy. (Since we assume the wires to have negligible resistance, no thermal energy will appear in them.) During the same interval, a charge $dq = i\, dt$ will have moved through battery B, and the battery will have done work on this charge, according to Eq. 28-1, equal to

$$dW = \mathcal{E}\, dq = \mathcal{E}i\, dt.$$

From the principle of conservation of energy, the work done by the battery must equal the thermal energy that appears in the resistor:

$$\mathcal{E}i\, dt = i^2R\, dt.$$

This gives us

$$\mathcal{E} = iR,$$

which in words means the following: emf $\mathcal{E}$ is the energy per unit charge transferred to the moving charges by the battery. The quantity iR is the energy per unit charge transferred *from* the moving charges to thermal energy within the resistor. The energy per unit charge transferred to the moving charges is equal to the energy per unit charge transferred from them. Solving for i, we find

$$i = \frac{\mathcal{E}}{R}. \tag{28-2}$$

Potential Method

Suppose we start at any point in the circuit of Fig. 28-3 and mentally proceed around the circuit in either direction, adding algebraically the potential differences that we encounter. When we arrive at our starting point, we must have returned to our starting potential. Before actually doing so, we shall formalize this idea in a statement that holds not only for single-loop circuits such as that of Fig. 28-3 but for any complete loop in a *multiloop* circuit, as we shall discuss in Section 28-6:

LOOP RULE: The algebraic sum of the changes in potential encountered in a complete traversal of any loop of a circuit must be zero.

This is often referred to as *Kirchhoff's loop rule* (or *Kirchhoff's voltage law*), after German physicist Gustav Robert Kirchhoff. This rule is equivalent to saying that any point on the side of a mountain must have a unique elevation above sea level. If you start from any point and return to it after walking around the mountain, the algebraic sum of the changes in elevation that you encounter must be zero.

In Fig. 28-3, let us start at point a, whose potential is V_a, and mentally walk clockwise around the circuit until we are back at a, keeping track of potential changes as we move. Our starting point is at the low-potential terminal of the battery. Since the battery is ideal, the potential difference between its terminals is equal to $\mathcal{E}$. So when we pass through the battery to the high-potential terminal, the change in potential is $+\mathcal{E}$.

As we walk along the top wire to the top end of the resistor, there is no potential change because the wire has negligible resistance: it is at the same potential as the high-potential terminal of the battery. So too is the top end of the resistor. When we pass through the resistor, however, the change in potential is $-iR$.

We return to point a along the bottom wire. Since this wire also has negligible resistance, we again find no potential change. Back at point a, the potential is again V_a. Because we traversed a complete loop, our initial potential, as modified for potential changes along the way, must be equal to our final potential; that is,

$$V_a + \mathcal{E} - iR = V_a.$$

The value of V_a cancels from this equation, which becomes

$$\mathcal{E} - iR = 0.$$

Solving this equation for i gives us the same result, $i = \mathcal{E}/R$, as the energy method (Eq. 28-2).

If we apply the loop rule to a complete *counterclockwise* walk around the circuit, the rule gives us

$$-\mathcal{E} + iR = 0$$

and we again find that $i = \mathcal{E}/R$. Thus you may mentally circle a loop in either direction to apply the loop rule.

To prepare for circuits more complex than that of Fig. 28-3, let us set down two rules for finding potential differences as we move around a loop:

RESISTANCE RULE: For a move through a resistance in the direction of the current, the change in potential is $-iR$; in the opposite direction it is $+iR$.

EMF RULE: For a move through an ideal emf device in the direction of the emf arrow, the change in potential is $+\mathcal{E}$; in the opposite direction it is $-\mathcal{E}$.

CHECKPOINT **1:** The figure shows the current i in a single-loop circuit with a battery B and a resistance R (and wires of negligible resistance). (a) Should the emf arrow at B be drawn leftward or rightward? At points a, b, and c, rank (b) the magnitude of the current, (c) the electric potential, and (d) the electric potential energy of the charge carriers, greatest first.

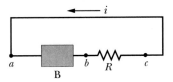

28-4 OTHER SINGLE-LOOP CIRCUITS

In this section we extend the simple circuit of Fig. 28-3 in two ways.

Internal Resistance

Figure 28-4a shows a real battery, with an internal resistance r, wired to an external resistor of resistance R. The internal resistance of the battery is the electrical resistance of the conducting materials of the battery and thus is an unremovable feature of the battery. In Fig. 28-4a, however, the battery is drawn as if it could be separated into an ideal battery with emf $\mathscr{E}$ and a resistor of resistance r. The order in which the symbols for these separated parts are drawn does not matter.

If we apply the loop rule clockwise beginning at point a, we obtain

$$\mathscr{E} - ir - iR = 0. \tag{28-3}$$

Solving for the current, we find

$$i = \frac{\mathscr{E}}{R + r}. \tag{28-4}$$

Note that this equation reduces to Eq. 28-2 if the battery is ideal, that is, if $r = 0$.

Figure 28-4b shows graphically the changes in electric potential around the circuit. (To better link Fig. 28-4b with the *closed circuit,* imagine curling the graph into a cylinder with point a at the left overlapping point a at the right.) Note how traversing the circuit is like walking around a (potential) mountain and returning to your starting point —you also return to the starting elevation.

In this book, when a battery is not described as real or if no internal resistance is indicated, you can generally assume that it is ideal. But, of course, in the real world, batteries are always real and have internal resistance.

Resistances in Series

Figure 28-5a shows three resistances connected **in series** to an ideal battery of emf $\mathscr{E}$. The battery applies a potential difference $V = \mathscr{E}$ across the three-resistance combination.

> Connected resistances are said to be in series when a potential difference that is applied across their combination is the sum of the resulting potential differences across all the resistances.

In less formal language, this definition often means that the resistances occur one after another along a single path for the current, as in Fig. 28-5a.

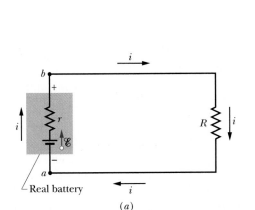

(a)

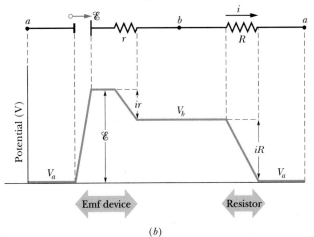

(b)

FIGURE 28-4 (a) A single-loop circuit containing a real battery having internal resistance r and emf $\mathscr{E}$. (b) The circuit is shown spread out at the top. The potentials encountered in traversing the circuit clockwise from a are shown in the graph. The potential V_a is arbitrarily assigned a value of zero, and other potentials in the circuit are graphed relative to V_a.

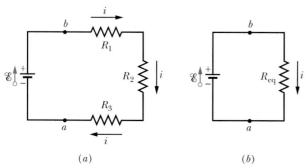

(a)   *(b)*

FIGURE 28-5 *(a)* Three resistors are connected in series between points *a* and *b*. *(b)* An equivalent circuit, with the three resistors replaced with their equivalent resistance R_{eq}.

We seek the single resistance R_{eq} that is equivalent to the three-resistance series combination of Fig. 28-5*a*. By *equivalent*, we mean that R_{eq} can replace the combination without changing the current *i* through the combination or the potential difference between *a* and *b*. Let us apply the loop rule, starting at terminal *a* and going clockwise around the circuit. We find

$$\mathcal{E} - iR_1 - iR_2 - iR_3 = 0,$$

or

$$i = \frac{\mathcal{E}}{R_1 + R_2 + R_3}. \qquad (28\text{-}5)$$

If we replaced the three resistances with a single equivalent resistance R_{eq}, we would have (Fig. 28-5*b*)

$$i = \frac{\mathcal{E}}{R_{eq}}. \qquad (28\text{-}6)$$

Comparison of Eqs. 28-5 and 28-6 shows that

$$R_{eq} = R_1 + R_2 + R_3.$$

The extension to *n* resistances is straightforward and is

$$R_{eq} = \sum_{j=1}^{n} R_j \qquad (n \text{ resistances in series}). \quad (28\text{-}7)$$

Note that when resistances are in series, their equivalent resistance is greater than any of the individual resistances.

28-5 POTENTIAL DIFFERENCES

We often want to find the potential difference between two points in a circuit. In Fig. 28-4*a*, for example, what is the potential difference between points *b* and *a*? To find out, let us start at point *b* and traverse the circuit clockwise to point *a*, passing through resistor *R*. If V_a and V_b are the potentials at *a* and *b*, respectively, we have

$$V_b - iR = V_a$$

because (according to our resistance rule) we experience a decrease in potential in going through a resistance in the direction of the current. We rewrite this as

$$V_b - V_a = +iR, \qquad (28\text{-}8)$$

which tells us that point *b* is at greater potential than point *a*. Combining Eq. 28-8 with Eq. 28-4, we have

$$V_b - V_a = \mathcal{E} \frac{R}{R + r}, \qquad (28\text{-}9)$$

where again *r* is the internal resistance of the emf device.

> To find the potential difference between any two points in a circuit, start at one point and traverse the circuit to the other, following any path, and add algebraically the changes in potential that you encounter.

Let us again calculate $V_b - V_a$, starting again at point *b* but this time proceeding counterclockwise to *a* through the battery. We have

$$V_b + ir - \mathcal{E} = V_a$$

or

$$V_b - V_a = \mathcal{E} - ir. \qquad (28\text{-}10)$$

Combining this with Eq. 28-4 again leads to Eq. 28-9.

The quantity $V_b - V_a$ in Fig. 28-4 is the potential difference of the battery across the battery terminals. As noted earlier, $V_b - V_a$ is equal to the emf $\mathcal{E}$ of the battery only if the battery has no internal resistance ($r = 0$ in Eq. 28-9) or if the circuit is open ($i = 0$ in Eq. 28-10).

Suppose that in Fig. 28-4, $\mathcal{E} = 12$ V, $R = 10\ \Omega$, and $r = 2.0\ \Omega$. Then Eq. 28-9 tells us that the potential across the battery's terminals is

$$V_b - V_a = 12 \text{ V} \frac{10\ \Omega}{10\ \Omega + 2.0\ \Omega} = 10 \text{ V}.$$

In "pumping" charge through itself, the battery (via electrochemical reactions) does work per unit charge of $\mathcal{E} = 12$ J/C, or 12 V. However, because of the internal resistance of the battery, it produces a potential difference of only 10 J/C, or 10 V, across its terminals.

Power, Potential, and Emf

When a battery or some other type of emf device does work on the charge carriers of a current *i*, it transfers energy from its source of energy (such as the chemical source in a battery) to the charge carriers. Because a real emf device has an internal resistance *r*, it also transfers energy to internal thermal energy via resistive dissipation as discussed in Section 27-7. Let us relate these transfers.

The net rate P of energy transfer from the emf device to the charge carriers is given by Eq. 27-21:

$$P = iV, \qquad (28\text{-}11)$$

where V is the potential across the terminals of the emf device. From Eq. 28-10, we can substitute $V = \mathscr{E} - ir$ into Eq. 28-11 to find

$$P = i(\mathscr{E} - ir) = i\mathscr{E} - i^2r. \qquad (28\text{-}12)$$

We see that the term i^2r in Eq. 28-12 is the rate P_r of energy transfer to thermal energy within the emf device:

$$P_r = i^2r \qquad \text{(internal dissipation rate).} \quad (28\text{-}13)$$

Then the term $i\mathscr{E}$ in Eq. 28-12 must be the rate P_{emf} at which the emf source transfers energy to *both* the charge carriers and to internal thermal energy. Thus,

$$P_{\text{emf}} = i\mathscr{E} \qquad \text{(power of emf device).} \quad (28\text{-}14)$$

If a battery is being *recharged*, with the current in the "wrong way" through it, the energy transfer is then *from* the charge carriers *to* the battery—both to the battery's chemical energy and to the energy dissipated in the internal resistance r. The rate of change of the chemical energy is given by Eq. 28-14; the rate of dissipation is given by Eq. 28-13; and the rate at which the carriers supply energy is given by Eq. 28-11.

CHECKPOINT **2**: In Fig. 28-5a, if $R_1 > R_2 > R_3$, rank the three resistors according to (a) the current through them and (b) the potential difference across them, greatest first.

SAMPLE PROBLEM 28-1

What is the current in the circuit of Fig. 28-6a? The emfs and the resistances have the following values:

$$\mathscr{E}_1 = 4.4 \text{ V}, \quad \mathscr{E}_2 = 2.1 \text{ V},$$

$$r_1 = 2.3 \text{ } \Omega, \quad r_2 = 1.8 \text{ } \Omega, \quad R = 5.5 \text{ } \Omega.$$

SOLUTION: The two batteries are connected so that they oppose each other, but $\mathscr{E}_1$, because it is larger than $\mathscr{E}_2$, controls the direction of the current in the circuit, which is clockwise. The loop rule, applied counterclockwise from point a, yields

$$-\mathscr{E}_1 + ir_1 + iR + ir_2 + \mathscr{E}_2 = 0.$$

Check that this equation also results from applying the loop rule clockwise or from starting at some point other than a.

Also, compare this equation term by term with Fig. 28-6b, which shows the potential changes graphically (with the potential at point a arbitrarily taken to be zero).

Solving the above loop equation for the current i, we obtain

$$i = \frac{\mathscr{E}_1 - \mathscr{E}_2}{R + r_1 + r_2} = \frac{4.4 \text{ V} - 2.1 \text{ V}}{5.5 \text{ } \Omega + 2.3 \text{ } \Omega + 1.8 \text{ } \Omega}$$

$$= 0.2396 \text{ A} \approx 240 \text{ mA}. \qquad \text{(Answer)}$$

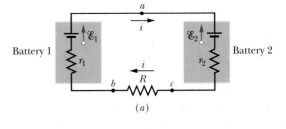

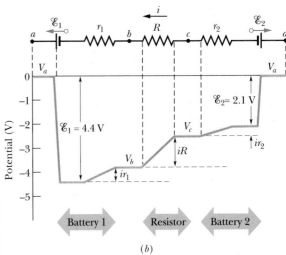

FIGURE 28-6 Sample Problems 28-1 and 28-2. (*a*) A single-loop circuit containing two real batteries and a resistor. The batteries oppose each other; that is, they tend to send current in opposite directions through the resistor. (*b*) A graph of the potentials encountered in traversing this circuit counterclockwise from point a, with the potential at a arbitrarily taken to be zero. (To better link the circuit with the graph, mentally cut the circuit at a and then unfold the left side of the circuit toward the left and the right side of the circuit toward the right.) As battery 1 is traversed from the higher-potential terminal to the lower-potential terminal against the current, the potential decreases by $\mathscr{E}_1$ and increases by ir_1. As the resistor R is traversed against the current, the potential increases by iR. As battery 2 is traversed from the lower-potential terminal to the higher-potential terminal against the current, the potential increases by ir_2 and by $\mathscr{E}_2$.

PROBLEM SOLVING TACTICS

TACTIC 1: *Assuming the Direction of a Current*

In solving circuit problems, you do not need to know the direction of a current in advance. Instead, you can just assume its direction. To show this, assume the current in Fig. 28-6a is counterclockwise; that is, reverse the direction of the current arrows shown. Applying the loop rule counterclockwise from point *a* now yields

$$-\mathscr{E}_1 - ir_1 - iR - ir_2 + \mathscr{E}_2 = 0$$

or

$$i = -\frac{\mathscr{E}_1 - \mathscr{E}_2}{R + r_1 + r_2}.$$

Substituting numerical values (see above) yields $i = -240$ mA for the current. The minus sign is a signal that the current is opposite the direction we initially assumed.

SAMPLE PROBLEM 28-2

(a) What is the potential difference between the terminals of battery 1 in Fig. 28-6a?

SOLUTION: Let us start at point *b* (effectively the negative terminal of battery 1) and travel through battery 1 to point *a* (effectively the positive terminal), keeping track of potential changes. We find that

$$V_b - ir_1 + \mathscr{E}_1 = V_a,$$

which gives us

$$\begin{aligned}V_a - V_b &= -ir_1 + \mathscr{E}_1 \\ &= -(0.2396 \text{ A})(2.3 \ \Omega) + 4.4 \text{ V} \\ &= +3.84 \text{ V} \approx 3.8 \text{ V}. \quad \text{(Answer)}\end{aligned}$$

We can verify this result by starting at point *b* in Fig. 28-6a and traversing the circuit counterclockwise to point *a*. For this different path we find

$$V_b + iR + ir_2 + \mathscr{E}_2 = V_a$$

or $\quad V_a - V_b = i(R + r_2) + \mathscr{E}_2$

$$\begin{aligned}&= (0.2396 \text{ A})(5.5 \ \Omega + 1.8 \ \Omega) + 2.1 \text{ V} \\ &= +3.84 \text{ V} \approx 3.8 \text{ V}, \quad \text{(Answer)}\end{aligned}$$

exactly as before. The potential difference between two points has the same value for all paths connecting those points.

(b) What is the potential difference between the terminals of battery 2 in Fig. 28-6a?

SOLUTION: Let us start at point *c* (the negative terminal of battery 2) and travel through battery 2 to point *a* (the positive terminal), keeping track of potential changes. We find

$$V_c + ir_2 + \mathscr{E}_2 = V_a$$

or $\qquad V_a - V_c = ir_2 + \mathscr{E}_2$

$$\begin{aligned}&= (0.2396 \text{ A})(1.8 \ \Omega) + 2.1 \text{ V} \\ &= +2.5 \text{ V}. \quad \text{(Answer)}\end{aligned}$$

Here the potential difference (2.5 V) between the terminals of the battery is *larger* than the emf (2.1 V) of the battery because charge is being forced through the battery in a direction opposite the direction in which it would normally move.

C**HECKPOINT 3:** A battery has an emf of 12 V and an internal resistance of 2 Ω. Is the terminal-to-terminal potential difference greater than, less than, or equal to 12 V if the current in the battery is (a) from the negative to the positive terminal, (b) from the positive terminal to the negative terminal, and (c) zero?

28-6 MULTILOOP CIRCUITS

Figure 28-7 shows a circuit containing more than one loop. For simplicity, we assume the batteries are ideal. There are two *junctions* in this circuit, at *b* and *d*, and there are three *branches* connecting these junctions. The branches are the left branch (*bad*), the right branch (*bcd*), and the central branch (*bd*). What are the currents in the three branches?

We arbitrarily label the currents, using a different symbol for each branch. Current i_1 has the same value everywhere in branch *bad*; i_2 has the same value everywhere in branch *bcd*; and i_3 is the current through branch *bd*. The directions of the currents are chosen arbitrarily.

Consider junction *d*. Charge comes into the junction via incoming currents i_1 and i_3, and it leaves via outgoing current i_2; there is no increase or decrease of charge at the junction. This condition means that

$$i_1 + i_3 = i_2. \quad (28\text{-}15)$$

You can easily check that applying this condition to junction *b* leads to exactly the same equation. Equation 28-15 suggests a general principle:

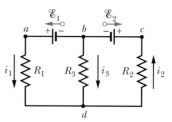

FIGURE 28-7 A multiloop circuit consisting of three branches: left-hand branch *bad,* right-hand branch *bcd,* and central branch *bd.* The circuit also consists of three loops: left-hand loop *badb,* right-hand loop *bcdb,* and big loop *badcb.*

This rule is often called *Kirchhoff's junction rule* (or *Kirchhoff's current law*). It is simply a statement of the conservation of charge for a steady flow of charge—there is neither a build-up nor a depletion of charge at at junction. Thus our basic tools for solving complex circuits are the *loop rule* (based on the conservation of energy) and the *junction rule* (based on the conservation of charge).

Equation 28-15 is a single equation involving three unknowns. To solve the problem completely (that is, to find all three currents), we need two more equations involving those same unknowns. We obtain them by applying the loop rule twice. In the circuit of Fig. 28-7, we have three loops from which to choose: the left-hand loop *(badb)*, the right-hand loop *(bcdb)*, and the big loop *(badcb)*. Which two loops we choose does not matter— let's choose the left-hand loop and the right-hand loop.

If we traverse the left-hand loop in a counterclockwise direction from point *b*, the loop rule gives us

$$\mathscr{E}_1 - i_1 R_1 + i_3 R_3 = 0. \tag{28-16}$$

If we traverse the right-hand loop in a counterclockwise direction from point *b*, the loop rule gives us

$$-i_3 R_3 - i_2 R_2 - \mathscr{E}_2 = 0. \tag{28-17}$$

We now had three equations (Eqs. 28-15, 28-16, and 28-17) in the three unknown currents, and they can be solved by a variety of techniques.

If we had applied the loop rule to the big loop, we would have obtained (moving counterclockwise from *b*) the equation

$$\mathscr{E}_1 - i_1 R_1 - i_2 R_2 - \mathscr{E}_2 = 0.$$

This equation may look like fresh information, but in fact it is only the sum of Eqs. 28-16 and 28-17. (It would, however, yield the proper results when used with Eq. 28-15 and either 28-16 or 28-17.)

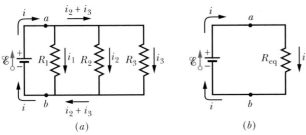

(a)   (b)

FIGURE 28-8 (*a*) Three resistors connected in parallel across points *a* and *b*. (*b*) An equivalent circuit, with the three resistors replaced with their equivalent resistance R_{eq}.

Resistances in Parallel

Figure 28-8*a* shows three resistances connected **in parallel** to an ideal battery of emf $\mathscr{E}$. The battery applies a potential difference $V = \mathscr{E}$ across each resistor in this parallel combination.

Connected resistances are said to be in parallel when a potential difference that is applied across their combination results in that same potential difference across each resistance.

We seek the single resistance R_{eq} that is equivalent to this parallel combination; R_{eq} is then the resistance that can replace the combination without changing the current *i* through the combination or the potential difference *V* applied across the combination.

The currents in the three branches of Fig. 28-8*a* are

$$i_1 = \frac{V}{R_1}, \quad i_2 = \frac{V}{R_2}, \quad \text{and} \quad i_3 = \frac{V}{R_3},$$

where *V* is the potential difference between *a* and *b*. If we apply the junction rule at point *a* and then substitute these values, we find

$$i = i_1 + i_2 + i_3 = V \left(\frac{1}{R_1} + \frac{1}{R_2} + \frac{1}{R_3} \right). \tag{28-18}$$

If we replaced the parallel combination with the equivalent resistance R_{eq} (Fig. 28-8*b*), we would have

$$i = \frac{V}{R_{\text{eq}}}. \tag{2819}$$

Comparing Eqs. 28-18 and 28-19 leads to

$$\frac{1}{R_{\text{eq}}} = \frac{1}{R_1} + \frac{1}{R_2} + \frac{1}{R_3}. \tag{28-20}$$

Extending this result to the case of *n* resistances, we have

$$\frac{1}{R_{\text{eq}}} = \sum_{j=1}^{n} \frac{1}{R_j} \qquad (n \text{ resistances in parallel}). \tag{28-21}$$

For the case of two resistances, the equivalent resistance is their product divided by their sum. That is,

$$R_{\text{eq}} = \frac{R_1 R_2}{R_1 + R_2}. \tag{28-22}$$

If you accidentally took the equivalent resistance to be the sum divided by the product, you would notice at once that this result would be dimensionally incorrect.

Note that when two or more resistances are connected in parallel, the equivalent resistance is smaller than any of the combining resistances. Table 28-1 summarizes the equivalence relations for resistors and capacitors in series and in parallel.

TABLE 28-1 **SERIES AND PARALLEL RESISTORS AND CAPACITORS**

SERIES	PARALLEL
Resistors	
$R_{eq} = \sum_{j=1}^{n} R_j$	$\dfrac{1}{R_{eq}} = \sum_{j=1}^{n} \dfrac{1}{R_j}$
Eq. 28-7	Eq. 28-21
Same current through all resistors	Same potential difference across all resistors
Capacitors	
$\dfrac{1}{C_{eq}} = \sum_{j=1}^{n} \dfrac{1}{C_j}$	$C_{eq} = \sum_{j=1}^{n} C_j$
Eq. 26-20	Eq. 26-19
Same charge on all capacitors	Same potential difference across all capacitors

$\mathbb{C}$HECKPOINT 4: A battery, with potential V across it and current i through it, is connected to a combination of two identical resistors. What are the potential difference across and the current through either resistor if the resistors are (a) in series and (b) in parallel?

SAMPLE PROBLEM 28-3

Figure 28-9a shows a multiloop circuit containing one ideal battery and four resistors with the following values:

$$R_1 = 20 \ \Omega, \quad R_2 = 20 \ \Omega, \quad \mathscr{E} = 12 \ \text{V},$$
$$R_3 = 30 \ \Omega, \quad R_4 = 8.0 \ \Omega.$$

(a) What is the current through the battery?

SOLUTION: The current through the battery is also the current through R_1. So, to find the current, we need to write an equation for a loop through R_1; either the left-hand loop or the big loop will do. Noting that since the emf arrow of the battery points upward and the current the battery supplies is clockwise, we might consider applying the loop rule to the left-hand loop clockwise from point a, getting

$$+\mathscr{E} - iR_1 - iR_2 - iR_4 = 0 \quad \text{(incorrect)}.$$

However, this equation is incorrect because it assumes that R_1, R_2, and R_4 all have the same current i. Resistors R_1 and R_4 do have the same current, because the current passing through R_4 must pass through the battery and then through R_1 with no change in value. But that current splits at junction point b—only part passes through R_2, the rest through R_3.

To distinguish the several currents in the circuit, we must label them individually as in Fig. 28-9b. Then, circling clockwise from a, we can write the loop rule for the left-hand loop as

$$+\mathscr{E} - i_1R_1 - i_2R_2 - i_1R_4 = 0.$$

Unfortunately, this equation contains two unknown, i_1 and i_2; we would need at least one more equation to find them.

A second, much easier option is to simplify the circuit of Fig. 28-9b by finding equivalent resistances. Note carefully that R_1 and R_2 are *not* in series and thus cannot be replaced with an equivalent resistance. However, R_2 and R_3 are in parallel; so we can use either Eq. 28-21 or Eq. 28-22 to find their equivalent resistance R_{23}. From the latter,

$$R_{23} = \frac{R_2R_3}{R_2 + R_3} = \frac{(20 \ \Omega)(30 \ \Omega)}{50 \ \Omega} = 12 \ \Omega.$$

We can now redraw the circuit as in Fig. 28-9c; note that the current through R_{23} must be i_1 because the current i_1 through R_1 and R_4 must continue through R_{23}. For this simple one-loop circuit, the loop rule (applied clockwise from point a) yields

$$+\mathscr{E} - i_1R_1 - i_1R_{23} - i_1R_4 = 0.$$

Substituting the given data, we find

$$12 \ \text{V} - i_1(20 \ \Omega) - i_1(12 \ \Omega) - i_1(8.0 \ \Omega) = 0,$$

which gives us

$$i_1 = \frac{12 \ \text{V}}{40 \ \Omega} = 0.30 \ \text{A}. \quad \text{(Answer)}$$

(b) What is the current i_2 through R_2?

SOLUTION: Look again at Fig. 28-9c. From it and the preceding answer we know that the current through R_{23} is $i_1 =$

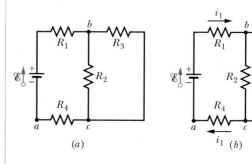

(a)

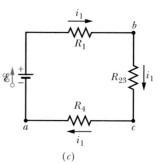

(c)

FIGURE 28-9 Sample Problem 28-3. (a) A multiloop circuit with an ideal battery of emf $\mathscr{E}$ and four resistors. (b) Assumed currents through the resistors. (c) A simplification of the circuit, with resistances R_2 and R_3 replaced with their equivalent resistance R_{23}. The current through R_{23} is equal to that through R_1 and R_4.

0.30 A. Then we can use Eq. 27-8 ($R = V/i$) to find the potential difference V_{23} across R_{23}, which is

$$V_{23} = i_1 R_{23} = (0.30 \text{ A})(12 \text{ }\Omega) = 3.6 \text{ V.}$$

This is also the potential difference across R_2 (and across R_3). Thus, applying Eq. 27-8 now to R_2, we can write

$$i_2 = \frac{V_2}{R_2} = \frac{3.6 \text{ V}}{20 \text{ }\Omega} = 0.18 \text{ A.} \qquad \text{(Answer)}$$

(c) What is the current i_3 through R_3?

SOLUTION: From Fig. 28-9b and the earlier results, application of the junction rule at point b gives us

$$i_3 = i_1 - i_2 = 0.30 \text{ A} - 0.18 \text{ A}$$
$$= 0.12 \text{ A.} \qquad \text{(Answer)}$$

SAMPLE PROBLEM 28-4

Figure 28-10 shows a circuit whose elements have the following values:

$$\mathcal{E}_1 = 3.0 \text{ V}, \qquad \mathcal{E}_2 = 6.0 \text{ V},$$
$$R_1 = 2.0 \text{ }\Omega, \qquad R_2 = 4.0 \text{ }\Omega.$$

The three batteries are ideal batteries. Find the magnitude and direction of the current in each of the three branches.

SOLUTION: It is not worthwhile to try to simplify this circuit, because no two resistors are in parallel, and the resistors that are in series (those in the right branch or those in the left branch) present no problem. So, we shall apply the junction and loop rules, and then solve some simultaneous equations.

Using arbitrarily chosen directions for the currents as shown in Fig. 28-10, we apply the junction rule at point a by writing

$$i_3 = i_1 + i_2. \qquad (28\text{-}23)$$

An application of the junction rule at junction b gives only the same equation. So we next apply the loop rule to any two of the three loops of the circuit. We first arbitrarily choose the left-hand loop, arbitrarily start at point a, and arbitrarily traverse the loop in the counterclockwise direction, obtaining

$$-i_1 R_1 - \mathcal{E}_1 - i_1 R_1 + \mathcal{E}_2 + i_2 R_2 = 0.$$

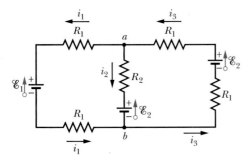

FIGURE 28-10 Sample Problem 28-4. A multiloop circuit with three ideal batteries and five resistors.

Substituting the given data and simplifying yield

$$i_1(4.0 \text{ }\Omega) - i_2(4.0 \text{ }\Omega) = 3.0 \text{ V.} \qquad (28\text{-}24)$$

For our second application of the loop rule, we arbitrarily choose to traverse the right-hand loop clockwise from point a, finding

$$+i_3 R_1 - \mathcal{E}_2 + i_3 R_1 + \mathcal{E}_2 + i_2 R_2 = 0.$$

Substituting the given data and simplifying yield

$$i_2(4.0 \text{ }\Omega) + i_3(4.0 \text{ }\Omega) = 0. \qquad (28\text{-}25)$$

Using Eq. 28-23 to eliminate i_3 from Eq. 28-25 and simplifying give us

$$i_1(4.0 \text{ }\Omega) + i_2(8.0 \text{ }\Omega) = 0. \qquad (28\text{-}26)$$

We now have a system of two equations (Eqs. 28-24 and 28-26) in two unknowns (i_1 and i_2) to solve either "by hand" (which is easy enough here) or with a "math package." (One solution technique is Cramer's rule, given in Appendix E.) We find

$$i_2 = -0.25 \text{ A.}$$

(The minus sign signals that our arbitrary choice of direction for i_2 in Fig. 28-10 is wrong; i_2 should point up through $\mathcal{E}_2$ and R_2.) Substituting $i_2 = -0.25$ A into Eq. 28-26 and solving for i_1 then give us

$$i_1 = 0.50 \text{ A.} \qquad \text{(Answer)}$$

With Eq. 28-23 we then find that

$$i_3 = i_1 + i_2 = 0.25 \text{ A.} \qquad \text{(Answer)}$$

The positive answers we obtained for i_1 and i_3 signal that our choices of directions for these currents are correct. We can now correct the direction for i_2 and write its magnitude as

$$i_2 = 0.25 \text{ A.} \qquad \text{(Answer)}$$

SAMPLE PROBLEM 28-5

Electric fish generate current with biological cells called *electroplaques*, which are physiological emf devices. The electroplaques in the South American eel shown in the photograph that opens this chapter are arranged in 140 rows, each row stretching horizontally along the body and each containing 5000 electroplaques. The arrangement is suggested in Fig. 28-11a; each electroplaque has an emf $\mathcal{E}$ of 0.15 V and an internal resistance r of 0.25 Ω.

(a) If the water surrounding the eel has resistance $R_w = 800$ Ω, how much current can the eel produce in the water, from near its head to its tail?

SOLUTION: To answer, we simplify the circuit of Fig. 28-11a, first considering a single row. The total emf $\mathcal{E}_{\text{row}}$ along a row of 5000 electroplaques is the sum of the emfs:

$$\mathcal{E}_{\text{row}} = 5000\mathcal{E} = (5000)(0.15 \text{ V}) = 750 \text{ V.}$$

The total resistance R_{row} along a row is the sum of the internal

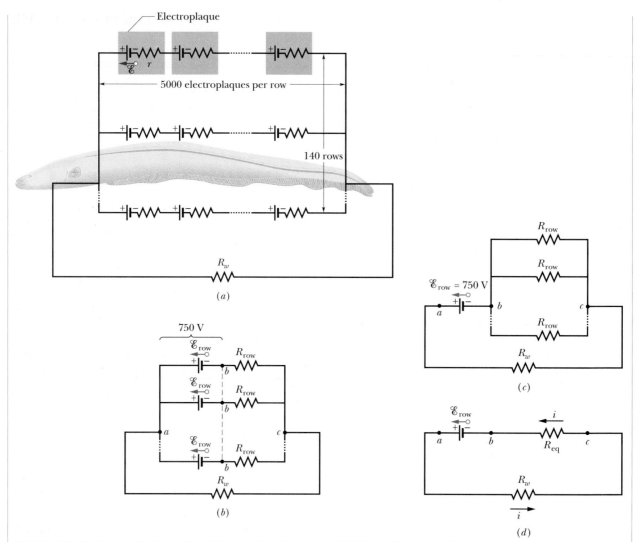

(a)

(b)

(c)

(d)

FIGURE 28-11 Sample Problem 28-5. (*a*) A model of the electric circuit of an eel in water. Each electroplaque of the eel has an emf $\mathscr{E}$ and internal resistance r. Along each of 140 rows extending from the head to the tail of the eel, there are 5000 electroplaques. The surrounding water has resistance R_w.

(*b*) The emf $\mathscr{E}_{\text{row}}$ and resistance R_{row} of each row. (*c*) The emf between points a and b is $\mathscr{E}_{\text{row}}$. Between points b and c are 140 parallel resistances R_{row}. (*d*) The simplified circuit, with R_{eq} replacing the parallel combination.

resistances of the 5000 electroplaques:

$$R_{\text{row}} = 5000r = (5000)(0.25\ \Omega) = 1250\ \Omega.$$

We can now represent each of the 140 identical rows as having a single emf $\mathscr{E}_{\text{row}}$ and a single resistance R_{row}, as shown in Fig. 28-11*b*.

In Fig. 28-11*b*, the emf between point a and point b on any row is $\mathscr{E}_{\text{row}} = 750$ V. Because the rows are identical and because they are all connected together at the left in Fig. 28-11*b*, all points b in that figure are at the same electric potential. Thus we can consider them to be connected so that there is only a single point b. The emf between point a and this single point b is $\mathscr{E}_{\text{row}} = 750$ V, so we can draw the circuit as shown in Fig. 28-11*c*.

Between points b and c in Fig. 28-11*c*, 140 resistances of $R_{\text{row}} = 1250\ \Omega$ are in parallel. The equivalent resistance R_{eq}

of this combination is given by Eq. 28-21 as

$$\frac{1}{R_{\text{eq}}} = \sum_{j=1}^{140} \frac{1}{R_j} = 140\ \frac{1}{R_{\text{row}}},$$

or

$$R_{\text{eq}} = \frac{R_{\text{row}}}{140} = \frac{1250\ \Omega}{140} = 8.93\ \Omega.$$

Replacing the parallel combination with R_{eq}, we obtain the simplified circuit of Fig. 28-11*d*. Applying the loop rule to this circuit counterclockwise from point b, we have

$$\mathscr{E}_{\text{row}} - iR_w - iR_{\text{eq}} = 0.$$

Solving for i and substituting the known data, we find

$$i = \frac{\mathscr{E}_{\text{row}}}{R_w + R_{\text{eq}}} = \frac{750\ \text{V}}{800\ \Omega + 8.93\ \Omega}$$

$$= 0.927\ \text{A} \approx 0.93\ \text{A}. \qquad \text{(Answer)}$$

If the head or tail of the eel is near a fish, much of this current could pass along a narrow path through the fish, stunning or killing it.

(b) How much current i_{row} travels through each row of Fig. 28-11a?

SOLUTION: Since the rows are identical, the current into and out of the eel is evenly divided among them:

$$i_{\text{row}} = \frac{i}{140} = \frac{0.927 \text{ A}}{140} = 6.6 \times 10^{-3} \text{ A}. \quad \text{(Answer)}$$

Thus the current through each row is small, about two orders of magnitude smaller than the current through the water. This means that the eel need not stun or kill itself when it stuns or kills a fish.

PROBLEM SOLVING TACTICS

TACTIC 2: *Solving Circuits of Batteries and Resistors*

Here are two general techniques for solving circuits for unknown currents or potential differences.

1. If a circuit can be simplified by replacing resistors in series or in parallel with their equivalents, do so. If you can reduce the circuit to a single loop, then you can find the current through the battery with that loop, as in Sample Problem 28-3a. You may then have to "work backward," undoing the resistor simplification process, to find the current or potential difference for any particular resistor, as in Sample Problem 28-3b.
2. If a circuit cannot be simplified to a single loop, use the junction rule and the loop rule to write a set of simultaneous equations, as in Sample Problem 28-4. You need have only as many independent equations as there are unknowns in those equations. If you have to find the current or potential difference for a particular resistor, you can ensure that its current or potential difference appears in the equations by having at least one of the loops pass through the designated resistor.

TACTIC 3: *Arbitrary Choices in Solving Circuit Problems*

In Sample Problem 28-4, we made several arbitrary choices. (1) We assumed directions for the currents in Fig. 28-10 arbitrarily. (2) We chose which of the three possible loops to write equations for arbitrarily. (3) We chose the direction in which to traverse each loop arbitrarily. (4) We chose the starting and ending point for each traversal arbitrarily.

Such arbitrariness often worries a beginning circuit solver, but an experienced circuit solver knows that it does not matter. Just keep two rules firmly in mind. First, make sure you traverse each chosen loop completely. Second, once you have chosen a direction for a current, stick with it until you get numerical values for all the currents. If you were wrong about

a direction, the algebra will signal you with a minus sign. Then you can make a correction by simply erasing the minus sign and reversing the arrow representing that current in the circuit diagram. However, *you should not make this correction until you have completed all the required calculations for the circuit,* as we did in Sample Problem 28-4.

28-7 THE AMMETER AND THE VOLTMETER

An instrument used to measure currents is called an *ammeter*. To measure the current in a wire, you usually have to break or cut the wire and insert the ammeter so that the current to be measured passes through the meter, as shown in Fig. 28-12.

It is essential that the resistance R_A of the ammeter be very small compared to other resistances in the circuit. Otherwise, the very presence of the meter will change the current to be measured.

A meter used to measure potential differences is called a *voltmeter*. To find the potential difference between any two points in the circuit, the voltmeter terminals are connected between those points, without breaking or cutting the wire (Fig. 28-12).

It is essential that the resistance R_V of a voltmeter be very large compared to the resistance of any circuit element across which the voltmeter is connected. Otherwise, the meter itself becomes an important circuit element and alters the potential difference that is to be measured.

Often a single meter is packaged so that, by means of a switch, it can be made to serve as either an ammeter or a voltmeter—and usually also as an *ohmmeter*, designed to measure the resistance of any element connected between its terminals. Such a versatile unit is called a *multimeter*.

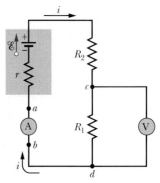

FIGURE 28-12 A single-loop circuit, showing how to connect an ammeter (A) and a voltmeter (V).

28-8 *RC* CIRCUITS

In preceding sections we dealt only with circuits in which the currents did not vary with time. Here we begin a discussion of time-varying currents.

Charging a Capacitor

The capacitor of capacitance C in Fig. 28-13 is initially uncharged. To charge it, we close switch S on point a. This completes an *RC series circuit* consisting of the capacitor, an ideal battery of emf $\mathcal{E}$, and a resistance R.

From Section 26-2, we already know that as soon as the circuit is complete, charge begins to flow (current exists) between a capacitor plate and a battery terminal on each side of the capacitor. This current increases the charge q on the plates and the potential difference V_C ($= q/C$) across the capacitor. When that potential difference equals the potential difference across the battery (which here is equal to the emf $\mathcal{E}$), the current is zero. From Eq. 26-1 ($q = CV$), the *equilibrium* (final) *charge* on the then fully charged capacitor is equal to $C\mathcal{E}$.

Here we want to examine the charging process. In particular we want to know how the charge $q(t)$ on the capacitor plates, the potential difference $V_C(t)$ across the capacitor, and the current $i(t)$ in the circuit vary with time during the charging process. We begin by applying the loop rule to the circuit, traversing it clockwise from the negative terminal of the battery. We find

$$\mathcal{E} - iR - \frac{q}{C} = 0. \qquad (28\text{-}27)$$

The last term on the left side represents the potential difference across the capacitor. The term is negative because the capacitor's top plate, which is connected to the battery's positive terminal, is at a higher potential than the lower plate. So, there is a drop in potential as we move down through the capacitor.

We cannot immediately solve Eq. 28-27 because it contains two variables, i and q. However, those variables are not independent but are related by

$$i = \frac{dq}{dt}. \qquad (28\text{-}28)$$

Substituting this for i in Eq. 28-27 and rearranging, we find

$$R\frac{dq}{dt} + \frac{q}{C} = \mathcal{E} \qquad \text{(charging equation).} \quad (28\text{-}29)$$

This differential equation describes the time variation of the charge q on the capacitor in Fig. 28-13. To solve it, we need to find the function $q(t)$ that satisfies this equation and also satisfies the condition that the capacitor be initially uncharged: $q = 0$ at $t = 0$.

We shall show below that the solution to Eq. 28-29 is

$$q = C\mathcal{E}(1 - e^{-t/RC}) \qquad \begin{array}{l}\text{(charging a}\\ \text{capacitor).}\end{array} \quad (28\text{-}30)$$

(Here e is the exponential base, $2.718 \cdots$, and not the elementary charge.) Note that Eq. 28-30 does indeed satisfy our required initial condition, because at $t = 0$ the term $e^{-t/RC}$ is unity; so the equation gives $q = 0$. Note also that at $t = \infty$ (that is, a long time later), the term $e^{-t/RC}$ is zero; so the equation gives the proper value for the full (equilibrium) charge on the capacitor, namely, $q = C\mathcal{E}$. A plot of $q(t)$ for the charging process is given in Fig. 28-14a.

The derivative of $q(t)$ is the current $i(t)$ charging the capacitor:

$$i = \frac{dq}{dt} = \left(\frac{\mathcal{E}}{R}\right)e^{-t/RC} \qquad \begin{array}{l}\text{(charging a}\\ \text{capacitor).}\end{array} \quad (28\text{-}31)$$

A plot of $i(t)$ for the charging process is given in Fig. 28-14b. Note that the current has the initial value $\mathcal{E}/R$ and that it decreases to zero as the capacitor becomes fully charged. Note also that the initial value $\mathcal{E}/R$ implies that at

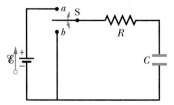

FIGURE 28-13 When switch S is closed on a, the capacitor C is *charged* through the resistor R. When the switch is afterward closed on b, the capacitor *discharges* through R.

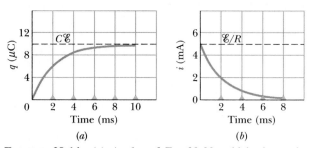

FIGURE 28-14 (*a*) A plot of Eq. 28-30, which shows the buildup of charge on the capacitor of Fig. 28-13. (*b*) A plot of Eq. 28-31, which shows the decline of the charging current in the circuit of Fig. 28-13. The curves are plotted for $R = 2000 \ \Omega$, $C = 1 \ \mu F$, and $\mathcal{E} = 10 \ V$; the small triangles represent successive intervals of one time constant.

$t = 0$, the capacitor acts as if it were a wire with negligible resistance.

By combining Eq. 26-1 ($q = CV$) and Eq. 28-30, we find that the potential difference $V_C(t)$ across the capacitor during the charging process is

$$V_C = \frac{q}{C} = \mathcal{E}(1 - e^{-t/RC}) \qquad \text{(charging a capacitor).} \qquad (28\text{-}32)$$

This tells us that $V_C = 0$ at $t = 0$ and that $V_C = \mathcal{E}$ when the capacitor is fully charged at $t = \infty$.

The Time Constant

The product RC that appears in Eqs. 28-30, 28-31, and 28-32 has the dimensions of time (because the argument of an exponential must be dimensionless); in fact, $1.0 \; \Omega \times 1.0 \; \text{F} = 1.0 \; \text{s}$. RC is called the **capacitive time constant** of the circuit and is represented with the symbol τ:

$$\tau = RC \qquad \text{(time constant).} \qquad (28\text{-}33)$$

From Eq. 28-30, we can now see that at time $t = \tau (= RC)$, the charge on the initially uncharged capacitor of Fig. 28-13 has increased from zero to

$$q = C\mathcal{E}(1 - e^{-1}) = 0.63 C\mathcal{E}. \qquad (28\text{-}34)$$

In words, during the first time constant τ the charge has increased from zero to 63% of its final value $C\mathcal{E}$. In Fig. 28-14, the small triangles along the times axes mark successive intervals of one time constant during the charging of the capacitor. The charging times for RC circuits are often stated in terms of τ: the greater τ is, the greater the charging time.

Discharging a Capacitor

Assume now that the capacitor of Fig. 28-13 is fully charged to a potential V_0 equal to the emf $\mathcal{E}$ of the battery. At a new time $t = 0$ the switch S is thrown from a to b so that the capacitor can *discharge* through resistance R. How do the charge $q(t)$ on the capacitor and the current $i(t)$ through the discharge loop of capacitor and resistance now vary with time?

The differential equation describing $q(t)$ is like Eq. 28-29 except that now, with no battery in the discharge loop, $\mathcal{E} = 0$. Thus,

$$R\frac{dq}{dt} + \frac{q}{C} = 0 \qquad \text{(discharging equation).} \qquad (28\text{-}35)$$

The solution to this differential equation is

$$q = q_0 e^{-t/RC} \qquad \text{(discharging a capacitor),} \qquad (28\text{-}36)$$

where $q_0 (= CV_0)$ is the initial charge on the capacitor. You can verify by substitution that Eq. 28-36 is indeed a solution of Eq. 28-35.

Equation 28-36 tells us that q decreases exponentially with time, at a rate that is set by the capacitive time constant $\tau = RC$. At time $t = \tau$, the capacitor's charge has been reduced to $q_0 e^{-1}$, or about 37% of the initial value. Note that a greater τ means a greater discharge time.

Differentiating Eq. 28-36 gives us the current $i(t)$:

$$i = \frac{dq}{dt} = -\left(\frac{q_0}{RC}\right)e^{-t/RC} \qquad \text{(discharging a capacitor).} \qquad (28\text{-}37)$$

This tells us that the current also decreases exponentially with time, at a rate set by τ. The initial current i_0 is equal to q_0/RC. Note that you can find i_0 by simply applying the loop rule to the circuit at $t = 0$; just then the capacitor's initial potential V_0 is connected across the resistance R, so the current must be $i_0 = V_0/R = (q_0/C)/R = q_0/RC$. The minus sign in Eq. 38-37 can be ignored; it merely means that the capacitor's charge q is decreasing.

Derivation of Eq. 28-30

To solve Eq. 28-29, we first rewrite it as

$$\frac{dq}{dt} + \frac{q}{RC} = \frac{\mathcal{E}}{R}. \qquad (28\text{-}38)$$

The general solution to this differential equation is of the form

$$q = q_p + Ke^{-at}, \qquad (28\text{-}39)$$

where q_p is a *particular solution* of the differential equation, K is a constant to be evaluated from the initial conditions, and $a = 1/RC$ is the coefficient of q in Eq. 28-38. To find q_p, we set $dq/dt = 0$ in Eq. 28-38 (corresponding to the final condition of no further charging) and solve, obtaining

$$q_p = C\mathcal{E}. \qquad (28\text{-}40)$$

To evaluate K, we first substitute this into Eq. 28-39 to get

$$q = C\mathcal{E} + Ke^{-at}.$$

Then substituting the initial conditions $q = 0$ and $t = 0$ yields

$$0 = C\mathcal{E} + K,$$

or $K = -C\mathcal{E}$. Finally, with the values of q_p, a, and K inserted, Eq. 28-39 becomes

$$q = C\mathcal{E} - C\mathcal{E}e^{-t/RC}$$

which, with a slight modification, is Eq. 28-30.

CHECKPOINT 5: The table gives four sets of values for the circuit elements in Fig. 28-13. Rank the sets according to (a) the initial current (as the switch is closed on *a*) and (b) the time required for the current to decrease to half its initial value, greatest first.

	1	2	3	4
$\mathscr{E}$ (V)	12	12	10	10
R (Ω)	2	3	10	5
C (μF)	3	2	0.5	2

SAMPLE PROBLEM 28-6

A capacitor of capacitance C is discharging through a resistor of resistance R.

(a) In terms of the time constant $\tau = RC$, when will the charge on the capacitor be half its initial value?

SOLUTION: The charge on the capacitor varies according to Eq. 28-36,

$$q = q_0 e^{-t/RC},$$

in which q_0 is the initial charge. We are asked to find the time t at which $q = \frac{1}{2}q_0$, or at which

$$\tfrac{1}{2}q_0 = q_0 e^{-t/RC}. \qquad (28\text{-}41)$$

After canceling q_0, we realize that the time t we seek is "buried" inside an exponential function. To expose the symbol t in Eq. 28-41, we take the natural logarithms of both sides of the equation. (The natural logarithm is the inverse function of the exponential function.) We find

$$\ln \tfrac{1}{2} = \ln(e^{-t/RC}) = -\frac{t}{RC}$$

or $\qquad t = (-\ln \tfrac{1}{2})RC = 0.69RC = 0.69\tau.$ (Answer)

(b) When will the energy stored in the capacitor be half its initial value?

SOLUTION: The energy stored in the capacitor is, from Eqs. 26-21 and 28-36,

$$U = \frac{q^2}{2C} = \frac{q_0^2}{2C} e^{-2t/RC} = U_0 e^{-2t/RC}, \qquad (28\text{-}42)$$

in which U_0 is the initial stored energy. We are asked to find the time at which $U = \frac{1}{2}U_0$, or at which

$$\tfrac{1}{2}U_0 = U_0 e^{-2t/RC}.$$

Canceling U_0 and taking the natural logarithms of both sides, we obtain

$$\ln \tfrac{1}{2} = -\frac{2t}{RC}$$

or $\qquad t = -RC\,\frac{\ln \tfrac{1}{2}}{2} = 0.35RC = 0.35\tau.$ (Answer)

It takes longer (0.69τ versus 0.35τ) for the *charge* to fall to half its initial value than for the *stored energy* to fall to half its initial value. Doesn't this result surprise you?

(c) At what rate P_R is thermal energy produced in the resistor during the discharging process? At what rate P_C is stored energy lost by the capacitor during the discharging process?

SOLUTION: The current through the resistor during the discharging is given by Eq. 28-37. From Eq. 27-22 ($P = i^2 R$), we then have

$$P_R = i^2 R = \left[-\frac{q_0}{RC} e^{-t/RC} \right]^2 R$$

$$= \frac{q_0^2}{RC^2} e^{-2t/RC}. \qquad (Answer)$$

Stored energy is lost by the capacitor at the rate $P_C = dU/dt$, where U is the energy stored there. From Eq. 28-42, we then have

$$P_C = \frac{dU}{dt} = \frac{d}{dt}(U_0 e^{-2t/RC}) = -\frac{2U_0}{RC} e^{-2t/RC}.$$

Substituting $q_0^2/2C$ for U_0 gives us

$$P_C = -\frac{q_0^2}{RC^2} e^{-2t/RC}. \qquad (Answer)$$

Note that $P_C + P_R = 0$. In words, stored energy lost by the capacitor is transferred completely to thermal energy of the resistor.

SAMPLE PROBLEM 28-7

The circuit in Fig. 28-15 consists of an ideal battery with emf $\mathscr{E} = 12$ V, two resistors with resistances $R_1 = 4.0\ \Omega$ and $R_2 = 6.0\ \Omega$, and an initially uncharged capacitor with capacitance $C = 6.0\ \mu$C. The circuit is completed when switch S is closed at time $t = 0$.

(a) At time $t = 2.0\tau$, what is the potential difference across the capacitor?

SOLUTION: In Fig. 28-15, the capacitor is being charged through R_1 by an emf $\mathscr{E}$ connected across them, just as in Fig. 28-13. (Resistance R_2 does not change this fact.) So, we can use Eq. 28-32,

$$V_C = \mathscr{E}(1 - e^{-t/RC}),$$

to find the potential difference V_C across the capacitor, except here resistance R is R_1. Substituting $t = 2.0\tau = 2.0R_1 C$ and given data, we then find

$$V_C = (12\text{ V})(1 - e^{-2.0 R_1 C/R_1 C})$$

$$= (12\text{ V})(1 - e^{-2.0}) = 10\text{ V}. \qquad (Answer)$$

(b) At time $t = 2.0\tau$, what are the potential differences V_{R_1} and V_{R_2} across the two resistors? Do those potential differences increase, decrease, or remain the same while the capacitor is being charged?

SOLUTION: If we apply the loop rule to the big loop of Fig. 28-15, clockwise from the negative terminal of the battery, we find

$$\mathcal{E} - V_C - V_{R_1} = 0. \qquad (28\text{-}43)$$

We now know that at $t = 2.0\tau$, the potential difference V_C is equal to 10 V. Substituting this and $\mathcal{E} = 12$ V into Eq. 28-43 yields

$$V_{R_1} = 2.0 \text{ V.} \qquad \text{(Answer)}$$

During the charging of the capacitor, the battery's emf $\mathcal{E}$ is constant and the potential difference V_C across the capacitor increases. By writing Eq. 28-43 as $V_{R_1} = \mathcal{E} - V_C$, we see that V_{R_1} must decrease during the charging process.

If we apply the loop rule to the left-hand loop of Fig.

28-15, again clockwise from the negative terminal, we find

$$\mathcal{E} - V_{R_2} = 0$$

and

$$V_{R_2} = \mathcal{E} = 12 \text{ V.} \qquad \text{(Answer)}$$

Thus V_{R_2} does not change during the charging process.

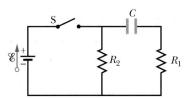

FIGURE 28-15 Sample Problem 28-7. When switch S is closed, the circuit is complete and the battery begins to charge the capacitor.

REVIEW & SUMMARY

Emf

An **emf device** does work on charges to maintain a potential difference between its output terminals. If dW is the work the device does to force positive charge dq from the negative to the positive terminal, then the **emf** (work per unit charge) of the device is

$$\mathcal{E} = \frac{dW}{dq} \qquad \text{(definition of } \mathcal{E}\text{).} \qquad (28\text{-}1)$$

The volt is the SI unit of emf as well as of potential difference. An **ideal emf device** is one that lacks any internal resistance. The potential difference between its terminals is equal to the emf. A **real emf device** has internal resistance. The potential difference between its terminals is equal to the emf only if there is no current through the device.

Analyzing Circuits

The change in potential in traversing a resistance R in the direction of the current is $-iR$; in the opposite direction it is $+iR$. The change in potential in traversing an ideal emf device in the direction of the emf arrow is $+\mathcal{E}$; in the opposite direction it is $-\mathcal{E}$. Conservation of energy leads to the loop rule:

Loop Rule. *The algebraic sum of the changes in potential encountered in a complete traversal of any loop of a circuit must be zero.*

Conservation of charge gives us the junction rule:

Junction Rule. *The sum of the currents entering any junction must be equal to the sum of the currents leaving that junction.*

Single-Loop Circuits

The current in a single-loop circuit containing a single resistance R and an emf device with emf $\mathcal{E}$ and internal resistance r is

$$i = \frac{\mathcal{E}}{R + r}, \qquad (28\text{-}4)$$

which reduces to $i = \mathcal{E}/R$ for an ideal emf device with $r = 0$.

Power

When a real battery of emf $\mathcal{E}$ and internal resistance r does work on the charge carriers in a current i through it, the rate P of energy transfer to the charge carriers is

$$P = iV, \qquad (28\text{-}11)$$

where V is the potential across the terminals of the battery. The rate P_r of energy transfer to thermal energy within the battery is

$$P_r = i^2 r. \qquad (28\text{-}13)$$

And the rate P_{emf} at which the chemical energy within the battery changes is

$$P_{\text{emf}} = i\mathcal{E}. \qquad (28\text{-}14)$$

Series Resistances

Resistances are in **series** if the sum of their individual potential differences is equal to the potential difference applied across the combination. The equivalent resistance of the series combination is

$$R_{\text{eq}} = \sum_{j=1}^{n} R_j \qquad (n \text{ resistances in series).} \qquad (28\text{-}7)$$

Other circuit elements may also be connected in series.

Parallel Resistances

Resistances are in **parallel** if their individual potential differences are equal to the applied potential difference. The equivalent resistance of the parallel combination is

$$\frac{1}{R_{\text{eq}}} = \sum_{j=1}^{n} \frac{1}{R_j} \qquad (n \text{ resistances in parallel).} \qquad (28\text{-}21)$$

Other circuit elements may also be connected in parallel.

RC *Circuits*

When an emf $\mathscr{E}$ is applied to a resistance R and capacitance C in series, as in Fig. 28-13 with the switch at a, the charge on the capacitor increases according to

$$q = C\mathscr{E}(1 - e^{-t/RC}) \quad \text{(charging capacitor),} \quad (28\text{-}30)$$

in which $C\mathscr{E} = q_0$ is the equilibrium (final) charge and $RC = \tau$ is the **capacitive time constant** of the circuit. During the charging, the current is

$$i = \frac{dq}{dt} = \left(\frac{\mathscr{E}}{R}\right)e^{-t/RC} \quad \begin{array}{c}\text{(charging}\\\text{capacitor).}\end{array} \quad (28\text{-}31)$$

When a capacitor discharges through a resistance R, the charge on the capacitor decays according to

$$q = q_0 e^{-t/RC} \quad \text{(discharging capacitor).} \quad (28\text{-}36)$$

During the discharging, the current is

$$i = \frac{dq}{dt} = -\left(\frac{q_0}{RC}\right)e^{-t/RC} \quad \begin{array}{c}\text{(discharging}\\\text{capacitor).}\end{array} \quad (28\text{-}37)$$

QUESTIONS

1. Figure 28-16 shows current i passing through a battery. The following table gives four sets of values for i and the battery's emf $\mathscr{E}$ and internal resistance r; it also gives the *polarity* (orientation of the terminals) of the battery. Rank the sets according to the rate at which energy is transferred between the battery and the charge carriers, greatest transfer *to* the carriers first and greatest transfer *from* the carriers last.

	$\mathscr{E}$	r	i	POLARITY
(1)	$15\mathscr{E}_1$	0	i_1	+ at left
(2)	$10\mathscr{E}_1$	0	$2i_1$	+ at left
(3)	$10\mathscr{E}_1$	0	$2i_1$	− at left
(4)	$10\mathscr{E}_1$	r_1	$2i_1$	− at left

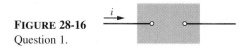

FIGURE 28-16
Question 1.

2. For each circuit in Fig. 28-17, are the resistors connected in series, in parallel, or neither?

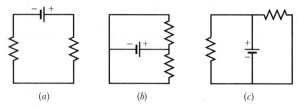

FIGURE 28-17 Question 2.

3. (a) In Fig. 28-18a, are resistors R_1 and R_3 in series? (b) Are resistors R_1 and R_2 in parallel? (c) Rank the equivalent resistances of the four circuits shown in Fig. 28-18, greatest first.

4. What is the equivalent resistance of three resistors, each of resistance R, if they are connected to an ideal battery (a) in series with one another and (b) in parallel with one another? (c) Is the potential difference across the series arrangement greater than, less than, or equal to that across the parallel arrangement?

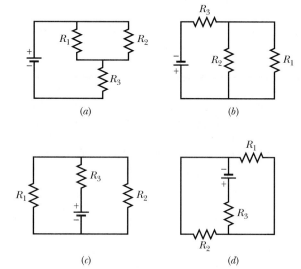

FIGURE 28-18 Questions 3 and 7.

5. You are to connect resistors R_1 and R_2, with $R_1 > R_2$, to a battery, first individually, then in series, and then in parallel. Rank those arrangements according to the amount of current through the battery, greatest first.

6. Two resistors are wired to a battery. (a) In which arrangement, parallel or series, are the potential differences across each resistor and across the equivalent resistance all equal? (b) In which arrangement are the currents through each resistor and through the equivalent resistance all equal?

7. (a) In Fig. 28-18a, with $R_1 > R_2$, is the potential difference across R_2 more than, less than, or equal to that across R_1? (b) Is the current through resistor R_2 more than, less than, or equal to that through resistor R_1?

8. Initially, a single resistor R_1 is wired to a battery. Then resistor R_2 is added in parallel. Are (a) the potential difference across R_1 and (b) the current i_1 through R_1 now more than, less than, or the same as previously? (c) Is the equivalent resistance R_{12} of R_1 and R_2 more than, less than, or equal to R_1? (d) Is the total current through R_1 and R_2 together more than, less than, or equal to the current through R_1 previously?

9. A resistor R_1 is wired to a battery; then resistor R_2 is added in series. Are (a) the potential difference across R_1 and (b) the current i_1 through R_1 now more than, less than, or the same as previously? (c) Is the equivalent resistance R_{12} of R_1 and R_2 more than, less than, or equal to R_1?

10. (a) In Fig. 28-19, when the branch with R_2 is added as indicated, does the rate at which electrical energy is transferred to thermal energy in R_1 increase, decrease, or stay the same? (b) Does the rate at which electrical energy is supplied by the battery increase, decrease, or stay the same? (c) Repeat (a) and (b) if, instead, R_2 is added in series with R_1.

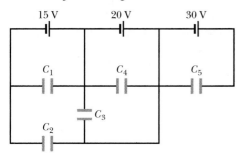

FIGURE 28-19 Question 10.

11. Without written calculation, determine the potential difference across each capacitor in Fig. 28-20.

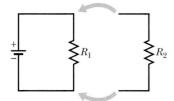

FIGURE 28-20 Question 11.

12. *Res-monster maze.* In Fig. 28-21, all the resistors have a resistance of $4.0\ \Omega$ and all the (ideal) batteries have an emf of 4.0 V. What is the current through resistor R? (If you can find the proper loop through this maze, you can answer the question with a few seconds of mental calculation.)

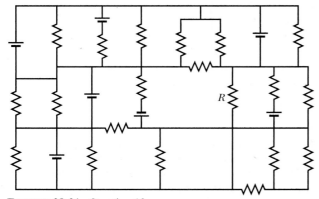

FIGURE 28-21 Question 12.

13. *Cap-monster maze.* In Fig. 28-22, all the capacitors have a capacitance of $6.0\ \mu$F, and all the batteries have an emf of 10 V. What is the charge on capacitor C? (If you can find the proper loop through this maze, you can answer the question with a few seconds of mental calculation.)

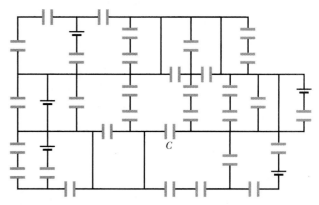

FIGURE 28-22 Question 13.

14. You are to connect n identical real batteries in series between circuit points a and b in Fig. 28-23a, and you have three choices: $n = 14$, $n = 12$, $n = 16$. Rank the choices according to (a) the total emf between a and b and (b) the total resistance between a and b, greatest first. Next, you are to connect the batteries in parallel between circuit points c and d in Fig. 28-23b. Rank the choices of n according to (c) the total emf between c and d and (d) the total resistance between c and d, greatest first.

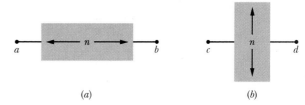

(a)   (b)

FIGURE 28-23 Question 14.

15. Figure 28-24 shows plots of $V(t)$ for three capacitors that discharge (separately) through the same resistor. Rank the plots according to the capacitances of the capacitors, greatest first.

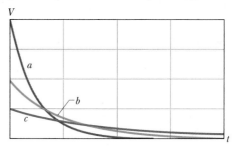

FIGURE 28-24 Question 15.

16. A capacitor is discharged from different initial charges and across different resistances in the three situations listed below. Rank the situations according to (a) the current through the resist-

ance at the start of the discharge and (b) the time required for the current to decrease to half its starting value, greatest first.

	1	2	3
Initial charge	$12q$	$12q$	$6q$
Resistance	$2R$	$3R$	R

17. Figure 28-25 shows three sections of circuit that are to be connected in turn to the same battery via a switch as in Fig. 28-13. The resistors are all identical; so are the capacitors. Rank the sections according to (a) the final (equilibrium) charge on the capacitor and (b) the time required for the capacitor to reach 50% of its final charge, greatest first.

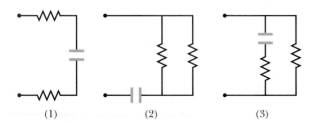

FIGURE 28-25 Question 17.

18. The five sections of circuit in Fig. 28-26 are to be connected, in turn, to the same 12 V battery via a switch as in Fig. 28-13. The resistors are all identical; so are the capacitors. Rank the sections according to the time required for the capacitors to reach 50% of their final potential, greatest first.

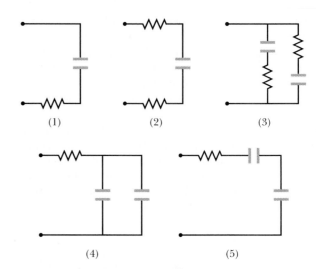

FIGURE 28-26 Questions 18 and 19.

19. Rank the five sections of Question 18 according to the potential across any resistor in the section when the potential across any capacitor in the section reaches 4 V, greatest first.

20. (a) Does the time required for the charge q on a capacitor in an RC circuit to build up to a certain fraction of its equilibrium value depend on the value of the applied emf? (b) Does the time required for the charge to change by a certain amount Δq depend on the applied emf? (c) Does the amount of charge for the fully charged capacitor depend on the internal resistance of the battery charging it?

EXERCISES & PROBLEMS

SECTION 28-5 Potential Differences

1E. A standard flashlight battery can deliver about 2.0 W·h of energy before it runs down. (a) If a battery costs 80¢, what is the cost of operating a 100 W lamp for 8.0 h using batteries? (b) What is the cost if power provided by an electric utility company, at 6¢ per kilowatt-hour, is used?

2E. (a) How much work does an ideal battery with a 12.0 V emf do on an electron that passes through the battery from the positive to the negative terminal? (b) If 3.4×10^{18} electrons pass through each second, what is the power of the battery?

3E. A 5.0 A current is set up in a circuit for 6.0 min by a rechargeable battery with a 6.0 V emf. By how much is the chemical energy of the battery reduced?

4E. A certain car battery with a 12 V emf has an initial charge of 120 A·h. Assuming that the potential across the terminals stays constant until the battery is completely discharged, for how many hours can it deliver energy at the rate of 100 W?

5E. In Fig. 28-27, $\mathscr{E}_1 = 12$ V and $\mathscr{E}_2 = 8$ V. (a) What is the direction of the current in the resistor? (b) Which battery is doing positive work? (c) Which point, A or B, is at the higher potential?

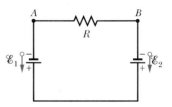

FIGURE 28-27 Exercise 5.

6E. Assume that the batteries in Fig. 28-28 have negligible internal resistance. Find (a) the current in the circuit, (b) the power dissipated in each resistor, and (c) the power of each battery, stating whether energy is supplied to or absorbed by it.

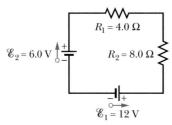

FIGURE 28-28   Exercise 6.

7E. A wire of resistance 5.0 Ω is connected to a battery whose emf $\mathscr{E}$ is 2.0 V and whose internal resistance is 1.0 Ω. In 2.0 min, (a) how much energy is transferred from chemical to electrical form? (b) How much energy appears in the wire as thermal energy? (c) Account for the difference between (a) and (b).

8E. In Fig. 28-4a, put $\mathscr{E}$ = 2.0 V and r = 100 Ω. Plot (a) the current and (b) the potential difference across R, as functions of R over the range 0 to 500 Ω. Make both plots on the same graph. (c) Make a third plot by multiplying together, for various values of R, the corresponding values on the two plotted curves. What is the physical significance of this third plot?

9E. A car battery with a 12 V emf and an internal resistance of 0.040 Ω is being charged with a current of 50 A. (a) What is the potential difference across its terminals? (b) At what rate is energy being dissipated as thermal energy in the battery? (c) At what rate is electrical energy being converted to chemical energy? (d) What are the answers to (a) and (b) when the battery is used to supply 50 A to the starter motor?

10E. In Fig. 28-29, if the potential at point P is 100 V, what is the potential at point Q?

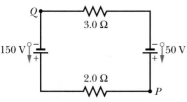

FIGURE 28-29   Exercise 10.

11E. In Fig. 28-30, the section of circuit AB absorbs 50 W of power when a current i = 1.0 A passes through it in the indicated direction. (a) What is the potential difference between A and B? (b) Emf device C does not have internal resistance. What is its emf? (c) What is its *polarity* (the orientation of its positive and negative terminals)?

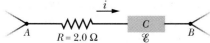

FIGURE 28-30   Exercise 11.

12E. In Fig. 28-5a calculate the potential difference across R_2, assuming $\mathscr{E}$ = 12 V, R_1 = 3.0 Ω, R_2 = 4.0 Ω, and R_3 = 5.0 Ω.

13E. In Fig. 28-6a calculate the potential difference between a and c by considering a path that contains R, r_2, and $\mathscr{E}_2$. (See Sample Problem 28-2.)

14E. A gasoline gauge for an automobile is shown schematically in Fig. 28-31. The indicator (on the dashboard) has a resistance of 10 Ω. The tank unit is simply a float connected to a variable resistor whose resistance is 140 Ω when the tank is empty, is 20 Ω when the tank is full, and varies linearly with the volume of gasoline. Find the current in the circuit when the tank is (a) empty, (b) half-full, (c) and full.

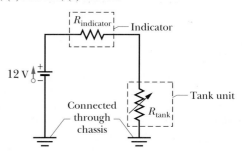

FIGURE 28-31   Exercise 14.

15P. A 10 km long underground cable extends east to west and consists of two parallel wires, each of which has resistance 13 Ω/km. A short develops at distance x from the west end when a conducting path of resistance R connects the wires (Fig. 28-32). The resistance of the wires and the short is then 100 Ω when the measurement is made from the east end, and 200 Ω when it is made from the west end. What are (a) x and (b) R?

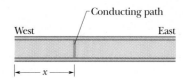

FIGURE 28-32   Problem 15.

16P. (a) In Fig. 28-33 what value must R have if the current in the circuit is to be 1.0 mA? Take $\mathscr{E}_1$ = 2.0 V, $\mathscr{E}_2$ = 3.0 V, and $r_1 = r_2$ = 3.0 Ω. (b) What is the rate at which thermal energy appears in R?

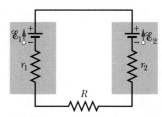

FIGURE 28-33   Problem 16.

17P. The current in a single-loop circuit with one resistance R is 5.0 A. When an additional resistance of 2.0 Ω is inserted in series with R, the current drops to 4.0 A. What is R?

18P. Thermal energy is to be generated in a 0.10 Ω resistor at the rate of 10 W by connecting the resistor to a battery whose emf is 1.5 V. (a) What potential difference must exist across the resistor? (b) What must be the internal resistance of the battery?

19P. Power is supplied by a device of emf $\mathcal{E}$ to a transmission line with resistance R. Find the ratio of the power dissipated in the line for $\mathcal{E} = 110{,}000$ V to that dissipated for $\mathcal{E} = 110$ V, assuming the power supplied is the same for the two cases.

20P. Wires A and B, having equal lengths of 40.0 m and equal diameters of 2.60 mm, are connected in series. A potential difference of 60.0 V is applied between the ends of the composite wire. The resistances of the wires are 0.127 and 0.729 Ω, respectively. Determine (a) the current density in each wire and (b) the potential difference across each wire. (c) Identify the wire materials. See Table 27-1.

21P. The starting motor of an automobile is turning too slowly, and the mechanic has to decide whether to replace the motor, the cable, or the battery. The manufacturer's manual says that the 12 V battery should have no more than 0.020 Ω internal resistance, the motor no more than 0.200 Ω resistance, and the cable no more than 0.040 Ω resistance. The mechanic turns on the motor and measures 11.4 V across the battery, 3.0 V across the cable, and a current of 50 A. Which part is defective?

22P. Two batteries having the same emf $\mathcal{E}$ but different internal resistances r_1 and r_2 ($r_1 > r_2$) are connected in series to an external resistance R. (a) Find the value of R that makes the potential difference zero between the terminals of one battery. (b) Which battery is it?

23P. A solar cell generates a potential difference of 0.10 V when a 500 Ω resistor is connected across it, and a potential difference of 0.15 V when a 1000 Ω resistor is substituted. What are (a) the internal resistance and (b) the emf of the solar cell? (c) The area of the cell is 5.0 cm^2, and the rate per unit area at which it receives energy from light is 2.0 mW/cm^2. What is the efficiency of the cell for converting the light energy to thermal energy in the 1000 Ω external resistor?

24P. (a) In Fig. 28-4a, show that the rate at which energy is dissipated in R as thermal energy is a maximum when $R = r$. (b) Show that this maximum power is $P = \mathcal{E}^2/4r$.

25P. A battery of emf $\mathcal{E} = 2.00$ V and internal resistance $r = 0.500$ Ω is driving a motor. The motor is lifting a 2.00 N mass at constant speed $v = 0.500$ m/s. Assuming no energy losses, find (a) the current i in the circuit and (b) the potential difference V across the terminals of the motor. (c) Discuss the fact that there are two solutions to this problem.

26P. A temperature-stable resistor is made by connecting a resistor made of silicon in series with one made of iron. If the required total resistance is 1000 Ω in a wide temperature range around 20°C, what should be the resistances of the two resistors? See Table 27-1.

SECTION 28-6   Multiloop Circuits

27E. Four 18.0 Ω resistors are connected in parallel across a 25.0 V ideal battery. What is the current through the battery?

28E. A total resistance of 3.00 Ω is to be produced by connecting an unknown resistance to a 12.0 Ω resistance. What must be the value of the unknown resistance and should it be connected in series or in parallel?

29E. By using only two resistors—singly, in series, or in parallel—you are able to obtain resistances of 3.0, 4.0, 12, and 16 Ω. What are the two resistances?

30E. In Fig. 28-34, find the equivalent resistance between points (a) A and B, (b) A and C, and (c) B and C. (*Hint:* Imagine that a battery is connected between points A and C.)

FIGURE 28-34   Exercise 30.

31E. In Fig. 28-35, find the equivalent resistance between points D and E. (*Hint:* Imagine that a battery is connected between points D and E.)

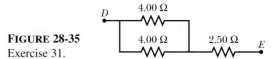

FIGURE 28-35
Exercise 31.

32E. In Fig. 28-36 find the current in each resistor and the potential difference between a and b. Put $\mathcal{E}_1 = 6.0$ V, $\mathcal{E}_2 = 5.0$ V, $\mathcal{E}_3 = 4.0$ V, $R_1 = 100$ Ω, and $R_2 = 50$ Ω.

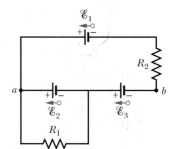

FIGURE 28-36
Exercise 32.

33E. Figure 28-37 shows a circuit containing three switches, labeled S_1, S_2, and S_3. Find the current at a for all possible combinations of switch settings. Put $\mathcal{E} = 120$ V, $R_1 = 20.0$ Ω, and $R_2 = 10.0$ Ω. Assume that the battery has no resistance.

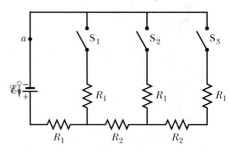

FIGURE 28-37   Exercise 33.

34E. Two lightbulbs, one of resistance R_1 and the other of resistance R_2, where $R_1 > R_2$, are connected to a battery (a) in parallel and (b) in series. Which bulb is brighter (dissipates more energy) in each case?

35E. In Fig. 28-7, calculate the potential difference between points c and d by as many paths as possible. Assume that $\mathcal{E}_1 = 4.0$ V, $\mathcal{E}_2 = 1.0$ V, $R_1 = R_2 = 10$ Ω, and $R_3 = 5.0$ Ω.

36E. Nine copper wires of length *l* and diameter *d* are connected in parallel to form a single composite conductor of resistance *R*. What must be the diameter *D* of a single copper wire of length *l* if it is to have the same resistance?

37E. A 120 V power line is protected by a 15 A fuse. What is the maximum number of 500 W lamps that can be simultaneously operated in parallel on this line without "blowing" the fuse because of an excess of current?

38E. A circuit containing five resistors connected to a battery with a 12.0 V emf is shown in Fig. 28-38. What is the potential difference across the 5.0 Ω resistor?

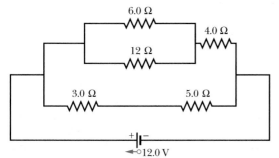

FIGURE 28-38 Exercise 38.

39P. In Fig. 28-39, find the equivalent resistance between points (a) *F* and *H* and (b) *F* and *G*. (*Hint:* For each pair of points, imagine that a battery is connected across the pair.)

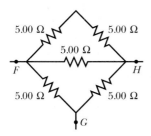

FIGURE 28-39 Problem 39.

40P. Two resistors R_1 and R_2 may be connected either in series or in parallel across an ideal battery with emf $\mathscr{E}$. We desire the rate of electrical energy dissipation of the parallel combination to be five times that of the series combination. If $R_1 = 100 \ \Omega$, what is R_2? (*Hint:* There are two answers.)

41P. You are given a number of 10 Ω resistors, each capable of dissipating only 1.0 W without being destroyed. What is the minimum number of such resistors that you need to combine in series or parallel to make a 10 Ω resistance that is capable of dissipating at least 5.0 W?

42P. Two batteries of emf $\mathscr{E}$ and internal resistance *r* are connected in parallel across a resistor *R*, as in Fig. 28-40*a*. (a) For what value of *R* is the rate of electrical energy dissipation by the resistor a maximum? (b) What is the maximum energy dissipation rate?

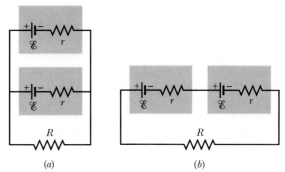

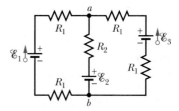

FIGURE 28-40 Problems 42 and 44.

43P. (a) Calculate the current through each ideal battery in Fig. 28-41. Assume that $R_1 = 1.0 \ \Omega$, $R_2 = 2.0 \ \Omega$, $\mathscr{E}_1 = 2.0$ V, and $\mathscr{E}_2 = \mathscr{E}_3 = 4.0$ V. (b) Calculate $V_a - V_b$.

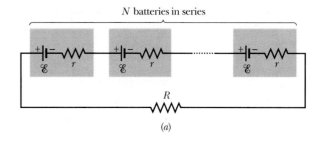

FIGURE 28-41 Problem 43.

44P. You are given two batteries of emf $\mathscr{E}$ and internal resistance *r*. They may be connected either in parallel (Fig. 28-40*a*) or in series (Fig. 28-40*b*) and are used to establish a current in a resistor *R*. (a) Derive expressions for the current in *R* for both arrangements. Which will yield the larger current (b) when $R > r$ and (c) when $R < r$?

45P. A group of *N* identical batteries of emf $\mathscr{E}$ and internal resistance *r* may be connected all in series (Fig. 28-42*a*) or all in parallel (Fig. 28-42*b*) and then across a resistor *R*. Show that both arrangements will give the same current in *R* if $R = r$.

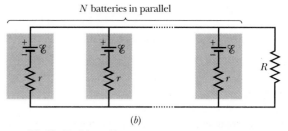

FIGURE 28-42 Problem 45.

46P. A three-way 120 V lamp bulb that contains two filaments is rated for 100-200-300 W. One filament burns out. Afterward, the bulb operates at the same intensity (dissipates energy at the same rate) on its lowest and its highest switch positions but does not operate at all on the middle position. (a) How are the two filaments wired to the three switch positions? (b) Calculate the resistances of the filaments.

47P. (a) In Fig. 28-43, what is the equivalent resistance of the network shown? (b) What is the current in each resistor? Put $R_1 = 100 \, \Omega$, $R_2 = R_3 = 50 \, \Omega$, $R_4 = 75 \, \Omega$, and $\mathscr{E} = 6.0$ V; assume the battery is ideal.

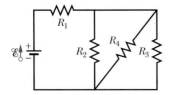

FIGURE 28-43
Problem 47.

48P. In Fig. 28-44, $\mathscr{E}_1 = 3.00$ V, $\mathscr{E}_2 = 1.00$ V, $R_1 = 5.00 \, \Omega$, $R_2 = 2.00 \, \Omega$, $R_3 = 4.00 \, \Omega$, and both batteries are ideal. (a) What is the rate at which energy is dissipated in R_1? In R_2? In R_3? (b) What is the power of battery 1? Of battery 2?

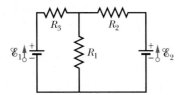

FIGURE 28-44
Problem 48.

49P. In the circuit of Fig. 28-45, for what value of R will the ideal battery transfer energy to the resistors (a) at a rate of 60.0 W, (b) at the maximum possible rate, and (c) at the minimum possible rate? (d) For (b) and (c), what are those rates?

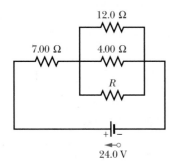

FIGURE 28-45
Problem 49.

50P. In the circuit of Fig. 28-46, $\mathscr{E}$ has a constant value but R can be varied. Find the value of R that results in the maximum heating in that resistor. The battery is ideal.

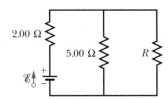

FIGURE 28-46
Problem 50.

51P. A copper wire of radius $a = 0.250$ mm has an aluminum jacket of outer radius $b = 0.380$ mm. (a) There is a current $i =$

2.00 A in the composite wire. Using Table 27-1, calculate the current in each material. (b) What is the length of the composite wire if a potential difference $V = 12.0$ V between the ends maintains the current?

52P. Figure 28-47 shows a battery connected across a uniform resistor R_0. A sliding contact can move across the resistor from $x = 0$ at the left to $x = 10$ cm at the right. Moving the contact changes how much resistance is to the left of the contact and how much is to the right. Find an expression for the power dissipated in resistor R as a function of x. Plot the function for $\mathscr{E} = 50$ V, $R = 2000 \, \Omega$, and $R_0 = 100 \, \Omega$.

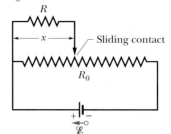

FIGURE 28-47
Problem 52.

SECTION 28-7 The Ammeter and the Voltmeter

53E. A simple ohmmeter is made by connecting a 1.50 V flashlight battery in series with a resistance R and an ammeter that reads from 0 to 1.00 mA, as shown in Fig. 28-48. Resistance R is adjusted so that when the clip leads are shorted together, the meter deflects to its full-scale value of 1.00 mA. What external resistance across the leads results in a deflection of (a) 10%, (b) 50%, and (c) 90% of full scale? (d) If the ammeter has a resistance of 20.0 Ω and the internal resistance of the battery is negligible, what is the value of R?

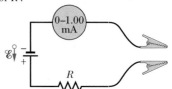

FIGURE 28-48
Exercise 53.

54E. For sensitive manual control of current in a circuit, you can use a parallel combination of variable resistors of the sliding contact type, as in Fig. 28-49. (Moving the contact changes how much resistance is in the circuit.) Suppose the full resistance R_1 of resistor A is 20 times the full resistance R_2 of resistor B. (a) What procedure should be used to adjust the current i to the desired value? (b) Why is the parallel combination better than a single variable resistor?

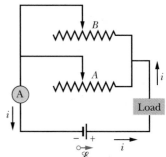

FIGURE 28-49
Exercise 54.

55P. (a) In Fig. 28-50, determine what the ammeter will read, assuming $\mathscr{E} = 5.0$ V (for the ideal battery), $R_1 = 2.0\ \Omega$, $R_2 = 4.0\ \Omega$, and $R_3 = 6.0\ \Omega$. (b) The ammeter and the source of emf are now physically interchanged. Show that the ammeter reading remains unchanged.

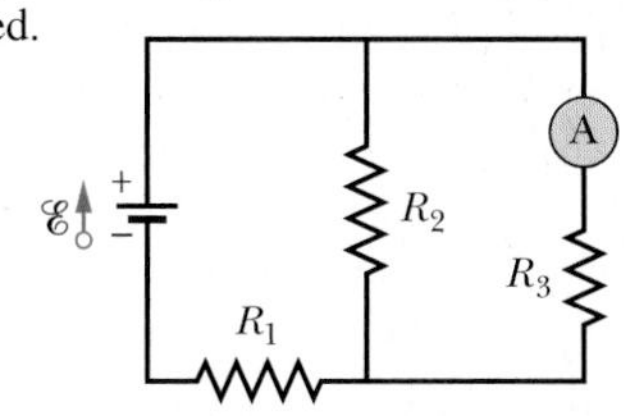

FIGURE 28-50 Problem 55.

56P. What current, in terms of $\mathscr{E}$ and R, does the ammeter in Fig. 28-51 read? Assume that it has zero resistance and that the battery is ideal.

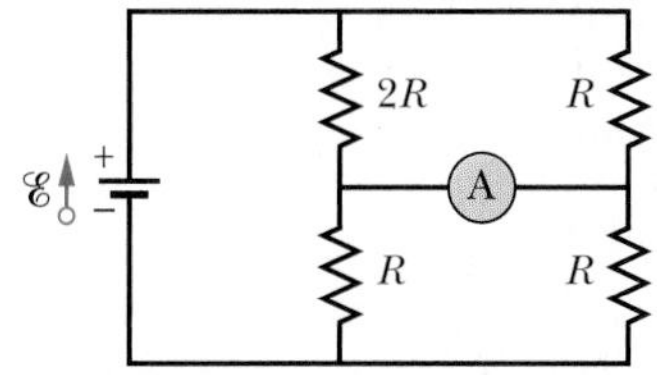

FIGURE 28-51 Problem 56.

57P. When the lights of an automobile are switched on, an ammeter in series with them reads 10 A and a voltmeter connected across them reads 12 V. See Fig. 28-52. When the electric starting motor is turned on, the ammeter reading drops to 8.0 A and the lights dim somewhat. If the internal resistance of the battery is 0.050 Ω and that of the ammeter is negligible, what are (a) the emf of the battery and (b) the current through the starting motor when the lights are on?

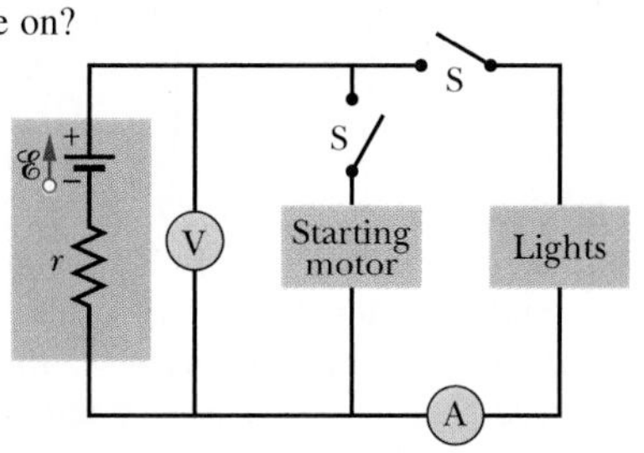

FIGURE 28-52 Problem 57.

58P. In Fig. 28-12, assume that $\mathscr{E} = 3.0$ V, $r = 100\ \Omega$, $R_1 = 250\ \Omega$, and $R_2 = 300\ \Omega$. If the voltmeter resistance $R_V = 5.0$ kΩ, what percent error is made in reading the potential difference across R_1? Ignore the presence of the ammeter.

59P. In Fig. 28-12, assume that $\mathscr{E} = 5.0$ V, $r = 2.0\ \Omega$, $R_1 = 5.0\ \Omega$, and $R_2 = 4.0\ \Omega$. If the ammeter resistance $R_A = 0.10\ \Omega$, what percent error is made in reading the current? Assume that the voltmeter is not present.

60P. A voltmeter (resistance R_V) and an ammeter (resistance R_A) are connected to measure a resistance R, as in Fig. 28-53*a*. The resistance is given by $R = V/i$, where V is the voltmeter reading and i is the current in the resistor R. Some of the current (i') registered by the ammeter goes through the voltmeter so that the ratio of the meter readings ($= V/i'$) gives only an *apparent* resistance reading R'. Show that R and R' are related by

$$\frac{1}{R} = \frac{1}{R'} - \frac{1}{R_V}.$$

Note that as $R_V \to \infty$, $R' \to R$.

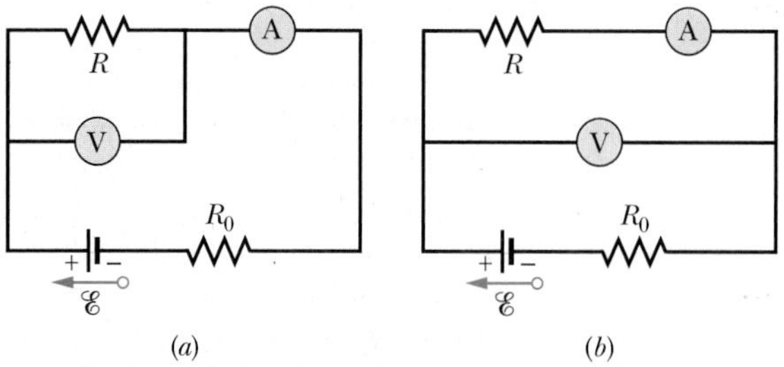

FIGURE 28-53 Problems 60, 61, and 62.

61P. (See Problem 60.) If meters are used to measure resistance, they may also be connected as in Fig. 28-53*b*. Again the ratio of the meter readings gives only an apparent resistance R'. Show that now R' is related to R by

$$R = R' - R_A,$$

in which R_A is the ammeter resistance. Note that as $R_A \to 0$, $R' \to R$.

62P. (See Problems 60 and 61.) In Fig. 28-53 the ammeter and voltmeter resistances are 3.00 and 300 Ω, respectively. Take $\mathscr{E} = 12.0$ V for the ideal battery and $R_0 = 100\ \Omega$. If $R = 85.0\ \Omega$, (a) what will the meters read for the two different connections? (b) What apparent resistance R' will be computed in each case?

63P. In Fig. 28-54, R_s is to be adjusted in value by moving the sliding contact across it until points a and b are brought to the same potential. (One tests for this condition by momentarily connecting a sensitive ammeter between a and b; if these points are at the same potential, the ammeter will not deflect.) Show that when this adjustment is made, the following relation holds:

$$R_x = R_s\left(\frac{R_2}{R_1}\right).$$

An unknown resistance (R_x) can be measured in terms of a standard (R_s) using this device, which is called a Wheatstone bridge.

64P. (a) If points a and b in Fig. 28-54 are connected by a wire of resistance r, show that the current in the wire is

$$i = \frac{\mathscr{E}(R_s - R_x)}{(R + 2r)(R_s + R_x) + 2R_sR_x},$$

where $\mathscr{E}$ is the emf of the ideal battery and $R = R_1 = R_2$. Assume that R_0 equals zero. (b) Is this formula consistent with the result of Problem 63?

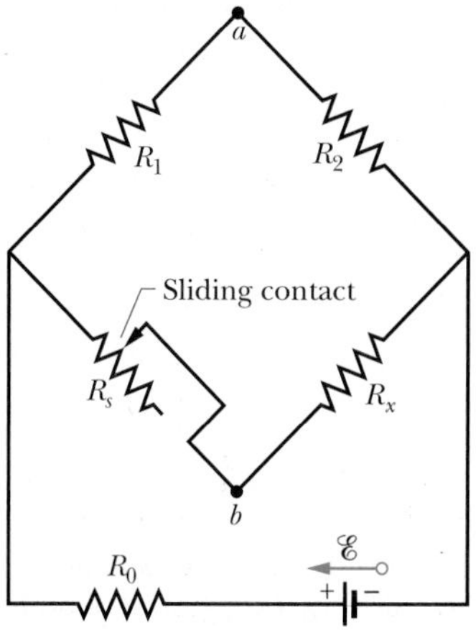

FIGURE 28-54 Problems 63 and 64.

SECTION 28-8 *RC* Circuits

65E. A capacitor with initial charge q_0 is discharged through a resistor. In terms of the time constant τ, how long is required for the capacitor to lose (a) the first one-third of its charge and (b) two-thirds of its charge?

66E. In an *RC* series circuit, $\mathscr{E} = 12.0$ V, $R = 1.40$ MΩ, and $C = 1.80$ μF. (a) Calculate the time constant. (b) Find the maximum charge that will appear on the capacitor during charging. (c) How long does it take for the charge to build up to 16.0 μC?

67E. How many time constants must elapse for an initially uncharged capacitor in an *RC* series circuit to be charged to 99.0% of its equilibrium charge?

68E. A 15.0 kΩ resistor and a capacitor are connected in series and then a 12.0 V potential difference is suddenly applied across them. The potential difference across the capacitor rises to 5.00 V in 1.30 μs. (a) Calculate the time constant of the circuit. (b) Find the capacitance of the capacitor.

69P. A 3.00 MΩ resistor and a 1.00 μF capacitor are connected in series with an ideal battery of $\mathscr{E} = 4.00$ V. At 1.00 s after the connection is made, what are the rates at which (a) the charge of the capacitor is increasing, (b) energy is being stored in the capacitor, (c) thermal energy is appearing in the resistor, and (d) energy is being delivered by the battery?

70P. A capacitor with an initial potential difference of 100 V is discharged through a resistor when a switch between them is closed at $t = 0$. At $t = 10.0$ s, the potential difference across the capacitor is 1.00 V. (a) What is the time constant of the circuit? (b) What is the potential difference across the capacitor at $t = 17.0$ s?

71P. Figure 28-55 shows the circuit of a flashing lamp, like those attached to barrels at highway construction sites. The fluorescent lamp L (of negligible capacitance) is connected in parallel across the capacitor C of an *RC* circuit. There is a current through the lamp only when the potential difference across it reaches the breakdown voltage V_L; in this event, the capacitor discharges completely through the lamp and the lamp flashes briefly. Suppose that two flashes per second are needed. For a lamp with breakdown voltage $V_L = 72.0$ V, a 95.0 V ideal battery, and a 0.150 μF capacitor, what should be the resistance R of the resistor?

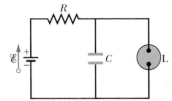

FIGURE 28-55
Problem 71.

72P. A 1.0 μF capacitor with an initial stored energy of 0.50 J is discharged through a 1.0 MΩ resistor. (a) What is the initial charge on the capacitor? (b) What is the current through the resistor when the discharge starts? (c) Determine V_C, the potential difference across the capacitor, and V_R, the potential difference across the resistor, as functions of time. (d) Express the production rate of thermal energy in the resistor as a function of time.

73P. The potential difference between the plates of a leaky (meaning that charge leaks from one plate to the other) 2.0 μF capacitor drops to one-fourth its initial value in 2.0 s. What is the equivalent resistance between the capacitor plates?

74P. An initially uncharged capacitor C is fully charged by a device of constant emf $\mathscr{E}$ connected in series with a resistor R. (a) Show that the final energy stored in the capacitor is half the energy supplied by the emf device. (b) By direct integration of i^2R over the charging time, show that the thermal energy dissipated by the resistor is also half the energy supplied by the emf device.

75P. A controller on an electronic arcade game consists of a variable resistor connected across the plates of a 0.220 μF capacitor. The capacitor is charged to 5.00 V, then discharged through the resistor. The time for the potential difference across the plates to decrease to 0.800 V is measured by a clock inside the game. If the range of discharge times that can be handled effectively is from 10.0 μs to 6.00 ms, what should be the resistance range of the resistor?

76P. The circuit of Fig. 28-56 shows a capacitor C, two ideal batteries, two resistors, and a switch S. Initially S has been open for a long time. If it is then closed for a long time, by how much does the charge on the capacitor change? Assume $C = 10$ μF, $\mathscr{E}_1 = 1.0$ V, $\mathscr{E}_2 = 3.0$ V, $R_1 = 0.20$ Ω, and $R_2 = 0.40$ Ω.

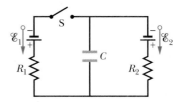

FIGURE 28-56
Problem 76.

77P*. In the circuit of Fig. 28-57, $\mathscr{E} = 1.2$ kV, $C = 6.5$ μF, $R_1 = R_2 = R_3 = 0.73$ MΩ. With C completely uncharged, switch S is suddenly closed (at $t = 0$). (a) Determine the current through each resistor for $t = 0$ and $t = \infty$. (b) Draw qualitatively a graph of the potential difference V_2 across R_2 from $t = 0$ to $t = \infty$. (c) What are the numerical values of V_2 at $t = 0$ and $t = \infty$? (d) Give the physical meaning of "$t = \infty$" in this case.

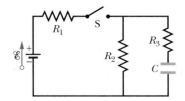

FIGURE 28-57
Problem 77.

Electronic Computation

78. Figure 28-58 shows a portion of a circuit. The rest of the circuit draws current i at the connections A and B, as indicated. Take $\mathscr{E}_1 = 10$ V, $\mathscr{E}_2 = 15$ V, $R_1 = R_2 = 5.0$ Ω, $R_3 = R_4 = 8.0$ Ω, and $R_5 = 12$ Ω. (a) For each of four values of i—0, 4.0, 8.0, and 12 A—find the current through each ideal battery and tell if the battery is charging or discharging. Also find the potential difference V_{AB}. (b) The portion of the circuit not shown consists of an emf and a resistor in series. What are their values?

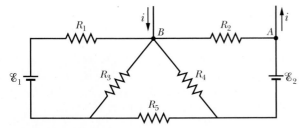

FIGURE 28-58 Problem 78.

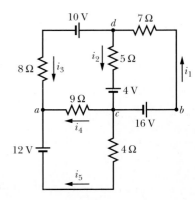

FIGURE 28-59
Problems 80 and 81.

79. The following table gives the electric potential difference V_T across the terminals of a battery as a function of current i being drawn from the battery. (a) Write an equation that represents the relationship between the terminal potential difference V_T and the current i. Enter the data into your graphing calculator and perform a linear regression fit of V_T versus i. (b) From the parameters of the fit, find (b) the battery's emf and (c) its internal resistance.

i (A):	50	75	100	125	150	175	200
V_T (V):	10.7	9.0	7.7	6.0	4.8	3.0	1.7

80. Consider the circuit in Fig. 28-59. (a) Apply the junction rule to junctions d and a and the loop rule to the three loops to produce five simultaneous, linearly independent equations. (b) Represent the five linear equations by the matrix equation $[A][B] = [C]$, where

$$[B] = \begin{bmatrix} i_1 \\ i_2 \\ i_3 \\ i_4 \\ i_5 \end{bmatrix}.$$

What are the matrices $[A]$ and $[C]$? (c) Have the calculator perform $[A]^{-1}[C]$ to find the values of i_1, i_2, i_3, i_4, and i_5.

81. For the same situation as in Problem 80 and having already solved for the five unknown currents, do the following. (a) Find the electric potential difference across the 9 Ω resistor. (b) Find the rate at which work is being done on the 7 Ω resistor. (c) Find the rate at which the 12 V battery is doing work on the circuit. (d) Find the rate at which the 4 V battery is doing work on the circuit. (e) Of the points in the circuit labeled a and c, which is at the higher electric potential?

82. A capacitor with capacitance C_0, after having been connected to a battery with emf $\mathcal{E}_0$ for a long time, is discharged through a 200,000 Ω resistor at time $t = 0$. The potential difference across the capacitor is then measured as a function of time for a brief time interval; the results are recorded below. (a) Write an equation that describes the potential difference across the capacitor as a function of time. Enter the data into your calculator and have the calculator perform a linear regression fit of $\ln V_C$ versus t. From the parameters of the fit, determine (b) the emf $\mathcal{E}_0$ of the battery and (c) the time constant τ for the circuit. (d) Finally, determine the value of C_0.

V_C (V):	9.9	7.2	5.7	4.4	3.4	2.7	2.0
t (s):	0.2	0.4	0.6	0.8	1.0	1.2	1.4

29
Magnetic Fields

If you are outside on a dark night in the middle to high latitudes, you might be able to see an aurora, a ghostly "curtain" of light that hangs down from the sky. This curtain is not just local: it may be several hundred kilometers high and several thousand kilometers long, stretching around Earth in an arc. However, it is less than 1 km thick. What produces this huge display, and what makes it so thin?

29-1 THE MAGNETIC FIELD

We have discussed how a charged plastic rod produces a vector field—the electric field **E**—at all points in the space around it. Similarly, a magnet produces a vector field—the **magnetic field B**—at all points in the space around it. You get a hint of that magnetic field whenever you attach a note to a refrigerator door with a small magnet, or accidentally erase a computer disk by bringing it near a magnet. The magnet acts on the door or disk *by means of* its magnetic field.

In a familiar type of magnet, a wire coil is wound around an iron core and a current is sent through the coil; the strength of the magnetic field is determined by the size of the current. In industry, such **electromagnets** are used for sorting scrap iron (Fig. 29-1) among many other things. You are probably more familiar with **permanent magnets**—magnets, like the refrigerator-door type, that do not need current to have a magnetic field.

FIGURE 29-1 Scrap metal collected by an electromagnet at a steel mill.

In Chapter 23 we saw that an *electric charge* sets up an electric field that can then affect other electric charges. Here, we might reasonably expect that a *magnetic charge* sets up a magnetic field that can then affect other magnetic charges. Although such magnetic charges, called *magnetic monopoles*, are predicted by certain theories, their existence has not been confirmed.

So then, how are magnetic fields set up? There are two ways. (1) Moving electrically charged particles, such as a current in a wire, create magnetic fields. (2) Elementary particles such as electrons have an intrinsic magnetic field around them; that is, these fields are a basic characteristic of the particles, just as are their mass and electric charge (or lack of charge). As we shall discuss in Chapter 32, the magnetic fields of the electrons in certain materials add together to give a net magnetic field around the material. This is true for the material in permanent magnets (which is good, because they can then hold notes to a refrigerator door). In other materials, the magnetic fields of all the electrons cancel out, giving no net magnetic field surrounding the material. This is true for the material in your body (which is also good, because otherwise you might be slammed up against a refrigerator door every time you passed one).

Experimentally we find that when a charged particle (either alone or part of a current) moves through a magnetic field, a force due to the field can act on the particle. In this chapter we focus on the relation between the magnetic field and this force.

29-2 THE DEFINITION OF B

We determined the electric field **E** at a point by putting a test particle of charge q at rest at that point and measuring the electric force $\mathbf{F}_E$ acting on the particle. We then defined **E** as

$$\mathbf{E} = \frac{\mathbf{F}_E}{q}. \tag{29-1}$$

If a magnetic monopole were available, we could define **B** in a similar way. Because such particles have not been found, we must define **B** in another way, in terms of the magnetic force $\mathbf{F}_B$ exerted on a moving electrically charged test particle.

In principle, we do this by firing a charged particle through the point where **B** is to be defined, using various directions and speeds for the particle and determining the force $\mathbf{F}_B$ that acts on the particle at that point. After many such trials we would find that when the particle's velocity **v** is along a particular axis through the point, force $\mathbf{F}_B$ is zero. For all other directions of **v**, the magnitude of $\mathbf{F}_B$ is always proportional to $v \sin \phi$, where ϕ is the angle between the zero-force axis and the direction of **v**. Furthermore, the direction of $\mathbf{F}_B$ is always perpendicular to the direction of $\mathbf{v}$. (These results suggest that a cross product is involved.)

We can then define a magnetic field $\mathbf{B}$ to be a vector quantity that is directed along the zero-force axis. We can next measure the magnitude of $\mathbf{F}_B$ when $\mathbf{v}$ is directed perpendicular to that axis and then define the magnitude of $\mathbf{B}$ in terms of that force magnitude:

$$B = \frac{F_B}{|q|v},$$

where q is the charge of the particle.

We can summarize all these results with the following vector equation:

$$\mathbf{F}_B = q\mathbf{v} \times \mathbf{B}. \qquad (29\text{-}2)$$

That is, the force $\mathbf{F}_B$ on the particle is equal to the charge q times the cross product of its velocity $\mathbf{v}$ and the magnetic field $\mathbf{B}$. Using Eq. 3-20 to evaluate the cross product, we can write the magnitude of $\mathbf{F}_B$ as

$$F_B = |q|vB \sin \phi, \qquad (29\text{-}3)$$

where ϕ is the angle between the directions of velocity $\mathbf{v}$ and magnetic field $\mathbf{B}$.

Finding the Magnetic Force on a Particle

Equation 29-3 tells us that the magnitude of the force $\mathbf{F}_B$ acting on a particle in a magnetic field is proportional to the charge q and speed v of the particle. Thus, the force is equal to zero if the charge is zero or if the particle is stationary. Equation 29-3 also tells us that the magnitude of the force is zero if $\mathbf{v}$ and $\mathbf{B}$ are either parallel ($\phi = 0°$) or antiparallel ($\phi = 180°$), and the force is at its maximum when $\mathbf{v}$ and $\mathbf{B}$ are perpendicular to each other.

Equation 29-2 tells us all this plus the direction of $\mathbf{F}_B$. From Section 3-7, we know that the cross product $\mathbf{v} \times \mathbf{B}$ in

Eq. 29-2 is a vector that is perpendicular to the two vectors $\mathbf{v}$ and $\mathbf{B}$. The right-hand rule (Fig. 29-2a) tells us that the thumb of the right hand points in the direction of $\mathbf{v} \times \mathbf{B}$ when the fingers sweep $\mathbf{v}$ into $\mathbf{B}$. If q is positive, then (by Eq. 29-2) the force $\mathbf{F}_B$ has the same sign as $\mathbf{v} \times \mathbf{B}$ and thus must be in the same direction. That is, for positive q, $\mathbf{F}_B$ points along the thumb as in Figs. 29-2b. If q is negative, then the force $\mathbf{F}_B$ and the cross product $\mathbf{v} \times \mathbf{B}$ have opposite signs and thus must be in opposite directions. So, for negative q, $\mathbf{F}_B$ points opposite the thumb as in Fig. 29-2c.

Regardless of the sign of the charge, however,

The force $\mathbf{F}_B$ acting on a charged particle moving with velocity $\mathbf{v}$ through a magnetic field $\mathbf{B}$ is *always* perpendicular to $\mathbf{v}$ and $\mathbf{B}$.

Thus $\mathbf{F}_B$ *never* has a component parallel to $\mathbf{v}$. This means that $\mathbf{F}_B$ cannot change the particle's speed v (and thus it cannot change the particle's kinetic energy). The force can change only the direction of $\mathbf{v}$ (and thus the direction of travel); only in this sense does $\mathbf{F}_B$ accelerate the particle.

To develop a feeling for Eq. 29-2, consider Fig. 29-3, which shows some tracks left by charged particles moving rapidly through a *bubble chamber* at the Lawrence Berkeley Laboratory. The chamber, which is filled with liquid hydrogen, is immersed in a strong uniform magnetic field that points out of the plane of the figure. At the left in Fig. 29-3 an incoming gamma ray—which leaves no track because it is uncharged—transforms into an electron (spiral track marked e⁻) and a positron (track marked e⁺) while it knocks an electron out of a hydrogen atom (long track marked e⁻). Check with Eq. 29-2 and Fig. 29-2 that the three tracks made by these two negative particles and one positive particle curve in the proper directions.

The SI unit for $\mathbf{B}$ that follows from Eqs. 29-2 and 29-3 is the newton per coulomb-meter per second. For conve-

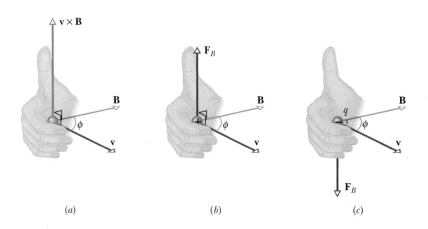

(a)   *(b)*   *(c)*

FIGURE 29-2 (*a*) The right-hand rule (in which $\mathbf{v}$ is swept into $\mathbf{B}$ through the smaller angle ϕ between them) gives the direction of $\mathbf{v} \times \mathbf{B}$ as the direction of the thumb. (*b*) If q is positive, then the direction of $\mathbf{F}_B = q\mathbf{v} \times \mathbf{B}$ is in the direction of $\mathbf{v} \times \mathbf{B}$. (*c*) If q is negative, then the direction of $\mathbf{F}_B$ is opposite that of $\mathbf{v} \times \mathbf{B}$.

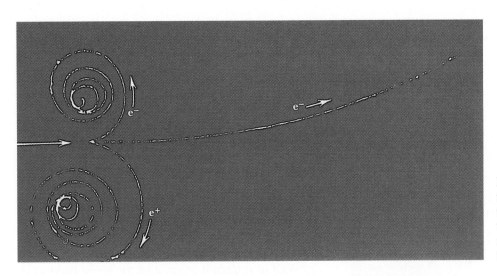

FIGURE 29-3 The tracks of two electrons (e⁻) and a positron (e⁺) in a bubble chamber that is immersed in a uniform magnetic field that points out of the plane of the page.

nience, this is called the **tesla** (T):

$$1 \text{ tesla} = 1 \text{ T} = 1 \frac{\text{newton}}{(\text{coulomb})(\text{meter/second})}.$$

Recalling that a coulomb per second is an ampere, we have

$$1 \text{ T} = 1 \frac{\text{newton}}{(\text{coulomb/second})(\text{meter})}$$

$$= 1 \frac{\text{N}}{\text{A} \cdot \text{m}}. \tag{29-4}$$

An earlier (non-SI) unit for **B**, still in common use, is the *gauss* (G), and

$$1 \text{ tesla} = 10^4 \text{ gauss}. \tag{29-5}$$

Table 29-1 lists the magnetic fields that occur in a few situations. Note that Earth's magnetic field near the planet's surface is about 10^{-4} T (= 100 μT or 1 gauss).

TABLE 29-1   SOME APPROXIMATE MAGNETIC FIELDS

At the surface of a neutron star	10^8 T
Near a big electromagnet	1.5 T
Near a small bar magnet	10^{-2} T
At Earth's surface	10^{-4} T
In interstellar space	10^{-10} T
Smallest value in a magnetically shielded room	10^{-14} T

Magnetic Field Lines

We can represent magnetic fields with field lines, just as we did for electric fields. Similar rules apply. That is, (1) the direction of the tangent to a magnetic field line at any point gives the direction of **B** at that point, and (2) the spacing of the lines represents the magnitude of **B**—the magnetic field is stronger where the lines are closer together, and conversely.

Figure 29-4*a* shows how the magnetic field near a *bar magnet* (a permanent magnet in the shape of a bar) can be represented by magnetic field lines. The lines all pass through the magnet, and they form closed loops (even those that are not shown closed in the figure). The external magnetic effects of a bar magnet are strongest near its ends, where the field lines are most closely spaced. Thus the magnetic field of the bar magnet in Fig. 29-4*b* collects the iron filings near the two ends of the magnet.

Because a magnetic field has direction, the (closed) field lines enter one end of a magnet and exit the other end. The end of a magnet from which the field lines emerge is called the *north pole* of the magnet; the other end, where field lines enter the magnet, is called the *south pole*. The magnets we use to fix notes on refrigerators are short bar

---

CHECKPOINT **1:** The figure shows three situations in which a charged particle with velocity **v** travels through a uniform magnetic field **B**. In each situation, what is the direction of the magnetic force $\mathbf{F}_B$ on the particle?

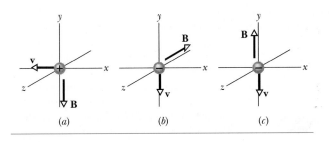

(a)   (b)   (c)

(a)   (b)

FIGURE 29-4 (a) The magnetic field lines for a bar magnet. (b) A "cow magnet": a bar magnet that is intended to be slipped down into the rumen of a cow to prevent accidentally ingested bits of scrap iron from reaching the cow's intestines.

magnets. Figure 29-5 shows two other common shapes for magnets: a *horseshoe magnet* and a magnet that has been bent around into the shape of a C so that the *pole faces* are facing each other. (The magnetic field between the pole faces can then be approximately uniform.) Regardless of the shape of the magnets, if we place two of them near each other we find:

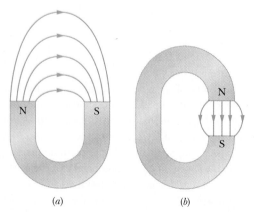

(a)   (b)

FIGURE 29-5 (a) A horseshoe magnet and (b) a C-shaped magnet. (Only some of the external field lines are shown.)

Opposite magnetic poles attract each other, and like magnetic poles repel each other.

Earth has a magnetic field that is produced in its core by still unknown mechanisms. On Earth's surface, we can detect this magnetic field with a compass, which is essentially a slender bar magnet on a low-friction pivot. This bar magnet, or this needle, turns because its north-pole end is attracted toward the Arctic region of Earth. Thus, the *south* pole of Earth's magnetic field must be located toward the Arctic. Logically, we then should call the pole there a south pole. However, because we call that direction north, we are trapped into the statement that Earth has a *geomagnetic north pole* in that direction.

With more careful observation we would find that in the northern hemisphere, the magnetic field lines of Earth generally point down into Earth and toward the Arctic. And in the southern hemisphere, they generally point up out of Earth and away from the Antarctic, that is, away from Earth's *geomagnetic south pole*.

SAMPLE PROBLEM 29-1

A uniform magnetic field **B**, with magnitude 1.2 mT, points vertically upward throughout the volume of a laboratory chamber. A proton with kinetic energy 5.3 MeV enters the chamber, moving horizontally from south to north. What magnetic deflecting force acts on the proton as it enters the chamber? The proton mass is 1.67×10^{-27} kg.

SOLUTION: The magnetic deflecting force depends on the speed of the proton, which we can find from $K = \frac{1}{2}mv^2$. Solving for v, we find

$$v = \sqrt{\frac{2K}{m}} = \sqrt{\frac{(2)(5.3 \text{ MeV})(1.60 \times 10^{-13} \text{ J/MeV})}{1.67 \times 10^{-27} \text{ kg}}}$$

$$= 3.2 \times 10^7 \text{ m/s}.$$

Equation 29-3 then yields

$$F_B = |q|vB \sin \phi$$
$$= (1.60 \times 10^{-19} \text{ C})(3.2 \times 10^7 \text{ m/s})$$
$$\times (1.2 \times 10^{-3} \text{ T})(\sin 90°)$$
$$= 6.1 \times 10^{-15} \text{ N}. \qquad \text{(Answer)}$$

This may seem like a small force, but it acts on a particle of small mass, producing a large acceleration, namely,

$$a = \frac{F_B}{m} = \frac{6.1 \times 10^{-15} \text{ N}}{1.67 \times 10^{-27} \text{ kg}} = 3.7 \times 10^{12} \text{ m/s}^2.$$

It remains to find the direction of $\mathbf{F}_B$. We know that **v** points horizontally from south to north and **B** points vertically up. The right-hand rule (see Fig. 29-2b) shows us that the deflecting force $\mathbf{F}_B$ must point horizontally from west to east, as Fig. 29-6 shows. (The array of dots in the figure represents a magnetic field pointing directly out of the plane of the figure. An array of Xs would have represented a magnetic field pointing directly into that plane.)

If the charge of the particle were negative, the magnetic deflecting force would point in the opposite direction, that is, horizontally from east to west. This is predicted automatically by Eq. 29-2, if we substitute $-e$ for q.

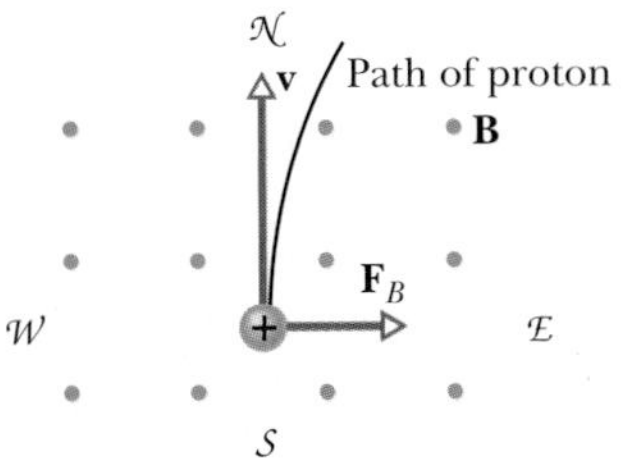

FIGURE 29-6 Sample Problem 29-1. An overhead view of a proton moving from south to north with velocity **v** in a chamber. A magnetic field points vertically upward in the chamber, as represented by the array of dots (which resemble the tips of arrows). The proton is deflected toward the east.

PROBLEM SOLVING TACTICS

TACTIC 1: *Classical and Relativistic Formulas for Kinetic Energy*

In Sample Problem 29-1, we used the (approximate) classical expression ($K = \frac{1}{2}mv^2$) for the kinetic energy of the proton rather than the (exact) relativistic expression (see Eq. 7-51). The criterion for when the classical expression may safely be used is that $K \ll mc^2$, where mc^2 is the rest energy of the particle. In this case, $K = 5.3$ MeV and the rest energy of a proton is 938 MeV. This proton passes the test and we were justified in treating it as "slow," that is, in using the classical $K = \frac{1}{2}mv^2$ formula for the kinetic energy. That is not always the case in dealing with energetic particles.

29-3 CROSSED FIELDS: DISCOVERY OF THE ELECTRON

Both an electric field **E** and a magnetic field **B** can produce a force on a charged particle. When the two fields are perpendicular to each other, they are said to be *crossed fields*. Here we shall examine what happens to charged particles, namely, electrons, as they move through crossed fields. We use as our example the experiment that led to the discovery of the electron in 1897 by J. J. Thomson at Cambridge University.

Figure 29-7 shows a modern, simplified version of Thomson's experimental apparatus—a *cathode ray tube* (which is like the "picture tube" in a standard television set). Charged particles (which we now know as electrons) are emitted by a hot filament at the rear of the evacuated tube and are accelerated by an applied potential difference V. After they pass through a slit in screen C, they form a narrow beam. They then pass through a region of crossed **E** and **B** fields, headed toward a fluorescent screen S, where they will produce a spot of light (on a television screen the spot would be part of the picture). The forces on the charged particles in the crossed-fields region can deflect them from the center of the screen. By controlling the magnitudes and directions of the fields, Thomson could thus control where the spot of light appeared on the screen. For the particular field arrangement of Fig. 29-7, electrons are forced up the page by the electric field **E** and down the page by the magnetic field **B**—that is, the forces are *in opposition*. Thomson's procedure was equivalent to the following series of steps.

1. Set $E = 0$ and $B = 0$ and note the position of the spot on screen S due to the undeflected beam.

2. Turn on **E** and measure the resulting beam deflection.

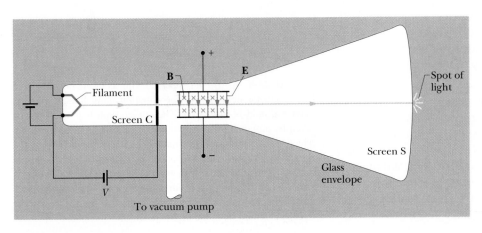

FIGURE 29-7 A modern version of J. J. Thomson's apparatus for measuring the ratio of mass to charge for the electron. The electric field **E** is established by connecting a battery across the deflecting plate terminals. The magnetic field **B** is set up by means of a current in a system of coils (not shown). The magnetic field shown is into the plane of the figure, as represented by the array of Xs (which resemble the feathered ends of arrows).

3. Maintaining **E**, now turn on **B** and adjust its value until the beam returns to the undeflected position. (With the forces in opposition, they can be made to cancel.)

We discussed the deflection of a charged particle moving through an electric field **E** between two plates (step 2 here) in Sample Problem 23-8. We found that the deflection of the particle at the far end of the plates is

$$y = \frac{qEL^2}{2mv^2}, \qquad (29\text{-}6)$$

where v is the particle's speed, m its mass, and q its charge, and L is the length of the plates. We can apply this same equation to the beam of electrons in Fig. 29-7 by measuring the deflection of the beam on screen S and then working back to calculate the deflection y at the end of the plates. (Because the direction of the deflection is set by the sign of the particle's charge, Thomson was able to show that the particles that were lighting up his screen were negatively charged.)

When the two fields in Fig. 29-7 are adjusted so that the two deflecting forces cancel (step 3), we have from Eqs. 29-1 and 29-3

$$|q|E = |q|vB \sin(90°) = |q|vB,$$

or

$$v = \frac{E}{B}. \qquad (29\text{-}7)$$

Thus the crossed fields allow us to measure the speed of the charged particles passing through them. Substituting Eq. 29-7 for v in Eq. 29-6 and rearranging yield

$$\frac{m}{q} = \frac{B^2L^2}{2yE}, \qquad (29\text{-}8)$$

in which all quantities on the right can be measured. Thus, the crossed fields allow us to measure the ratio m/q of the particles moving through Thomson's apparatus.

Thomson claimed that these particles are found in all matter. He also claimed that they are lighter than the lightest known atom (hydrogen) by a factor of more than 1000. (The exact ratio proved later to be 1836.15.) His m/q measurement, coupled with the boldness of his two claims, is considered to be the "discovery of the electron."

$\mathbb{C}$HECKPOINT **2:** The figure shows four directions for the velocity vector **v** of a positively charged particle moving through a uniform electric field **E** (directed out of the page and represented by an encircled dot) and a uniform magnetic field **B**. (a) Rank directions 1, 2, and 3 according to the magnitude of the net force on the particle, greatest first. (b) Of all four directions, which might result in a net force of zero?

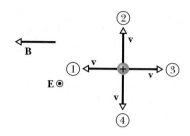

29-4 CROSSED FIELDS: THE HALL EFFECT

As we just discussed, a beam of electrons in a vacuum can be deflected by a magnetic field. Can the drifting conduction electrons in a copper wire also be deflected by a magnetic field? In 1879, Edwin H. Hall, then a 24-year-old graduate student at the Johns Hopkins University, showed that they can. This **Hall effect** allows us to find out whether the charge carriers in a conductor are positively or negatively charged. Beyond that, we can measure the number of such carriers per unit volume of the conductor.

Figure 29-8a shows a copper strip of width d, carrying a current i whose conventional direction is from the top of

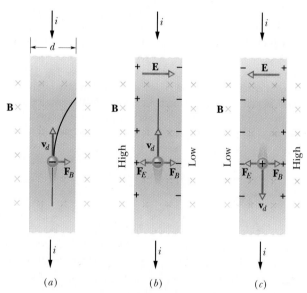

(a)   (b)   (c)

FIGURE 29-8 A strip of copper carrying a current i is immersed in a magnetic field **B**. (a) The situation immediately after the magnetic field is turned on. The curved path that will then be taken by an electron is shown. (b) The situation at equilibrium, which quickly follows. Note that negative charges pile up on the right side of the strip, leaving uncompensated positive charges on the left. Thus the left side is at a higher potential than the right side. (c) For the same current direction, if the charge carriers were positively charged, *they* would pile up on the right side, and the right side would be at the higher potential.

the figure to the bottom. The charge carriers are electrons and, as we know, they drift (with drift speed v_d) in the opposite direction, from bottom to top. At the instant shown in Fig. 29-8a, an external magnetic field **B**, pointing into the plane of the figure, has just been turned on. From Eq. 29-2 we see that a magnetic deflecting force $\mathbf{F}_B$ will act on each drifting electron, pushing it toward the right edge of the strip.

As time goes on, electrons will move to the right, mostly piling up on the right edge of the strip, leaving uncompensated positive charges in fixed positions at the left edge. The separation of positive and negative charges produces an electric field **E** within the strip, pointing from left to right in Fig. 29-8b. This field will exert an electric force $\mathbf{F}_E$ on each electron, tending to push it to the left.

An equilibrium quickly develops in which the electric force on each electron builds up until it just cancels the magnetic force. When this happens, as Fig. 29-8b shows, the force due to **B** and the force due to **E** are in balance. The drifting electrons then move along the strip toward the top of the page with no further collection of electrons on the right edge of the strip and thus no further increase in the electric field **E.**

A *Hall potential difference V* is associated with the electric field across strip width d. From Eq. 25-42, the magnitude of that potential difference is

$$V = Ed. \tag{29-9}$$

By connecting a voltmeter across the width, we can measure the potential difference between the two edges of the strip. Moreover, the voltmeter can tell us which edge is at higher potential. For the situation of Fig. 29-8a, we would find that the left edge is at higher potential, which is consistent with our assumption that the charge carriers are negatively charged.

For a moment, let us make the opposite assumption, that the charge carriers in current i are positively charged (Fig. 29-8c). Convince yourself that as these charge carriers moved from top to bottom in the strip, they would be pushed to the right edge by $\mathbf{F}_B$ and thus that the *right* edge would be at higher potential. Because that last statement is contradicted by our voltmeter reading, the charge carriers must be negatively charged.

Now for the quantitative part. When the electric and magnetic forces are in balance (Fig. 29-8b), Eqs. 29-1 and 29-3 give us

$$eE = ev_dB. \tag{29-10}$$

From Eq. 27-7, the drift speed v_d is

$$v_d = \frac{J}{ne} = \frac{i}{neA}, \tag{29-11}$$

in which $J\ (= i/A)$ is the current density in the strip, A is the cross-sectional area of the strip, and n is the *number density* of charge carriers (their number per unit volume).

In Eq. 29-10, substituting for E with Eq. 29-9 and substituting for v_d with Eq. 29-11, we obtain

$$n = \frac{Bi}{Vle}, \tag{29-12}$$

in which $l\ (= A/d)$ is the thickness of the strip. Thus we can find n in terms of quantities that we can measure.

It is also possible to use the Hall effect to measure directly the drift speed v_d of the charge carriers, which you may recall is of the order of centimeters per hour. In this clever experiment, the metal strip is moved mechanically through the magnetic field in a direction opposite that of the drift velocity of the charge carriers. The speed of the moving strip is then adjusted until the Hall potential difference vanishes. At this condition, with no Hall effect, the velocity of the charge carriers *with respect to the magnetic field* must be zero. So the velocity of the strip must be equal in magnitude but opposite in direction to the velocity of the negative charge carriers.

SAMPLE PROBLEM 29-2

Figure 29-9 shows a solid metal cube, of edge length $d = 1.5$ cm, moving in the positive y direction at a constant velocity **v** of magnitude 4.0 m/s. The cube moves through a uniform magnetic field **B** of magnitude 0.050 T and pointing in the positive z direction.

(a) Which cube face is at a lower electric potential and which is at a higher electric potential because of the motion through the field?

SOLUTION: When the cube first began to move through the magnetic field, the conduction electrons within the cube also began to move through the field. Because of their motion, they experienced a force $\mathbf{F}_B$ given by Eq. 29-2. In Fig. 29-9, $\mathbf{F}_B$ acts in the negative direction of the x axis. This means that some of the electrons were deflected by $\mathbf{F}_B$ to the (hidden) left cube face, making that face negatively charged and the right face positively charged. This charge separation produces an electric field **E** directed from the right face toward the left face.

FIGURE 29-9 Sample Problem 29-2. A solid metal cube of edge length d moves at constant velocity **v** through a uniform magnetic field **B**.

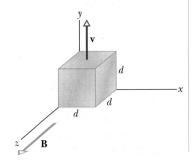

Thus, the left face is at lower potential and the right face is at higher potential.

(b) What is the potential difference V between the faces of higher and lower electric potential?

SOLUTION: The electric field E that is produced by the charge separation causes a force F_E to act on the electrons; F_E is directed toward the right cube face, in the direction opposite that of force F_B. Equilibrium, in which $F_E = F_B$, is reached quickly after the cube begins to move through the magnetic field. From Eqs. 29-1 and 29-3, we then have

$$eE = evB.$$

Substituting for E with Eq. 29-9 ($V = Ed$) then yields

$$V = dvB. \qquad (29\text{-}13)$$

Substituting the given data, we now find

$$V = (0.015 \text{ m})(4.0 \text{ m/s})(0.050 \text{ T})$$
$$= 0.0030 \text{ V} = 3.0 \text{ mV}. \qquad \text{(Answer)}$$

CHECKPOINT 3: The figure shows a metallic, rectangular solid that is to move at a certain speed v through the uniform magnetic field **B**. Its dimensions are multiples of d, as shown. You have six choices for the direction of the velocity of the solid: it can be parallel to x, y, or z, in either the positive or negative direction. (a) Rank the six choices according to the potential set up across the solid, greatest first. (b) For which choice is the front face at lower potential?

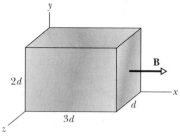

29-5 A CIRCULATING CHARGED PARTICLE

If a particle moves in a circle at constant speed, we can be sure that the net force acting on the particle is constant in magnitude and points toward the center of the circle, always perpendicular to the particle's velocity. Think of a stone tied to a string and whirled in a circle on a smooth horizontal surface, or of a satellite moving in a circular orbit around Earth. In the first case, the tension in the string provides the necessary force and centripetal acceleration. In the second case, Earth's gravitational attraction provides the force and acceleration.

Figure 29-10 shows another example: a beam of electrons is projected into a chamber by an *electron gun* G. The electrons enter in the plane of the page with velocity **v** and move in a region of uniform magnetic field **B** directed out of the plane of the figure. As a result, a magnetic force $\mathbf{F}_B = q\mathbf{v} \times \mathbf{B}$ continually deflects the electrons, and because **v** and **B** are perpendicular to each other, this deflection causes the electrons to follow a circular path. The path is visible in the photo because atoms of gas in the chamber emit light when some of the circulating electrons collide with them.

We would like to determine the parameters that characterize the circular motion of these electrons, or of any particle of charge magnitude q and mass m moving perpendicular to a uniform magnetic field **B** at speed v. From Eq. 29-3, the force acting on the particle has a magnitude of qvB. So from Newton's second law applied to uniform circular motion (Eq. 6-20),

$$F = ma = \frac{mv^2}{r}, \qquad (29\text{-}14)$$

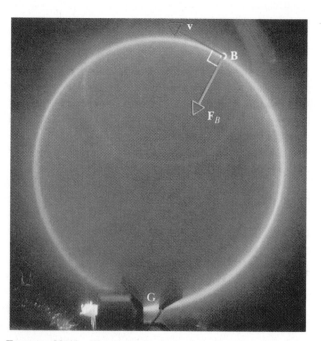

FIGURE 29-10 Electrons circulating in a chamber containing gas at low pressure (their path is the glowing circle). A uniform magnetic field **B**, pointing directly out of the plane of the page, fills the chamber. Note the radially directed magnetic force $\mathbf{F}_B$: for circular motion to occur, $\mathbf{F}_B$ *must* point toward the center of the circle. Use the right-hand rule for cross products to confirm that $\mathbf{F}_B = q\mathbf{v} \times \mathbf{B}$ gives $\mathbf{F}_B$ the proper direction.

we have

$$qvB = \frac{mv^2}{r}. \qquad (29\text{-}15)$$

Solving for r, we find the radius of the circular path as

$$r = \frac{mv}{qB} \quad \text{(radius)}. \qquad (29\text{-}16)$$

The period T (the time for one full revolution) is equal to the circumference divided by the speed:

$$T = \frac{2\pi r}{v} = \frac{2\pi}{v}\frac{mv}{qB} = \frac{2\pi m}{qB} \quad \text{(period)}. \qquad (29\text{-}17)$$

The frequency f is

$$f = \frac{1}{T} = \frac{qB}{2\pi m} \quad \text{(frequency)}. \qquad (29\text{-}18)$$

The angular frequency ω of the motion is then

$$\omega = 2\pi f = \frac{qB}{m} \quad \text{(angular frequency)}. \qquad (29\text{-}19)$$

The quantities T, f, and ω do not depend on the speed of the particle (provided that speed is much less than the speed of light). Fast particles move in large circles and slow ones in small circles, but all particles with the same charge-to-mass ratio q/m take the same time T (the period) to complete one round trip. Using Eq. 29-2, you can show that if you are looking in the direction of **B,** the direction of rotation for a positive particle is always counterclockwise; that for a negative particle is always clockwise.

Helical Paths

If the velocity of a charged particle has a component parallel to the (uniform) magnetic field, the particle will move in a helical path about the direction of the field vector. Figure 29-11a, for example, shows the velocity vector **v** of such a particle resolved into two components, one parallel to **B** and one perpendicular to it:

$$v_\| = v\cos\phi \quad \text{and} \quad v_\perp = v\sin\phi. \qquad (29\text{-}20)$$

The parallel component determines the *pitch p* of the helix, that is, the distance between adjacent turns (Fig. 29-11b). The perpendicular component determines the radius of the helix and is the quantity to be substituted for v in Eq. 29-16.

Figure 29-11c shows a charged particle spiraling in a nonuniform magnetic field. The more closely spaced field lines at the left and right sides indicate that the magnetic field is stronger there. When the field at an end is strong enough, the particle "reflects" from that end. If the particle reflects from both ends, it is said to be trapped in a *magnetic bottle.*

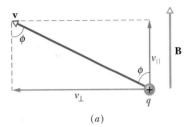
(a)

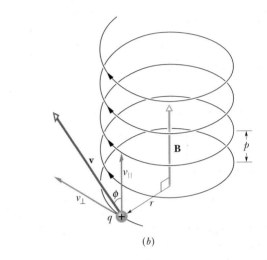
(b)

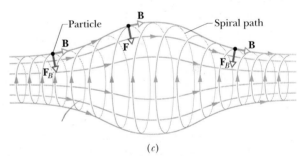
(c)

FIGURE 29-11 (a) A charged particle moves in a magnetic field, its velocity making an angle ϕ with the field direction. (b) The particle follows a helical path, of radius r and pitch p. (c) A charged particle spiraling in a nonuniform magnetic field. (The particle can become trapped, spiraling back and forth between the strong field regions at either end.) Note that the magnetic force vectors at the left and right sides have a component pointing toward the center of the figure.

Electrons and protons are trapped in this way by the terrestrial magnetic field, forming the *Van Allen radiation belts,* which loop well above Earth's atmosphere, between Earth's north and south geomagnetic poles. The trapped particles bounce back and forth, from end to end of the magnetic bottle, within a few seconds.

When a large solar flare shoots additional energetic electrons and protons into the radiation belts, an electric

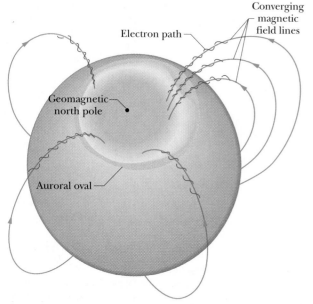

FIGURE 29-12 The auroral oval surrounding Earth's geomagnetic north pole (in northwestern Greenland). Magnetic field lines converge toward that pole. Electrons moving toward Earth are "caught by" and spiral around these field lines, entering the terrestrial atmosphere at high latitudes and producing aurora within the oval.

FIGURE 29-13 A false-color image of aurora inside the north auroral oval, recorded by the satellite *Dynamic Explorer*, using ultraviolet light emitted by oxygen atoms excited in the aurora. The sun-lit portion of Earth is the crescent at the left.

field is produced in the region where electrons normally reflect. This field eliminates the reflection and drives electrons down into the atmosphere, where they collide with atoms and molecules of air, causing that air to emit light. This light forms the aurora—a curtain of light that hangs down to an altitude of about 100 km. Green light is emitted by oxygen atoms, and pink light is emitted by nitrogen molecules, but often the light is so dim that we perceive only white light.

An auroral display extends in an arc above Earth in a region called the *auroral oval* (Figs. 29-12 and 29-13). Although the display is long, it is less than 1 km thick (north to south) because the paths of the electrons producing it converge as the electrons spiral down the converging magnetic field lines (Fig. 29-12).

$\mathbb{C}$HECKPOINT 4: The figure shows the circular paths of two particles that travel at the same speed in a uniform magnetic field **B**, which points into the page. One particle is a proton; the other is an electron (which is less massive). (a) Which particle follows the smaller circle, and (b) does that particle travel clockwise or counterclockwise?

$$\underset{\textbf{B}}{\otimes}$$

SAMPLE PROBLEM 29-3

Figure 29-14 shows the essentials of a *mass spectrometer*, which can be used to measure the mass of an ion: an ion of mass m (to be measured) and charge q is produced in source S. The initially stationary ion is accelerated by the electric field due to a potential difference V. The ion leaves S and enters a separator chamber in which a uniform magnetic field **B** is perpendicular to the path of the ion. The magnetic field causes the ion to move in a semicircle, striking (and thus altering) a photographic plate at distance x from the entry slit. Suppose that in a certain trial $B = 80.000$ mT and $V = 1000.0$ V and ions of charge $q = +1.6022 \times 10^{-19}$ C strike the plate at $x = 1.6254$ m. What is the mass m of the individual ions, in unified atomic mass units (1 u = 1.6605×10^{-27} kg)?

SOLUTION: We need to relate the ion mass m to the measured distance x in Fig. 29-14. To do so, we first note that $x = 2r$, where r is the radius of the semicircular path taken by the ion. Then we note that r is related to mass m by $r = mv/qB$ (Eq. 29-16), where v is the speed of the ion upon entering and then moving through the magnetic field.

We can relate the speed v to the accelerating potential V by applying the law of conservation of energy to the ion: its kinetic energy $\frac{1}{2}mv^2$ at the end of the acceleration is equal to its potential energy qV at the start of the acceleration. Thus

$$\tfrac{1}{2}mv^2 = qV$$

and

$$v = \sqrt{\frac{2qV}{m}}. \qquad (29\text{-}21)$$

Substituting this into Eq. 29-16 gives us

$$r = \frac{mv}{qB} = \frac{m}{qB}\sqrt{\frac{2qV}{m}} = \frac{1}{B}\sqrt{\frac{2mV}{q}}.$$

Thus,
$$x = 2r = \frac{2}{B}\sqrt{\frac{2mV}{q}}.$$

Solving this for m and substituting the given data yield

$$m = \frac{B^2 q x^2}{8V}$$

$$= \frac{(0.080000 \text{ T})^2 (1.6022 \times 10^{-19} \text{ C})(1.6254 \text{ m})^2}{8(1000.0 \text{ V})}$$

$$= 3.3863 \times 10^{-25} \text{ kg} = 203.93 \text{ u.} \qquad \text{(Answer)}$$

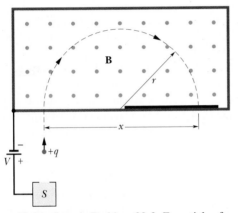

FIGURE 29-14 Sample Problem 29-3. Essentials of a mass spectrometer. A positive ion, after being accelerated from its source S by potential difference V, enters a chamber of uniform magnetic field **B**. There it travels through a semicircle of radius r and strikes a photographic plate at a distance x from where it entered the chamber.

SAMPLE PROBLEM 29-4

An electron with a kinetic energy of 22.5 eV moves into a region of uniform magnetic field **B** of magnitude 4.55×10^{-4} T. The angle between the directions of **B** and the electron's velocity **v** is 65.5°. What is the pitch of the helical path taken by the electron?

SOLUTION: The pitch p is the distance the electron travels parallel to the magnetic field **B** during one period T of revolution. That distance is $v_{\parallel}T$, where $v_{\parallel}$ is the electron's speed parallel to **B**. Using Eqs. 29-20 and 29-17, we find that

$$p = v_{\parallel}T = (v \cos \phi)\frac{2\pi m}{qB}. \qquad (29\text{-}22)$$

We can calculate the electron's speed v from its kinetic energy as we did for the proton in Sample Problem 29-1. (The kinetic energy of 22.5 eV is much less than the electron's rest energy

of 5.11×10^5 eV, so we need not use the relativistic formula for the kinetic energy.) We find that $v = 2.81 \times 10^6$ m/s. Substituting this and known data in Eq. 29-22 gives us

$$p = (2.81 \times 10^6 \text{ m/s})(\cos 65.5°)$$

$$\times \frac{2\pi (9.11 \times 10^{-31} \text{ kg})}{(1.60 \times 10^{-19} \text{ C})(4.55 \times 10^{-4} \text{ T})}$$

$$= 9.16 \text{ cm.} \qquad \text{(Answer)}$$

29-6 CYCLOTRONS AND SYNCHROTRONS

What is the structure of matter on the smallest scale? This question has always intrigued physicists. One way of getting at the answer is to allow an energetic charged particle (a proton, for example) to slam into a solid target. Better yet, allow two such energetic protons to collide head-on. Then analyze the debris from many such collisions to learn the nature of the subatomic particles of matter. The Nobel Prizes in physics for 1976 and 1984 were awarded for just such studies.

How can we give a proton enough kinetic energy for such an experiment? The direct approach is to allow the proton to "fall" through a potential difference V, thereby increasing its kinetic energy by eV. As we want higher and higher energies, however, it becomes more and more difficult to establish the necessary potential difference.

A better way is to arrange for the proton to circulate in a magnetic field, and to give it a modest electrical "kick" once per revolution. For example, if a proton circulates 100 times in a magnetic field and receives an energy boost of 100 keV every time it completes an orbit, it will end up with a kinetic energy of (100)(100 keV) or 10 MeV. Two very useful devices are based on this principle.

The Cyclotron

Figure 29-15 is a top view of the region of a *cyclotron* in which the particles (protons, say) circulate. The two hollow D-shaped objects (open on their straight edges) are made of sheet copper. These *dees*, as they are called, form part of an electrical oscillator, which establishes an alternating potential difference across the gap between them. The dees are immersed in a magnetic field ($B = 1.5$ T) whose direction is out of the plane of the page and which is set up by a large electromagnet.

Suppose that a proton, injected by source S at the center of the cyclotron in Fig. 29-15, initially moves toward a negatively charged dee. It will accelerate toward this dee and enter it. Once inside, it is shielded from electric fields by the copper walls of the dee; that is, the electric

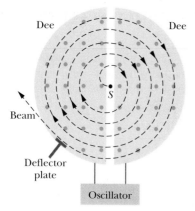

FIGURE 29-15 The elements of a cyclotron, showing the particle source S and the dees. A uniform magnetic field emerges from the plane of the figure. Circulating protons spiral outward within the hollow dees, gaining energy every time they cross the gap between the dees.

field does not enter the dee. The magnetic field, however, is not screened by the (nonmagnetic) copper dee, so the proton moves in a circular path whose radius, which depends on its speed, is given by Eq. 29-16, $r = mv/qB$.

Let us assume that at the instant the proton emerges into the center gap from the first dee, the potential difference between the dees has been reversed. Thus the proton *again* faces a negatively charged dee and is *again* accelerated. This process continues, the circulating proton always being in step with the oscillations of the dee potential, until the proton spirals out to the edge of the dee system.

The key to the operation of the cyclotron is that the frequency f at which the proton circulates in the field (and which does *not* depend on its speed) must be equal to the fixed frequency f_{osc} of the electrical oscillator, or

$$f = f_{osc} \quad \text{(resonance condition).} \quad (29-23)$$

This *resonance condition* says that, if the energy of the circulating proton is to increase, energy must be fed to it at a frequency f_{osc} that is equal to the natural frequency f at which the proton circulates in the magnetic field.

Combining Eqs. 29-18 and 29-23 allows us to write the resonance condition as

$$qB = 2\pi m f_{osc}. \quad (29-24)$$

For the proton, q and m are fixed. The oscillator (we assume) is designed to work at a single fixed frequency f_{osc}. We then "tune" the cyclotron by varying B until Eq. 29-24 is satisfied and a beam of energetic protons appears.

The Proton Synchrotron

At proton energies above 50 MeV, the conventional cyclotron begins to fail because one of the assumptions of its design—that the frequency of revolution of a charged particle circulating in a magnetic field is independent of the particle's speed—is true only for speeds that are much less than the speed of light. At greater proton speeds, we must treat the problem relativistically.

According to relativity theory, as the speed of a circulating proton approaches that of light, the proton takes a longer and longer time to make the trip around its orbit. This means that the frequency of revolution of the circulating proton decreases steadily. Thus the protons get out of step with the cyclotron's oscillator—whose frequency remains fixed at f_{osc}—and eventually the energy of the circulating proton stops increasing.

There is another problem. For a 500 GeV proton in a magnetic field of 1.5 T, the path radius is 1.1 km. The magnet for a conventional cyclotron of the proper size would be impossibly expensive, the area of its pole faces being about 4×10^6 m².

The *proton synchrotron* is designed to meet these two difficulties. The magnetic field B and the oscillator frequency f_{osc}, instead of having fixed values as in the conventional cyclotron, are made to vary with time during the accelerating cycle. When this is done properly, (1) the frequency of the circulating protons remains in step with the oscillator at all times, and (2) the protons follow a circular—not a spiral—path. Thus the magnet need extend only along that circular path, not over some 4×10^6 m². The circular path, however, still must be large if high energies are to be achieved. The proton synchrotron at the Fermi National Accelerator Laboratory (Fermilab) in Illinois (Fig. 29-16) has a circumference of 6.3 km and can produce protons with energies of about 1 TeV (= 10^{12} eV).

FIGURE 29-16 An aerial view of Fermilab.

SAMPLE PROBLEM 29-5

Suppose a cyclotron is operated at an oscillator frequency of 12 MHz and has a dee radius $R = 53$ cm.

(a) What is the magnitude of the magnetic field needed for deuterons to be accelerated in the cyclotron?

SOLUTION: A deuteron has the same charge as a proton but approximately twice the mass ($m = 3.34 \times 10^{-27}$ kg). From Eq. 29-24,

$$B = \frac{2\pi m f_{osc}}{q} = \frac{(2\pi)(3.34 \times 10^{-27} \text{ kg})(12 \times 10^6 \text{ s}^{-1})}{1.60 \times 10^{-19} \text{ C}}$$

$$= 1.57 \text{ T} \approx 1.6 \text{ T}. \qquad \text{(Answer)}$$

Note that, to allow protons to be accelerated, B would have to be reduced by a factor of 2, assuming that the oscillator frequency remained fixed at 12 MHz.

(b) What is the resulting kinetic energy of the deuterons?

SOLUTION: From Eq. 29-16, the speed of a deuteron circulating with a radius equal to the dee radius R is given by

$$v = \frac{RqB}{m} = \frac{(0.53 \text{ m})(1.60 \times 10^{-19} \text{ C})(1.57 \text{ T})}{3.34 \times 10^{-27} \text{ kg}}$$

$$= 3.99 \times 10^7 \text{ m/s}.$$

This speed corresponds to a kinetic energy of

$$K = \tfrac{1}{2}mv^2$$

$$= \tfrac{1}{2}(3.34 \times 10^{-27} \text{ kg})(3.99 \times 10^7 \text{ m/s})^2$$

$$\times (1 \text{ MeV}/1.60 \times 10^{-13} \text{ J})$$

$$= 16.6 \text{ MeV} \approx 17 \text{ MeV}. \qquad \text{(Answer)}$$

29-7 MAGNETIC FORCE ON A CURRENT-CARRYING WIRE

We have already seen (in connection with the Hall effect) that a magnetic field exerts a sideways force on moving electrons in a wire. This force must be transmitted to the wire itself, because the conduction electrons cannot escape sideways out of the wire.

In Fig. 29-17a, a vertical wire, carrying no current and fixed in place at both ends, extends through the gap between the vertical pole faces of a magnet. The magnetic field between the faces points outward from the page. In Fig. 29-17b, a current is sent upward through the wire; the wire deflects to the right. In Fig. 29-17c, we reverse the direction of the current and the wire deflects to the left.

Figure 29-18 shows what happens inside the wire of Fig. 29-17. We see one of the conduction electrons, drifting downward with an assumed drift speed v_d. Equation 29-3, in which we must put $\phi = 90°$, tells us that a force

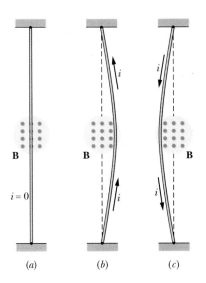

FIGURE 29-17 A flexible wire passes between the pole faces of a magnet (only the farther pole face is shown). (a) Without current in the wire, the wire is straight. (b) With upward current, the wire is deflected rightward. (c) With downward current, the deflection is leftward. The connections for getting the current into the wire at one end and out of it at the other end are not shown.

$\mathbf{F}_B$ of magnitude ev_dB must act on each such electron. From Eq. 29-2 we see that this force must point to the right. We expect then that the wire as a whole will experience a force to the right, in agreement with Fig. 29-17b.

If, in Fig. 29-18, we were to reverse *either* the direction of the magnetic field *or* the direction of the current, the force on the wire would reverse, pointing now to the left. Note too that it does not matter whether we consider negative charges drifting downward in the wire (the actual case) or positive charges drifting upward. The direction of the deflecting force on the wire is the same. We are safe then in dealing with the conventional direction of current, which assumes positive charge carriers.

Consider a length L of the wire in Fig. 29-18. The conduction electrons in this section of wire will drift past plane xx in Fig. 29-18 in a time $t = L/v_d$. Thus in that time a charge given by

$$q = it = i\frac{L}{v_d}$$

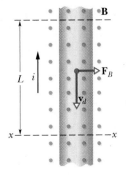

FIGURE 29-18 A close-up view of a section of the wire of Fig. 29-17b. The current direction is upward, which means that electrons drift downward. A magnetic field that emerges from the plane of the page causes the electrons and the wire to be deflected to the right.

will pass through that plane. Substituting this into Eq. 29-3 yields

$$F_B = q v_d B \sin \phi$$

$$= \frac{iL}{v_d} v_d B \sin 90°$$

or $\qquad F_B = iLB.$ $\qquad$ (29-25)

This equation gives the force that acts on a segment of a straight wire of length L, carrying a current i and immersed in a magnetic field **B** that is perpendicular to the wire.

If the magnetic field is *not* perpendicular to the wire, as in Fig. 29-19, the magnetic force is given by a generalization of Eq. 29-25:

$$\mathbf{F}_B = i\mathbf{L} \times \mathbf{B} \qquad \text{(force on a current).} \quad (29\text{-}26)$$

Here **L** is a *length vector* that points along the wire segment in the direction of the (conventional) current.

Equation 29-26 is equivalent to Eq. 29-2 in that either can be taken as the defining equation for **B**. In practice, we define **B** from Eq. 29-26. It is much easier to measure the magnetic force acting on a wire than that on a single moving charge.

If a wire is not straight, we can imagine it broken up into small straight segments and apply Eq. 29-26 to each segment. The force on the wire as a whole is then the vector sum of all the forces on the segments that make it up. In the differential limit, we can write

$$d\mathbf{F}_B = i\, d\mathbf{L} \times \mathbf{B}, \qquad (29\text{-}27)$$

and we can find the resultant force on any given arrangement of currents by integrating Eq. 29-27 over that arrangement.

In using Eq. 29-27, bear in mind that there is no such thing as an isolated current-carrying wire segment of length dL. There must always be a way to introduce the current into the segment at one end and take it out at the other end.

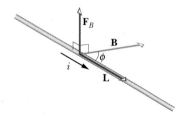

FIGURE 29-19 A wire carrying current i makes an angle ϕ with magnetic field **B**. The wire has length L in the field and length vector **L** (in the direction of the current). A magnetic force $\mathbf{F}_B = i\mathbf{L} \times \mathbf{B}$ acts on the wire.

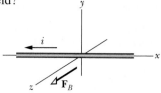

SAMPLE PROBLEM 29-6

A straight, horizontal stretch of copper wire has a current $i = 28$ A through it. What are the magnitude and direction of the minimum magnetic field **B** needed to suspend the wire, that is, to balance its weight? Its linear density is 46.6 g/m.

SOLUTION: Figure 29-20 shows the situation for a section of wire of length L, with the current out of the page. If the field is to be minimal, the force $\mathbf{F}_B$ that it exerts on the section must be upward, as shown. Equation 29-26 then requires that the field **B** be horizontal and, for the situation of Fig. 29-20, directed to the right.

In order to balance the weight of the section, the force $\mathbf{F}_B$ must have the magnitude $F_B = mg$, where m is the mass of the section. From Eq. 29-25, we then have

$$iLB = mg.$$

Solving for B and substituting known data yield

$$B = \frac{(m/L)g}{i} = \frac{(46.6 \times 10^{-3} \text{ kg/m})(9.8 \text{ m/s}^2)}{28 \text{ A}}$$

$$= 1.6 \times 10^{-2} \text{ T.} \qquad \text{(Answer)}$$

This is about 160 times the strength of Earth's magnetic field.

FIGURE 29-20 Sample Problem 29-6. A current-carrying wire (shown in cross section) can be made to "float" in a magnetic field. The current in the wire emerges from the plane of the page, and the magnetic field points to the right.

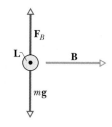

SAMPLE PROBLEM 29-7

Figure 29-21 shows a length of wire with a central semicircular arc, placed in a uniform magnetic field **B** that points out of the plane of the figure. If the wire carries a current i, what resultant magnetic force **F** acts on it?

SOLUTION: The force that acts on each straight section has the magnitude, from Eq. 29-25,

$$F_1 = F_3 = iLB$$

and points down, as shown by F_1 and F_3 in Figure 29-21.

A segment of the central arc of length dL has a force dF acting on it, whose magnitude is given by

$$dF = iB \, dL = iB(R \, d\theta)$$

and whose direction is radially toward point O, the center of the arc. Only the downward component $dF \sin \theta$ of this force element is effective. The horizontal component is canceled by an oppositely directed horizontal component associated with a symmetrically located segment on the opposite side of the arc.

Thus the total force on the central arc points down and is given by

$$F_2 = \int_0^\pi dF \sin \theta = \int_0^\pi (iBR \, d\theta)\sin \theta$$

$$= iBR \int_0^\pi \sin \theta \, d\theta = 2iBR.$$

The resultant force on the entire wire is then

$$F = F_1 + F_2 + F_3 = iLB + 2iBR + iLB$$

$$= 2iB(L + R). \qquad \text{(Answer)}$$

Note that this force is equal to the force that would act on a straight wire of length $2(L + R)$. This would be true no matter what the shape of the central segment.

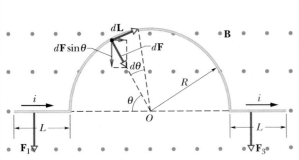

FIGURE 29-21 Sample Problem 29-7. A wire segment carrying a current i is immersed in a magnetic field. The resultant force on the wire is directed downward.

SAMPLE PROBLEM 29-8

Figure 29-22a shows a wire carrying a current $i = 6.0$ A in the positive direction of the x axis and lying in a *nonuniform* magnetic field given by $\mathbf{B} = (2.0 \text{ T/m})x\mathbf{i} + (2.0 \text{ T/m})x\mathbf{j}$, with $\mathbf{B}$ in teslas and x in meters. What is the net magnetic force $\mathbf{F}_B$ on the section of the wire between $x = 0$ and $x = 2.0$ m?

SOLUTION: Because the field varies along the section of wire, we cannot just substitute the data into Eq. 29-26, which holds only for a uniform magnetic field $\mathbf{B}$. Instead, we must

mentally divide the wire into differential lengths and then use Eq. 29-26 to find the differential force $d\mathbf{F}_B$ on each length. Then we can sum these differential forces to find the net magnetic force $\mathbf{F}_B$ on the full section of wire.

Figure 29-22b shows a differential length vector $d\mathbf{L}$ along the wire in the direction of the current; the vector has length dx and points in the positive direction of the x axis. Thus we can write this vector $d\mathbf{L}$ as

$$d\mathbf{L} = dx \, \mathbf{i}. \qquad (29\text{-}28)$$

(Be careful not to confuse the unit vector $\mathbf{i}$ with the current i.) Now, by Eq. 29-26, the differential force $d\mathbf{F}_B$ on the length dx of the wire is

$$d\mathbf{F}_B = i \, d\mathbf{L} \times \mathbf{B}$$

$$= i(dx \, \mathbf{i}) \times (2.0x\mathbf{i} + 2.0x\mathbf{j})$$

$$= i \, dx[2.0x(\mathbf{i} \times \mathbf{i}) + 2.0x(\mathbf{i} \times \mathbf{j})]$$

$$= i \, dx[0 + 2.0x\mathbf{k}] = 2.0ix \, dx \, \mathbf{k}, \qquad (29\text{-}29)$$

where the constant 2.0 has the unit teslas per meter. From this result we see that the magnetic force does not depend on the x component of $\mathbf{B}$ (because that component is along the direction of the current). We also see that the magnetic force $d\mathbf{F}_B$ on length dx of the wire is in the positive direction of the z axis (out of the page in Fig. 29-22c) and has magnitude $dF_B = (2.0 \text{ T/m})ix \, dx$.

Because the direction of the force $d\mathbf{F}_B$ is the same for all the differential lengths dx of the wire, we can find the magnitude of the total force by summing all the differential force magnitudes dF_B. To do so, we integrate dF_B from $x = 0$ to $x = 2.0$ m and then substitute the given data. We get

$$F_B = \int dF_B = \int_0^{2.0 \text{ m}} (2.0 \text{ T/m})ix \, dx$$

$$= (2.0 \text{ T/m})i \left[\tfrac{1}{2}x^2 \right]_0^{2.0 \text{ m}} = (2.0 \text{ T/m})(6.0 \text{ A})(\tfrac{1}{2})(2.0 \text{ m})^2$$

$$= 24 \text{ (T·A·m)} = 24 \text{ N}. \qquad \text{(Answer)}$$

This force is directed along the positive direction of the z axis.

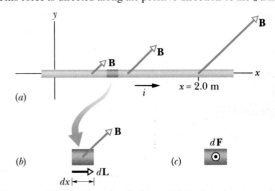

FIGURE 29-22 Sample Problem 29-8. (*a*) A wire with current i lies in a nonuniform magnetic field $\mathbf{B}$. (*b*) An element of the wire, with differential length vector $d\mathbf{L}$ and length dx. (*c*) The differential force $d\mathbf{F}$ acting on the element of (*b*) due to the magnetic field; the force is directed out of the page.

29-8 TORQUE ON A CURRENT LOOP

Much of the world's work is done by electric motors. The forces behind this work are the magnetic forces that we studied in the preceding section, that is, the forces that a magnetic field exerts on a wire that carries a current.

Figure 29-23a shows a simple motor, consisting of a single current-carrying loop immersed in a magnetic field **B**. The two magnetic forces **F** and −**F** combine to exert a torque on the loop, tending to rotate it about its central axis. Although many essential details have been omitted, the figure does suggest how the action of a magnetic field, exerting a torque on a current loop, produces the rotary motion of the electric motor. Let us analyze the action.

Figure 29-24a shows a rectangular loop of sides a and b, carrying a current i and immersed in a uniform magnetic field **B**. We place it in the field so that its long sides, labeled 1 and 3, are perpendicular to the field direction

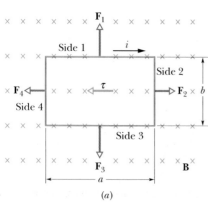

FIGURE 29-23 The elements of an electric motor. A rectangular loop of wire, carrying a current and free to rotate about a fixed axis, is placed in a magnetic field. A commutator (not shown) reverses the direction of the current every half-revolution so that the magnetic torque always acts in the same direction.

(which is into the page), but its short sides, labeled 2 and 4, are not. Wires to lead the current into and out of the loop are needed but, for simplicity, they are not shown.

To define the orientation of the loop in the magnetic field, we use a normal vector **n** that is perpendicular to the plane of the loop. Figure 29-24b shows a right-hand rule for finding the direction of **n**. Point or curl the fingers of your right hand in the direction of the current at any point on the loop. Your extended thumb then points in the direction of the normal vector **n**.

The normal vector of the loop is at an angle θ to the direction of the magnetic field **B**, as shown in Fig. 29-24c. We wish to find the net force and net torque acting on the loop in this orientation.

The net force is the vector sum of the forces acting on each of the four sides of the loop. For side 2 the vector **L** in Eq. 29-26 points in the direction of the current and has magnitude b. The angle between **L** and **B** for side 2 (see Fig. 29-24c) is 90° − θ. Thus the magnitude of the force acting on this side is

$$F_2 = ibB \sin(90° - θ) = ibB \cos θ. \quad (29\text{-}30)$$

You can show that the force **F**$_4$ acting on side 4 has the same magnitude as **F**$_2$ but points in the opposite direction. Thus **F**$_2$ and **F**$_4$ cancel out exactly. Their net force is zero and, because their common line of action is through the center of the loop, their net torque is also zero.

The situation is different for sides 1 and 3. Here the common magnitude of **F**$_1$ and **F**$_3$ is iaB, and the two forces point in opposite directions so that they do not tend to move the loop up or down. However, as Fig. 29-24c shows, these two forces do not share the same line of action so they do produce a net torque. The torque tends to rotate the loop so as to align its normal vector **n** with the direction

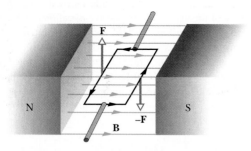

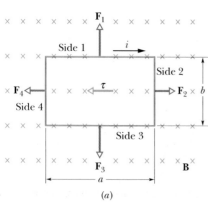

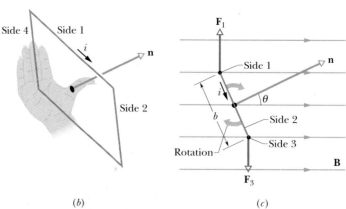

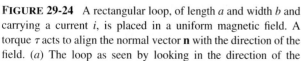

FIGURE 29-24 A rectangular loop, of length a and width b and carrying a current i, is placed in a uniform magnetic field. A torque τ acts to align the normal vector **n** with the direction of the field. (a) The loop as seen by looking in the direction of the magnetic field. (b) A perspective of the loop showing how a right-hand rule gives the direction of **n**, which is perpendicular to the plane of the loop. (c) A side view of the loop, from side 2. The loop rotates as indicated.

of the magnetic field **B**. That torque has moment arm $(b/2)$ $\sin \theta$ about the center of the loop. The magnitude τ' of the torque due to forces F_1 and F_3 is (see Fig. 29-24c)

$$\tau' = \left(iaB \frac{b}{2} \sin \theta \right) + \left(iaB \frac{b}{2} \sin \theta \right)$$
$$= iabB \sin \theta.$$

Suppose we replace the single loop of current with a *coil* of N loops, or *turns*. Further, suppose that the turns are wound tightly enough that they can be approximated as having the same dimensions and lying in a plane. Then the turns form a *flat coil* and the torque τ' derived above acts on each of them. The total torque is then

$$\tau = N\tau' = NiabB \sin \theta = (NiA)B \sin \theta, \quad (29\text{-}31)$$

in which $A \ (= ab)$ is the area enclosed by the coil. The quantities in parentheses (NiA) are grouped together because they are all properties of the coil: its number of turns, its area, and the current it carries. This equation holds for all flat coils, no matter what their shape, provided the magnetic field is uniform.

Instead of focusing on the motion of the coil, it is simpler to keep track of the vector **n**, which is normal to the plane of the coil. Equation 29-31 tells us that a current-carrying flat coil placed in a magnetic field will tend to rotate so that **n** points in the field direction.

SAMPLE PROBLEM 29-9

Analog voltmeters and ammeters work by measuring the torque exerted by a magnetic field on a current-carrying coil. The reading is displayed by means of the deflection of a pointer over a scale. Figure 29-25 shows the basic *galvanometer*, on which both analog ammeters and analog voltmeters are based. The coil is 2.1 cm high and 1.2 cm wide; it has 250 turns and is mounted so that it can rotate about an axis (into the page) in a uniform radial magnetic field with $B = 0.23$ T. For any orientation of the coil, the net magnetic field through the coil is perpendicular to the normal vector of the coil. A spring Sp provides a countertorque that balances the magnetic torque, so that a given steady current i in the coil results in a steady angular deflection ϕ. If a current of 100 μA produces an angular deflection of 28°, what must be the torsional constant κ of the spring, as used in Eq. 16-24 ($\tau = -\kappa\phi$)?

SOLUTION: Setting the magnetic torque (Eq. 29-31) equal to the spring torque and using absolute magnitudes yield

$$\tau = NiAB \sin \theta = \kappa\phi, \quad (29\text{-}32)$$

in which ϕ is the angular deflection of the coil and pointer, and $A \ (= 2.52 \times 10^{-4} \text{ m}^2)$ is the area encircled by the coil. Since the net magnetic field through the coil is always perpen-

dicular to the normal vector of the coil, $\theta = 90°$ for any orientation of the pointer.

Solving Eq. 29-32 for κ, we find

$$\kappa = \frac{NiAB \sin \theta}{\phi}$$
$$= (250)(100 \times 10^{-6} \text{ A})(2.52 \times 10^{-4} \text{ m}^2)$$
$$\times \frac{(0.23 \text{ T})(\sin 90°)}{28°}$$
$$= 5.2 \times 10^{-8} \text{ N·m/degree.} \quad \text{(Answer)}$$

Many modern ammeters and voltmeters are of the digital, direct-reading type and operate in a way that does not involve a moving coil.

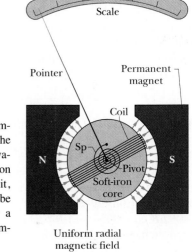

FIGURE 29-25 Sample Problem 29-9. The elements of a galvanometer. Depending on the external circuit, this device can be wired up as either a voltmeter or an ammeter.

29-9 THE MAGNETIC DIPOLE

We can describe the current-carrying coil of the preceding section with a single vector μ, its **magnetic dipole moment**. We take the direction of μ to be that of the normal vector **n** to the plane of the coil, as in Fig. 29-24c. We define the magnitude of μ as

$$\mu = NiA \quad \text{(magnetic moment).} \quad (29\text{-}33)$$

Thus Eq. 29-31 becomes

$$\tau = \mu B \sin \theta, \quad (29\text{-}34)$$

in which θ is the angle between the vectors μ and **B**.

We can generalize this to the vector relation

$$\tau = \mu \times B, \quad (29\text{-}35)$$

which reminds us very much of the corresponding equation for the torque exerted by an *electric* field on an *electric* dipole, namely Eq. 23-34:

$$\tau = p \times E.$$

FIGURE 29-26 The orientations of highest and lowest energy of a magnetic dipole in an external magnetic field **B**. The direction of the current i gives the direction of the magnetic dipole moment μ via the right-hand rule shown for **n** in Fig. 29–24b.

Highest energy   Lowest energy

In each case the torque exerted by the external field—either magnetic or electric—is equal to the vector product of the corresponding dipole moment and the field vector.

While an external magnetic field is exerting a torque on a magnetic dipole—such as a current-carrying coil—work must be done to change the orientation of the dipole. The magnetic dipole must then have a **magnetic potential energy** that depends on the dipole's orientation in the field. For electric dipoles we have shown (Eq. 23-38) that

$$U(\theta) = -\mathbf{p} \cdot \mathbf{E}.$$

In strict analogy, we can write for the magnetic case

$$U(\theta) = -\boldsymbol{\mu} \cdot \mathbf{B}. \qquad (29\text{-}36)$$

A magnetic dipole has its lowest energy ($= -\mu B \cos 0 = -\mu B$) when its dipole moment $\boldsymbol{\mu}$ is lined up with the magnetic field (Fig. 29-26). And it has its highest energy ($= -\mu B \cos 180° = +\mu B$) when $\boldsymbol{\mu}$ points in a direction opposite the field. The difference in energy between these two orientations is

$$\Delta U = (+\mu B) - (-\mu B) = 2\mu B. \qquad (29\text{-}37)$$

This much work must be done by an external agent (something other than the magnetic field) to turn a magnetic dipole through 180°, starting when the dipole is lined up with the magnetic field.

So far, we have identified only a current-carrying coil as a magnetic dipole. However, a simple bar magnet is also a magnetic dipole. So is a rotating sphere of charge. Earth itself is a magnetic dipole. Finally, most subatomic particles, including the electron, the proton, and the neutron, have magnetic dipole moments. As you will see in Chapter 32, all these quantities can be viewed as current loops. For comparison, some approximate magnetic dipole moments are shown in Table 29-2.

TABLE 29-2   SOME MAGNETIC DIPOLE MOMENTS

A small bar magnet	5 J/T
Earth	8.0×10^{22} J/T
A proton	1.4×10^{-26} J/T
An electron	9.3×10^{-24} J/T

CHECKPOINT **6:** The figure shows four orientations, at angle θ, of a magnetic dipole moment $\boldsymbol{\mu}$ in a magnetic field. Rank the orientations according to (a) the magnitude of the torque on the dipole and (b) the potential energy of the dipole, greatest first.

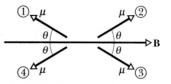

SAMPLE PROBLEM 29-10

(a) What is the magnetic dipole moment of the coil of Sample Problem 29-9, assuming that it carries a current of 100 μA?

SOLUTION: The *magnitude* of the magnetic dipole moment of the coil, whose area A is 2.52×10^{-4} m², is

$$\mu = NiA$$
$$= (250)(100 \times 10^{-6} \text{ A})(2.52 \times 10^{-4} \text{ m}^2)$$
$$= 6.3 \times 10^{-6} \text{ A} \cdot \text{m}^2 = 6.3 \times 10^{-6} \text{ J/T}. \quad \text{(Answer)}$$

You can show that these two sets of units are identical. The second set of units follows logically from Eq. 29-36.

The *direction* of $\boldsymbol{\mu}$, as inspection of Fig. 29-25 shows, is that of the pointer. You can verify this by showing that, if we assume $\boldsymbol{\mu}$ to be in the pointer direction, the torque predicted by Eq. 29-35 is such that it would indeed move the pointer clockwise across the scale.

(b) The magnetic dipole moment of the galvanometer coil is lined up with an external magnetic field whose strength is 0.85 T. How much work would be required to turn the coil end for end?

SOLUTION: The required work is equal to the increase in potential energy; that is, from Eq. 29-37,

$$W = \Delta U = 2\mu B = 2(6.3 \times 10^{-6} \text{ J/T})(0.85 \text{ T})$$
$$= 10.7 \times 10^{-6} \text{ J} \approx 11 \ \mu\text{J}. \quad \text{(Answer)}$$

This is about equal to the work needed to lift an aspirin tablet through a vertical height of 3 mm.

(c) What is the magnitude of the maximum torque τ that the external field **B** can exert on the magnetic dipole moment?

SOLUTION: From Eq. 29–34, the maximum torque occurs when the magnitude of $\sin \theta$ is 1. Thus, we have

$$\tau = \mu B \sin \theta$$
$$= (6.3 \times 10^{-6} \text{ J/T})(0.85 \text{ T})(1)$$
$$= 5.4 \times 10^{-6} \text{ N} \cdot \text{m}. \quad \text{(Answer)}$$

REVIEW & SUMMARY

Magnetic Field B

A **magnetic field B** is defined in terms of the force $\mathbf{F}_B$ acting on a test particle with charge q moving through the field with velocity $\mathbf{v}$:

$$\mathbf{F}_B = q\mathbf{v} \times \mathbf{B}. \qquad (29\text{-}2)$$

The SI unit for $\mathbf{B}$ is the **tesla** (T): $1 \text{ T} = 1 \text{ N/(A} \cdot \text{m)} = 10^4$ gauss.

The Hall Effect

When a conducting strip of thickness l carrying a current i is placed in a magnetic field $\mathbf{B}$, some charge carriers (with charge e) build up on the sides of the conductor, creating a potential difference V across the strip. The polarity of V gives the sign of the charge carriers; the number density n of charge carriers can be calculated with

$$n = \frac{Bi}{Vle}. \qquad (29\text{-}12)$$

A Charged Particle Circulating in a Magnetic Field

A charged particle with mass m and charge magnitude q moving with velocity $\mathbf{v}$ perpendicular to a magnetic field $\mathbf{B}$ will travel in a circle. Applying Newton's second law to the circular motion yields

$$qvB = \frac{mv^2}{r}, \qquad (29\text{-}15)$$

from which we find the radius r of the circle to be

$$r = \frac{mv}{qB}. \qquad (29\text{-}16)$$

The frequency of revolution f, the angular frequency ω, and the period of the motion T are given by

$$f = \frac{\omega}{2\pi} = \frac{1}{T} = \frac{qB}{2\pi m}. \qquad (29\text{-}19, 29\text{-}18, 29\text{-}17)$$

Cyclotrons and Synchrotrons

A cyclotron is a particle accelerator that uses a magnetic field to hold a charged particle in a circular orbit of increasing radius so that a modest accelerating potential may act on the particle repeatedly, providing it with high energy. Because the moving particle gets out of step with the oscillator as its speed approaches that of light, there is an upper limit to the energy attainable with the cyclotron. A synchrotron avoids this difficulty. Here both B and the oscillator frequency f_{osc} are programmed to change cyclically so that the particle not only can go to high energies but can do so at a constant orbital radius.

Magnetic Force on a Current-Carrying Wire

A straight wire carrying a current i in a uniform magnetic field experiences a sideways force

$$\mathbf{F}_B = i\mathbf{L} \times \mathbf{B}. \qquad (29\text{-}26)$$

The force acting on a current element $i\,d\mathbf{L}$ in a magnetic field is

$$d\mathbf{F}_B = i\,d\mathbf{L} \times \mathbf{B}. \qquad (29\text{-}27)$$

The direction of the length vector $\mathbf{L}$ or $d\mathbf{L}$ is that of the current i.

Torque on a Current-Carrying Coil

A coil (of area A and carrying current i, with N turns) in a uniform magnetic field $\mathbf{B}$ will experience a torque $\boldsymbol{\tau}$ given by

$$\boldsymbol{\tau} = \boldsymbol{\mu} \times \mathbf{B}. \qquad (29\text{-}35)$$

Here $\boldsymbol{\mu}$ is the **magnetic dipole moment** of the coil, with magnitude $\mu = NiA$ and direction given by a right-hand rule.

Orientation Energy of a Magnetic Dipole

The **magnetic potential energy** of a magnetic dipole in a magnetic field is

$$U(\theta) = -\boldsymbol{\mu} \cdot \mathbf{B}. \qquad (29\text{-}36)$$

QUESTIONS

1. Figure 29-27 shows four directions for the velocity vector $\mathbf{v}$ of a negatively charged particle moving at angle θ to a uniform magnetic field $\mathbf{B}$. (a) Rank the directions according to the magnitude of the magnetic force on the particle, greatest first. (b) Which gives a magnetic force out of the plane of the page?

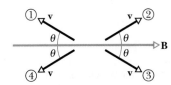

FIGURE 29-27
Question 1.

2. Here are four situations in which a proton has a velocity $\mathbf{v}$ while moving through a uniform magnetic field $\mathbf{B}$:

(a) $\mathbf{v} = 2\mathbf{i} - 3\mathbf{j}$ and $\mathbf{B} = 4\mathbf{k}$

(b) $\mathbf{v} = 3\mathbf{i} + 2\mathbf{j}$ and $\mathbf{B} = -4\mathbf{k}$

(c) $\mathbf{v} = 3\mathbf{j} - 2\mathbf{k}$ and $\mathbf{B} = 4\mathbf{i}$

(d) $\mathbf{v} = 20\mathbf{i}$ and $\mathbf{B} = -4\mathbf{i}$.

Without written calculation, rank the situations according to the magnitude of the magnetic force on the proton, greatest first.

3. Figure 29-28 shows three situations in which a positive particle of velocity $\mathbf{v}$ moves through a uniform magnetic field $\mathbf{B}$ and experiences a magnetic force $\mathbf{F}_B$. In each situation, determine whether the orientations of the vectors are physically reasonable.

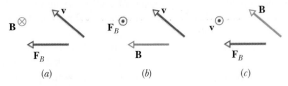

FIGURE 29-28 Question 3.

4. Figure 29-29 shows the path of an electron in a region of uniform magnetic field. The path consists of two straight sections, each between a pair of uniformly charged plates, and two half-circles. Which plate is at the higher electric potential in (a) the top pair of plates and (b) the bottom pair? (c) What is the direction of the magnetic field?

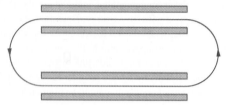

FIGURE 29-29 Question 4.

5. In Section 29-3, we discussed a charged particle moving through crossed fields with the forces $\mathbf{F}_E$ and $\mathbf{F}_B$ in opposition. We found that the particle moves in a straight line (that is, neither force dominates the motion) if its speed is given by Eq. 29-7 ($v = E/B$). Which of the two forces dominates if the speed of the particle is, instead, (a) $v < E/B$ and (b) $v > E/B$?

6. Figure 29-30 shows crossed and uniform electric and magnetic fields $\mathbf{E}$ and $\mathbf{B}$ and, at a certain instant, the velocity vectors of the 10 charged particles listed in Table 29-3. (The vectors are not drawn to scale.) The table gives the signs of the charges and the speeds of the particles; the speeds are given as either less than or greater than E/B (see Question 5). Which particles will move out of the page toward you after the instant of Fig. 29-30?

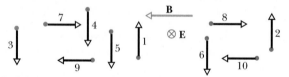

FIGURE 29-30 Question 6.

TABLE 29-3 **QUESTION 6**

PARTICLE	CHARGE	SPEED	PARTICLE	CHARGE	SPEED
1	+	Less	6	−	Greater
2	+	Greater	7	+	Less
3	+	Less	8	+	Greater
4	+	Greater	9	−	Less
5	−	Less	10	−	Greater

7. An airplane flies due west over Massachusetts, where Earth's magnetic field is directed downward and to the north. (a) On which wing, left or right, are some of the conduction electrons moved to the wingtip by the magnetic force on them? (b) Which wingtip gets the conduction electrons if the flight is eastward?

8. Figure 29-31 shows the cross section of a solid conductor carrying a current perpendicular to the page. (a) Which pair of the four terminals (*a, b, c, d*) should be used to measure the Hall

voltage if the magnetic field is in the positive direction of the *x* axis, the charge carriers are negative, and they move out of the page? Which terminal of the pair is at the higher potential? (b) Repeat for a magnetic field in the negative direction of the *y* axis and positive charge carriers moving out of the page. (c) Discuss the situation if the magnetic field is in the positive *z* direction.

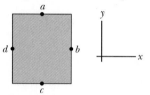

FIGURE 29-31 Question 8.

9. In Fig. 29-32, a charged particle enters a uniform magnetic field **B** with speed v_0, moves through a half-circle in time T_0, and then leaves the field. (a) Is the charge positive or negative? (b) Is the final speed of the particle greater than, smaller than, or equal to v_0? (c) If the initial speed had been $0.5v_0$, would the time spent in field **B** have been greater than, less than, or equal to T_0? (d) Would the path have been a half-circle, more than a half-circle, or less than a half-circle?

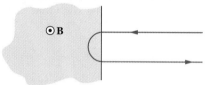

FIGURE 29-32 Question 9.

10. Figure 29-33 shows the path of a particle through six regions of uniform magnetic field, where the path is either a half-circle or a quarter-circle. Upon leaving the last region, the particle travels between two charged, parallel plates and is deflected toward the plate of higher potential. What are the directions of the magnetic fields in the six regions?

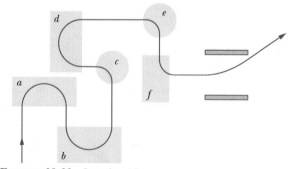

FIGURE 29-33 Question 10.

11. Figure 29-34 shows the path of an electron that passes through two regions containing uniform magnetic fields of magnitudes B_1 and B_2. Its path in each region is a half-circle. (a) Which field is stronger? (b) What are the directions of the two fields? (c) Is the time spent by the electron in the B_1 region greater than, less than, or the same as the time spent in the B_2 region?

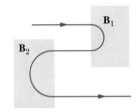

FIGURE 29-34
Question 11.

12. Particle Roundabout. Figure 29-35 shows 11 paths through a region of uniform magnetic field. One path is a straight line; the rest are half-circles. Table 29-4 gives the masses, charges, and speeds of 11 particles that take these paths through the field. Which path corresponds to which particle?

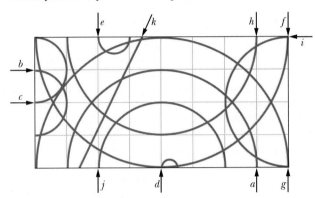

FIGURE 29-35 Question 12.

TABLE 29-4 **QUESTION 12**

PARTICLE	MASS	CHARGE	SPEED
1	$2m$	q	v
2	m	$2q$	v
3	$m/2$	q	$2v$
4	$3m$	$3q$	$3v$
5	$2m$	q	$2v$
6	m	$-q$	$2v$
7	m	$-4q$	v
8	m	$-q$	v
9	$2m$	$-2q$	$3v$
10	m	$-2q$	$8v$
11	$3m$	0	$3v$

13. Figure 29-36 shows three situations in which a charged particle moves in a spiral path through a uniform magnetic field. In which is the particle negatively charged?

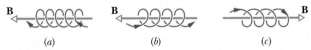

FIGURE 29-36 Question 13.

14. Figure 29-37 shows four views of a horseshoe magnet and a straight wire in which electrons are flowing out of the page, perpendicular to the plane of the magnet. In which case will the magnetic force on the wire point toward the top of the page?

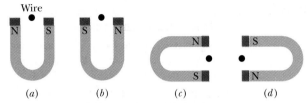

FIGURE 29-37 Question 14.

15. A wire carries a current i in the negative x direction, through a magnetic field **B.** Without written calculation, rank the following choices for **B** according to the magnitude of the magnetic forces they exert on the wire, greatest first: $\mathbf{B}_1 = 2\mathbf{i} + 3\mathbf{j}$, $\mathbf{B}_2 = 4\mathbf{i} - 3\mathbf{j}$, $\mathbf{B}_3 = 6\mathbf{i} + 3\mathbf{k}$, and $\mathbf{B}_4 = -8\mathbf{i} - 3\mathbf{k}$.

16. The dead-quiet ''caterpillar drive'' for submarines in the movie *The Hunt for Red October* is based on a *magnetohydrodynamic* (MHD) drive: as the ship moves forward, seawater flows through multiple channels in a structure built around the rear of the hull. Figure 29-38 shows the essentials of a channel. Magnets, positioned along opposite sides of the channel with opposite poles facing each other, create a magnetic field within the channel. Electrodes (not shown) create an electric field across the channel. The electric field drives a current across the channel and through the water; the magnetic force on the current propels the water toward the rear of the channel, thus propelling the ship forward. In Fig. 29-38, should the electric field be directed upward, downward, leftward, rightward, frontward, or rearward?

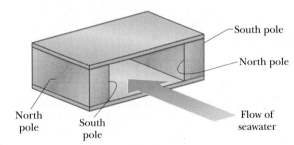

FIGURE 29-38 Question 16.

17. (a) In Checkpoint 6, if the dipole moment $\boldsymbol{\mu}$ rotates from orientation 1 to orientation 2, is the work done on the dipole *by the magnetic field* positive, negative, or zero? (b) Rank the work done on the dipole by the magnetic field for rotations from orientation 1 to (1) orientation 2, (2) orientation 3, and (3) orientation 4, greatest first.

EXERCISES & PROBLEMS

SECTION 29-2 The Definition of B

1E. Express the unit of a magnetic field B in terms of the dimensions M, L, T, and Q (mass, length, time, and charge).

2E. An alpha particle travels at a velocity $\mathbf{v}$ of magnitude 550 m/s through a uniform magnetic field $\mathbf{B}$ of magnitude 0.045 T. (An alpha particle has a charge of $+3.2 \times 10^{-19}$ C and a mass of 6.6×10^{-27} kg.) The angle between $\mathbf{v}$ and $\mathbf{B}$ is 52°. What are the magnitudes of (a) the force $\mathbf{F}_B$ acting on the particle due to the field and (b) the acceleration of the particle due to $\mathbf{F}_B$? (c) Does the speed of the particle increase, decrease, or remain equal to 550 m/s?

3E. An electron in a TV camera tube is moving at 7.20×10^6 m/s in a magnetic field of strength 83.0 mT. (a) Without knowing the direction of the field, what can you say about the greatest and least magnitudes of the force acting on the electron due to the field? (b) At one point the acceleration of the electron is 4.90×10^{14} m/s². What is the angle between the electron's velocity and the magnetic field?

4E. A proton traveling at 23.0° with respect to a magnetic field of strength 2.60 mT experiences a magnetic force of 6.50×10^{-17} N. Calculate (a) the proton's speed and (b) its kinetic energy in electron-volts.

5P. Each of the electrons in the beam of a television tube has a kinetic energy of 12.0 keV. The tube is oriented so that the electrons move horizontally from geomagnetic south to geomagnetic north. The vertical component of Earth's magnetic field points down and has a magnitude of 55.0 μT. (a) In what direction will the beam deflect? (b) What is the acceleration of a single electron due to the magnetic field? (c) How far will the beam deflect in moving 20.0 cm through the television tube?

6P. An electron that has velocity $\mathbf{v} = (2.0 \times 10^6 \text{ m/s})\mathbf{i} + (3.0 \times 10^6 \text{ m/s})\mathbf{j}$ moves through a magnetic field $\mathbf{B} = (0.030 \text{ T})\mathbf{i} - (0.15 \text{ T})\mathbf{j}$. (a) Find the force on the electron. (b) Repeat your calculation for a proton having the same velocity.

7P. An electron that is moving through a uniform magnetic field has a velocity $\mathbf{v} = (40 \text{ km/s})\mathbf{i} + (35 \text{ km/s})\mathbf{j}$ when it experiences a force $\mathbf{F} = -(4.2 \text{ fN})\mathbf{i} + (4.8 \text{ fN})\mathbf{j}$ due to the magnetic field. If $B_x = 0$, calculate the magnetic field $\mathbf{B}$.

SECTION 29-3 Crossed Fields: Discovery of the Electron

8E. A proton travels through uniform magnetic and electric fields. The magnetic field is $\mathbf{B} = -2.5\mathbf{i}$ mT. At one instant the velocity of the proton is $\mathbf{v} = 2000\mathbf{j}$ m/s. At that instant, what is the magnitude of the net force acting on the proton if the electric field is (a) $4.0\mathbf{k}$ V/m, (b) $-4.0\mathbf{k}$ V/m, and (c) $4.0\mathbf{i}$ V/m?

9E. An electron with kinetic energy 2.5 keV moves horizontally into a region of space in which there is a downward-directed electric field of magnitude 10 kV/m. (a) What are the magnitude

and direction of the (smallest) magnetic field that will cause the electron to continue to move horizontally? Ignore the gravitational force, which is rather small. (b) Is it possible for a proton to pass through this combination of fields undeflected? If so, under what circumstances?

10E. An electric field of 1.50 kV/m and a magnetic field of 0.400 T act on a moving electron to produce no net force. (a) Calculate the minimum speed v of the electron. (b) Draw the vectors $\mathbf{E}$, $\mathbf{B}$, and $\mathbf{v}$.

11P. An electron has an initial velocity of $(12.0 \mathbf{j} + 15.0 \mathbf{k})$ km/s and a constant acceleration of $(2.00 \times 10^{12} \text{ m/s}^2)\mathbf{i}$ in a region in which uniform electric and magnetic fields are present. If $\mathbf{B} = (400 \ \mu\text{T})\mathbf{i}$, find the electric field $\mathbf{E}$.

12P. An electron is accelerated through a potential difference of 1.0 kV and directed into a region between two parallel plates separated by 20 mm with a potential difference of 100 V between them. The electron is moving perpendicular to the electric field when it enters the region between the plates. What magnetic field is necessary perpendicular to both the electron path and the electric field so that the electron travels in a straight line?

13P. An ion source is producing ions of ^{6}Li (mass = 6.0 u), each with a charge of $+e$. The ions are accelerated by a potential difference of 10 kV and pass horizontally into a region in which there is a uniform vertical magnetic field of magnitude $B = 1.2$ T. Calculate the strength of the smallest electric field, to be set up over the same region, that will allow the ^{6}Li ions to pass through undeflected.

SECTION 29-4 Crossed Fields: The Hall Effect

14E. A strip of copper 150 μm wide is placed in a uniform magnetic field $\mathbf{B}$ of magnitude 0.65 T, with $\mathbf{B}$ perpendicular to the strip. A current $i = 23$ A is then sent through the strip such that a Hall potential difference V appears across the width. Calculate V. (The number of charge carriers per unit volume for copper is 8.47×10^{28} electrons/m³.)

15E. Show that, in terms of the Hall electric field E and the current density J, the number of charge carriers per unit volume is given by $n = JB/eE$.

16P. In a Hall-effect experiment, a current of 3.0 A sent lengthwise through a conductor 1.0 cm wide, 4.0 cm long, and 10 μm thick produces a transverse (across the width) Hall voltage of 10 μV when a magnetic field of 1.5 T is passed perpendicularly through the thickness of the conductor. From these data, find (a) the drift velocity of the charge carriers and (b) the number density of charge carriers. (c) Show on a diagram the polarity of the Hall voltage with assumed current and magnetic field directions, assuming also that the charge carriers are electrons.

17P. (a) In Fig. 29-8, show that the ratio of the Hall electric field E to the electric field E_C responsible for moving charge (the current) along the length of the strip is

$$\frac{E}{E_C} = \frac{B}{ne\rho},$$

where ρ is the resistivity of the material and n is the number density of the charge carriers. (b) Compute this ratio numerically for Exercise 14. (See Table 27-1.)

18P. A metal strip 6.50 cm long, 0.850 cm wide, and 0.760 mm thick moves with constant velocity **v** through a magnetic field $B = 1.20$ mT pointing perpendicular to the strip, as shown in Fig. 29-39. A potential difference of 3.90 μV is measured between points x and y across the strip. Calculate the speed v.

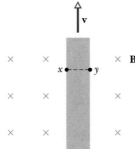

FIGURE 29-39
Problem 18.

SECTION 29-5 A Circulating Charged Particle

19E. An electron is accelerated from rest by a potential difference of 350 V. It then enters a uniform magnetic field of magnitude 200 mT with its velocity perpendicular to the field. Calculate (a) the speed of the electron and (b) the radius of its path in the magnetic field.

20E. What uniform magnetic field, applied perpendicular to a beam of electrons moving at 1.3×10^6 m/s, is required to make the electrons travel in a circular arc of radius 0.35 m?

21E. (a) In a magnetic field with $B = 0.50$ T, for what path radius will an electron circulate at 10% the speed of light? (b) What will be its kinetic energy in electron-volts? Ignore the small relativistic effects.

22E. What uniform magnetic field must be set up in space to permit a proton of speed 1.0×10^7 m/s to move in a circle the size of Earth's equator?

23E. An electron with kinetic energy 1.20 keV circles in a plane perpendicular to a uniform magnetic field. The orbit radius is 25.0 cm. Find (a) the speed of the electron, (b) the magnetic field, (c) the frequency of circling, and (d) the period of the motion.

24E. Physicist S. A. Goudsmit devised a method for measuring accurately the masses of heavy ions by timing their periods of revolution in a known magnetic field. A singly charged ion of iodine makes 7.00 rev in a field of 45.0 mT in 1.29 ms. Calculate its mass, in unified atomic mass units. Actually, the mass measurements are carried out to much greater accuracy than these approximate data suggest.

25E. An alpha particle ($q = +2e$, $m = 4.00$ u) travels in a circular path of radius 4.50 cm in a magnetic field with $B = 1.20$ T. Calculate (a) its speed, (b) its period of revolution, (c) its kinetic energy in electron-volts, and (d) the potential difference through which it would have to be accelerated to achieve this energy.

26E. (a) Find the frequency of revolution of an electron with an energy of 100 eV in a magnetic field of 35.0 μT. (b) Calculate the radius of the path of this electron if its velocity is perpendicular to the magnetic field.

27E. A beam of electrons whose kinetic energy is K emerges from a thin-foil "window" at the end of an accelerator tube. There is a metal plate a distance d from this window and perpendicular to the direction of the emerging beam (Fig. 29-40). Show that we can prevent the beam from hitting the plate if we apply a magnetic field B such that

$$B \geq \sqrt{\frac{2mK}{e^2 d^2}},$$

in which m and e are the electron mass and charge. How should **B** be oriented?

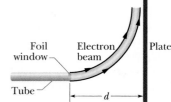

FIGURE 29-40
Exercise 27.

28P. A source injects an electron of speed $v = 1.5 \times 10^7$ m/s into a uniform magnetic field of magnitude $B = 1.0 \times 10^{-3}$ T. The velocity of the electron makes an angle $\theta = 10°$ with the direction of the magnetic field. Find the distance d from the point of injection at which the electron next crosses the field line that passes through the injection point.

29P. In a nuclear experiment a proton with kinetic energy 1.0 MeV moves in a circular path in a uniform magnetic field. What energy must (a) an alpha particle ($q = +2e$, $m = 4.0$ u) and (b) a deuteron ($q = +e$, m = 2.0 u) have if they are to circulate in the same orbit?

30P. A proton, a deuteron, and an alpha particle (see Problem 29), accelerated through the same potential difference, enter a region of uniform magnetic field **B**, moving perpendicular to **B**. (a) Compare their kinetic energies. If the radius of the proton's circular path is 10 cm, what are the radii of (b) the deuteron path and (c) the alpha-particle path?

31P. A proton, a deuteron, and an alpha particle (see Problem 29) with the same kinetic energies enter a region of uniform magnetic field **B**, moving perpendicular to **B**. Compare the radii of their circular paths.

32P. A proton of charge $+e$ and mass m enters a uniform magnetic field $\mathbf{B} = B\mathbf{i}$ with an initial velocity $\mathbf{v} = v_{0x}\mathbf{i} + v_{0y}\mathbf{j}$. Find an expression in unit-vector notation for its velocity **v** at any later time t.

33P. Two types of singly ionized atom having the same charge q but masses that differ by a small amount Δm are introduced into the mass spectrometer described in Sample Problem 29-3. (a) Calculate the difference in mass in terms of V, q, m (of either), B, and the distance Δx between the spots on the photographic plate. (b) Calculate Δx for a beam of singly ionized chlorine atoms of masses 35 and 37 u if $V = 7.3$ kV and $B = 0.50$ T.

34P. In a commercial mass spectrometer (see Sample Problem 29-3), uranium ions of mass 3.92×10^{-25} kg and charge 3.20×10^{-19} C are separated from related species. The ions are accelerated through a potential difference of 100 kV and then pass into a magnetic field, where they are bent in a path of radius 1.00 m. After traveling through 180° and passing through a slit of width 1.00 mm and height 1.00 cm, they are collected in a cup. (a) What is the magnitude of the (perpendicular) magnetic field in the separator? If the machine is used to separate out 100 mg of material per hour, calculate (b) the current of the desired ions in the machine and (c) the thermal energy produced in the cup in 1.00 h.

35P. Bainbridge's mass spectrometer, shown in Fig. 29-41, separates ions having the same velocity. The ions, after entering through slits S_1 and S_2, pass through a velocity selector composed of an electric field produced by the charged plates P and P′, and a magnetic field **B** perpendicular to the electric field and the ion path. The ions that pass undeviated through the crossed **E** and **B** fields enter into a region where a second magnetic field **B**′ exists, where they are made to follow circular paths. A photographic plate registers their arrival. Show that, for the ions, $q/m = E/rBB'$, where r is the radius of the circular orbit.

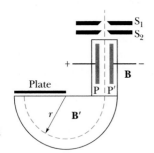

FIGURE 29-41
Problem 35.

36P. A positron with kinetic energy 2.0 keV is projected into a uniform magnetic field **B** of 0.10 T with its velocity vector making an angle of 89° with **B**. Find (a) the period, (b) the pitch p, and (c) the radius r of its helical path.

37P. A neutral particle is at rest in a uniform magnetic field of magnitude B. At time $t = 0$ it decays into two charged particles, each of mass m. (a) If the charge of one of the particles is $+q$, what is the charge of the other? (b) The two particles move off in separate paths, both of which lie in the plane perpendicular to **B**. At a later time the particles collide. Express the time from decay until collision in terms of m, B, and q.

38P. (a) What speed would a proton need to circle Earth at the equator, if Earth's magnetic field is everywhere horizontal there and directed along longitudinal lines? Relativistic effects must be taken into account. Take the magnitude of Earth's magnetic field to be 41 μT at the equator. (*Hint:* Replace the momentum mv in Eq. 29-16 with the relativistic momentum given in Eq. 9-24.) (b) Draw the velocity and magnetic field vectors corresponding to this situation.

39P. In Fig. 29-42, an electron of mass m, charge e, and negligible speed enters the region between two plates of potential difference V and plate separation d, initially headed directly toward the higher-potential top plate in the figure. A uniform magnetic field

of magnitude B is directed perpendicular to the plane of the figure. Find the minimum value of B at which the electron will not strike the top plate.

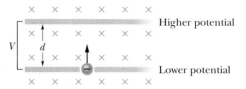

FIGURE 29-42 Problem 39.

SECTION 29-6 Cyclotrons and Synchrotrons

40E. In a certain cyclotron a proton moves in a circle of radius 0.50 m. The magnitude of the magnetic field is 1.2 T. (a) What is the cyclotron frequency? (b) What is the kinetic energy of the proton, in electron-volts?

41E. A physicist is designing a cyclotron to accelerate protons to one-tenth the speed of light. The magnet used will produce a field of 1.4 T. Calculate (a) the radius of the cyclotron and (b) the corresponding oscillator frequency. Relativity considerations are not significant.

42P. The oscillator frequency of the cyclotron in Sample Problem 29-5 has been adjusted to accelerate deuterons. (a) If protons are injected instead of deuterons, to what kinetic energy can the protons be accelerated, using the same oscillator frequency? (b) What magnetic field would be required? (c) What kinetic energy for protons could be produced if the magnetic field were left at the value used for deuterons? (d) What oscillator frequency would then be required? (e) Answer the same questions for alpha particles ($q = +2e$, $m = 4.0$ u).

43P. A deuteron in a cyclotron is moving in a magnetic field with $B = 1.5$ T and an orbit radius of 50 cm. Because of a grazing collision with a target, the deuteron breaks up, with negligible loss of kinetic energy, into a proton and a neutron. Discuss the subsequent motion of each. Assume that the deuteron energy is shared equally by the proton and neutron at breakup.

44P. Estimate the total path length traversed by a deuteron in the cyclotron of Sample Problem 29-5 during the acceleration process. Assume an accelerating potential between the dees of 80 kV.

SECTION 29-7 Magnetic Force on a Current-Carrying Wire

45E. A horizontal conductor in a power line carries a current of 5000 A from south to north. Earth's magnetic field (60.0 μT) is directed toward the north and is inclined downward at 70° to the horizontal. Find the magnitude and direction of the magnetic force on 100 m of the conductor due to Earth's field.

46E. A wire of 62.0 cm length and 13.0 g mass is suspended by a pair of flexible leads in a magnetic field of 0.440 T (Fig. 29-43). What are the magnitude and direction of the current required to remove the tension in the supporting leads?

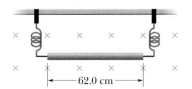

FIGURE 29-43
Exercise 46.

47E. A wire 1.80 m long carries a current of 13.0 A and makes an angle of 35.0° with a uniform magnetic field $B = 1.50$ T. Calculate the magnetic force on the wire.

48P. A wire 50 cm long lying along the x axis carries a current of 0.50 A in the positive x direction, through a magnetic field $\mathbf{B} = (0.0030\ \text{T})\mathbf{j} + (0.010\ \text{T})\mathbf{k}$. Find the force on the wire.

49P. A metal wire of mass m slides without friction on two horizontal rails spaced a distance d apart, as in Fig. 29-44. The track lies in a vertical uniform magnetic field $\mathbf{B}$. There is a constant current i through generator G, along one rail, across the wire, and back down the other rail. Find the speed and direction of the wire's motion as a function of time, assuming it to be stationary at $t = 0$.

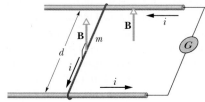

FIGURE 29-44 Problem 49.

50P. Figure 29-45 shows a wire of arbitrary shape carrying a current i between points a and b. The wire lies in a plane at right angles to a uniform magnetic field $\mathbf{B}$. (a) Prove that the force on the wire is the same as that on a straight wire carrying a current i directly from a to b. (*Hint:* Replace the wire with a series of "steps" parallel and perpendicular to the straight line joining a and b.) (b) Prove that the force on the wire becomes zero when points a and b are brought together so that the wire is a complete loop whose plane is perpendicular to $\mathbf{B}$.

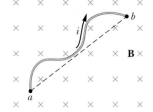

FIGURE 29-45
Problem 50.

51P. A long, rigid conductor, lying along the x axis, carries a current of 5.0 A in the negative direction. A magnetic field $\mathbf{B}$ is present, given by $\mathbf{B} = 3.0\mathbf{i} + 8.0x^2\mathbf{j}$, with x in meters and $\mathbf{B}$ in milliteslas. Calculate the force on the 2.0 m segment of the conductor that lies between $x = 1.0$ m and $x = 3.0$ m.

52P. Consider the possibility of a new design for an electric train. The engine is driven by the force on a conducting axle due

to the vertical component of Earth's magnetic field. Current is down one rail, through a conducting wheel, through the axle, through another conducting wheel, and then back to the source via the other rail. (a) What current is needed to provide a modest 10 kN force? Take the vertical component of Earth's field to be 10 μT and the length of the axle to be 3.0 m. (b) How much power would be lost for each ohm of resistance in the rails? (c) Is such a train totally unrealistic or just marginally unrealistic?

53P. A 1.0 kg copper rod rests on two horizontal rails 1.0 m apart and carries a current of 50 A from one rail to the other. The coefficient of static friction between rod and rails is 0.60. What is the smallest magnetic field (not necessarily vertical) that would cause the rod to slide?

SECTION 29-8 Torque on a Current Loop

54E. A single-turn current loop, carrying a current of 4.00 A, is in the shape of a right triangle with sides 50.0, 120, and 130 cm. The loop is in a uniform magnetic field of magnitude 75.0 mT whose direction is parallel to the current in the 130 cm side of the loop. (a) Find the magnitude of the magnetic force on each of the three sides of the loop. (b) Show that the total magnetic force on the loop is zero.

55E. Figure 29-46 shows a rectangular, 20-turn coil of wire, 10 cm by 5.0 cm. It carries a current of 0.10 A and is hinged along one long side. It is mounted in the xy plane, at an angle of 30° to the direction of a uniform magnetic field of 0.50 T. Find the magnitude and direction of the torque acting on the coil about the hinge line.

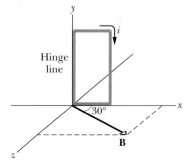

FIGURE 29-46
Exercise 55.

56E. A length L of wire carries a current i. Show that if the wire is formed into a circular coil, then the maximum torque in a given magnetic field is developed when the coil has one turn only and that maximum torque has the magnitude $\tau = (1/4\pi)L^2iB$.

57P. Prove that the relation $\tau = NiAB \sin\theta$ holds for closed loops of arbitrary shape and not only for rectangular loops as in Fig. 29-24. (*Hint:* Replace the loop of arbitrary shape with an assembly of adjacent long, thin, approximately rectangular loops that are nearly equivalent to the loop of arbitary shape as far as the distribution of current is concerned.)

58P. A closed wire loop with current i is in a uniform magnetic field $\mathbf{B}$, with the plane of the loop at angle θ to the direction of $\mathbf{B}$. Show that the total magnetic force on the loop is zero. Does your proof also hold for a nonuniform magnetic field?

59P. A particle of charge q moves in a circle of radius a with speed v. Treating the circular path as a current loop with constant current equal to its average current, find the maximum torque exerted on the loop by a uniform magnetic field of magnitude B.

60P. Figure 29-47 shows a wire ring of radius a that is perpendicular to the general direction of a radially symmetric diverging magnetic field. The magnetic field at the ring is everywhere of the same magnitude B, and its direction at the ring everywhere makes an angle θ with a normal to the plane of the ring. The twisted lead wires have no effect on the problem. Find the magnitude and direction of the force the field exerts on the ring if the ring carries a current i.

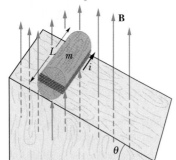

FIGURE 29-47 Problem 60.

61P. A certain galvanometer has a resistance of 75.3 Ω; its needle experiences a full-scale deflection when a current of 1.62 mA passes through its coil. (a) Determine the value of the auxiliary resistance required to convert the galvanometer to a voltmeter that reads 1.00 V at full-scale deflection. How is this resistance to be connected? (b) Determine the value of the auxiliary resistance required to convert the galvanometer to an ammeter that reads 50.0 mA at full-scale deflection. How is this resistance to be connected?

62P. Figure 29-48 shows a wooden cylinder with mass $m = 0.250$ kg and length $L = 0.100$ m, with $N = 10.0$ turns of wire wrapped around it longitudinally, so that the plane of the wire coil contains the axis of the cylinder. What is the least current i through the coil that will prevent the cylinder from rolling down a plane inclined at an angle θ to the horizontal, in the presence of a vertical, uniform magnetic field of 0.500 T, if the plane of the windings is parallel to the inclined plane?

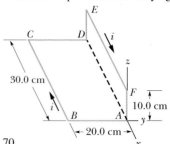

FIGURE 29-48
Problem 62.

SECTION 29-9 The Magnetic Dipole

63E. A circular coil of 160 turns has a radius of 1.90 cm. (a) Calculate the current that results in a magnetic dipole moment of 2.30 A·m². (b) Find the maximum torque that the coil, carrying this current, can experience in a uniform 35.0 mT magnetic field.

64E. The magnetic dipole moment of Earth is 8.00×10^{22} J/T. Assume that this is produced by charges flowing in Earth's mol-

ten outer core. If the radius of their circular path is 3500 km, calculate the current they produce.

65E. A circular wire loop whose radius is 15.0 cm carries a current of 2.60 A. It is placed so that the normal to its plane makes an angle of 41.0° with a uniform magnetic field of 12.0 T. (a) Calculate the magnetic dipole moment of the loop. (b) What torque acts on the loop?

66E. A current loop, carrying a current of 5.0 A, is in the shape of a right triangle with sides 30, 40, and 50 cm. The loop is in a uniform magnetic field of magnitude 80 mT whose direction is parallel to the current in the 50 cm side of the loop. Find the magnitude of (a) the magnetic dipole moment of the loop and (b) the torque on the loop.

67E. A stationary circular wall clock has a face with a radius of 15 cm. Six turns of wire are wound around its perimeter; the wire carries a current of 2.0 A in the clockwise direction. The clock is located where there is a constant, uniform external magnetic field of 70 mT (but the clock still keeps perfect time). At exactly 1:00 P.M., the hour hand of the clock points in the direction of the external magnetic field. (a) After how many minutes will the minute hand point in the direction of the torque on the winding due to the magnetic field? (b) Find the torque magnitude.

68E. Two concentric circular loops of radii 20.0 and 30.0 cm, located in the xy plane, each carry a clockwise current of 7.00 A (Fig. 29-49). (a) Find the net magnetic dipole moment of this system. (b) Repeat for reversed current in the inner loop.

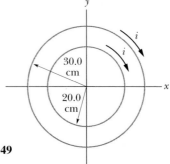

FIGURE 29-49
Exercise 68.

69P. A circular loop of wire having a radius of 8.0 cm carries a current of 0.20 A. A unit vector parallel to the dipole moment $\boldsymbol{\mu}$ of the loop is given by $0.60\mathbf{i} - 0.80\mathbf{j}$. If the loop is located in a magnetic field given by $\mathbf{B} = (0.25 \text{ T})\mathbf{i} + (0.30 \text{ T})\mathbf{k}$, find (a) the torque on the loop (in unit-vector notation) and (b) the magnetic potential energy of the loop.

70P. Figure 29-50 shows a current loop $ABCDEFA$ carrying a

FIGURE 29-50 Problem 70.

current $i = 5.00$ A. The sides of the loop are parallel to the coordinate axes, with $AB = 20.0$ cm, $BC = 30.0$ cm, and $FA = 10.0$ cm. Calculate the magnitude and direction of the magnetic dipole moment of this loop. (*Hint:* Imagine equal and opposite currents i in the line segment AD; then treat the two rectangular loops $ABCDA$ and $ADEFA$.)

Electronic Computation

71. A particle with mass m and charge q moves in a uniform electric field **E,** in the positive y direction, and a uniform magnetic field **B,** in the positive z direction. The force on the particle is $\mathbf{F} = q[\mathbf{E} + \mathbf{v} \times \mathbf{B}]$ and the acceleration of the particle is $\mathbf{a} = (q/m)[\mathbf{E} + \mathbf{v} \times \mathbf{B}]$. If **v** is in the xy plane, then the components of the acceleration are $a_x = (qB/m)v_y$ and $a_y = (qE/m) - (qB/m)v_x$. These components can be integrated twice to obtain expressions for the coordinates of the particle. If the particle starts at the origin with an initial velocity v_0 in the positive x direction, then

and

$$x = \frac{E}{B}t - \frac{1}{\omega}\left[\frac{E}{B} - v_0\right]\sin(\omega t)$$

$$y = -\frac{1}{\omega}\left[\frac{E}{B} - v_0\right][1 - \cos(\omega t)],$$

where $\omega = qB/m$. (a) By direct substitution, verify that these equations satisfy Newton's second law. Also verify that they lead to the given initial conditions. (b) Take $B = 1.2$ T, $E = 1.0 \times 10^4$ V/m, and $v_0 = 5.0 \times 10^4$ m/s and plot the trajectory of the particle for the first 4.0×10^{-6} s after it leaves the origin. The orbit can be described as a circle that translates in the positive x direction. Explain qualitatively why the motion is along the x axis when the electric field is along the y axis. Graph the trajectory for (b) $v_0 = 3.0 \times 10^4$ m/s, (c) $v_0 = 6.0 \times 10^4$ m/s, and (d) $v_0 = 9.0 \times 10^4$ m/s. (e) Why do some of the trajectories cross back over themselves while others do not? Why is one of the trajectories a straight line?

30
Magnetic Fields Due to Currents

This is the way we presently launch materials into space. But when we begin mining the Moon and the asteroids, where we will not have a source of fuel for such conventional rockets, we shall need a more effective way. Electromagnetic launchers may be the answer. A small prototype, the <u>electromagnetic rail gun</u>, can presently accelerate a projectile from rest to a speed of 10 km/s (2000 mi/h) within 1 ms. How can such rapid acceleration possibly be accomplished?

30-1 CALCULATING THE MAGNETIC FIELD DUE TO A CURRENT

As we discussed in Section 29-1, one way to produce a magnetic field is with moving charges, that is, with a current. Our goal in this chapter is to calculate the magnetic field that is produced by a given distribution of currents. We shall use the same basic procedure we used in Chapter 23 to calculate the electric field produced by a given distribution of charged particles.

Let us quickly review that basic procedure. We first mentally divide the charge distribution into charge elements dq, as is done for a charge distribution of arbitrary shape in Fig. 30-1a. We then calculate the field $d\mathbf{E}$ produced by a typical charge element at some point P. Because the electric fields contributed by different elements can be superimposed, we calculate the net field $\mathbf{E}$ at P by summing, via integration, the contributions $d\mathbf{E}$ from all the elements.

Recall that we express the magnitude of $d\mathbf{E}$ as

$$dE = \frac{1}{4\pi\epsilon_0}\frac{dq}{r^2}, \qquad (30\text{-}1)$$

in which r is the distance from the charge element dq to point P. For a positively charged element, the direction of $d\mathbf{E}$ is that of $\mathbf{r}$, where $\mathbf{r}$ is the vector that extends from the charge element dq to the point P. Using $\mathbf{r}$, we can rewrite Eq. 30-1 in vector form as

$$d\mathbf{E} = \frac{1}{4\pi\epsilon_0}\frac{dq}{r^3}\mathbf{r}, \qquad (30\text{-}2)$$

which indicates that the direction of the vector $d\mathbf{E}$ produced by a positively charged element is in the direction of the vector $\mathbf{r}$. Note that Eq. 30-2 is an inverse-square law

($d\mathbf{E}$ depends on inverse r^2) in spite of the exponent 3 in the denominator. That exponent is in the equation only because we added a factor of magnitude r in the numerator.

Now let us use the same basic procedure to calculate the magnetic field due to a current. Figure 30-1b shows a wire of arbitrary shape carrying a current i. We want to find the magnetic field $\mathbf{B}$ at a nearby point P. We first mentally divide the wire into differential elements $d\mathbf{s}$ that have length ds and are everywhere tangent to the wire and in the direction of the current. We can then define a differential *current-length element* to be $i\,d\mathbf{s}$; we wish to calculate the field $d\mathbf{B}$ produced at P by a typical current-length element. From experiment we find that magnetic fields, like electric fields, can be superimposed to find a net field. So, we can calculate the net field $\mathbf{B}$ at P by summing, via integration, the contributions $d\mathbf{B}$ from all the current-length elements. However, this summation is more challenging than the process associated with electric fields because of a complexity: whereas a charge element dq producing an electric field is a scalar, a current-length element $i\,d\mathbf{s}$ producing a magnetic field is the product of a scalar and a vector.

The magnitude of the field $d\mathbf{B}$ produced at point P by a current-length element $i\,d\mathbf{s}$ turns out to be

$$dB = \frac{\mu_0}{4\pi}\frac{i\,ds\,\sin\theta}{r^2}. \qquad (30\text{-}3)$$

Here μ_0 is a constant, called the *permeability constant*, whose value is defined to be exactly

$$\mu_0 = 4\pi \times 10^{-7}\ \text{T}\cdot\text{m/A}$$
$$\approx 1.26 \times 10^{-6}\ \text{T}\cdot\text{m/A}. \qquad (30\text{-}4)$$

The direction of $d\mathbf{B}$, shown as being into the page in Fig. 30-1b, is that of the cross product $d\mathbf{s} \times \mathbf{r}$, where $\mathbf{r}$ is now a vector that extends from the current element to point P. We can therefore write Eq. 30-3 in vector form as

$$d\mathbf{B} = \frac{\mu_0}{4\pi}\frac{i\,d\mathbf{s} \times \mathbf{r}}{r^3} \qquad \text{(Biot–Savart law).} \quad (30\text{-}5)$$

This vector equation and its scalar form, Eq. 30-3, are known as the **law of Biot and Savart** (rhymes with "Leo and bazaar"). The law is an inverse-square law (the exponent in the denominator of Eq. 30-5 is 3 only because of the factor $\mathbf{r}$ in the numerator). We shall use this law to calculate the net magnetic field $\mathbf{B}$ produced at a point by various distributions of current.

Magnetic Field Due to a Current in a Long Straight Wire

Shortly we shall use the law of Biot and Savart to prove that the magnitude of the magnetic field at a perpendicular

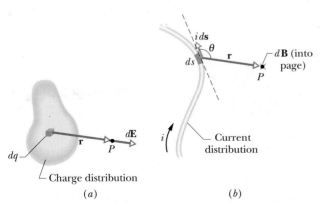

FIGURE 30-1 (a) A charge element dq produces a differential electric field $d\mathbf{E}$ at point P. (b) A current-length element $i\,d\mathbf{s}$ produces a differential magnetic field $d\mathbf{B}$ at point P. The green $\times$ (the tail of an arrow) at the dot for point P indicates that $d\mathbf{B}$ points *into* the page there.

distance r from a long (infinite) straight wire carrying a current i is given by

$$B = \frac{\mu_0 i}{2\pi r} \qquad \text{(long straight wire).} \qquad (30\text{-}6)$$

(Note carefully that in this specialized equation r is the *perpendicular* distance between the wire and a point at which B is to be evaluated. However, in Eqs. 30-3 and 30-5—which are fundamental—r is the distance between a current-length element in the wire and that point.)

The field magnitude B in Eq. 30-6 depends only on the current and the perpendicular distance r from the wire. We shall show in our derivation that the field lines of **B** form concentric circles around the wire, as Fig. 30-2 shows and as the iron filings in Fig. 30-3 suggest. The increase in the spacing of the lines in Fig. 30-2 with increasing distance from the wire represents the $1/r$ decrease in the magnitude of **B** predicted by Eq. 30-6.

Here is a simple right-hand rule for finding the direction of the magnetic field set up by a current-length element, such as a section of a long wire:

> Grasp the element in your right hand with your extended thumb pointing in the direction of the current. Your fingers will then naturally curl around in the direction of the magnetic field lines due to that element.

The result of applying this right-hand rule to the current in the straight wire of Fig. 30-2 is shown in a side view in Fig. 30-4a. Note that the fingers curl around the wire as the magnetic field lines do in Fig. 30-2. To determine the direction of **B** at any particular point, place your right hand around the wire so that your fingertips pass through that point, as in Figs. 30-4a and 30-4b. The direction of the fingertips at the point then gives the direction of **B** there.

Proof of Equation 30-6

Figure 30-5, which is just like Fig. 30-1b except that now the wire is straight, illustrates the task at hand: we seek the field **B** at point P, a perpendicular distance R from the wire. The magnitude of the differential magnetic field produced at P by the current-length element $i\,d\mathbf{s}$ located a distance r from P is given by Eq. 30-3:

$$dB = \frac{\mu_0}{4\pi} \frac{i\,ds\,\sin\theta}{r^2}.$$

The direction of $d\mathbf{B}$ in Fig. 30-5 is that of the vector $d\mathbf{s} \times \mathbf{r}$, namely, directly into the page.

Note that $d\mathbf{B}$ at point P has this same direction for all the current-length elements into which the wire can be di-

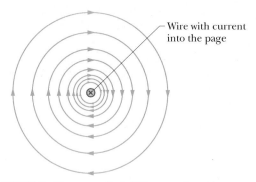

FIGURE 30-2 The magnetic field lines produced by a current in a long straight wire form concentric circles around the wire. Here the current is into the page, as indicated by the ×.

FIGURE 30-3 Iron filings that have been sprinkled onto cardboard collect in concentric circles when current is sent through the central wire. The alignment, which is along magnetic field lines, is caused by the magnetic field produced by the current.

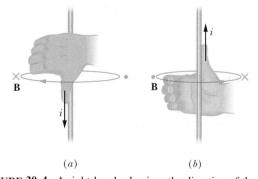

(a)   *(b)*

FIGURE 30-4 A right-hand rule gives the direction of the magnetic field due to a current in a wire. (a) The situation of Fig. 30-2, seen from the side. The magnetic field **B** at any point to the left of the wire points into the page, in the direction of the fingertips, as indicated by the ×. (b) If the current is reversed, **B** at any point to the left points out of the page, as indicated by the dot.

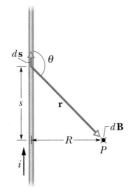

FIGURE 30-5 Calculating the magnetic field produced by a current i in a long straight wire. The field $d\mathbf{B}$ at P associated with the current-length element i $d\mathbf{s}$ points into the page, as shown.

vided. Thus we can find the magnitude of the magnetic field produced at P by the current-length elements in the upper half of the (infinitely) long wire by integrating dB in Eq. 30-3 from 0 to ∞. By Eq. 30-5, the magnetic field produced at P by a symmetrically located current-length element in the lower half of the wire has the same magnitude and direction as that from i $d\mathbf{s}$ in Fig. 30-5. Moreover, the magnetic field produced by the lower half is exactly the same as that produced by the upper half. So, to find the magnitude of the *total* magnetic field $\mathbf{B}$ at P, we need only multiply the result of our integration by 2. We get

$$B = 2 \int_0^\infty dB = \frac{\mu_0 i}{2\pi} \int_0^\infty \frac{\sin\theta\, ds}{r^2}. \qquad (30\text{-}7)$$

The variables θ, s, and r in this equation are not independent but (see Fig. 30-5) are related by

$$r = \sqrt{s^2 + R^2}$$

and $\qquad \sin\theta = \sin(\pi - \theta) = \dfrac{R}{\sqrt{s^2 + R^2}}.$

With these substitutions and integral 19 in Appendix E, Eq. 30-7 becomes

$$B = \frac{\mu_0 i}{2\pi} \int_0^\infty \frac{R\, ds}{(s^2 + R^2)^{3/2}}$$

$$= \frac{\mu_0 i}{2\pi R}\left[\frac{s}{(s^2 + R^2)^{1/2}}\right]_0^\infty = \frac{\mu_0 i}{2\pi R}. \qquad (30\text{-}8)$$

With a small change in notation, Eq. 30-8 becomes Eq. 30-6, the relation we set out to prove. Note that the magnetic field at P due to either the lower half or the upper half of the infinite wire in Fig. 30-5 is half this value; that is,

$$B = \frac{\mu_0 i}{4\pi R} \qquad \text{(semi-infinite straight wire).} \qquad (30\text{-}9)$$

Magnetic Field Due to a Current in a Circular Arc of Wire

To find the magnetic field produced at a point by a current in a curved wire, we would again use Eq. 30-3 to write the magnitude of the field produced by a single current-length element. And we would again integrate to find the net field produced by all the current-length elements. That integration can be difficult, depending on the shape of the wire; it is fairly straightforward, however, when the wire is a circular arc and the point is the center of curvature.

Figure 30-6a shows such an arc-shaped wire with central angle ϕ, radius R, and center C, carrying current i. At C, each current-length element i $d\mathbf{s}$ of the wire produces a magnetic field of magnitude dB given by Eq. 30-3. Moreover, as Fig. 30-6b shows, no matter where the element is located on the wire, the angle θ between the vectors $d\mathbf{s}$ and $\mathbf{r}$ is 90°; also, $r = R$. Thus, by substituting R for r and 90° for θ, we obtain from Eq. 30-3,

$$dB = \frac{\mu_0}{4\pi}\frac{i\, ds \sin 90°}{R^2} = \frac{\mu_0}{4\pi}\frac{i\, ds}{R^2}. \qquad (30\text{-}10)$$

The field at C due to each current-length element in the circular arc has this same magnitude.

An application of the right-hand rule anywhere along the wire (as in Fig. 30-6c) will show that all the differential fields $d\mathbf{B}$ have the same direction at C: directly out of the page. So the total field at C is simply the sum (via integration) of all the fields $d\mathbf{B}$. We use the identity $ds = R\, d\phi$ to change the variable of integration from ds to $d\theta$ and obtain, from Eq. 30-10,

$$B = \int dB = \int_0^\phi \frac{\mu_0}{4\pi}\frac{iR\, d\phi}{R^2} = \frac{\mu_0 i}{4\pi R}\int_0^\phi d\phi.$$

Integrating, we find that

$$B = \frac{\mu_0 i\phi}{4\pi R} \qquad \text{(at center of circular arc).} \qquad (30\text{-}11)$$

Note that this equation gives us the magnetic field *only* at the center of curvature of a circular arc of current. When you insert data into the equation, you must be careful to express ϕ in radians rather than degrees.

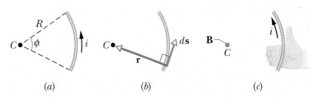

(a)   *(b)*   *(c)*

FIGURE 30-6 (a) A wire in the shape of a circular arc with center C carries current i. (b) For any element of wire along the arc, the angle between the directions of $d\mathbf{s}$ and $\mathbf{r}$ is 90°. (c) Determining the direction of the magnetic field at the center C due to the current in the wire; the field is out of the page, in the direction of the fingertips, as indicated by the colored dot at C.

SAMPLE PROBLEM 30-1

The wire in Fig. 30-7a carries a current i and consists of a circular arc of radius R and central angle $\pi/2$ rad, and two straight sections whose extensions intersect the center C of the arc. What magnetic field $\mathbf{B}$ does the current produce at C?

SOLUTION: To answer, we mentally divide the wire into three sections: (1) the straight section at left, (2) the straight section at right, and (3) the circular arc. Then we apply Eq. 30-3 to each section.

For any current-length element in section 1, the angle θ between $d\mathbf{s}$ and $\mathbf{r}$ is zero (Fig. 30-7b). So Eq. 30-3 gives us

$$dB_1 = \frac{\mu_0}{4\pi} \frac{i\,ds\,\sin\theta}{r^2} = \frac{\mu_0}{4\pi} \frac{i\,ds\,\sin 0}{r^2} = 0.$$

Thus the current along the entire length of wire in straight section 1 contributes no magnetic field at C:

$$B_1 = 0.$$

The same situation prevails in straight section 2, where the angle θ between $d\mathbf{s}$ and $\mathbf{r}$ for any current-length element is $180°$. Thus

$$B_2 = 0.$$

Because curved section 3 is a circular arc and we are to find the magnetic field at the center of curvature, we can use Eq. 30-11. Substituting $\pi/2$ rad for ϕ, we obtain

$$B_3 = \frac{\mu_0 i (\pi/2)}{4\pi R} = \frac{\mu_0 i}{8R}.$$

By applying the right-hand rule as in Fig. 30-7c, we see that B_3 points into the page at C.

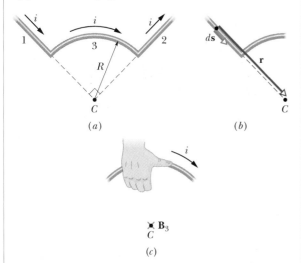

FIGURE 30-7 Sample Problem 30-1. (a) A wire consists of two straight sections (1 and 2) and a circular arc (3), and carries current i. (b) For a current-length element in section 1, the angle between $d\mathbf{s}$ and $\mathbf{r}$ is zero. (c) Determining the direction of magnetic field B_3 at C due to the current in the circular arc; the field is into the page there.

Thus the total magnetic field $\mathbf{B}$ produced at point C by the current in the wire has magnitude

$$B = B_1 + B_2 + B_3 = 0 + 0 + \frac{\mu_0 i}{8R} = \frac{\mu_0 i}{8R} \quad \text{(Answer)}$$

and points into the plane of the page.

CHECKPOINT **1:** The figure shows three circuits consisting of concentric circular arcs (either half- or quarter-circles of radii r, $2r$, and $3r$) and radial lengths. The circuits carry the same curent. Rank them according to the magnitude of the magnetic field produced at the center of curvature (the dot), greatest first.

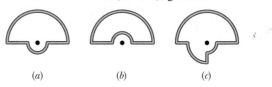

(a)   (b)   (c)

PROBLEM SOLVING TACTICS

TACTIC 1: *Right-Hand Rules*

To help you sort out the right-hand rules you have now seen (and the ones coming up), here is a review.

Right-Hand Rule for Cross Products. Introduced in Section 3-7, this is a way to determine the direction of the vector that results from a cross product. You point the fingers of your right hand so as to sweep the first vector expressed in the product into the second vector, through the smaller angle between the two vectors. Your outstretched thumb gives you the direction of the vector resulting from the cross product. In Chapter 12, we used this right-hand rule to find the directions of torque and angular momentum vectors; in Chapter 29, we used it to find the direction of the force on a current-carrying wire in a magnetic field.

Curled–Straight Right-Hand Rules for Magnetism. In many situations involving magnetism, you need to relate a "curled" element and a "straight" element. You can do so with the (curled) fingers and the (straight) thumb on your right hand. You have already seen an example in Section 29-8, in which we related the current around a loop (curled element) to the normal vector $\mathbf{n}$ (straight element) of the loop: you curl the fingers of your right hand around in the direction of the current along the loop; your outstretched thumb then gives the direction of $\mathbf{n}$. This is also the direction of the magnetic dipole moment $\boldsymbol{\mu}$ of the loop.

In this section you were introduced to a second curled–straight right-hand rule. To determine the direction of the magnetic field lines around a current-length element, you point the outstretched thumb of your right hand in the direction of the current. The fingers then curl around the current-length element in the direction of the field lines.

30-2 TWO PARALLEL CURRENTS

Two long parallel wires carrying currents exert forces on each other. Figure 30-8 shows two such wires, separated by a distance d and carrying currents i_a and i_b. Let us analyze the forces that these wires exert on each other.

We seek first the force on wire b in Fig. 30-8 due to the current in wire a. That current produces a magnetic field $\mathbf{B}_a$, and it is this magnetic field that actually causes the force we seek. To find the force, then, we need the magnitude and direction of the field $\mathbf{B}_a$ *at the site of wire* b. The magnitude of $\mathbf{B}_a$ at every point of wire b is, from Eq. 30-6,

$$B_a = \frac{\mu_0 i_a}{2\pi d}. \qquad (30\text{-}12)$$

A (curled–straight) right-hand rule tells us that the direction of $\mathbf{B}_a$ at wire b is down, as Fig. 30-8 shows.

Now that we have the field, we can find the force it exerts on wire b. Equation 29-27 tells us that the force $\mathbf{F}_{ba}$ exerted on a length L of wire b by the external magnetic field $\mathbf{B}_a$ is

$$\mathbf{F}_{ba} = i_b \mathbf{L} \times \mathbf{B}_a. \qquad (30\text{-}13)$$

In Fig. 30-8, vectors $\mathbf{L}$ and $\mathbf{B}_a$ are perpendicular. So, with Eq. 30-12, we can write

$$F_{ba} = i_b L B_a \sin 90° = \frac{\mu_0 L i_a i_b}{2\pi d}. \qquad (30\text{-}14)$$

The direction of $\mathbf{F}_{ba}$ is the direction of the cross product $\mathbf{L} \times \mathbf{B}_a$. Applying the right-hand rule for cross products to $\mathbf{L}$ and $\mathbf{B}_a$ in Fig. 30-8, we find that $\mathbf{F}_{ba}$ points directly toward wire a, as shown. The general procedure for finding the force on a current-carrying wire is this:

> To find the force on a current-carrying wire due to a second current-carrying wire, first find the field due to the second wire at the site of the first wire. Then find the force on the first wire due to that field.

We could now use this procedure to compute the force on wire a due to the current in wire b. We would find that the force would point directly toward wire b; hence the two wires with parallel currents attract each other. Similarly, if the two currents were antiparallel, we could show that the two wires repel each other. Thus,

> Parallel currents attract, and antiparallel currents repel.

The force acting between currents in parallel wires is the basis for the definition of the ampere, which is one of the seven SI base units. The definition, adopted in 1946, is this: The ampere is that constant current which, if maintained in two straight, parallel conductors of infinite length, of negligible circular cross section, and placed 1 m apart in vacuum, would produce on each of these conductors a force of 2×10^{-7} newton per meter of length.

Rail Gun

The basics of a rail gun are shown in Fig. 30-9a. A large current is sent out along one of two parallel conducting rails, across a conducting "fuse" (such as a narrow piece of copper) between the rails, and then back to the current source along the second rail. The projectile to be fired lies on the far side of the fuse and fits loosely between the rails. Immediately after the current begins, the fuse element melts and vaporizes, creating a conducting gas between the rails where the fuse had been.

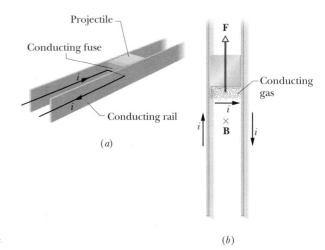

(a)

(b)

FIGURE 30-9 (a) A rail gun, as a current i is set up in it. The current rapidly causes the conducting fuse to vaporize. (b) The current produces a magnetic field $\mathbf{B}$ between the rails, and the field causes a force $\mathbf{F}$ to act on the conducting gas, which is part of the current path. The gas propels the projectile along the rails, launching it.

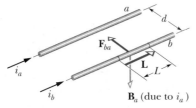

FIGURE 30-8 Two parallel wires carrying currents in the same direction attract each other. $\mathbf{B}_a$ is the magnetic field at wire b produced by the current in wire a. $\mathbf{F}_{ba}$ is the resulting force acting on wire b because it carries current in field $\mathbf{B}_a$.

The curled–straight right-hand rule of Fig. 30-4 reveals that the currents in the rails of Fig. 30-9a produce magnetic fields that are directed downward between the rails. The net magnetic field **B** exerts a force **F** on the gas due to the current i through the gas (Fig. 30-9b). With Eq. 30-13 and the right-hand rule for cross products, we find that **F** points outward along the rails. As the gas is forced outward along the rails, it pushes the projectile, accelerating it by as much as $5 \times 10^6 g$, and then launches it with a speed of 10 km/s, all within 1 ms.

SAMPLE PROBLEM 30-2

Two long parallel wires a distance $2d$ apart carry equal currents i in opposite directions, as shown in Fig. 30-10a. Derive an expression for $B(x)$, the magnitude of the resultant magnetic field for points at a distance x from the midpoint of a line joining the wires.

SOLUTION: Study of Fig. 30-10a and the use of the right-hand rule show that the fields set up by the currents in the individual wires point in the same direction for all points be-

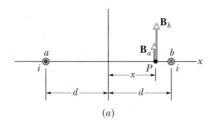

(a)

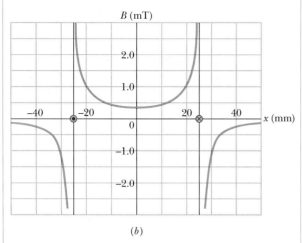

(b)

FIGURE 30-10 Sample Problem 30-2. (a) Two parallel wires carry currents of the same magnitude in opposite directions (out of and into the page). At points between the wires, such as P, the magnetic fields due to the separate currents point in the same direction. (b) A plot of $B(x)$ for $i = 25$ A and a wire separation of 50 mm.

---

tween the wires. From Eq. 30-6 we then have, at any point P between the wires,

$$B(x) = B_a(x) + B_b(x) = \frac{\mu_0 i}{2\pi(d+x)} + \frac{\mu_0 i}{2\pi(d-x)}$$

$$= \frac{\mu_0 id}{\pi(d^2 - x^2)}. \qquad \text{(Answer)} \quad (30\text{-}15)$$

Inspection of this relation shows that between the wires (1) $B(x)$ is symmetric about the midpoint ($x = 0$); (2) $B(x)$ has its minimum value (= $\mu_0 i/\pi d$) at this point; and (3) $B(x) \to \infty$ as $x \to \pm d$. At $x = \pm d$, the point P in Fig. 30-10a is within the wires on their axes. Our derivation of Eq. 30-6, however, is valid only for points outside the wires, so Eq. 30-15 holds only up to the surface of the wires.

Figure 30-10b plots Eq. 30-15 for $i = 25$ A and $2d = 50$ mm. We leave it as an exercise to show what the plot suggests: that Eq. 30-15 holds also for points beyond the wires, that is, for points with $|x| > d$.

SAMPLE PROBLEM 30-3

Figure 30-11a shows two long parallel wires carrying currents i_1 and i_2 in opposite directions. What are the magnitude and direction of the resultant magnetic field at point P? Assume the following values: $i_1 = 15$ A, $i_2 = 32$ A, and $d = 5.3$ cm.

SOLUTION: Figure 30-11b shows the individual magnetic fields **B$_1$** and **B$_2$** set up by currents i_1 and i_2, respectively, at P. (Verify that their directions are correct, as given by the appropriate right-hand rule.) The magnitudes of these fields at P are given by Eq. 30-6 as

$$B_1 = \frac{\mu_0 i_1}{2\pi R} = \frac{\mu_0 i_1}{2\pi(d/\sqrt{2})} = \frac{\sqrt{2}\mu_0}{2\pi d}\, i_1$$

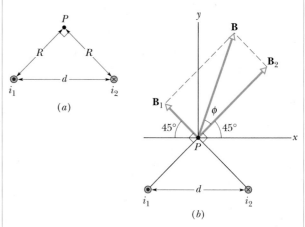

FIGURE 30-11 Sample Problem 30-3. (a) Two wires carry currents i_1 and i_2 in opposite directions (out of and into the page). Note the right angles at P. (b) The separate fields **B$_1$** and **B$_2$** are combined vectorially to yield the resultant field **B**.

and

$$B_2 = \frac{\mu_0 i_2}{2\pi R} = \frac{\mu_0 i_2}{2\pi (d/\sqrt{2})} = \frac{\sqrt{2}\mu_0}{2\pi d} i_2,$$

in which we have replaced R with its equal, $d/\sqrt{2}$, by noting that $R/d = \sin 45° = \sqrt{2}/2$.

The magnitude of the resultant magnetic field **B** is

$$B = \sqrt{B_1^2 + B_2^2} = \frac{\sqrt{2}\mu_0}{2\pi d}\sqrt{i_1^2 + i_2^2}$$

$$= \frac{(\sqrt{2})(4\pi \times 10^{-7}\ \text{T·m/A})\ \sqrt{(15\ \text{A})^2 + (32\ \text{A})^2}}{(2\pi)(5.3 \times 10^{-2}\ \text{m})}$$

$$= 1.89 \times 10^{-4}\ \text{T} \approx 190\ \mu\text{T}. \qquad \text{(Answer)}$$

The angle ϕ between **B** and $\mathbf{B}_2$ in Fig. 30-11b follows from

$$\phi = \tan^{-1} \frac{B_1}{B_2},$$

which, with B_1 and B_2 given above, yields

$$\phi = \tan^{-1} \frac{i_1}{i_2} = \tan^{-1} \frac{15\ \text{A}}{32\ \text{A}} = 25°.$$

The angle between **B** and the x axis is then

$$\phi + 45° = 25° + 45° = 70°. \qquad \text{(Answer)}$$

C HECKPOINT **2:** The figure shows three long, straight, parallel, equally spaced wires with identical currents either into or out of the page. Rank the wires according to the magnitude of the force on each due to the currents in the other two wires, greatest first.

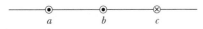

30-3 AMPERE'S LAW

We can find the net electric field due to *any* distribution of charges with the inverse-square law for the differential field $d\mathbf{E}$ (Eq. 30-2), but if the distribution is complicated, we may have to use a computer. Recall, however, that if the distribution has planar, cylindrical, or spherical symmetry, we can apply Gauss' law to find the net electric field with considerably less effort.

Similarly, we can find the net magnetic field due to *any* distribution of currents with the inverse-square law for the differential field $d\mathbf{B}$ (Eq. 30-5), but again we may have to use a computer for a complicated distribution. However, if the distribution has some symmetry, we may be able to apply **Ampere's law** to find the magnetic field with considerably less effort. This law was first advanced by André Marie Ampère (1775–1836), for whom the SI unit of current is named.

Ampere's law is

$$\oint \mathbf{B} \cdot d\mathbf{s} = \mu_0 i_{\text{enc}} \qquad \text{(Ampere's law). (30-16)}$$

The circle on the integral sign means that the scalar (or dot) product $\mathbf{B} \cdot d\mathbf{s}$ is to be integrated around a *closed* loop, called an *Amperian loop*. The current i_{enc} on the right is the *net* current encircled by that loop.

To see the meaning of the scalar product $\mathbf{B} \cdot d\mathbf{s}$ and its integral, let us first apply Ampere's law to the general situation of Fig. 30-12. The figure shows the cross sections of three long straight wires that carry currents i_1, i_2, and i_3 either directly into or directly out of the page. An arbitrary Amperian loop lying in the plane of the page encircles two of the currents but not the third. The counterclockwise direction marked on the loop indicates the arbitrarily chosen direction of integration for Eq. 30-16.

To apply Ampere's law, we mentally divide the loop into differential elements $d\mathbf{s}$ that are everywhere directed along the tangent to the loop in the direction of integration. At the location of the element $d\mathbf{s}$ shown in Fig. 30-12, the net magnetic field due to the three currents is **B**. Because the wires are perpendicular to the page, we know that the magnetic field at $d\mathbf{s}$ due to each current is in the plane of Fig. 30-12; thus their net magnetic field **B** at $d\mathbf{s}$ must also be in that plane. However, we do not know the orientation of **B** in the plane. In Fig. 30-12, **B** is arbitrarily drawn at an angle θ to the direction of $d\mathbf{s}$.

The scalar product $\mathbf{B} \cdot d\mathbf{s}$ on the left of Eq. 30-16 is then equal to $B \cos \theta\ ds$. Thus Ampere's law can be written as

$$\oint \mathbf{B} \cdot d\mathbf{s} = \oint B \cos \theta\ ds = \mu_0 i_{\text{enc}}. \qquad (30\text{-}17)$$

We can now interpret the scalar product $\mathbf{B} \cdot d\mathbf{s}$ as being the product of a length ds of the Amperian loop and the field component $B \cos \theta$ that is tangent to the loop. Then we can interpret the integration as being the summation of all such products around the entire loop.

When we can actually perform this integration, we do not need to know the direction of **B** before integrating.

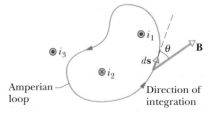

FIGURE 30-12 Ampere's law applied to an arbitrary Amperian loop that encircles two long straight wires but excludes a third wire. Note the directions of the currents.

Instead, we arbitrarily assume **B** to be generally in the direction of integration (as in Fig. 30-12). Then we use the following curled–straight right-hand rule to assign a plus sign or a minus sign to each of the currents that make up the net encircled current i_{enc}:

Curl the fingers of your right hand around the Amperian loop, with them pointing in the direction of integration. A current passing through the loop in the general direction of your outstretched thumb is assigned a plus sign, and a current moving generally in the opposite direction is assigned a minus sign.

Finally, we solve Eq. 30-17 for the magnitude of **B**. If B turns out positive, then the direction we assumed for **B** is correct. If it turns out negative, we neglect the minus sign and redraw **B** in the opposite direction.

In Fig. 30-13 we apply the curled–straight rule for Ampere's law to the situation of Fig. 30-12. With the indicated counterclockwise direction of integration, the net current encircled by the loop is

$$i_{enc} = i_1 - i_2.$$

(Current i_3 is not encircled by the loop.) So, we can rewrite Eq. 30-17 as

$$\oint B \cos \theta \, ds = \mu_0(i_1 - i_2). \qquad (30\text{-}18)$$

You might wonder why, since current i_3 contributes to the magnetic-field magnitude B on the left side of Eq. 30-18, it is not needed on the right side. The answer is that the contributions of current i_3 to the magnetic field cancel out because the integration in Eq. 30-18 is made around the full loop. In contrast, the contributions of an encircled current to the magnetic field do not cancel out.

We cannot solve Eq. 30-18 for the magnitude B of the magnetic field, because for the situation of Fig. 30-12 we do not have enough information to simplify and solve the integral. However, we do know the outcome of the integration: it must be equal to the value of $\mu_0(i_1 - i_2)$, which is set by the net current passing through the loop.

We shall now apply Ampere's law to two situations in which symmetry does allow us to simplify and solve the integral, hence to find the magnetic field.

The Magnetic Field Outside a Long Straight Wire with Current

Figure 30-14 shows a long straight wire that carries current i directly out of the page. Equation 30-6 tells us that the magnetic field **B** produced by the current has the same magnitude at all points that are the same distance r from the wire. That is, the field **B** has cylindrical symmetry about the wire. We can take advantage of that symmetry to simplify the integral in Ampere's law (Eq. 30-16) if we encircle the wire with a concentric circular Amperian loop of radius r, as in Fig. 30-14. The magnetic field **B** then has the same magnitude B at every point on the loop. We shall integrate counterclockwise, and we assume that **B** points in the same direction as the element ds in Fig. 30-14.

We can further simplify the quantity $B \cos \theta$ in Eq. 30-17 by noting that **B** is tangent to the loop at every point along the loop. Thus at every point the angle θ between ds and **B** is $0°$, so $\cos \theta = \cos 0° = 1$. The integral in Eq. 30-17 then becomes

$$\oint \mathbf{B} \cdot d\mathbf{s} = \oint B \cos \theta \, ds = B \oint ds = B(2\pi r).$$

Note that $\oint ds$ above is the summation of all the line segment lengths ds around the circular loop; that is, it simply gives the circumference $2\pi r$ of the loop.

Our right-hand rule gives us a plus sign for the current of Fig. 30-14. So the right side of Ampere's law becomes $+\mu_0 i$ and we then have

$$B(2\pi r) = \mu_0 i$$

or

$$B = \frac{\mu_0 i}{2\pi r}. \qquad (30\text{-}19)$$

This is precisely Eq. 30-6, which we derived earlier—with considerably more effort—using the law of Biot and Savart. In addition, because the magnitude B turned out positive, we know that the correct direction of **B** must be the one shown in Fig. 30-14.

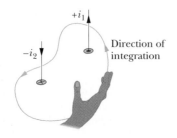

FIGURE 30-13 A right-hand rule for Ampere's law, to determine the signs for currents encircled by an Amperian loop. The situation is that of Fig. 30-12.

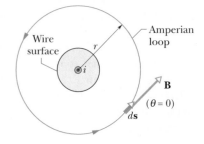

FIGURE 30-14 Using Ampere's law to find the magnetic field produced by a current i in a long straight wire. The Amperian loop is a concentric circle that lies outside the wire.

The Magnetic Field Inside a Long Straight Wire with Current

Figure 30-15 shows the cross section of a long straight wire of radius R that carries a uniformly distributed current i directly out of the page. Because the current is uniformly distributed about the center of the wire, the magnetic field **B** that it produces must be cylindrically symmetrical. So, to find the magnetic field at points inside the wire, we can again use an Amperian loop of radius r, as shown in Fig. 30-15, where now $r < R$. Symmetry again suggests that **B** is tangent to the loop, as shown. So the left side of Ampere's law again yields

$$\oint \mathbf{B} \cdot d\mathbf{s} = B \oint ds = B(2\pi r). \quad (30\text{-}20)$$

To find the right side of Ampere's law, we note that because the current is uniformly distributed, the current i_{enc} encircled by the loop is proportional to the area encircled by the loop. That is,

$$i_{enc} = i \frac{\pi r^2}{\pi R^2}. \quad (30\text{-}21)$$

Our right-hand rule tells us that i_{enc} gets a plus sign. Then Ampere's law gives us

$$B(2\pi r) = \mu_0 i \frac{\pi r^2}{\pi R^2}.$$

or

$$B = \left(\frac{\mu_0 i}{2\pi R^2}\right) r. \quad (30\text{-}22)$$

Thus, inside the wire, the magnitude B of the magnetic field is proportional to r; that magnitude is zero at the center and a maximum at the surface, where $r = R$. Note that Eqs. 30-19 and 30-22 give the same (maximum) value for B at $r = R$; that is, the expressions for the magnetic field outside the wire and inside the wire yield the same result at the surface of the wire.

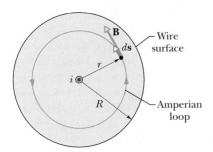

FIGURE 30-15 Using Ampere's law to find the magnetic field that a current i produces inside a long straight wire of circular cross section. The current is uniformly distributed over the cross section and emerges from the page. An Amperian loop is drawn inside the wire.

CHECKPOINT **3:** The figure shows three parallel and equal currents i and four Amperian loops. Rank the loops according to the magnitude of $\oint \mathbf{B} \cdot d\mathbf{s}$ along each, greatest first.

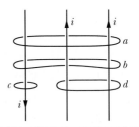

SAMPLE PROBLEM 30-4

Figure 30-16a shows the cross section of a long conducting cylinder with inner radius $a = 2.0$ cm and outer radius $b = 4.0$ cm. The cylinder carries a current out of the page, and the current density in the cross section is given by $J = cr^2$, with $c = 3.0 \times 10^6$ A/m^2 and r in meters. What is the magnetic field **B** at a point that is 3.0 cm from the central axis of the cylinder?

SOLUTION: Because the current distribution (hence the magnetic field) has cylindrical symmetry about the central axis of the cylinder, we can apply Ampere's law to the cross section to find the magnetic field. As shown in Fig. 30-16b, we draw an Amperian loop concentric with the cylinder and with radius $r = 3.0$ cm because we want to compute B at that distance from the central axis.

To apply Ampere's law, we must compute the current i_{enc} that is encircled by the Amperian loop. We cannot set up a proportionality as in Eq. 30-21, however, because here the current is not uniformly distributed. Instead, following the procedure of Sample Problem 27-2b, we must integrate the current density from the cylinder's inner radius a to the loop radius r:

$$i_{enc} = \int J\, dA = \int_a^r cr^2\, (2\pi r\, dr)$$

$$= 2\pi c \int_a^r r^3\, dr = 2\pi c \left[\frac{r^4}{4}\right]_a^r$$

$$= \frac{\pi c(r^4 - a^4)}{2}.$$

The direction of integration indicated in Fig. 30-16b is (arbitrarily) clockwise. Applying the right-hand rule for Ampere's law to that loop, we find that we should take i_{enc} as negative because the current is directed out of the page but our thumb is directed into the page.

We next evaluate the left side of Ampere's law exactly as we did in Fig. 30-15, and we again obtain Eq. 30-20. So, Ampere's law,

$$\oint \mathbf{B} \cdot d\mathbf{s} = \mu_0 i_{enc},$$

gives us

$$B(2\pi r) = -\frac{\mu_0 \pi c}{2}(r^4 - a^4).$$

Solving for B and substituting known data yield

$$B = -\frac{\mu_0 c}{4r}(r^4 - a^4)$$

$$= -\frac{(4\pi \times 10^{-7}\ \text{T·m/A})(3.0 \times 10^6\ \text{A/m}^2)}{4(0.030\ \text{m})}$$

$$\times\ [(0.030\ \text{m})^4 - (0.020\ \text{m})^4]$$

$$= -2.0 \times 10^{-5}\ \text{T}.$$

Thus, the magnetic field **B** has the magnitude

$$B = 2.0 \times 10^{-5}\ \text{T} \qquad \text{(Answer)}$$

and is directed opposite our direction of integration, hence counterclockwise in Fig. 30-16*b*.

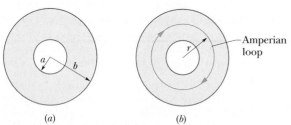

(a)   (b)

FIGURE 30-16 Sample Problem 30-4. (*a*) Cross section of a conducting cylinder of inner radius *a* and outer radius *b*. (*b*) An Amperian loop of radius *r* is added to compute the magnetic field at points that are a distance *r* from the central axis.

30-4 SOLENOIDS AND TOROIDS

Magnetic Field of a Solenoid

We now turn our attention to another situation in which Ampere's law proves useful. It concerns the magnetic field produced by the current in a long, tightly wound helical coil of wire. Such a coil is called a **solenoid** (Fig. 30-17). We assume that the length of the solenoid is much greater than the diameter.

Figure 30-18 shows a section through a portion of a "stretched-out" solenoid. The solenoid's magnetic field is the vector sum of the fields produced by the individual

FIGURE 30-17   A solenoid carrying current *i*.

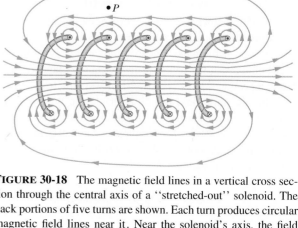

FIGURE 30-18   The magnetic field lines in a vertical cross section through the central axis of a "stretched-out" solenoid. The back portions of five turns are shown. Each turn produces circular magnetic field lines near it. Near the solenoid's axis, the field lines combine into a net magnetic field that is directed along the axis. The closely spaced field lines there indicate a strong magnetic field. Outside the solenoid the field lines are widely spaced: the field there is very weak.

turns. For points very close to each turn, the wire behaves magnetically almost like a long straight wire, and the lines of **B** there are almost concentric circles. Figure 30-18 suggests that the field tends to cancel between adjacent turns. It also suggests that, at points inside the solenoid and reasonably far from the wire, **B** is approximately parallel to the (central) solenoid axis. In the limiting case of an *ideal solenoid*, which is infinitely long and consists of tightly packed (*close-packed*) turns of square wire, the field inside the coil is uniform and parallel to the solenoid axis.

At points above the solenoid, such as *P* in Fig. 30-18, the field set up by the upper parts of the solenoid turns (marked ⊙) points to the left (as drawn near *P*) and tends to cancel the field set up by the lower parts of the turns (marked ⊗), which points to the right (not drawn). In the limiting case of an ideal solenoid, the magnetic field outside the solenoid is zero. Taking the external field to be zero is an excellent assumption for a real solenoid if its length is much greater than its diameter and if we consider external points such as point *P*. The direction of the magnetic field along the solenoid axis is given by a curled–straight right-hand rule: grasp the solenoid with your right hand so that your fingers follow the direction of the current in the windings; your extended right thumb then points in the direction of the axial magnetic field.

Figure 30-19 shows the lines of **B** for a real solenoid. The spacing of the lines of **B** in the central region shows that the field inside the coil is fairly strong and uniform over the cross section of the coil. The external field, however, is relatively weak.

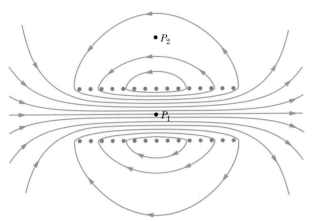

FIGURE 30-19 Magnetic field lines for a real solenoid of finite length. The field is strong and uniform at interior points such as P_1 but relatively weak at external points such as P_2.

Let us apply Ampere's law,

$$\oint \mathbf{B} \cdot d\mathbf{s} = \mu_0 i_{\text{enc}}, \qquad (30\text{-}23)$$

to the rectangular Amperian loop $abcd$ in the ideal solenoid of Fig. 30-20, where $\mathbf{B}$ is uniform within the solenoid and zero outside it. We write the integral $\oint \mathbf{B} \cdot d\mathbf{s}$ as the sum of four integrals, one for each path segment:

$$\oint \mathbf{B} \cdot d\mathbf{s} = \int_a^b \mathbf{B} \cdot d\mathbf{s} + \int_b^c \mathbf{B} \cdot d\mathbf{s}$$
$$+ \int_c^d \mathbf{B} \cdot d\mathbf{s} + \int_d^a \mathbf{B} \cdot d\mathbf{s}. \qquad (30\text{-}24)$$

The first integral on the right of Eq. 30-24 is Bh, where B is the magnitude of the uniform field $\mathbf{B}$ inside the solenoid and h is the (arbitrary) length of the path from a to b. The second and fourth integrals are zero because for every element of these paths $\mathbf{B}$ either is perpendicular to the path or is zero, and thus $\mathbf{B} \cdot d\mathbf{s}$ is zero. The third integral, which is along a path that lies outside the solenoid, is zero because $B = 0$ at all external points. Thus $\oint \mathbf{B} \cdot d\mathbf{s}$ for the entire rectangular path has the value Bh.

The net current i_{enc} encircled by the rectangular Amperian loop in Fig. 30-20 is not the same as the current i in

the solenoid windings because the windings pass more than once through this loop. Let n be the number of turns per unit length of the solenoid; then

$$i_{\text{enc}} = i(nh).$$

Ampere's law then gives us

$$Bh = \mu_0 i n h,$$

or

$$B = \mu_0 i n \qquad \text{(ideal solenoid).} \qquad (30\text{-}25)$$

Although we derived Eq. 30-25 for an infinitely long ideal solenoid, it holds quite well for actual solenoids if we apply it only at interior points, well away from the solenoid ends. Equation 30-25 is consistent with the experimental fact that B does not depend on the diameter or the length of the solenoid and that B is constant over the solenoidal cross section. A solenoid provides a practical way to set up a known uniform magnetic field for experimentation, just as a parallel-plate capacitor provides a practical way to set up a known uniform electric field.

Magnetic Field of a Toroid

Figure 30-21a shows a **toroid,** which we may describe as a solenoid bent into the shape of a doughnut. What magnetic field $\mathbf{B}$ is set up at its interior points (within the "tube" of the doughnut)? We can find out from Ampere's law and the symmetry.

From the symmetry, we see that the lines of $\mathbf{B}$ form concentric circles inside the toroid, directed as shown in Fig. 30-21b. Let us choose a concentric circle of radius r as an Amperian loop and traverse it in the clockwise direction. Ampere's law (Eq. 30-16) yields

$$(B)(2\pi r) = \mu_0 i N,$$

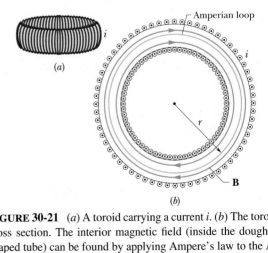

FIGURE 30-21 (a) A toroid carrying a current i. (b) The toroid's cross section. The interior magnetic field (inside the doughnut-shaped tube) can be found by applying Ampere's law to the Amperian loop shown.

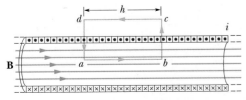

FIGURE 30-20 An application of Ampere's law to a section of a long ideal solenoid carrying a current i. The Amperian loop is the rectangle $abcd$.

where i is the current in the toroid windings (and is positive) and N is the total number of turns. This gives

$$B = \frac{\mu_0 i N}{2\pi} \frac{1}{r} \quad \text{(toroid).} \quad (30\text{-}26)$$

In contrast to the situation for a solenoid, B is not constant over the cross section of a toroid. It is easy to show, with Ampere's law, that $B = 0$ for points outside an ideal toroid (as if the toroid were made from an ideal solenoid).

The direction of the magnetic field within a toroid follows from our curled–straight right-hand rule: grasp the toroid with the fingers of your right hand curled in the direction of the current in the windings; your extended right thumb points in the direction of the magnetic field.

SAMPLE PROBLEM 30-5

A solenoid has length $L = 1.23$ m and inner diameter $d = 3.55$ cm, and it carries a current $i = 5.57$ A. It consists of five close-packed layers, each with 850 turns along length L. What is B at its center?

SOLUTION: From Eq. 30-25

$$B = \mu_0 i n = (4\pi \times 10^{-7} \text{ T·m/A})(5.57 \text{ A}) \frac{5 \times 850 \text{ turns}}{1.23 \text{ m}}$$

$$= 2.42 \times 10^{-2} \text{ T} = 24.2 \text{ mT.} \quad \text{(Answer)}$$

Note that Eq. 30-25 applies even though the solenoid has more than one layer of windings because the diameter of the windings does not enter into the equation.

30-5 A CURRENT-CARRYING COIL AS A MAGNETIC DIPOLE

So far we have examined the magnetic fields produced by a long straight wire, a solenoid, and a toroid. We turn our attention here to the field produced by a coil carrying current. You saw in Section 29-9 that such a coil behaves as a magnetic dipole in that, if we place it in an external magnetic field $\mathbf{B}$, a torque τ given by

$$\boldsymbol{\tau} = \boldsymbol{\mu} \times \mathbf{B} \quad (30\text{-}27)$$

acts on it. Here $\boldsymbol{\mu}$ is the magnetic dipole moment of the coil and has the magnitude NiA, where N is the number of turns (or loops), i is the current in each turn, and A is the area enclosed by each turn.

Recall that the direction of $\boldsymbol{\mu}$ is given by a curled–straight right-hand rule: grasp the coil so that the fingers of

your right hand curl around it in the direction of the current; your extended thumb then points in the direction of the dipole moment $\boldsymbol{\mu}$.

Magnetic Field of a Coil

We turn now to the other aspect of a coil as a magnetic dipole. What magnetic field does *it* produce at a point in the surrounding space? The problem does not have enough symmetry to make Ampere's law useful, so we must turn to the law of Biot and Savart. For simplicity, we first consider only a coil with a single circular loop and only points on its central axis, which we take to be a z axis. We shall show that the magnitude of the magnetic field is

$$B(z) = \frac{\mu_0 i R^2}{2(R^2 + z^2)^{3/2}}, \quad (30\text{-}28)$$

in which R is the radius of the circular loop and z is the distance of the point in question from the center of the loop. Furthermore, the direction of the magnetic field $\mathbf{B}$ is the same as the direction of the magnetic dipole moment $\boldsymbol{\mu}$ of the loop.

For axial points far from the loop, we have $z \gg R$ in Eq. 30-28. With that approximation, this equation reduces to

$$B(z) \approx \frac{\mu_0 i R^2}{2z^3}.$$

Recalling that πR^2 is the area A of the loop and extending our result to include a coil of N turns, we can write this equation as

$$B(z) = \frac{\mu_0}{2\pi} \frac{NiA}{z^3}$$

or, since $\mathbf{B}$ and $\boldsymbol{\mu}$ have the same direction, we can write the equation in vector form, substituting from the identity $\mu = NiA$:

$$\mathbf{B}(z) = \frac{\mu_0}{2\pi} \frac{\boldsymbol{\mu}}{z^3} \quad \text{(current-carrying coil).} \quad (30\text{-}29)$$

Thus we have two ways in which we can regard a current-carrying coil as a magnetic dipole: (1) it experiences a torque when we place it in an external magnetic field; (2) it generates its own intrinsic magnetic field, given, for distant points along its axis, by Eq. 30-29. Figure 30-22 shows the magnetic field of a current loop; one side of the loop acts as a north pole (in the direction of $\boldsymbol{\mu}$) and the other side as a south pole, as suggested by the ghosted magnet in the figure.

CHECKPOINT **4:** The figure shows four arrangements of circular loops of radius r or $2r$, centered on vertical axes and carrying identical currents in the directions indicated. Rank the arrangements according to the magnitude of the net magnetic field at the dot, midway between the loops on the central axis, greatest first.

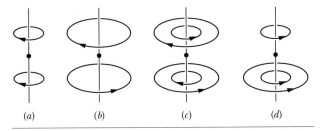

(a)   (b)   (c)   (d)

Proof of Equation 30-28

Figure 30-23 shows the back half of a circular loop of radius R carrying a current i. Consider a point P on the axis of the loop, a distance z from its plane. Let us apply the law of Biot and Savart to a length element located at the left side of the loop. The vector $d\mathbf{s}$ for this element points perpendicularly out of the page. The angle θ between $d\mathbf{s}$ and the vector $\mathbf{r}$ in Fig. 30-23 is 90°; the plane formed by these two vectors is perpendicular to the plane of the figure and contains both $\mathbf{r}$ and $d\mathbf{s}$. From the law of Biot and Savart (and the right-hand rule), the differential field $d\mathbf{B}$ produced at point P by the current in this element is perpendicular to this plane and thus lies in the plane of the figure, perpendicular to $\mathbf{r}$, as indicated in Fig. 30-23.

Let us resolve $d\mathbf{B}$ into two components: $dB_{\parallel}$ along the axis of the loop, and $dB_{\perp}$ perpendicular to this axis. From the symmetry, the vector sum of all the perpendicular

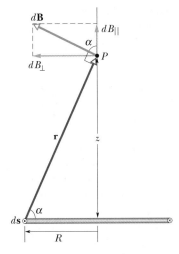

FIGURE 30-23 A current loop of radius R. The plane of the loop is perpendicular to the page and only the back half of the loop is shown. We use the law of Biot and Savart to find the magnetic field at point P on the central axis of the loop.

components $dB_{\perp}$ due to all the loop elements is zero. This leaves only the axial components $dB_{\parallel}$ and we have

$$B = \int dB_{\parallel}.$$

For the element $d\mathbf{s}$ in Fig. 30-23, the law of Biot and Savart (Eq. 30-3) gives

$$dB = \frac{\mu_0}{4\pi} \frac{i\,ds\,\sin 90°}{r^2}.$$

We also have

$$dB_{\parallel} = dB \cos \alpha.$$

Combining these two relations, we obtain

$$dB_{\parallel} = \frac{\mu_0 i \cos \alpha\, ds}{4\pi r^2}. \tag{30-30}$$

Figure 30-23 shows that r and α are not independent but are related to each other. Let us express each in terms of the variable z, the distance between point P and the center of the loop. The relations are

$$r = \sqrt{R^2 + z^2} \tag{30-31}$$

and

$$\cos \alpha = \frac{R}{r} = \frac{R}{\sqrt{R^2 + z^2}}. \tag{30-32}$$

Substituting Eqs. 30-31 and 30-32 into Eq. 30-30, we find

$$dB_{\parallel} = \frac{\mu_0 iR}{4\pi(R^2 + z^2)^{3/2}}\,ds.$$

Note that i, R, and z have the same values for all elements $d\mathbf{s}$ around the loop. So when we integrate this equation, noting that $\int ds$ is simply the circumference $2\pi R$ of the

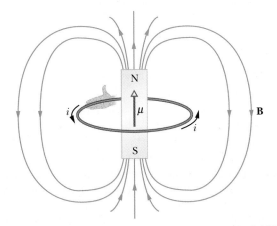

FIGURE 30-22 A current loop produces a magnetic field like that of a bar magnet and thus has associated north and south poles. The magnetic dipole moment $\boldsymbol{\mu}$ of the loop, given by a curled–straight right-hand rule, points from the south pole to the north pole, in the direction of the field **B** within the loop.

loop, we find that

$$B = \int dB_{\parallel} = \frac{\mu_0 iR}{4\pi(R^2 + z^2)^{3/2}} \int ds$$

or

$$B(z) = \frac{\mu_0 iR^2}{2(R^2 + z^2)^{3/2}},$$

which is Eq. 30-28, the relation we sought to prove.

REVIEW & SUMMARY

The Biot–Savart Law

The magnetic field set up by a current-carrying conductor can be found from the **Biot–Savart law.** This law asserts that the contribution $d\mathbf{B}$ to the field produced by a current-length element $i\,d\mathbf{s}$ at a point P, a distance r from the current element, is

$$d\mathbf{B} = \frac{\mu_0}{4\pi} \frac{i\,d\mathbf{s} \times \mathbf{r}}{r^3} \qquad \text{(Biot–Savart law).} \quad (30\text{-}5)$$

Here $\mathbf{r}$ is a vector that points from the element to the point in question. The quantity μ_0, called the permeability constant, has the value $4\pi \times 10^{-7}$ T·m/A $\approx 1.26 \times 10^{-6}$ T·m/A.

Magnetic Field of a Long Straight Wire

For a long straight wire carrying a current i, the Biot–Savart law gives, for the magnetic field at a distance r from the wire,

$$B = \frac{\mu_0 i}{2\pi r} \qquad \text{(long straight wire).} \quad (30\text{-}6)$$

Magnetic Field of a Circular Arc

The magnetic field at the center of a circular arc of wire of radius R that carries current i is

$$B = \frac{\mu_0 i\phi}{4\pi R} \qquad \text{(at center of circular arc).} \quad (30\text{-}11)$$

The Force Between Parallel Wires Carrying Currents

Parallel wires carrying currents in the same direction attract each other, whereas parallel wires carrying currents in opposite directions repel each other. The magnitude of the force on a length L of either wire is

$$F_{ba} = i_b LB_a \sin 90° = \frac{\mu_0 L i_a i_b}{2\pi d}, \quad (30\text{-}14)$$

where d is the wire separation, and i_a and i_b are the currents in the wires.

Ampere's Law

For some current distributions, **Ampere's law,**

$$\oint \mathbf{B} \cdot d\mathbf{s} = \mu_0 i_{enc} \qquad \text{(Ampere's law),} \quad (30\text{-}16)$$

can be used (instead of the Biot–Savart law) to calculate the magnetic field. The line integral in this equation is evaluated around a closed loop called an *Amperian loop*. The current i is the *net* current encircled by the loop.

Fields of a Solenoid and a Toroid

Inside a *long solenoid* carrying current i, at points not near its ends, the magnitude B of the magnetic field is

$$B = \mu_0 in \qquad \text{(ideal solenoid),} \quad (30\text{-}25)$$

where n is the number of turns per unit length. At a point inside a *toroid*, the magnitude B of the magnetic field is

$$B = \frac{\mu_0 iN}{2\pi} \frac{1}{r} \qquad \text{(toroid),} \quad (30\text{-}26)$$

where r is the distance from the center of the toroid to the point.

Field of a Magnetic Dipole

The magnetic field produced by a current-carrying coil, which is a *magnetic dipole*, at a point P located a distance z along the coil's central axis is parallel to the axis and is given by

$$\mathbf{B}(z) = \frac{\mu_0}{2\pi} \frac{\boldsymbol{\mu}}{z^3}, \quad (30\text{-}29)$$

where $\boldsymbol{\mu}$ is the dipole moment of the coil.

QUESTIONS

1. Figure 30-24 shows four arrangements in which long parallel wires carry equal currents directly into or out of the page at the corners of identical squares. Rank the arrangements according to the magnitude of the net magnetic field at the center of the square, greatest first.

2. Figure 30-25 shows cross sections of two long straight wires; the left-hand wire carries current i_1 directly out of the page. If the net magnetic field due to the two currents is to be zero at point P, (a) should the direction of current i_2 in the right-hand wire be directly into or out of the page and (b) should i_2 be greater than, less than, or equal to i_1?

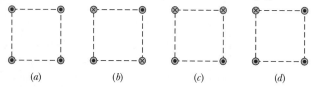

$(a) \qquad (b) \qquad (c) \qquad (d)$

FIGURE 30-24 Question 1.

FIGURE 30-25 Question 2.

3. In Fig. 30-26, two long straight wires just pass each other perpendicularly without touching. In which quadrants are there points at which the net magnetic field is zero?

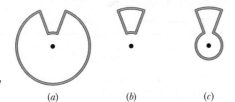

FIGURE 30-26
Question 3.

4. Figure 30-27 shows three circuits, each consisting of two concentric circular arcs, one of radius r and the other of a larger radius R, and two radial lengths. The circuits have the same current through them, and the radial lengths have the same angle between them. Rank the circuits according to the magnitude of the net magnetic field at the center, greatest first.

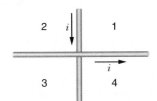

FIGURE 30-27
Question 4.   (a)   (b)   (c)

5. Figure 30-28 shows three sections of circuits, each section consisting of a wire curved along a circular arc (all with the same radius) and two long straight wires that are tangential to the arc. (In a, the straight wires pass each other without touching.) The sections carry equal currents. Rank the sections according to the magnitude of the magnetic field produced at the center of curvature, greatest first.

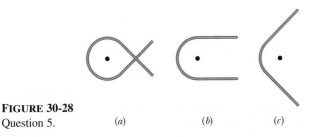

FIGURE 30-28
Question 5.   (a)   (b)   (c)

6. Figure 30-29 shows three sections of circuits, each consisting of a curved wire along a circular arc (all with the same radius and internal angle) and two long straight wires; the straight wires are radial to the arc in circuit a and tangential to the arcs in circuits b and c (in c, the straight wires pass each other without touching). The sections carry equal currents. The net magnetic field at the center of each arc is dominated by the contribution from the arc. Rank the sections according to the magnitude of the net magnetic field at the center, greatest first.

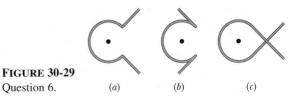

FIGURE 30-29
Question 6.   (a)   (b)   (c)

7. Figure 30-30 shows four arrangements in which long, parallel, equally spaced wires carry equal currents directly into or out of the page. Rank the arrangements according to the magnitude of the net force on the central wire due to the currents in the other wires, greatest first.

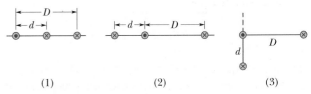

FIGURE 30-30   Question 7.

8. Figure 30-31 shows three arrangements of three long straight wires, carrying equal currents directly into or out of the page. (a) Rank the arrangements according to the magnitude of the net force on the wire with the current directed out of the page due to the currents in the other wires, greatest first. (b) In arrangement 3, is the angle between the net force on that wire and the dashed line equal to, less than, or more than 45°?

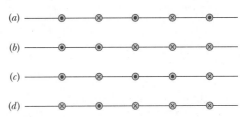

FIGURE 30-31   Question 8.

9. Figure 30-32 shows two arrangements in which two long straight wires with equal currents, directed into or out of the page, are at equal distances from a y axis. (a) For each arrangement, what is the direction of the net magnetic field at point P? (b) For each arrangement, if a third long straight wire, with a current directly out of the page, is placed at P, what is the direction of the net force on that wire due to the other currents?

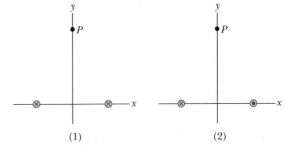

FIGURE 30-32 Question 9.

10. Figure 30-33 shows a long straight wire with current i to the right and three wire loops, all with the same clockwise current around them. The loops are all at the same distance from the long wire and have edge lengths of either L or $2L$. Rank the loops according to the magnitude of the net force on them due to the current in the long straight wire, greatest first.

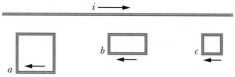

FIGURE 30-33 Question 10.

11. In Fig. 30-34, a messy loop of wire is placed on a slick table with points a and b fixed in place. If a current is then sent through the wire, will the wire be pushed outward into an arc or will it be pulled inward?

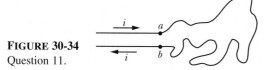

FIGURE 30-34
Question 11.

12. Figure 30-35 shows a uniform magnetic field **B** and four straight-line paths of equal lengths. Rank the paths according to the magnitude of $\int \mathbf{B} \cdot d\mathbf{s}$ taken along the paths, greatest first.

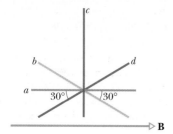

FIGURE 30-35
Question 12.

13. Figure 30-36*a* shows four circular Amperian loops concentric with a wire whose current is directed out of the page. The current is uniform across the wire's circular cross section. Rank the loops according to the magnitude of $\oint \mathbf{B} \cdot d\mathbf{s}$ around each, greatest first.

14. Figure 30-36*b* shows four circular Amperian loops and, in cross section, four long circular conductors, all of which are concentric. The currents in the conductors are, from smallest radius to largest radius, 4 A out of the page, 9 A into the page, 5 A out of the page, and 3 A into the page. Rank the loops according to the magnitude of $\oint \mathbf{B} \cdot d\mathbf{s}$ around each, greatest first.

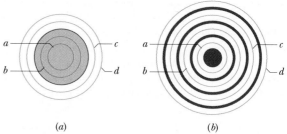

FIGURE 30-36 Questions 13 and 14.

15. Figure 30-37 shows four identical currents i and five Amperian paths encircling them. Rank the paths according to the value of $\oint \mathbf{B} \cdot d\mathbf{s}$ taken in the directions shown, most positive first and most negative last.

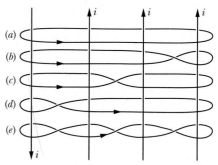

FIGURE 30-37 Question 15.

16. The following table gives the number of turns per unit length n and the current i through six ideal solenoids of different radii. You want to combine several of them concentrically to produce a net magnetic field of zero along the central axis. Can this be done with (a) two of them, (b) three of them, (c) four of them, and (d) five of them? If so, answer by listing which solenoids are to be used and indicate the directions of the currents.

Solenoid:	1	2	3	4	5	6
n:	5	4	3	2	10	8
i:	5	3	7	6	2	3

17. The position vector of a particle moving around a circle of radius r is **r**. What is the value of $\oint \mathbf{r} \cdot d\mathbf{s}$ around the circle?

18. What is the value of $\oint d\mathbf{s}$ around the perimeter of (a) a square with edge length a and (b) a equilateral triangle of edge length d?

EXERCISES & PROBLEMS

SECTION 30-1 Calculating the Magnetic Field Due to a Current

1E. The magnitude of the magnetic field 88.0 cm from the axis of a long straight wire is 7.30 μT. What is the current in the wire?

2E. A 10-gauge bare copper wire (2.6 mm in diameter) can carry a current of 50 A without overheating. For this current, what is the magnetic field at the surface of the wire?

3E. A surveyor is using a magnetic compass 20 ft below a power line in which there is a steady current of 100 A. (a) What is the magnetic field at the site of the compass due to the power line? (b) Will this interfere seriously with the compass reading? The horizontal component of Earth's magnetic field at the site is 20 μT.

4E. The electron gun in a TV tube fires electrons of kinetic energy 25 keV in a beam 0.22 mm in diameter at the screen; 5.6×10^{14} electrons arrive each second. Calculate the magnetic field produced by the beam at a point 1.5 mm from the beam axis.

5E. Figure 30-38 shows a 3.0 cm segment of wire, centered at the origin, carrying a current of 2.0 A in the positive y direction (as part of some complete circuit). To calculate the magnetic field **B** at a point several meters from the origin, one may use the Biot–Savart law in the form $B = (\mu_0/4\pi)i \, \Delta s \sin \theta/r^2$, in which $\Delta s = 3.0$ cm. This is because r and θ are essentially constant over the segment of wire. Calculate **B** (magnitude and direction) at the following (x, y, z) locations: (a) (0, 0, 5.0 m), (b) (0, 6.0 m, 0), (c) (7.0 m, 7.0 m, 0), (d) (−3.0 m, −4.0 m, 0).

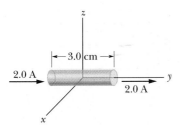

FIGURE 30-38
Exercise 5.

6E. A long wire carrying a current of 100 A is placed in a uniform external magnetic field of 5.0 mT. The wire is perpendicular to this magnetic field. Locate the points at which the resultant magnetic field is zero.

7E. At a position in the Philippines, Earth's magnetic field of 39 μT is horizontal and directed due north. Suppose the net field is zero exactly 8.0 cm above a long straight horizontal wire that carries a constant current. What are (a) the magnitude and (b) the direction of the current?

8E. A particle with positive charge q is a distance d from a long straight wire that carries a current i and is traveling with speed v perpendicular to the wire. What are the direction and magnitude of the force on the particle if it is moving (a) toward or (b) away from the wire?

9E. A straight conductor carrying a current i splits into identical semicircular arcs as shown in Fig. 30-39. What is the magnetic field at the center C of the resulting circular loop?

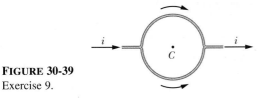

FIGURE 30-39
Exercise 9.

10E. A long straight wire carries a current of 50 A. An electron, traveling at 1.0×10^7 m/s, is 5.0 cm from the wire. What force acts on the electron if the electron velocity is directed (a) toward

the wire, (b) parallel to the wire in the direction of the current, and (c) perpendicular to both directions defined by (a) and (b)?

11P. The wire shown in Fig. 30-40 carries current i. What magnetic field **B** is produced at the center C of the semicircle by (a) each straight segment of length L, (b) the semicircular segment of radius R, and (c) the entire wire?

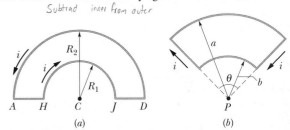

FIGURE 30-40
Problem 11.

12P. Use the Biot–Savart law to calculate the magnetic field **B** at C, the common center of the semicircular arcs AD and HJ in Fig. 30-41a. The two arcs, of radii R_2 and R_1, respectively, form part of the circuit $ADJHA$ carrying current i.

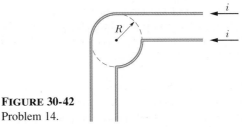

(a)   (b)

FIGURE 30-41 Problems 12 and 13.

13P. Consider the circuit of Fig. 30-41b. The curved segments are arcs of circles of radii a and b. The straight segments are along radii. Find the magnetic field **B** at point P (the center of curvature), assuming a current i in the circuit.

14P. Two infinitely long wires carry equal currents i. Each follows a 90° arc on the circumference of the same circle of radius R, in the configuration shown in Fig. 30-42. Show that **B** at the center of the circle is the same as the field **B** a distance R below an infinite straight wire carrying a current i to the left.

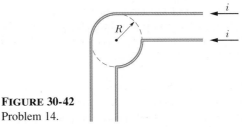

FIGURE 30-42
Problem 14.

15P. A long hairpin is formed by bending a very long wire as shown in Fig. 30-43. If the wire carries a 10 A current, what are the direction and magnitude of **B** at (a) point a and (b) midpoint b? Take $R = 5.0$ mm and the distance between a and b to be *much* larger than R (each straight section is "infinite").

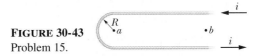

FIGURE 30-43
Problem 15.

16P. A wire carrying current i has the configuration shown in Fig. 30-44. Two semi-infinite straight sections, both tangent to the same circle, are connected by a circular arc, of central angle θ, along the circumference of the circle, with all sections lying in the same plane. What must θ be in order for B to be zero at the center of the circle?

FIGURE 30-44
Problem 16.

17P. In Fig. 30-45, a straight wire of length L carries current i. Show that the magnitude of the magnetic field **B** produced by this segment at P_1, a distance R from the segment along a perpendicular bisector, is

$$B = \frac{\mu_0 i}{2\pi R} \frac{L}{(L^2 + 4R^2)^{1/2}}.$$

Show that this expression reduces to an expected result as $L \to \infty$.

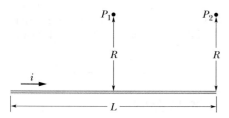

FIGURE 30-45 Problems 17 and 18.

18P. In Fig. 30-45, a straight wire of length L carries current i. Show that the magnitude of the magnetic field **B** produced by the wire at P_2, a perpendicular distance R from one end of the wire, is

$$B = \frac{\mu_0 i}{4\pi R} \frac{L}{(L^2 + R^2)^{1/2}}.$$

19P. A square loop of wire of edge length a carries current i. Using Problem 17, show that, at the center of the loop, the magnitude of the magnetic field produced by the current is

$$B = \frac{2\sqrt{2}\mu_0 i}{\pi a}.$$

20P. Using Problem 17, show that the magnitude of the magnetic field produced at the center of a rectangular loop of wire of length L and width W, carrying a current i, is

$$B = \frac{2\mu_0 i}{\pi} \frac{(L^2 + W^2)^{1/2}}{LW}.$$

Show that, for $L \gg W$, this expression reduces to a result consistent with the result of Sample Problem 30-2.

21P. A square loop of wire of edge length a carries current i. Using Problem 17, show that the magnitude of the magnetic field produced at a point on the axis of the loop and a distance x from its center is

$$B(x) = \frac{4\mu_0 i a^2}{\pi(4x^2 + a^2)(4x^2 + 2a^2)^{1/2}}.$$

Prove that this result is consistent with the result of Problem 19.

22P. Two wires, both of length L, are formed into a circle and a square, and each carries current i. Show that the square produces a greater magnetic field at its center than the circle produces at its center. (See Problem 19.)

23P. In Fig. 30-46, a current i is in a straight wire of length a. Show that the magnitude of the magnetic field produced by the current at point P is $B = \sqrt{2}\mu_0 i/8\pi a$.

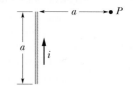

FIGURE 30-46
Problem 23.

24P. Find the magnetic field **B** at point P in Fig. 30-47 (See Problem 23.)

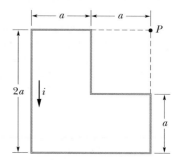

FIGURE 30-47
Problem 24.

25P. Find the magnetic field **B** at point P in Fig. 30-48 for $i = 10$ A and $a = 8.0$ cm. (See Problems 18 and 23.)

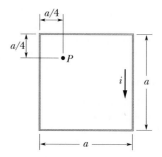

FIGURE 30-48
Problem 25.

26P. Figure 30-49 shows a cross section of a long, thin ribbon of width w that is carrying a uniformly distributed total current i into the page. Calculate the magnitude and direction of the magnetic field **B** at a point P in the plane of the ribbon at a distance d from its edge. (*Hint:* Imagine the ribbon to be constructed from many long, thin, parallel wires.)

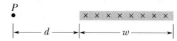

FIGURE 30-49 Problem 26.

SECTION 30-2 Two Parallel Currents

27E. Two long parallel wires are 8.0 cm apart. What equal currents must be in the wires if the magnetic field halfway between them is to have a magnitude of 300 μT? Consider both (a) parallel and (b) antiparallel currents.

28E. Two long parallel wires a distance d apart carry currents of i and $3i$ in the same direction. Locate the point or points at which their magnetic fields cancel.

29E. Two long straight parallel wires, separated by 0.75 cm, are perpendicular to the plane of the page as shown in Fig. 30-50. Wire 1 carries a current of 6.5 A into the page. What must be the current (magnitude and direction) in wire 2 for the resultant magnetic field at point P to be zero?

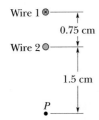

FIGURE 30-50
Exercise 29.

30E. Figure 30-51 shows five long parallel wires in the xy plane. Each wire carries a current $i = 3.00$ A in the positive x direction. The separation between adjacent wires is $d = 8.00$ cm. In unit-vector notation, what is the magnetic force per meter exerted on each of these five wires by the other wires?

FIGURE 30-51
Exercise 30.

31E. For the wires in Sample Problem 30-2, show that Eq. 30-15 holds for points beyond the wires, that is, for points with $|x| > d$.

32E. Each of two long straight parallel wires 10 cm apart carries a current of 100 A. Figure 30-52 shows a cross section, with the wires running perpendicular to the page and point P lying on the perpendicular bisector of the line between the wires. Find the magnitude and direction of the magnetic field at P when the current in the left-hand wire is out of the page and the current in the right-hand wire is (a) out of the page and (b) into the page.

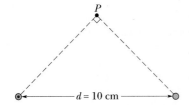

FIGURE 30-52
Exercise 32.

33P. Assume that both currents in Fig. 30-10a are in the same direction, out of the plane of the figure. Show that the magnetic field in the plane defined by the wires is

$$B(x) = \frac{\mu_0 ix}{\pi(x^2 - d^2)}.$$

Assume that $i = 10$ A and $d = 2.0$ cm in Fig. 30-10a, and plot $B(x)$ for the range -2 cm $< x < 2$ cm. Assume that the wire diameters are negligible.

34P. Four long copper wires are parallel to each other, their cross sections forming the corners of a square with sides $a = 20$ cm. A 20 A current exists in each wire in the direction shown in Fig. 30-53. What are the magnitude and direction of **B** at the center of the square?

FIGURE 30-53
Problems 34, 35, and 36.

35P. Suppose, in Fig. 30-53, that the identical currents i are all out of the page. What is the force per unit length (magnitude and direction) on any one wire?

36P. In Fig. 30-53 what is the force per unit length acting on the lower left wire, in magnitude and direction, with the current directions as shown? The currents are i.

37P. Two long wires a distance d apart carry equal antiparallel currents i, as in Fig. 30-54. (a) Show that the magnitude of the magnetic field at point P, which is equidistant from the wires, is given by

$$B = \frac{2\mu_0 id}{\pi(4R^2 + d^2)}.$$

(b) In what direction does **B** point?

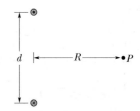

FIGURE 30-54
Problem 37.

38P. In Fig. 30-55, the long straight wire carries a current of 30 A and the rectangular loop carries a current of 20 A. Calculate the resultant force acting on the loop. Assume that $a = 1.0$ cm, $b = 8.0$ cm, and $L = 30$ cm.

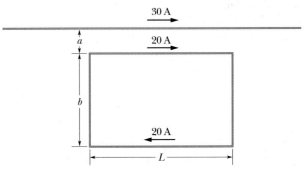

FIGURE 30-55 Problem 38.

39P. Figure 30-56 is an idealized schematic drawing of a rail gun. Projectile P sits between the two wide circular rails; a source of current sends current through the rails and through the (conducting) projectile itself (a fuse is not used). (a) Let w be the distance between the rails, R the radius of the rails, and i the current. Show that the force on the projectile is directed to the right along the rails and is given approximately by

$$F = \frac{i^2 \mu_0}{2\pi} \ln \frac{w + R}{R}.$$

(b) If the projectile starts from the left end of the rails at rest, find the speed v at which it is expelled at the right. Assume that $i = 450$ kA, $w = 12$ mm, $R = 6.7$ cm, $L = 4.0$ m, and the mass of the projectile is $m = 10$ g.

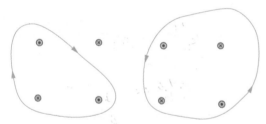

FIGURE 30-56   Problem 39.

SECTION 30-3   Ampere's Law

40E. Each of the eight conductors in Fig. 30-57 carries 2.0 A of current into or out of the page. Two paths are indicated for the line integral $\oint \mathbf{B} \cdot d\mathbf{s}$. What is the value of the integral for the path (a) at the left and (b) at the right?

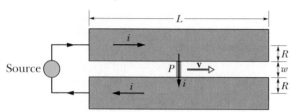

FIGURE 30-57   Exercise 40.

41E. Eight wires cut the page perpendicularly at the points shown in Fig. 30-58. A wire labeled with the integer k ($k = 1, 2, \ldots, 8$) carries the current ki. For those with odd k, the current is out of the page; for those with even k, it is into the page. Evaluate $\oint \mathbf{B} \cdot d\mathbf{s}$ along the closed path in the direction shown.

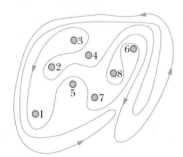

FIGURE 30-58
Exercise 41.

42E. In a certain region there is a uniform current density of 15 A/m² in the positive z direction. What is the value of $\oint \mathbf{B} \cdot d\mathbf{s}$ when the line integral is taken along the three straight-line segments from $(4d, 0, 0)$ to $(4d, 3d, 0)$ to $(0, 0, 0)$ to $(4d, 0, 0)$, where $d = 20$ cm?

43E. Figure 30-59 shows a cross section of a long cylindrical conductor of radius a, carrying a uniformly distributed current i. Assume that $a = 2.0$ cm and $i = 100$ A, and plot $B(r)$ over the range $0 < r < 6.0$ cm.

FIGURE 30-59
Exercise 43.

44P. Two square conducting loops carry currents of 5.0 and 3.0 A as shown in Fig. 30-60. What is the value of $\oint \mathbf{B} \cdot d\mathbf{s}$ for each of the two closed paths shown?

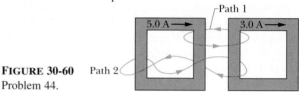

FIGURE 30-60   Path 2
Problem 44.

45P. Show that a uniform magnetic field $\mathbf{B}$ cannot drop abruptly to zero, as is suggested just to the right of point a in Fig. 30-61, as one moves perpendicular to $\mathbf{B}$, say along the horizontal arrow in the figure. (*Hint:* Apply Ampere's law to the rectangular path shown by the dashed lines.) In actual magnets "fringing" of the magnetic field lines always occurs, which means that $\mathbf{B}$ approaches zero in a gradual manner. Modify the field lines in the figure to indicate a more realistic situation.

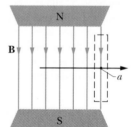

FIGURE 30-61
Problem 45.

46P. Figure 30-62a shows a cross section of a hollow cylindrical conductor of radii a and b, carrying a uniformly distributed current i. (a) Show that $B(r)$ for the range $b < r < a$ is given by

$$B = \frac{\mu_0 i}{2\pi(a^2 - b^2)} \frac{r^2 - b^2}{r}.$$

(b) Show that when $r = a$, this equation gives the magnetic field magnitude B for a long straight wire; when $r = b$, it gives zero magnetic field; and when $b = 0$, it gives the magnetic field inside a solid conductor. (c) Assume that $a = 2.0$ cm, $b = 1.8$ cm, and $i = 100$ A, and plot $B(r)$ for the range $0 < r < 6$ cm.

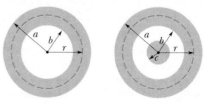

FIGURE 30-62
Problems 46 and 47.   (a)   (b)

47P. Figure 30-62*b* shows a cross section of a long conductor of a type called a coaxial cable and gives its radii (a, b, c). Equal but opposite currents i are uniformly distributed in the two conductors. Derive expressions for $B(r)$ in the ranges (a) $r < c$, (b) $c < r < b$, (c) $b < r < a$, and (d) $r > a$. (e) Test these expressions for all the special cases that occur to you. (f) Assume that $a = 2.0$ cm, $b = 1.8$ cm, $c = 0.40$ cm, and $i = 120$ A and plot the function $B(r)$ over the range $0 < r < 3$ cm.

48P. The current density inside a long, solid, cylindrical wire of radius a is in the direction of the central axis and varies linearly with radial distance r from the axis according to $J = J_0 r/a$. Find the magnetic field inside the wire.

49P. A long circular pipe with outside radius R carries a (uniformly distributed) current i into the page as shown in Fig. 30-63. A wire runs parallel to the pipe at a distance of $3R$ from center to center. Find the magnitude and direction of the current in the wire such that the resultant magnetic field at point P has the same magnitude as the resultant field at the center of the pipe but is in the opposite direction.

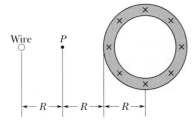

FIGURE 30-63
Problem 49.

50P. Figure 30-64 shows a cross section of a long cylindrical conductor of radius a containing a long cylindrical hole of radius b. The axes of the cylinder and hole are parallel and are a distance d apart; a current i is uniformly distributed over the tinted area. (a) Use superposition to show that the magnetic field at the center of the hole is

$$B = \frac{\mu_0 id}{2\pi(a^2 - b^2)}.$$

(b) Discuss the two special cases $b = 0$ and $d = 0$. (c) Use Ampere's law to show that the magnetic field in the hole is uniform. (*Hint:* Regard the cylindrical hole as filled with two equal currents moving in opposite directions, thus canceling each other. Assume that each of these currents has the same current density as that in the actual conductor. Then superimpose the fields due to two complete cylinders of current, of radii a and b, each cylinder having the same current density.)

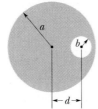

FIGURE 30-64
Problem 50.

51P. Figure 30-65 shows a cross section of an infinite conducting sheet with a current per unit x-length λ emerging perpendicularly out of the page. (a) Use the Biot–Savart law and symmetry to show that for all points P above the sheet, and all points P' below it, the magnetic field $\mathbf{B}$ is parallel to the sheet and directed as shown. (b) Use Ampere's law to prove that $B = \frac{1}{2}\mu_0\lambda$ at all points P and P'.

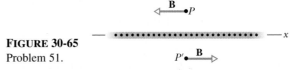

FIGURE 30-65
Problem 51.

52P*. The magnetic field in a certain region is given in milliteslas by $\mathbf{B} = 3.0\mathbf{i} + 8.0(x^2/d^2)\mathbf{j}$, where x is in meters and d is a constant with the unit of length. Some current must exist in the region to cause the specified $\mathbf{B}$ field. (a) Evaluate the integral $\oint \mathbf{B} \cdot d\mathbf{s}$ along the straight path from $(d, 0, 0)$ to $(d, d, 0)$. (b) Let $d = 0.50$ m in the expression for $\mathbf{B}$ and apply Ampere's law to determine what current flows perpendicularly through a square of side length 0.5 m that lies in the first quadrant of the xy plane, with one corner at the origin. (c) Is this current in the $+\mathbf{k}$ or $-\mathbf{k}$ direction?

SECTION 30-4 Solenoids and Toroids

53E. A 200 turn solenoid having a length of 25 cm and a diameter of 10 cm carries a current of 0.30 A. Calculate the magnitude of the magnetic field $\mathbf{B}$ inside the solenoid.

54E. A solenoid 95.0 cm long has a radius of 2.00 cm and a winding of 1200 turns; it carries a current of 3.60 A. Calculate the magnitude of the magnetic field inside the solenoid.

55E. A solenoid 1.30 m long and 2.60 cm in diameter carries a current of 18.0 A. The magnetic field inside the solenoid is 23.0 mT. Find the length of the wire forming the solenoid.

56E. A toroid having a square cross section, 5.00 cm on a side, and an inner radius of 15.0 cm has 500 turns and carries a current of 0.800 A. (It is made up of a square solenoid—instead of round as in Fig. 30-17—bent into a doughnut shape.) What is the magnetic field inside the toroid at (a) the inner radius and (b) the outer radius of the toroid?

57E. Show that if the thickness of a toroid is very small compared to its radius of curvature (a very skinny toroid), then Eq. 30-26 for the field inside a toroid reduces to Eq. 30-25 for the field inside a solenoid. Explain why this result is to be expected.

58P. Treat an ideal solenoid as a thin cylindrical conductor whose current per unit length, measured parallel to the cylinder axis, is λ. By doing so, show that the magnitude of the magnetic field inside an ideal solenoid can be written as $B = \mu_0\lambda$. This is the value of the *change* in B that you encounter as you move from inside the solenoid to outside, through the solenoid wall. Show that the same change occurs as you move through an infinite flat current sheet such as that of Fig. 30-65 (see Problem 51). Does this equality surprise you?

59P. In Section 30-4 we showed that the magnetic field at any radius r inside a toroid is given by

$$B = \frac{\mu_0 iN}{2\pi r}.$$

Show that as you move from a point just inside a toroid to a point just outside, the magnitude of the *change* in **B** that you encounter —at any radius r—is just $\mu_0\lambda$. Here λ is the current per unit length along a circumference of radius r within the toroid. Compare with the similar result found in Problem 58. Isn't the equality surprising?

60P. A long solenoid with 10.0 turns/cm and a radius of 7.00 cm carries a current of 20.0 mA. A current of 6.00 A exists in a straight conductor located along the axis of the solenoid. (a) At what radial distance from the axis will the direction of the resulting magnetic field be at $45.0°$ to the axial direction? (b) What is the magnitude of the magnetic field there?

61P. A long solenoid has 100 turns/cm and carries current i. An electron moves within the solenoid in a circle of radius 2.30 cm perpendicular to the solenoid axis. The speed of the electron is $0.0460c$ (c = speed of light). Find the current i in the solenoid.

SECTION 30-5 A Current-Carrying Coil as a Magnetic Dipole

62E. What is the magnetic dipole moment $\boldsymbol{\mu}$ of the solenoid described in Exercise 53?

63E. Figure 30-66a shows a length of wire carrying a current i and bent into a circular coil of one turn. In Fig. 30-66b the same length of wire has been bent more sharply, to give a coil of two turns, each of half the original radius. (a) If B_a and B_b are the magnitudes of the magnetic fields at the centers of the two coils, what is the ratio B_b/B_a? (b) What is the ratio of the dipole moments, μ_b/μ_a, of the coils?

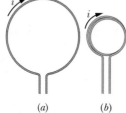

FIGURE 30-66
Exercise 63.

(a)   *(b)*

64E. Figure 30-67 shows an arrangement known as a Helmholtz coil. It consists of two circular coaxial coils, each of N turns and radius R, separated by a distance R. The two coils carry equal currents i in the same direction. Find the magnitude of the net magnetic field at P, midway between the coils.

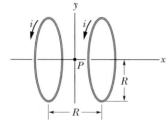

FIGURE 30-67   Exercise 64; Problems 68 and 72.

65E. A student makes a short electromagnet by winding 300 turns of wire around a wooden cylinder of diameter $d = 5.0$ cm. The coil is connected to a battery producing a current of 4.0 A in the wire. (a) What is the magnetic moment of this device? (b) At

what axial distance $z \gg d$ will the magnetic field of this dipole have the magnitude 5.0 μT (approximately one-tenth that of Earth's magnetic field)?

66E. The magnitude $B(x)$ of the magnetic field at points on the axis of a square current loop of side a is given in Problem 21. (a) Show that the axial magnetic field of this loop, for $x \gg a$, is that of a magnetic dipole (see Eq. 30-29). (b) What is the magnetic dipole moment of this loop?

67P. A length of wire is formed into a closed circuit with radii a and b, as shown in Fig. 30-68, and carries a current i. (a) What are the magnitude and direction of **B** at point P? (b) Find the magnetic dipole moment of the circuit.

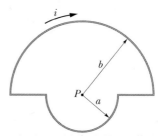

FIGURE 30-68
Problem 67.

68P. Two 300 turn coils of radius R each carry a current i. They are arranged a distance R apart, as in Fig. 30-67. For $R = 5.0$ cm and $i = 50$ A, plot the magnitude B of the net magnetic field as a function of distance x along the common axis over the range $x = -5$ cm to $x = +5$ cm, taking $x = 0$ at the midpoint P. (Such coils provide an especially uniform field **B** near point P.) (*Hint:* See Eq. 30-28.)

69P. A circular loop of radius 12 cm carries a current of 15 A. A coil of radius 0.82 cm, having 50 turns and a current of 1.3 A, is concentric with the loop. (a) What magnetic field **B** does the loop produce at its center? (b) What torque acts on the coil? Assume that the planes of the loop and coil are perpendicular and that the magnetic field due to the loop is essentially uniform throughout the volume occupied by the coil.

70P. (a) A long wire is bent into the shape shown in Fig. 30-69, without the wire actually touching itself at P. The radius of the circular section is R. Determine the magnitude and direction of **B** at the center C of the circular portion when the current i is as indicated. (b) Suppose the circular part of the wire is rotated without distortion about the indicated diameter, until the plane of the circle is perpendicular to the straight portion of the wire. The magnetic dipole moment associated with the circular loop is now in the direction of the current in the straight part of the wire. Determine **B** at C in this case.

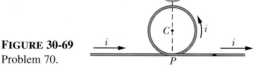

FIGURE 30-69
Problem 70.

71P. A conductor carries a current of 6.0 A along the closed path *abcdefgha* involving 8 of the 12 edges of a cube of side 10 cm as shown in Fig. 30-70. (a) Why can one regard this as the superposition of three square loops: *bcfgb*, *abgha*, and *cdefc*? (*Hint:*

Draw currents around those square loops.) (b) Use this superposition to find the magnetic dipole moment $\boldsymbol{\mu}$ (magnitude and direction) of the closed path. (c) Calculate **B** at the points $(x, y, z) = (0, 5.0 \text{ m}, 0)$ and $(5.0 \text{ m}, 0, 0)$.

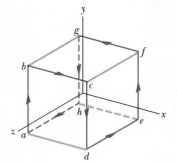

FIGURE 30-70
Problem 71.

72P. In Exercise 64 (Fig. 30-67), let the separation of the coils be a variable s (not necessarily equal to the coil radius R). (a) Show that the first derivative of the magnitude of the net magnetic field of the coils (dB/dx) vanishes at the midpoint P regardless of the value of s. Why would you expect this to be true from symmetry? (b) Show that the second derivative (d^2B/dx^2) also vanishes at P, provided $s = R$. This accounts for the uniformity of B near P for this particular coil separation.

Electronic Computation

73. The magnetic field of a circular loop with current i, at a point on the central axis through the loop, is parallel to that axis. The magnitude of the field is given by

$$B = \frac{\mu_0 i R^2}{2(R^2 + z^2)^{3/2}},$$

where R is the radius of the loop and z is the distance from the center of the loop. A solenoid can be constructed mathematically by using many such circular loops that are identical in radius and current, coaxial, and closely spaced. Suppose the solenoid has a length of 25.0 cm and a radius of 1.00 cm and consists of N equally spaced loops, each with a current of 1.00 A. For (a) $N = 11$, (b) $N = 21$, and (c) $N = 51$, compute the magnitude of the magnetic field at the center of the solenoid by summing the fields produced by the individual loops. For each value of N, compare the result with the value found using Eq. 30-25, which holds for a long solenoid with a large number of tightly spaced loops.

74. A computer can be used to demonstrate Ampere's law for a situation in which the Amperian loop does not coincide with a magnetic field line. Suppose that a square with edges of length a is centered at the origin of a coordinate system whose x and y axes are parallel to sides of the square. A long straight wire carrying current i is perpendicular to the plane of the square and crosses the x axis at $x = x'$. Evaluate $\oint \mathbf{B} \cdot d\mathbf{s}$ numerically. Divide a side of the square into N segments of equal length Δs and for each segment evaluate $\mathbf{B} \cdot \mathbf{u} \Delta s$. Here **B** is the magnetic field at the center of the segment and **u** is a unit vector that is parallel to the segment and is in the direction of integration. For different segments, **u** might be $\mathbf{i}$, $\mathbf{j}$, $-\mathbf{i}$, or $-\mathbf{j}$. The magnetic field at a point with coordinates x and y is given by

$$\mathbf{B} = \frac{\mu_0 i [-y\mathbf{i} + (x - x')\mathbf{j}]}{2\pi[(x - x')^2 + y^2]}.$$

For sides that are parallel to the x axis, take $y = a/2$ or $-a/2$; for sides that are parallel to the y axis, take $x = a/2$ or $-a/2$. Suppose that the length of a side is 1.00 m and the current is 1.00 A, and, for each of the following cases, calculate the sum over segments for each side of the square separately; then add the results to find the total for the square. Compare the total to $\mu_0 i_{enc}$. The value $N = 50$ should give you three-significant-figure accuracy: (a) $x' = 0$ (the wire is at the center of the square), (b) $x' = 0.200$ m (the wire passes inside the square at an off-center point), (c) $x' = 0.400$ m (the wire passes through the square near the center of a side), and (d) $x' = 0.600$ m (the wire is outside the square).

75. Two long parallel conductors carry currents parallel to the z axis. The conductors intersect the xy plane at points along the x axis: one intersects at $x = a$ and carries a current i_1 in the $+z$ direction; the other conductor intersects at $x = 0$ and carries a current i_2 that can be varied in both magnitude and direction. The current is considered to be positive if directed in the positive z direction and negative if directed in the negative z direction. (a) Write an equation that gives the net magnetic field **B** along the x axis for $x > a$. (b) Rewrite the equation for $x = 2a$ after substituting for i_2 with $i_2 = bi_1$, where b is a variable. (c) With this rewritten equation, graph **B** versus b for the range $3 > b > -3$. Positive **B** corresponds to the magnetic field being directed in the positive y direction, negative in the negative y direction.

31
Induction and Inductance

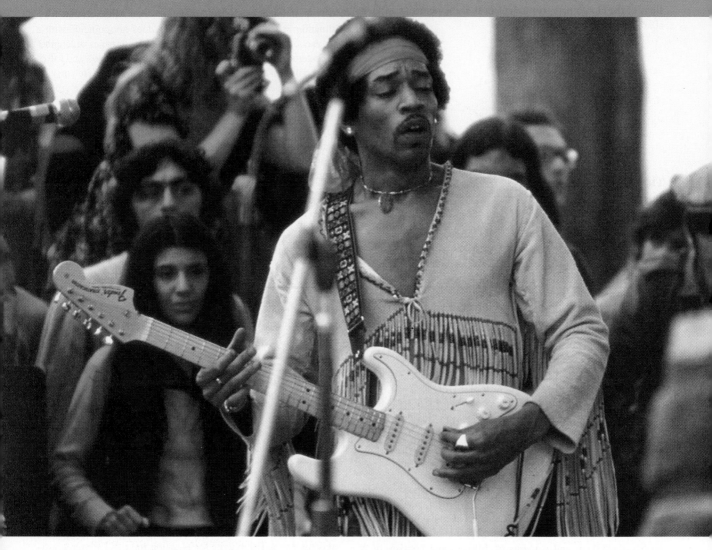

Soon after rock began in the mid-1950s, guitarists switched from acoustic guitars to electric guitars. But it was Jimi Hendrix who first understood the electric guitar as an electronic instrument. He exploded on the scene in the 1960s, ripping his pick along the strings, positioning himself and his guitar in front of a speaker to sustain feedback, and then laying down chords on top of the feedback. He shoved rock forward from the melodies of Buddy Holly into the psychedelia of the late 1960s and into the early heavy metal of Led Zeppelin in the 1970s, and his ideas continue to influence rock today. But what is it about an electric guitar that distinguishes it from an acoustic guitar and enabled Hendrix to make so much broader use of this electronic instrument?

31-1 TWO SYMMETRIC SITUATIONS

In Section 29-8, we saw that if we put a closed conducting loop in a magnetic field and then send current through the loop, forces due to the magnetic field create a torque to turn the loop:

$$\text{current loop} + \text{magnetic field} \Rightarrow \text{torque.} \quad (31\text{-}1)$$

Suppose that, instead, with the current off, we turn the loop by hand. Will the opposite of Eq. 31-1 occur? That is, will a current now appear in the loop:

$$\text{torque} + \text{magnetic field} \Rightarrow \text{current?} \quad (31\text{-}2)$$

The answer is yes—a current does appear. The situations of Eqs. 31-1 and 31-2 are symmetric. The physical law on which Eq. 31-2 depends is called *Faraday's law of induction*. Whereas Eq. 31-1 is the basis for the electric motor, Eq. 31-2 and Faraday's law are the basis for the electric generator. This chapter is concerned with that law and the process it describes.

31-2 TWO EXPERIMENTS

Let us examine two simple experiments to prepare for our discussion of Faraday's law of induction.

First Experiment. Figure 31-1 shows a conducting loop connected to a sensitive current meter. Since there is no battery or other source of emf included, there is no current in the circuit. However, if we move a bar magnet toward the loop, a current suddenly appears in the circuit. The current disappears when the magnet stops. If we then move the magnet away, a current again suddenly appears, but now in the opposite direction. If we experimented for a while, we would discover the following:

1. A current appears only if there is relative motion between the loop and the magnet (one must move relative to the other); the current disappears when the relative motion between them ceases.

2. Faster motion produces a greater current.

3. If moving the magnet's north pole toward the loop causes, say, clockwise current, then moving the north pole away causes counterclockwise current. And moving the south pole toward or away from the loop also causes currents, but in the reversed directions.

The current produced in the loop is called an **induced current,** the work done per unit charge in producing that current (in moving the conduction electrons constituting the current) is called an **induced emf,** and the process of producing the current and emf is called **induction.**

Second Experiment. For this experiment we use the apparatus of Fig. 31-2, with the two conducting loops close to each other but not touching. If we close switch S, to turn on a current in the right-hand loop, the meter suddenly and briefly registers a current—an induced current—in the left-hand loop. If we then open the switch, another sudden and brief induced current appears in the left-hand loop, but in the opposite direction. We get an induced current (and thus an induced emf) only when the current in the right-hand loop is changing (either turning on or turning off) and not when it is constant (even if it is large).

The induced emf and induced current in these experiments are apparently caused when something changes. But what is that "something"? Faraday knew.

31-3 FARADAY'S LAW OF INDUCTION

Faraday realized that an emf and a current can be induced in a loop, as in our two experiments, by changing the *amount of magnetic field* passing through the loop. He further realized that the "amount of magnetic field" can be visualized in terms of the magnetic field lines passing through the loop. **Faraday's law of induction,** stated in terms of our experiments, is:

> An emf is induced in the left-hand loop in Figs. 31-1 and 31-2 when the number of magnetic field lines that pass through the loop is changing.

FIGURE 31-2 The current meter registers a current in the left-hand wire loop just as switch S is closed (to turn on current in the right-hand wire loop) or opened (to turn off the current in the right-hand loop). No motion of the coils is involved.

FIGURE 31-1 A current meter registers a current in the wire loop when the magnet is moving with respect to the loop.

Most important, the actual number of field lines passing through the loop does not matter; the values of the induced emf and induced current are determined by the *rate* at which that number changes.

In our first experiment (Fig. 31-1), the magnetic field lines spread from the north pole of the magnet. So, as we bring the north pole toward the loop, the number of field lines passing through the loop increases. That increase apparently causes conduction electrons in the loop to move (the induced current) and provides energy (the induced emf) for their motion. When the magnet stops moving, the number of field lines through the loop no longer changes and the induced current and induced emf disappear.

In our second experiment (Fig. 31-2), when the switch is open (no current), there are no field lines. But when we turn on the current in the right-hand loop, the increasing current builds up a magnetic field around that loop and at the left-hand loop. While the field builds, the number of magnetic field lines through the left-hand loop increases. As in the first experiment, the increase in field lines through that loop apparently induces a current and an emf there. When the current in the right-hand loop reaches a final, steady value, the number of field lines through the left-hand loop no longer changes, and the induced current and induced emf disappear.

Faraday's law does not explain *why* a current and an emf are induced in either experiment; it is just a statement that helps us visualize the induction.

A Quantitative Treatment

To put Faraday's law to work, we need a way to calculate the *amount of magnetic field* that passes through a loop. In Chapter 24, in a similar situation, we needed to calculate the amount of an electric field that passes through a surface. There we defined an electric flux $\Phi_E = \int \mathbf{E} \cdot d\mathbf{A}$. So, here we define a magnetic flux. Suppose a loop enclosing an area A is placed in a magnetic field $\mathbf{B}$. Then the magnetic flux through the loop is

$$\Phi_B = \int \mathbf{B} \cdot d\mathbf{A} \qquad \begin{array}{l}\text{(magnetic flux}\\ \text{through area } A\text{).}\end{array} \qquad (31\text{-}3)$$

As in Chapter 24, $d\mathbf{A}$ is a vector of magnitude dA that is perpendicular to a differential area dA.

As a special case of Eq. 31-3, suppose that the loop lies in a plane and that the magnetic field is perpendicular to the plane of the loop. Then we can write the dot product in Eq. 31-3 as $B\,dA \cos 0° = B\,dA$. If the magnetic field is also uniform, then B can be brought out in front of the integral sign. The remaining $\int dA$ then gives just the area A of the loop. So, Eq. 31-3 reduces to

$$\Phi_B = BA \qquad (\mathbf{B} \perp \mathbf{A},\ \mathbf{B} \text{ uniform}). \qquad (31\text{-}4)$$

From Eqs. 31-3 and 31-4, we see that the SI unit for magnetic flux is the tesla–square meter, which is called the *weber* (abbreviated Wb):

$$1 \text{ weber} = 1 \text{ Wb} = 1 \text{ T} \cdot \text{m}^2. \qquad (31\text{-}5)$$

With the notion of magnetic flux, we can state Faraday's law in a more quantitative and useful way:

> The magnitude of the emf $\mathscr{E}$ induced in a conducting loop is equal to the rate at which the magnetic flux Φ_B through that loop changes with time.

As you will see in the next section, the induced emf $\mathscr{E}$ tends to oppose the flux change, so Faraday's law is formally written as

$$\mathscr{E} = -\frac{d\Phi_B}{dt} \qquad \text{(Faraday's law),} \qquad (31\text{-}6)$$

with the minus sign indicating that opposition. We often neglect the minus sign in Eq. 31-6, seeking only the magnitude of the induced emf.

If we change the magnetic flux through a coil of N turns, an induced emf appears in every turn and the total emf induced in the coil is the sum of these individual induced emfs. If the coil is tightly wound (*closely packed*), so that the same magnetic flux Φ_B passes through all the turns, the total emf induced in the coil is

$$\mathscr{E} = -N\frac{d\Phi_B}{dt} \qquad \text{(coil of } N \text{ turns).} \qquad (31\text{-}7)$$

Here are the general means by which we can change the magnetic flux through a coil:

1. Change the magnitude B of the magnetic field within the coil.

2. Change the area of the coil, or the portion of that area that happens to lie within the magnetic field (for example, by expanding the coil or sliding it out of the field).

3. Change the angle between the direction of the magnetic field $\mathbf{B}$ and the area of the coil (for example, by rotating the coil so that $\mathbf{B}$ is first perpendicular to the plane of the coil and then is along that plane).

SAMPLE PROBLEM 31-1

The long solenoid S of Fig. 31-3 has 220 turns/cm and carries a current $i = 1.5$ A; its diameter D is 3.2 cm. At its center we place a 130-turn close-packed coil C of diameter $d = 2.1$ cm. The current in the solenoid is reduced to zero at a steady rate in 25 ms. What emf is induced in coil C while the current in the solenoid is changing?

SOLUTION: Coil C is located in the magnetic field B produced by the current in the solenoid. As the current decreases, so does B. Thus, the magnetic flux through the coil decreases. During this decrease an emf is induced in the coil via Faraday's law. To find the induced emf, we first find the initial magnetic field B_i of the solenoid by substituting given data into Eq. 30-25:

$$B_i = \mu_0 i n$$
$$= (4\pi \times 10^{-7} \text{ T·m/A})$$
$$\times (1.5 \text{ A})(220 \text{ turns/cm})(100 \text{ cm/m})$$
$$= 4.15 \times 10^{-2} \text{ T}.$$

The area A of each turn of coil C is $\frac{1}{4}\pi d^2$, which is equal to 3.46×10^{-4} m². The solenoid's magnetic field is perpendicular to this area, and we assume it to be uniform across the area. So we can find the initial magnetic flux $\Phi_{B,i}$ through each turn of coil C by substituting known data into Eq. 31-4:

$$\Phi_{B,i} = BA = (4.15 \times 10^{-2} \text{ T})(3.46 \times 10^{-4} \text{ m}^2)$$
$$= 1.44 \times 10^{-5} \text{ Wb} = 14.4 \text{ } \mu\text{Wb}.$$

The final magnetic field B_f and magnetic flux $\Phi_{B,f}$ are both zero. Thus, the change in flux through each turn of coil C is $\Delta\Phi_B = 14.4$ μWb.

Because the solenoid's current decreases at a steady rate, so does the magnetic flux, and we can rewrite Faraday's law (Eq. 31-7) as

$$\mathcal{E} = N\frac{\Delta\Phi_B}{\Delta t},$$

where N is the number of turns (130) of the coil. (We ignore the minus sign in Eq. 31-7 because we are looking for the magnitude of $\mathcal{E}$ only.) Substituting known data then gives us

$$\mathcal{E} = (130 \text{ turns})\frac{14.4 \times 10^{-6} \text{ Wb}}{25 \times 10^{-3} \text{ s}}$$
$$= 7.5 \times 10^{-2} \text{ V} = 75 \text{ mV.} \qquad \text{(Answer)}$$

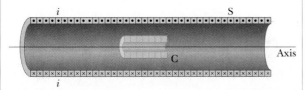

FIGURE 31-3 Sample Problem 31-1. A coil C is located inside a solenoid S, which carries current i.

CHECKPOINT 1: The graph gives the magnitude $B(t)$ of a uniform magnetic field that exists throughout a conducting loop, perpendicular to the plane of the loop.

Rank the five regions of the graph according to the magnitude of the emf induced in the loop, greatest first.

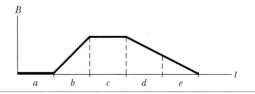

31-4 LENZ'S LAW

Soon after Faraday proposed his law of induction, Heinrich Friedrich Lenz devised a rule—now known as **Lenz's law**—for determining the direction of an induced current in a loop:

An induced current has a direction such that the magnetic field due to *the current* opposes the change in the magnetic field that induces the current.

Furthermore, the direction of an induced emf is that of the induced current.

To get a feel for Lenz's law, let us apply it in two different but equivalent ways to Fig. 31-4, where the north pole of a magnet is being moved toward a conducting loop.

1. *Opposition to Pole Movement.* The approach of the magnet's north pole in Fig. 31-4 increases the magnetic field in the loop and thereby induces a current in the loop. From Fig. 30-22, we know that the loop then acts as a magnetic dipole with a south pole and a north pole, and that its magnetic dipole moment μ points from south to north. To *oppose* the magnetic field increase being caused by the approaching magnet, the loop's north pole (and thus μ) must face *toward* the approaching north pole so as to repel it (Fig. 31-4). Then, the curled–straight right-hand rule for

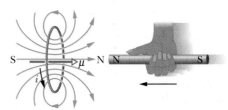

FIGURE 31-4 Lenz's law at work. As the magnet is moved toward the loop, a counterclockwise current is induced in the loop; the current produces its own magnetic field, with magnetic dipole moment μ, so as to oppose the motion of the magnet.

μ (Fig. 30-22) tells us that the current induced in the loop must be counterclockwise in Fig. 31-4.

If we next pull the magnet away from the loop, a current will again be induced in the loop. Now, however, the loop will have a south pole facing the retreating north pole of the magnet, so as to oppose the retreat. So the induced current will be clockwise.

2. *Opposition to Flux Change.* In Fig. 31-4, with the magnet initially distant, no magnetic flux passes through the loop. As the north pole of the magnet then nears the loop with its magnetic field **B** directed *toward the left,* the flux through the loop increases. To oppose this increase in flux, the induced current i must set up its own field $\mathbf{B}_i$ *directed toward the right* inside the loop, as shown in Fig. 31-5a; then the rightward flux of field $\mathbf{B}_i$ opposes the increasing leftward flux of field **B**. The curled–straight right-hand rule of Fig. 30-22 then tells us that i must be counterclockwise in Fig. 31-5a.

Note carefully that the flux of $\mathbf{B}_i$ always opposes the *change* in the flux of **B**, but that does not always mean that $\mathbf{B}_i$ points opposite **B**. For example, if we next pull the magnet away from the loop, the flux Φ_B from the magnet is still directed to the left through the loop, but it is now

decreasing. So the flux of $\mathbf{B}_i$ must now be to the left inside the loop, to oppose the *decrease* in Φ_B, as shown in Fig. 31-5b. Thus, $\mathbf{B}_i$ and **B** are now in the same direction.

Figures 31-5c and d show the situations in which the south pole of the magnet approaches and retreats from the loop, respectively.

Electric Guitars

Figure 31-6 shows a Fender Stratocaster, the type of electric guitar that was used by Jimi Hendrix and by many other musicians. Whereas an acoustic guitar depends for its sound on the acoustic resonance produced in the hollow body of the instrument by the oscillations of the strings, an electric guitar is a solid instrument, so there is no body resonance. Instead, the oscillations of the metal strings are sensed by electric "pickups" that send signals to an amplifier and a set of speakers.

The basic construction of a pickup is shown in Fig. 31-7. Wire connecting the instrument to the amplifier is coiled around a small magnet. The magnetic field of the magnet produces a north and south pole in the section of the metal string just above the magnet. That section then has its own magnetic field. When the string is plucked and thus made to oscillate, its motion relative to the coil changes the flux of its magnetic field through the coil, inducing a current in the coil. As the string oscillates toward and away from the coil, the induced current changes direction at the same frequency as the string's oscillations, thus relaying the frequency of oscillation to the amplifier and speaker.

On a Stratocaster, there are three groups of pickups, placed at the near end of the strings (on the wide part of the body). The group closest to the near end better detects the high-frequency oscillations of the strings; the group farthest from the near end better detects the low-frequency oscillations. By throwing a toggle switch on the guitar, the musician can select which group or which pair of groups will send signals to the amplifier and speakers.

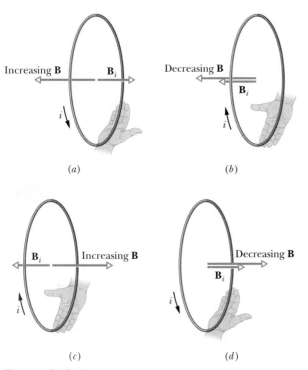

(a) (b)

(c) (d)

FIGURE 31-5 The current i induced in a loop has the direction such that the current's magnetic field $\mathbf{B}_i$ opposes the *change* in the magnetic field **B** inducing i. The field $\mathbf{B}_i$ is always directed opposite an increasing field **B** (a, c) and in the same direction as a decreasing **B** (b, d).

FIGURE 31-6 A Fender Stratocaster has three groups of six electric pickups each (within the wide part of the body). A toggle switch (at the bottom of the guitar) allows the musician to determine which group of pickups sends signals to an amplifier and thus to a speaker system.

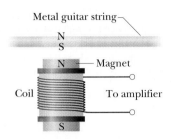

FIGURE 31-7 A side view of an electric guitar pickup. When the metal string (which acts like a magnet) is made to oscillate, the variation in magnetic flux induces a current in the coil.

To gain further control over his music, Hendrix sometimes rewrapped the wire in the pickup coils of his guitar to change the number of turns. In this way, he altered the amount of emf induced in the coils and thus their relative sensitivity. Even without this additional measure, you can see that the electric guitar offers far more control over the musical sound that is produced than can be obtained with an acoustic guitar.

CHECKPOINT 2: The figure shows three situations in which identical circular conducting loops are in uniform magnetic fields that are either increasing (Inc) or decreasing (Dec) in magnitude at identical rates. In each, the dashed line coincides with a diameter. Rank the situations according to the magnitude of the current induced in the loops, greatest first.

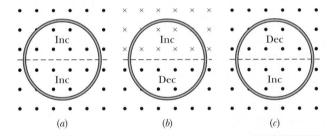

(a)   (b)   (c)

SAMPLE PROBLEM 31-2

Figure 31-8 shows a conducting loop consisting of a half-circle of radius $r = 0.20$ m and three straight sections. The half-circle lies in a uniform magnetic field **B** that is directed out of the page; the field magnitude is given by $B = 4.0t^2 + 2.0t + 3.0$, with B in teslas and t in seconds. An ideal battery with emf $\mathscr{E}_{bat} = 2.0$ V is connected to the loop. The resistance of the loop is $2.0\ \Omega$.

(a) What are the magnitude and direction of the emf $\mathscr{E}_{ind}$ induced along the loop by **B** at $t = 10$ s?

SOLUTION: Equation 31-6 tells us that the magnitude of $\mathscr{E}_{ind}$ is equal to the rate $d\Phi_B/dt$ at which the magnetic flux through the loop changes. Because the field is uniform and perpendicular to the plane of the loop, the flux is given by Eq. 31-4: $\Phi_B = BA$. Using this equation and realizing that only the field magnitude B changes in time (not the area A), we

rewrite Eq. 31-6 as

$$\mathscr{E}_{ind} = \frac{d\Phi_B}{dt} = \frac{d(BA)}{dt} = A\frac{dB}{dt}.$$

Because the flux penetrates the loop only within the half-circle, the area A in this equation is $\frac{1}{2}\pi r^2$. Substituting this and the given expression for B yields

$$\mathscr{E}_{ind} = A\frac{dB}{dt} = \frac{\pi r^2}{2}\frac{d}{dt}(4.0t^2 + 2.0t + 3.0)$$
$$= \frac{\pi r^2}{2}(8.0t + 2.0).$$

At $t = 10$ s, then,

$$\mathscr{E}_{ind} = \frac{\pi(0.20\text{ m})^2}{2}[8.0(10) + 2.0]$$
$$= 5.152\text{ V} \approx 5.2\text{ V.} \qquad\text{(Answer)}$$

In Fig. 31-8, the flux through the loop is out of the page and increasing. Then, according to Lenz's law, the flux of the induced field B_i (due to the induced current) must be *into* the page. Using the curled–straight right-hand rule (Fig. 30-7c), we find that the induced current, must be clockwise around the loop. The induced emf $\mathscr{E}_{ind}$ must then also be clockwise.

(b) What is the current in the loop at $t = 10$ s?

SOLUTION: The induced emf $\mathscr{E}_{ind}$ tends to drive a current clockwise around the loop; the battery's emf $\mathscr{E}_{bat}$ tends to drive a current counterclockwise. Because $\mathscr{E}_{ind}$ is greater than $\mathscr{E}_{bat}$, the net emf $\mathscr{E}_{net}$ is clockwise, and thus so is the current. To find the current at $t = 10$ s, we use Eq. 28-2 ($i = \mathscr{E}/R$):

$$i = \frac{\mathscr{E}_{net}}{R} = \frac{\mathscr{E}_{ind} - \mathscr{E}_{bat}}{R} = \frac{5.152\text{ V} - 2.0\text{ V}}{2.0\ \Omega}$$
$$= 1.58\text{ A} \approx 1.6\text{ A.} \qquad\text{(Answer)}$$

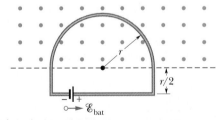

FIGURE 31-8 Sample Problem 31-2. A battery is connected to a conducting loop consisting of a half-circle of radius r that lies in a uniform magnetic field. The field is directed out of the page; its magnitude is changing.

SAMPLE PROBLEM 31-3

Figure 31-9 shows a rectangular loop of wire immersed in a nonuniform and varying magnetic field **B** that is perpendicular to and directed into the page. The field's magnitude is given by $B = 4t^2x^2$, with B in teslas, t in seconds, and x in meters. The loop has width $W = 3.0$ m and height $H = 2.0$ m. What are the magnitude and direction of the induced emf $\mathscr{E}$ along the loop at $t = 0.10$ s?

SOLUTION: The magnitude of the induced emf $\mathscr{E}$ is given by Faraday's law: $\mathscr{E} = d\Phi_B/dt$. But to use it, we need an expression for the flux Φ_B through the loop at any time t. Because B is *not* uniform over the area enclosed by the loop, we *cannot* use Eq. 31–4 ($\Phi_B = BA$) to find that expression; instead we must use Eq. 31-3 ($\Phi_B = \int \mathbf{B} \cdot d\mathbf{A}$).

In Fig. 31-9, $\mathbf{B}$ is perpendicular to the plane of the loop (hence parallel to the differential area vector $d\mathbf{A}$), so the dot product in Eq. 31-3 gives $B\,dA$. Because the magnetic field varies with the coordinate x but not with the coordinate y, we can take the differential area dA to be the area of a vertical strip of height H and width dx (as shown in Fig. 31-9). Then $dA = H\,dx$, and the flux through the loop is

$$\Phi_B = \int \mathbf{B} \cdot d\mathbf{A} = \int B\,dA = \int BH\,dx = \int 4t^2x^2H\,dx.$$

Treating t as a constant for this integration and inserting the integration limits $x = 0$ and $x = 3.0$ m, we obtain

$$\Phi_B = 4t^2H \int_0^{3.0} x^2\,dx = 4t^2H\left[\frac{x^3}{3}\right]_0^{3.0} = 72t^2,$$

where we have substituted $H = 2.0$ m and Φ_B is in webers. Now we can use Faraday's law to find the magnitude of $\mathscr{E}$ at any time t:

$$\mathscr{E} = \frac{d\Phi_B}{dt} = \frac{d(72t^2)}{dt} = 144t,$$

in which $\mathscr{E}$ is in volts. At $t = 0.10$ s,

$$\mathscr{E} = (144 \text{ V/s})(0.10 \text{ s}) \approx 14 \text{ V}. \qquad \text{(Answer)}$$

The flux of $\mathbf{B}$ through the loop is into the page in Fig. 31-9 and is increasing in magnitude because B is increasing in magnitude with time. According to Lenz's law, the field B_i of the induced current that opposes this increase is directed out of the page. The curled–straight right-hand rule of Fig. 31-5 tells us that the induced current is counterclockwise around the loop, and thus so is the induced emf $\mathscr{E}$.

31-5 INDUCTION AND ENERGY TRANSFERS

By Lenz's law, whether you move the magnet toward or away from the loop in Fig. 31-1, a force resists the motion, requiring your applied force to do positive work. At the same time, thermal energy is produced in the material of the loop because of the material's electrical resistance to the current that is induced. The energy you transfer to the closed *loop + magnet* system via your applied force ends up in this thermal energy. (For now, we neglect energy that is radiated away from the loop as electromagnetic waves during the induction.) The faster you move the magnet, the more rapidly your applied force does work, and the greater the rate of production of thermal energy in the loop.

Regardless of how current is induced in a loop, energy is always transferred to thermal energy during the process because of the electrical resistance of the loop (unless the loop is superconducting). For example, in Fig. 31-2, when switch S is closed and a current is briefly induced in the left-hand loop, energy is transferred from the battery to thermal energy in that loop.

Figure 31-10 shows another situation involving induced current. A rectangular loop of wire of width L has one end in a uniform external magnetic field that is directed perpendicularly into the plane of the loop. This field may be produced, for example, by a large electromagnet.

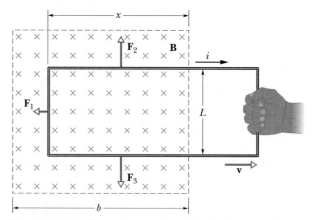

FIGURE 31-10 You pull a closed conducting loop out of a magnetic field at constant velocity **v**. While the loop is moving, a clockwise current i is induced in the loop, and the loop segments still within the magnetic field experience forces $\mathbf{F}_1$, $\mathbf{F}_2$, and $\mathbf{F}_3$.

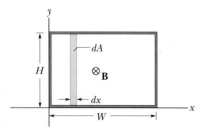

FIGURE 31-9 Sample Problem 31-3. A closed conducting loop, of width W and height H, lies in a nonuniform, varying magnetic field that points directly into the page. To apply Faraday's law, we use the vertical strip of height H, width dx, and area dA.

The dashed lines in Fig. 31-10 show the assumed limits of the magnetic field; the fringing of the field at its edges is neglected. You are asked to pull this loop to the right at a constant velocity **v.**

The situation of Fig. 31-10 does not differ in any essential way from that of Fig. 31-1. In each case a magnetic field and a conducting loop are in relative motion; in each case the flux of the field through the loop is changing with time. It is true that in Fig. 31-1 the flux is changing because **B** is changing and in Fig. 31-10 the flux is changing because the area of the loop still in the magnetic field is changing, but that difference is not important. The important difference between the two arrangements is that the arrangement of Fig. 31-10 makes calculations easier. Let us now calculate the rate at which you do mechanical work as you pull steadily on the loop in Fig. 31-10.

As you will see, if you are to pull the loop at a constant velocity **v**, you must apply a constant force **F** to the loop because a force of equal magnitude but opposite direction acts on the loop to oppose you. From Eq. 7-49, the rate at which you do work is then

$$P = Fv, \qquad (31\text{-}8)$$

where F is the magnitude of your force. We wish to find an expression for P in terms of the magnitude B of the magnetic field and the characteristics of the loop, namely, its resistance R to current and its dimension L.

As you move the loop to the right in Fig. 31-10, the portion of its area within the magnetic field decreases. Thus the flux through the loop also decreases and, according to Lenz's law, a current is produced in the loop. It is the presence of this current that causes the force that opposes your pull.

To find the current, we first apply Faraday's law. When x is the length of the loop still in the magnetic field, the area of the loop still in the field is Lx. Then from Eq. 31-4, the magnitude of the flux through the loop is

$$\Phi = BA = BLx. \qquad (31\text{-}9)$$

As x decreases, the flux decreases. Faraday's law tells us that with this flux decrease, an emf is induced in the loop. Dropping the minus sign in Eq. 31-6 and using Eq. 31-9, we can write the magnitude of this emf as

$$\mathcal{E} = \frac{d\Phi}{dt} = \frac{d}{dt} BLx = BL \frac{dx}{dt} = BLv, \quad (31\text{-}10)$$

in which we have replaced dx/dt with v, the speed at which the loop moves.

Figure 31-11 shows the circuit through which the charge flows: emf $\mathcal{E}$ is represented on the left, and the collective resistance R of the loop is represented on the

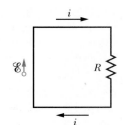

FIGURE 31-11 A circuit diagram for the loop of Fig. 31-10 while the loop is moving.

right. The direction of $\mathcal{E}$ is obtained as in Fig. 31-5b; the induced current i must have the same direction.

To find the magnitude of the induced current, we cannot apply the loop rule for potential differences in a circuit because, as you will see in Section 31-6, we cannot define a potential difference for an induced emf. However, we can apply the equation $i = \mathcal{E}/R$, as we did in Sample Problem 31-2. With Eq. 31-10, this becomes

$$i = \frac{BLv}{R}. \qquad (31\text{-}11)$$

Along the three segments of the loop in Fig. 31-10 where this current is in the magnetic field, sideways deflecting forces act on the loop. From Eq. 29-26 we know that such a deflecting force is, in general notation,

$$\mathbf{F}_d = i\mathbf{L} \times \mathbf{B}. \qquad (31\text{-}12)$$

In Fig. 31-10, the deflecting forces acting on the three segments of the loop are marked $\mathbf{F}_1$, $\mathbf{F}_2$, and $\mathbf{F}_3$. Note, however, that from the symmetry, $\mathbf{F}_2$ and $\mathbf{F}_3$ are equal in magnitude and cancel. This leaves only $\mathbf{F}_1$, which is directed opposite your force **F** on the loop and thus is the force that opposes you. So $\mathbf{F} = -\mathbf{F}_1$.

Using Eq. 31-12 to obtain the magnitude of $\mathbf{F}_1$ and noting that the angle between **B** and the length vector **L** for the left segment is 90°, we write

$$F = F_1 = iLB \sin 90° = iLB. \qquad (31\text{-}13)$$

Substituting Eq. 31-11 for i in Eq. 31-13 then gives us

$$F = \frac{B^2L^2v}{R}. \qquad (31\text{-}14)$$

Since B, L, and R are constants, the speed v at which you move the loop is constant if the magnitude F of the force you apply to the loop is also constant.

Substituting Eq. 31-14 into Eq. 31-8, we find the rate at which you do work on the loop as you pull it from the magnetic field:

$$P = Fv = \frac{B^2L^2v^2}{R} \qquad \text{(rate of doing work).} \quad (31\text{-}15)$$

To cook food on an *induction stove*, oscillating current is sent through a conducting coil that lies just below the cooking surface. The magnetic field produced by that current oscillates and induces an oscillating current in the conducting cooking pan. Because the pan has some resistance to that current, the electrical energy of the current is continuously transformed to thermal energy, resulting in a temperature increase of the pan and the food in it. The cooking surface itself might never get hot.

To complete our analysis, let us find the rate at which thermal energy appears in the loop as you pull it along at constant speed. We calculate it from Eq. 27-22,

$$P = i^2 R. \tag{31-16}$$

Substituting for i from Eq. 31-11, we find

$$P = \left(\frac{BLv}{R}\right)^2 R = \frac{B^2 L^2 v^2}{R} \qquad \begin{matrix}\text{(thermal}\\\text{energy rate),}\end{matrix} \tag{31-17}$$

which is exactly equal to the rate at which you are doing work on the loop (Eq. 31-15). Thus the work that you do in pulling the loop through the magnetic field appears as thermal energy in the loop, manifesting itself as a small increase in the temperature of the loop.

Eddy Currents

Suppose we replace the conducting loop of Fig. 31-10 with a solid conducting plate. If we then move the plate out of the magnetic field as we did the loop (Fig. 31-12a), the relative motion of the field and the conductor again induces a current in the conductor. Thus we again encounter an opposing force and must do work because of the induced current. With the plate, however, the conduction electrons making up the induced current do not follow one path as they do with the loop. Instead, the electrons swirl about within the plate as if they were caught in an eddy (or

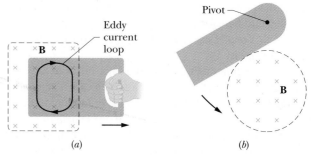

FIGURE 31-12 (*a*) As you pull a solid conducting plate out of a magnetic field, *eddy currents* are induced in the plate. A typical loop of eddy current is shown; it has the same clockwise sense of circulation as the current in the conducting loop of Fig. 31-10. (*b*) A conducting plate is allowed to swing like a pendulum about a pivot and into a region of magnetic field. As it enters and leaves the field, eddy currents are induced in the plate.

whirlpool) of water. Such a current is called an *eddy current* and can be represented as in Fig. 31-12a *as if* it followed a single path.

As with the conducting loop of Fig. 31-10, the current induced in the plate results in mechanical energy being dissipated as thermal energy. The dissipation is more apparent in the arrangement of Fig. 31-12b: a conducting plate, free to rotate about a pivot, swings down through a magnetic field like a pendulum. Each time the plate enters and leaves the field, a portion of its mechanical energy is transferred to its thermal energy. After several swings, no mechanical energy remains and the plate just hangs from its pivot.

SAMPLE PROBLEM 31-4

Figure 31-13a shows a rectangular conducting loop of resistance R, width L, and length b being pulled at constant speed v through a region of width d in which a uniform magnetic field **B** is produced by an electromagnet. Let $L = 40$ mm, $b = 10$ cm, $d = 15$ cm, $R = 1.6 \ \Omega$, $B = 2.0$ T, and $v = 1.0$ m/s.

(a) Plot the flux Φ_B through the loop as a function of the position x of the right side of the loop.

SOLUTION: The flux is zero when the loop is not in the field; it is BLb (= 8 mWb) when the loop is entirely in the field; it is BLx when the loop is entering the field; and then it is $BL[b - (x - d)]$ when the loop is leaving the field. These results lead to the plot of Fig. 31-13b, which you should verify.

(b) Plot the induced emf as a function of the position of the loop. Indicate the directions of the induced emf.

SOLUTION: From Eq. 31-6, the induced emf is $-d\Phi_B/dt$, which we can write as

$$\mathcal{E} = -\frac{d\Phi_B}{dt} = -\frac{d\Phi_B}{dx}\frac{dx}{dt} = -\frac{d\Phi_B}{dx} v,$$

where $d\Phi_B/dx$ is the slope of the curve of Fig. 31-13b. The emf is plotted as a function of x in Fig. 31-13c.

Lenz's law shows that when the loop is entering the field, the current and emf are counterclockwise in Fig. 31-13a; when the loop is leaving the field, the emf is clockwise in that figure. In Fig. 31-13c, a counterclockwise emf is plotted as a negative value, a clockwise emf as a positive value. There is *no* emf when the loop is either entirely out of the field or entirely in it because, in these two situations, the flux through the loop is not changing.

(c) Plot the rate of production of thermal energy in the loop as a function of the position of the loop.

SOLUTION: Substituting $i = \mathcal{E}/R$ in Eq. 31-16 gives us the rate of thermal energy production as

$$P = i^2R = \frac{\mathcal{E}^2}{R}.$$

We can calculate P by squaring the ordinate of the curve of Fig. 31-13c and dividing by R, being careful of powers of 10

in the units. The result is plotted in Fig. 31-13d. Note that thermal energy is produced only when the loop is entering or leaving the magnetic field.

In practice, the external magnetic field **B** cannot drop sharply to zero at its boundary but must approach zero smoothly. The result would be a rounding of the corners of the curves plotted in Fig. 31-13.

CHECKPOINT 3: The figure shows four wire loops, with edge lengths of either L or $2L$. All four loops will move through a region of uniform magnetic field **B** (directed out of the page) at the same constant velocity. Rank the four loops according to the maximum magnitude of the emf induced as they move through the field, greatest first.

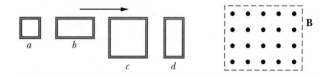

31-6 INDUCED ELECTRIC FIELDS

Let us place a copper ring of radius r in a uniform external magnetic field, as in Fig. 31-14a. The field—neglecting fringing—fills a cylindrical volume of radius R. Suppose that we increase the strength of this field at a steady rate, perhaps by increasing—in an appropriate way—the current in the windings of the electromagnet that produces the field. The magnetic flux through the ring will then change at a steady rate and—by Faraday's law—an induced emf and thus an induced current will appear in the ring. From Lenz's law we can deduce that the direction of the induced current is counterclockwise in Fig. 31-14a.

If there is a current in the copper ring, an electric field must be present along the ring; an electric field is needed to do the work of moving the conduction electrons. Moreover, the field must have been produced by the changing magnetic flux. This **induced electric field E** is just as real as an electric field produced by static charges; either field will exert a force $q_0\mathbf{E}$ on a particle of charge q_0. By this line of reasoning, we are led to a useful and informative restatement of Faraday's law of induction:

A changing magnetic field produces an electric field.

The striking feature of this statement is that the electric field is induced even if there is no copper ring.

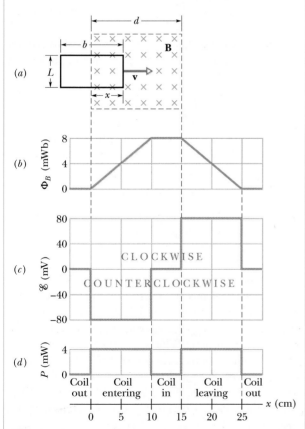

FIGURE 31-13 Sample Problem 31-4. (a) A closed conducting loop is pulled at constant velocity **v** completely through a magnetic field. (b) The flux through the loop as a function of the position x of the right side of the loop. (c) The induced emf as a function of x. (d) The rate at which thermal energy appears in the loop as a function of x.

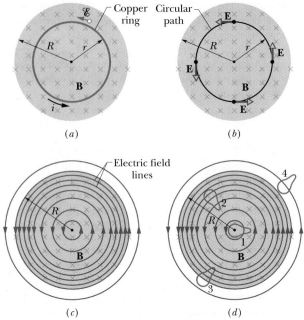

FIGURE 31-14 (*a*) If the magnetic field increases at a steady rate, a constant induced current appears, as shown, in the copper ring of radius r. (*b*) Induced electric fields appear at various points even when the ring is removed. (*c*) The complete picture of the induced electric fields, displayed as field lines. (*d*) Four similar closed paths that enclose identical areas. Equal emfs are induced around paths 1 and 2, which lie entirely within the region of changing magnetic field. A smaller emf is induced around path 3, which only partially lies in that region. No emf is induced around path 4, which lies entirely outside the magnetic field.

To fix these ideas, consider Fig. 31-14*b*, which is just like Fig. 31-14*a* except the copper ring has been replaced by a hypothetical circular path of radius r. We assume, as previously, that the magnetic field **B** is increasing in magnitude at a constant rate dB/dt. The electric field induced at various points around the circular path must—from the symmetry—be tangent to the circle, as Fig. 31-14*b* shows.* Hence the circular path is also an electric field line. There is nothing special about the circle of radius r, so the electric field lines produced by the changing magnetic field must be a set of concentric circles, as in Fig. 31-14*c*.

As long as the magnetic field is increasing with time, the electric field represented by the circular field lines in Fig. 31-14*c* will be present. If the magnetic field remains constant with time, there will be no induced electric field and thus no electric field lines. If the magnetic field is

decreasing with time (at a constant rate), the electric field lines will still be concentric circles as in Fig. 31-14*c*, but they will now have the opposite direction. All this is what we have in mind when we say: "A changing magnetic field produces an electric field."

A Reformulation of Faraday's Law

Consider a particle of charge q_0 moving around the circular path of Fig. 31-14*b*. The work W done on it in one revolution by the induced electric field is $\mathscr{E}q_0$, where $\mathscr{E}$ is the induced emf, that is, the work done per unit charge in moving the test charge around the path. From another point of view, the work is

$$\int \mathbf{F} \cdot d\mathbf{s} = (q_0 E)(2\pi r), \qquad (31\text{-}18)$$

where $q_0 E$ is the magnitude of the force acting on the test charge and $2\pi r$ is the distance over which that force acts. Setting these two expressions for W equal to each other and canceling q_0, we find that

$$\mathscr{E} = 2\pi r E. \qquad (31\text{-}19)$$

More generally, we can rewrite Eq. 31-18 to give the work done on a particle of charge q_0 moving along any closed path:

$$W = \oint \mathbf{F} \cdot d\mathbf{s} = q_0 \oint E \cdot d\mathbf{s}. \qquad (31\text{-}20)$$

(The circle indicates that the integral is to be taken around the closed path.) Substituting $\mathscr{E}q_0$ for W, we find that

$$\mathscr{E} = \oint \mathbf{E} \cdot d\mathbf{s}. \qquad (31\text{-}21)$$

This integral reduces at once to Eq. 31-19 if we evaluate it for the special case of Fig. 31-14*b*.

With Eq. 31-21, we can expand the meaning of induced emf. Previously, induced emf has meant the work per unit charge done in maintaining current due to a changing magnetic flux. Or it has meant the work done per unit charge on a charged particle that moves around a closed path in a changing magnetic flux. But with Eq. 31-21, we no longer actually need a current or a particle to speak of induced emf. An induced emf is the sum—via integration—of quantities $\mathbf{E} \cdot d\mathbf{s}$ around a closed path, where **E** is the electric field induced by a changing magnetic flux and $d\mathbf{s}$ is a differential length vector along the closed path.

If we combine Eq. 31-21 with Faraday's law in Eq. 31-6 ($\mathscr{E} = -d\Phi_B/dt$), we can rewrite Faraday's law as

$$\oint \mathbf{E} \cdot d\mathbf{s} = -\frac{d\Phi_B}{dt} \qquad \text{(Faraday's law).} \qquad (31\text{-}22)$$

---

*Arguments of symmetry would also permit the lines of **E** around the circular path to be *radial*, rather than tangential. However, such radial lines would imply that there are free charges, distributed symmetrically about the axis of symmetry, on which the electric field lines could begin or end; there are no such charges.

This equation says simply that a changing magnetic field induces an electric field. The changing magnetic field appears on the right side of this equation, the electric field on the left.

Faraday's law in the form of Eq. 31-22 can be applied to *any* closed path that can be drawn in a changing magnetic field. Figure 31-14d, for example, shows four such paths, all having the same shape and area but located in different positions in the changing field. For paths 1 and 2, the induced emfs $\mathscr{E}$ ($= \oint \mathbf{E} \cdot d\mathbf{s}$) are equal because these paths lie entirely in the magnetic field and thus have the same value of $d\Phi_B/dt$. This is true even though the electric field vectors around these paths are distributed differently, as indicated by the patterns of electric field lines. For path 3 the induced emf is smaller because the enclosed flux Φ_B (hence $d\Phi_B/dt$) is smaller, and for path 4 the induced emf is zero, even though the electric field is not zero at any point on the path.

A New Look at Electric Potential

Induced electric fields are produced not by static charges but by a changing magnetic flux. Although electric fields produced in either way exert forces on charged particles, there is an important difference between them. The simplest evidence of this difference is that the field lines of induced electric fields form closed loops, as in Fig. 31-14c. Field lines produced by static charges never do so but must start on positive charges and end on negative charges.

In a more formal sense, we can state the difference between electric fields produced by induction and those produced by static charges in these words:

Electric potential has meaning only for electric fields that are produced by static charges; it has no meaning for electric fields that are produced by induction.

You can understand this statement qualitatively by considering what happens to a charged particle that makes a single journey around the circular path in Fig. 31-14b. It starts at a certain point and, upon its return to that same point, has experienced an emf $\mathscr{E}$ of, let us say, 5 V. Its potential should have increased by this amount. This is impossible, however, because otherwise the same point in space would have two different values of potential. We must conclude that potential has no meaning for electric fields that are set up by changing magnetic fields.

We can take a more formal look by recalling Eq. 25-18, which defines the potential difference between two points *i* and *f*:

$$V_f - V_i = -\int_i^f \mathbf{E} \cdot d\mathbf{s}. \qquad (31\text{-}23)$$

In Chapter 25 we had not yet encountered Faraday's law of induction, so the electric fields involved in the derivation of Eq. 25-18 were those due to static charges. If *i* and *f* in Eq. 31-23 are the same point, the path connecting them is a closed loop, V_i and V_f are identical, and Eq. 31-23 reduces to

$$\oint \mathbf{E} \cdot d\mathbf{s} = 0. \qquad (31\text{-}24)$$

However, when a changing magnetic flux is present, this integral is *not* zero but is $-d\Phi_B/dt$, as Eq. 31-22 asserts. Again, we conclude that electric potential has no meaning for electric fields associated with induction.

SAMPLE PROBLEM 31-5

In Fig. 31-14b, take $R = 8.5$ cm and $dB/dt = 0.13$ T/s.

(a) Find an expression for the magnitude E of the induced electric field at points within the magnetic field, at radius r from the center of the magnetic field. Evaluate the expression for $r = 5.2$ cm.

SOLUTION: We can find an expression for E at radius r by applying Faraday's law in the form of Eq. 31-22, with the closed integration path a circle of radius r (as in Fig. 31-14b). We have assumed from the symmetry that $\mathbf{E}$ in Fig. 31-14b is tangent to the circular path at all points. The path vector $d\mathbf{s}$ is also always tangent to the circular path, so the dot product $\mathbf{E} \cdot d\mathbf{s}$ in Eq. 31-22 must have the magnitude $E\,ds$ at all points on the path. We can also assume from the symmetry that E has the same value at all points along the circular path. Then, dropping the minus sign, Eq. 31-22 gives us

$$\oint \mathbf{E} \cdot d\mathbf{s} = \oint E\,ds = E \oint ds = E(2\pi r) = \frac{d\Phi_B}{dt}. \qquad (31\text{-}25)$$

From Eq. 31-4, the magnetic flux through the circular path of integration is

$$\Phi_B = BA = B(\pi r^2). \qquad (31\text{-}26)$$

Substituting this into Eq. 31-25, we find that

$$E(2\pi r) = (\pi r^2)\frac{dB}{dt},$$

or

$$E = \frac{r}{2}\frac{dB}{dt}. \qquad \text{(Answer)} \qquad (31\text{-}27)$$

Equation 31-27 gives the magnitude of the electric field at any point for which $r < R$ (that is, within the magnetic field). Substituting given values yields, for the magnitude of $\mathbf{E}$ at $r = 5.2$ cm,

$$E = \frac{(5.2 \times 10^{-2}\text{ m})}{2}(0.13\text{ T/s})$$

$$= 0.0034\text{ V/m} = 3.4\text{ mV/m}. \qquad \text{(Answer)}$$

(b) Find an expression for the magnitude E of the induced electric field at points that are outside the magnetic field, at radius r. Evaluate the expression for $r = 12.5$ cm.

SOLUTION: Proceeding as in (a), we again obtain Eq. 31-25. However, we do not then obtain Eq. 31-26, because the closed path is now outside the magnetic field. The magnetic flux encircled by the closed path is now the area πR^2 of the magnetic field region. So,

$$\Phi_B = BA = B(\pi R^2). \qquad (31\text{-}28)$$

Substituting this into Eq. 31-25 and solving for E yield

$$E = \frac{R^2}{2r} \frac{dB}{dt}. \qquad \text{(Answer)} \qquad (31\text{-}29)$$

Since E is not zero here, we know that an electric field is induced even at points that are outside the changing magnetic field, an important result that (as you shall see in Section 33-11) makes transformers possible. With the given data, Eq. 31-29 yields the magnitude of $\mathbf{E}$ at $r = 12.5$ cm:

$$E = \frac{(8.5 \times 10^{-2} \text{ m})^2}{(2)(12.5 \times 10^{-2} \text{ m})} (0.13 \text{ T/s})$$

$$= 3.8 \times 10^{-3} \text{ V/m} = 3.8 \text{ mV/m}. \qquad \text{(Answer)}$$

Equations 31-27 and 31-29 give the same result, as they must, for $r = R$. Figure 31-15 shows a plot of $E(r)$ based on these two equations.

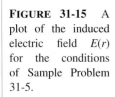

FIGURE 31-15 A plot of the induced electric field $E(r)$ for the conditions of Sample Problem 31-5.

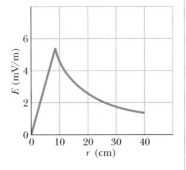

CHECKPOINT 4: The figure shows five lettered regions in which a uniform magnetic field extends either directly out of the page (as in region a) or into the page. The field is increasing in magnitude at the same steady rate in all five regions; the regions are identical in area. Also shown are four numbered paths along which $\oint \mathbf{E} \cdot d\mathbf{s}$ has the magnitudes given below in terms of a quantity mag. Determine whether the magnetic fields in regions b through e are directed into or out of the page.

Path:	1	2	3	4
$\oint \mathbf{E} \cdot d\mathbf{s}$:	mag	2(mag)	3(mag)	0

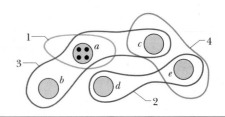

31-7 INDUCTORS AND INDUCTANCE

We found in Chapter 26 that a capacitor can be used to produce a desired electric field. We considered the parallel-plate arrangement as a basic type of capacitor. Similarly, an **inductor** (symbol ⦙⦙⦙⦙) can be used to produce a desired magnetic field. We shall consider a long solenoid (more specifically, a short length near the middle of a long solenoid) as our basic type of inductor.

If we establish a current i in the windings (or turns) of an inductor (a solenoid), the current produces a magnetic flux Φ through the central region of the inductor. The **inductance** of the inductor is then

$$L = \frac{N\Phi}{i} \qquad \text{(inductance defined),} \qquad (31\text{-}30)$$

in which N is the number of turns. The windings of the inductor are said to be *linked* by the shared flux, and the product $N\Phi$ is called the *magnetic flux linkage*. The inductance $\mathbf{L}$ is thus a measure of the flux linkage produced by the inductor per unit of current.

Because the SI unit of magnetic flux is the tesla–square meter, the SI unit of inductance is the tesla–square meter per ampere ($\text{T} \cdot \text{m}^2/\text{A}$). We call this the **henry** (H), after American physicist Joseph Henry, the codiscoverer of the law of induction and a contemporary of Faraday. Thus

$$1 \text{ henry} = 1 \text{ H} = 1 \text{ T} \cdot \text{m}^2/\text{A}. \qquad (31\text{-}31)$$

Through the rest of this chapter we assume that all inductors, no matter what their geometric arrangement, have no magnetic materials such as iron in their vicinity. Such materials would distort the magnetic field of an inductor.

Inductance of a Solenoid

Consider a long solenoid of cross-sectional area A. What is the inductance per unit length near its middle?

To use the defining equation for inductance (Eq. 31-30), we must calculate the flux linkage set up by a given current in the solenoid windings. Consider a length l near the middle of this solenoid. The flux linkage for this section of the solenoid is

$$N\Phi = (nl)(BA)$$

The crude inductors with which Michael Faraday discovered the law of induction. In those days amenities such as insulated wire were not commercially available. It is said that Faraday insulated his wires by wrapping them with strips from one of his wife's petticoats.

in which n is the number of turns per unit length of the solenoid and B is the magnetic field within the solenoid.

The magnitude B is given by Eq. 30-25,

$$B = \mu_0 in,$$

so from Eq. 31-30,

$$L = \frac{N\Phi}{i} = \frac{(nl)(BA)}{i} = \frac{(nl)(\mu_0 in)(A)}{i}$$
$$= \mu_0 n^2 lA. \quad (31\text{-}32)$$

Thus the inductance per unit length for a long solenoid near its center is

$$\frac{L}{l} = \mu_0 n^2 A \quad \text{(solenoid).} \quad (31\text{-}33)$$

Inductance—like capacitance—depends only on the geometry of the device. The dependence on the square of the number of turns per unit length is to be expected. If you, say, triple n, you triple not only the number of turns (N) but you also triple the flux ($\Phi = BA = \mu_0 inA$) through each turn, multiplying the flux linkage $N\Phi$ and thus the inductance L by a factor of 9.

If the solenoid is very much longer than its radius, then Eq. 31-32 gives its inductance to a good approximation. This approximation neglects the spreading of the magnetic field lines near the ends of the solenoid, just as the parallel-plate capacitor formula ($C = \epsilon_0 A/d$) neglects the fringing of the electric field lines near the edges of the capacitor plates.

From Eq. 31-32, and recalling that n is a number per unit length, we can see that an inductance can be written as a product of the permeability constant μ_0 and a quantity with the dimensions of a length. This means that μ_0 can be expressed in the unit henry per meter:

$$\mu_0 = 4\pi \times 10^{-7} \text{ T} \cdot \text{m/A}$$
$$= 4\pi \times 10^{-7} \text{ H/m.} \quad (31\text{-}34)$$

SAMPLE PROBLEM 31-6

Figure 31-16 shows a cross section, in the plane of the page, of a toroid of N turns like that in Fig. 30-21a but of rectangular cross section; its dimensions are as indicated.

(a) What is its inductance L?

SOLUTION: To use the definition of inductance, Eq. 31-30, we need to find the magnetic flux Φ due to a current i through the toroid. From Eq. 30-26, we already know the magnitude B of the magnetic field within the toroid (also due to i):

$$B = \frac{\mu_0 iN}{2\pi r}, \quad (31\text{-}35)$$

where r is the distance from the center of the toroid. This equation holds regardless of the shape or dimensions of the toroid's cross section.

Because B is *not* uniform over the cross section, we cannot use Eq. 31-4 ($\Phi_B = BA$) to find the flux Φ, but instead must use Eq. 31-3,

$$\Phi = \int \mathbf{B} \cdot d\mathbf{A}. \quad (31\text{-}36)$$

The field $\mathbf{B}$ is everywhere perpendicular to the cross section, as shown in Fig. 31-16; $\mathbf{B}$ is thus parallel to the differential cross-sectional area vector $d\mathbf{A}$, so the dot product in Eq. 31-36 gives $B\,dA$. For the differential area dA, we can use the area $h\,dr$ of the strip shown in Fig. 31-16. Substituting these quantities and Eq. 31-35 into Eq. 31-36 and integrating from $r = a$ to $r = b$ yield

$$\Phi = \int_a^b Bh\,dr = \int_a^b \frac{\mu_0 iN}{2\pi r} h\,dr$$
$$= \frac{\mu_0 iNh}{2\pi} \int_a^b \frac{dr}{r} = \frac{\mu_0 iNh}{2\pi} \ln \frac{b}{a}.$$

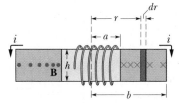

FIGURE 31-16 Sample Problem 31-6. A cross section of a toroid, showing the current in the windings and the associated magnetic field. See Fig. 30-21. The nonuniform magnetic field within the toroid is represented by nonuniformly spaced dots and ×s.

Equation 31-30 then gives us

$$L = \frac{N\Phi}{i} = \frac{N}{i} \frac{\mu_0 iNh}{2\pi} \ln \frac{b}{a},$$

so

$$L = \frac{\mu_0 N^2 h}{2\pi} \ln \frac{b}{a}. \quad \text{(Answer)} \quad (31\text{-}37)$$

(b) The toroid shown in Fig. 31-16 has $N = 1250$ turns, $a = 52$ mm, $b = 95$ mm, and $h = 13$ mm. What is its inductance?

SOLUTION: From Eq. 31-37

$$L = \frac{\mu_0 N^2 h}{2\pi} \ln \frac{b}{a}$$

$$= \frac{(4\pi \times 10^{-7} \text{ H/m})(1250)^2(13 \times 10^{-3} \text{ m})}{2\pi} \ln \frac{95 \text{ mm}}{52 \text{ mm}}$$

$$= 2.45 \times 10^{-3} \text{ H} \approx 2.5 \text{ mH}. \quad \text{(Answer)}$$

31-8 SELF-INDUCTION

If two coils—which we can now call inductors—are near each other, a current i in one coil produces a magnetic flux Φ through the second coil. We have seen that if we change this flux by changing the current, an induced emf appears in the second coil according to Faraday's law. An induced emf appears in the first coil as well.

> An induced emf $\mathcal{E}_L$ appears in any coil in which the current is changing.

This process (see Fig. 31-17) is called **self-induction,** and the emf that appears is called a **self-induced emf.** It obeys Faraday's law of induction just as other induced emfs do.

For any inductor, Eq. 31-30 tells us that

$$N\Phi = Li. \quad (31\text{-}38)$$

Faraday's law tell us that

$$\mathcal{E}_L = -\frac{d(N\Phi)}{dt}. \quad (31\text{-}39)$$

By combining Eqs. 31-38 and 31-39 we can write:

$$\mathcal{E}_L = -L \frac{di}{dt} \quad \text{(self-induced emf).} \quad (31\text{-}40)$$

Thus in any inductor (such as a coil, a solenoid, or a toroid) a self-induced emf appears whenever the current changes with time. The magnitude of the current has no influence on the magnitude of the induced emf; only the rate of change of the current counts.

You can find the *direction* of a self-induced emf from Lenz's law. The minus sign in Eq. 31-40 indicates that—as the law states—the self-induced emf acts to oppose the change that brings it about.

Suppose that, as in Fig. 31-18a, you set up a current i in a coil and arrange to have it increase with time at a rate di/dt. In the language of Lenz' law, this increase in the current is the "change" that the self-induction must oppose. For such opposition to occur, a self-induced emf must appear in the coil, pointing—as the figure shows—so as to oppose the increase in the current. If you cause the current to decrease with time, as in Fig. 31-18b, the self-induced emf must point in a direction that tends to oppose the decrease in the current, as the figure shows.

In Section 31-6 we saw that we cannot define an electric potential for an electric field (and thus for an emf) that is induced by a changing magnetic flux. This means that when a self-induced emf is produced in the inductor of Fig. 31-17, we cannot define an electric potential within the inductor itself, where the flux is changing. However, potentials can still be defined at points of the circuit outside this region, where the electric fields are due to charge distributions with associated electric potentials.

Moreover, we can define a potential difference V_L *across an inductor* (between its terminals, which we assume to be outside the region of changing flux). If the

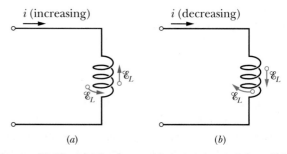

(a)   (b)

FIGURE 31-18   (a) The current i is increasing and the self-induced emf $\mathcal{E}_L$ appears along the coil in a direction such that it opposes the increase. The arrow representing $\mathcal{E}_L$ can be drawn along a turn of the coil or alongside the coil. Both are shown. (b) The current i is decreasing and the self-induced emf appears in a direction such that it opposes the decrease.

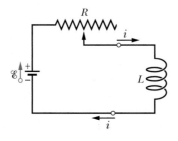

FIGURE 31-17   If the current in coil L is changed by varying the contact position on resistor R, a self-induced emf $\mathcal{E}_L$ will appear in the coil *while the current is changing.*

inductor is an *ideal inductor* (its wire has negligible resistance), the magnitude of V_L is equal to the magnitude of the self-induced emf $\mathcal{E}_L$.

If, instead, the wire in the inductor has resistance r, we mentally separate the inductor into a resistance r (which we take to be outside the region of changing flux) and an ideal inductor of self-induced emf $\mathcal{E}_L$. As with a real battery of emf $\mathcal{E}$ and internal resistance r, the potential difference across the terminals of a real inductor then differs from the emf. Unless otherwise indicated, we assume here that inductors are ideal.

CHECKPOINT **5:** The figure shows an emf $\mathcal{E}_L$ induced in a coil. Which of the following can describe the current through the coil: (a) constant and rightward, (b) constant and leftward, (c) increasing and rightward, (d) decreasing and rightward, (e) increasing and leftward, (f) decreasing and leftward?

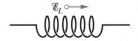

31-9 *RL* CIRCUITS

In Section 28-8 we saw that if you suddenly introduce an emf $\mathcal{E}$ into a single-loop circuit containing a resistor R and a capacitor C, the charge on the capacitor does not build up immediately to its final equilibrium value $C\mathcal{E}$ but approaches it in an exponential fashion:

$$q = C\mathcal{E}(1 - e^{-t/\tau_C}). \qquad (31\text{-}41)$$

The rate at which the charge builds up is determined by the capacitive time constant τ_C, defined in Eq. 28-33 as

$$\tau_C = RC. \qquad (31\text{-}42)$$

If you suddenly remove the emf from this same circuit, the charge does not immediately fall to zero but approaches zero in an exponential fashion:

$$q = q_0 e^{-t/\tau_C}. \qquad (31\text{-}43)$$

The time constant τ_C describes the fall of the charge as well as its rise.

An analogous slowing of the rise (or fall) of the current occurs if we introduce an emf $\mathcal{E}$ into (or remove it from) a single-loop circuit containing a resistor R and an inductor L. When the switch S in Fig. 31-19 is closed on a, for example, the current in the resistor starts to rise. If the inductor were not present, the current would rise rapidly to a steady value $\mathcal{E}/R$. Because of the inductor, however, a self-induced emf $\mathcal{E}_L$ appears in the circuit; from Lenz's law, this emf opposes the rise of the current, which means

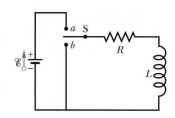

FIGURE 31-19 An *RL* circuit. When switch S is closed on a, the current rises and approaches a limiting value of $\mathcal{E}/R$.

that it opposes the battery emf $\mathcal{E}$ in polarity. Thus the resistor responds to the difference between two emfs, a constant one $\mathcal{E}$ due to the battery and a variable one $\mathcal{E}_L$ ($= -L\,di/dt$) due to self-induction. As long as $\mathcal{E}_L$ is present, the current in the resistor will be less than $\mathcal{E}/R$.

As time goes on, the rate at which the current increases becomes less rapid and the magnitude of the self-induced emf, which is proportional to di/dt, becomes smaller. Thus the current in the circuit approaches $\mathcal{E}/R$ asymptotically.

We can generalize these results as follows:

> Initially, an inductor acts to oppose changes in the current through it. A long time later, it acts like ordinary connecting wire.

Now let us analyze the situation quantitatively. With the switch S in Fig. 31-19 thrown to a, the circuit is equivalent to that of Fig. 31-20. Let us apply the loop rule, starting at x in this figure and moving clockwise around the loop. For the current direction shown, x will be higher in potential than y, which means that we encounter a potential change of $-iR$ as we traverse the resistor. Point y is higher in potential than point z because, for an increasing current, the self-induced emf will oppose the rise of the current by pointing as shown. Thus, as we traverse the inductor from y to z, we encounter a potential change of $\mathcal{E}_L = -L\,di/dt$. We encounter a rise in potential of $+\mathcal{E}$ in traversing the battery from z to x. The loop rule thus gives

$$-iR - L\frac{di}{dt} + \mathcal{E} = 0$$

or

$$L\frac{di}{dt} + Ri = \mathcal{E} \qquad (RL \text{ circuit}). \qquad (31\text{-}44)$$

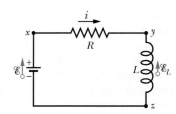

FIGURE 31-20 The circuit of Fig. 31-19 with the switch closed on a. We apply the loop rule clockwise, starting at x.

Equation 31-44 is a differential equation involving the variable i and its first derivative di/dt. We seek the function $i(t)$ such that when it and its first derivative are substituted in Eq. 31-44, the equation is satisfied and the initial condition $i(0) = 0$ is satisfied.

Equation 31-44 and its initial condition are of exactly the form of Eq. 28-29 for an RC circuit, with i replacing q, L replacing R, and R replacing $1/C$. The solution of Eq. 31-44 must then be of exactly the form of Eq. 28-30 with the same replacements. That solution is

$$i = \frac{\mathscr{E}}{R}(1 - e^{-Rt/L}), \qquad (31\text{-}45)$$

which we can rewrite as

$$i = \frac{\mathscr{E}}{R}(1 - e^{-t/\tau_L}) \qquad \text{(rise of current).} \quad (31\text{-}46)$$

Here τ_L, the **inductive time constant,** is given by

$$\tau_L = \frac{L}{R} \qquad \text{(time constant).} \quad (31\text{-}47)$$

Figure 31-21 shows how the potential differences V_R $(= iR)$ across the resistor and V_L $(= L\,di/dt)$ across the inductor vary with time for particular values of $\mathscr{E}$, L, and R. Compare this figure carefully with the corresponding figure for an RC circuit (Fig. 28-14).

To show that the quantity τ_L $(= L/R)$ has the dimensions of time, we put

$$1\,\frac{\text{H}}{\Omega} = 1\,\frac{\text{H}}{\Omega}\left(\frac{1\,\text{V}\cdot\text{s}}{1\,\text{H}\cdot\text{A}}\right)\left(\frac{1\,\Omega\cdot\text{A}}{1\,\text{V}}\right) = 1\,\text{s}.$$

The first quantity in parentheses is a conversion factor based on Eq. 31-40, and the second one is a conversion factor based on the relation $V = iR$.

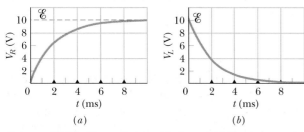

FIGURE 31-21 The variation with time of (a) V_R, the potential difference across the resistor in the circuit of Fig. 31-20, and (b) V_L, the potential difference across the inductor in that circuit. The small triangles represent successive intervals of one inductive time constant $\tau_L = L/R$. The figure is plotted for $R = 2000\ \Omega$, $L = 4.0$ H, and $\mathscr{E} = 10$ V.

The physical significance of the time constant follows from Eq. 31-46. If we put $t = \tau_L = L/R$ in this equation, it reduces to

$$i = \frac{\mathscr{E}}{R}(1 - e^{-1}) = 0.63\,\frac{\mathscr{E}}{R}.$$

Thus the time constant τ_L is the time it takes the current in the circuit to reach within $1/e$ (about 37%) of its final equilibrium value $\mathscr{E}/R$. Since the potential difference V_R across the resistor is proportional to the current i, the time dependence of the increasing current has the same shape as V_R, as plotted in Fig. 31-21a.

If the switch S in Fig. 31-19, having been closed on a long enough for the equilibrium current $\mathscr{E}/R$ to be established, is thrown to b, the effect will be to remove the battery from the circuit. (The connection to b must actually be made before the connection to a is broken. A switch that does this is called a *make-before-break* switch.)

The current through the resistor cannot drop immediately to zero but must decay to zero over time. The differential equation that governs the decay can be found by putting $\mathscr{E} = 0$ in Eq. 31-44:

$$L\frac{di}{dt} + iR = 0. \qquad (31\text{-}48)$$

By analogy with Eqs. 28-35 and 28-36, the solution of this differential equation that satisfies the initial condition $i(0) = i_0 = \mathscr{E}/R$ is

$$i = \frac{\mathscr{E}}{R}e^{-t/\tau_L} = i_0 e^{-t/\tau_L} \qquad \begin{matrix}\text{(decay of} \\ \text{current).}\end{matrix} \quad (31\text{-}49)$$

We see that both current rise (Eq. 31-46) and current decay (Eq. 31-49) in an RL circuit are governed by the same inductive time constant, τ_L.

We have used i_0 in Eq. 31-49 to represent the current at time $t = 0$. In our case that happened to be $\mathscr{E}/R$, but it could be any other initial value.

SAMPLE PROBLEM 31-7

Figure 31-22a shows a circuit that contains three identical resistors with resistance $R = 9.0\ \Omega$, two identical inductors with inductance $L = 2.0$ mH, and an ideal battery with emf $\mathscr{E} = 18$ V.

(a) What is the current i through the battery just after the switch is closed?

SOLUTION: Because the current through each inductor is zero before the switch is closed, it will also be zero just afterward. So, immediately after the switch is closed, the inductors act as broken wires, as indicated in Fig. 31-22b. We then have a single-loop circuit for which the loop rule gives us

$$\mathscr{E} - iR = 0.$$

Substituting given data, we find that

$$i = \frac{\mathscr{E}}{R} = \frac{18 \text{ V}}{9.0 \text{ }\Omega} = 2.0 \text{ A}. \qquad \text{(Answer)}$$

(b) What is the current i through the battery long after the switch has been closed?

SOLUTION: Long after the switch has been closed, the currents in the circuit have reached their equilibrium values. Then the inductors act as simple connecting wires, as indicated in Fig. 31-22c. We then have a circuit with three identical resistors in parallel; from Eq. 28-20, their equivalent resistance is $R_{eq} = R/3 = (9.0 \text{ }\Omega)/3 = 3.0 \text{ }\Omega$. The equivalent circuit in Fig. 31-22d then yields the loop equation $\mathscr{E} - iR_{eq} = 0$, or

$$i = \frac{\mathscr{E}}{R_{eq}} = \frac{18 \text{ V}}{3.0 \text{ }\Omega} = 6.0 \text{ A}. \qquad \text{(Answer)}$$

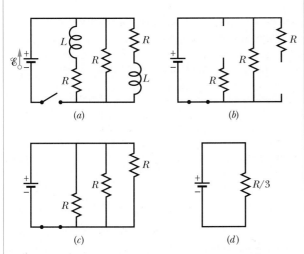

(a)

(b)

(c)

(d)

FIGURE 31-22   Sample Problem 31-7. (a) A multiloop RL circuit with an open switch. (b) The equivalent circuit just after the switch has been closed. (c) The equivalent circuit a long time later. (d) The single-loop circuit that is equivalent to circuit (c).

SAMPLE PROBLEM 31-8

A solenoid has an inductance of 53 mH and a resistance of 0.37 Ω. If it is connected to a battery, how long will the current take to reach half its final equilibrium value?

SOLUTION: The equilibrium value of the current is reached as $t \to \infty$; from Eq. 31-46 that value is $\mathscr{E}/R$. If the current has half this value at a particular time t_0, this equation becomes

$$\frac{1}{2} \frac{\mathscr{E}}{R} = \frac{\mathscr{E}}{R} (1 - e^{-t_0/\tau_L}).$$

We solve for t_0 by canceling $\mathscr{E}/R$, isolating the exponential, and taking the natural logarithm of each side. We find

$$t_0 = \tau_L \ln 2$$
$$= \frac{L}{R} \ln 2 = \frac{53 \times 10^{-3} \text{ H}}{0.37 \text{ }\Omega} \ln 2$$
$$= 0.10 \text{ s}. \qquad \text{(Answer)}$$

CHECKPOINT **6:** The figure shows three circuits with identical batteries, inductors, and resistors. Rank the circuits according to the current through the battery (a) just after the switch is closed and (b) a long time later, greatest first.

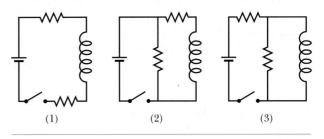

(1)

(2)

(3)

31-10   ENERGY STORED IN A MAGNETIC FIELD

When we pull two unlike charges apart, we say that the resulting electric potential energy is stored in the electric field of the charges. We get it back from the field by letting the charges move closer together again.

In the same way we can consider energy to be stored in a magnetic field. For example, two long, rigid, parallel wires carrying current in the same direction attract each other, and we must do work to pull them apart. In doing so, we store energy in the magnetic fields of the currents. We can get this stored energy back at any time by letting the wires move back to their original positions.

To derive a quantitative expression for the energy stored in a magnetic field, consider again Fig. 31-20, which shows a source of emf $\mathscr{E}$ connected to a resistor R and an inductor L. Equation 31-44, restated here for convenience,

$$\mathscr{E} = L \frac{di}{dt} + iR, \qquad (31\text{-}50)$$

is the differential equation that describes the growth of current in this circuit. We stress that this equation follows immediately from the loop rule and that the loop rule in turn is an expression of the principle of conservation of energy for single-loop circuits. If we multiply each side of Eq. 31-50 by i, we obtain

$$\mathscr{E}i = Li \frac{di}{dt} + i^2 R, \qquad (31\text{-}51)$$

which has the following physical interpretation in terms of work and energy:

1. If a charge dq passes through the battery of emf $\mathcal{E}$ in Fig. 31-20 in time dt, the battery does work on it in the amount $\mathcal{E}\,dq$. The rate at which the battery does work is $(\mathcal{E}\,dq)/dt$, or $\mathcal{E}i$. Thus the left side of Eq. 31-51 represents the rate at which the emf device delivers energy to the rest of the circuit.

2. The right-most term in Eq. 31-51 represents the rate at which energy appears as thermal energy in the resistor.

3. Energy that does not appear as thermal energy must, by our conservation-of-energy hypothesis, be stored in the magnetic field of the inductor. Since Eq. 31-51 represents a statement of the conservation of energy for RL circuits, the middle term must represent the rate dU_B/dt at which energy is stored in the magnetic field; so

$$\frac{dU_B}{dt} = Li\,\frac{di}{dt}. \qquad (31\text{-}52)$$

We can write this as

$$dU_B = Li\,di.$$

Integrating yields

$$\int_0^{U_B} dU_B = \int_0^i Li\,di$$

or

$$U_B = \tfrac{1}{2}Li^2 \quad \text{(magnetic energy)}, \qquad (31\text{-}53)$$

which represents the total energy stored by an inductor L carrying a current i. Note the similarity between this expression and the expression for the energy stored by a capacitor with capacitance C and charge q, namely,

$$U_E = \frac{q^2}{2C}. \qquad (31\text{-}54)$$

SAMPLE PROBLEM 31-9

A coil has an inductance of 53 mH and a resistance of 0.35 Ω.

(a) If a 12 V emf is applied across the coil, how much energy is stored in the magnetic field after the current has built up to its equilibrium value?

SOLUTION: The stored energy is given by Eq. 31-53,

$$U_B = \tfrac{1}{2}Li^2.$$

To find the equilibrium stored energy, we must substitute the equilibrium current in this expression. From Eq. 31-46 the equilibrium current is

$$i_\infty = \frac{\mathcal{E}}{R} = \frac{12\text{ V}}{0.35\ \Omega} = 34.3\text{ A}.$$

The substitution yields

$$\begin{aligned} U_{B\infty} = \tfrac{1}{2}Li_\infty^2 &= (\tfrac{1}{2})(53 \times 10^{-3}\text{ H})(34.3\text{ A})^2 \\ &= 31\text{ J}. \qquad\qquad \text{(Answer)}\end{aligned}$$

(b) After how many times constants will half this equilibrium energy be stored in the magnetic field?

SOLUTION: We are asked: At what time t will the relation

$$U_B = \tfrac{1}{2}U_{B\infty}$$

be satisfied? Equation 31-53 allows us to rewrite this condition as

$$\tfrac{1}{2}Li^2 = (\tfrac{1}{2})\tfrac{1}{2}Li_\infty^2$$

or

$$i = \left(\frac{1}{\sqrt{2}}\right)i_\infty.$$

But i is given by Eq. 31-46 and i_∞ (see above) is $\mathcal{E}/R$, so this equation becomes

$$\frac{\mathcal{E}}{R}(1 - e^{-t/\tau_L}) = \frac{\mathcal{E}}{\sqrt{2}R}.$$

By canceling $\mathcal{E}/R$ and rearranging, this can be written as

$$e^{-t/\tau_L} = 1 - \frac{1}{\sqrt{2}} = 0.293,$$

which yields

$$\frac{t}{\tau_L} = -\ln 0.293 = 1.23$$

or

$$t \approx 1.2\,\tau_L. \qquad\qquad \text{(Answer)}$$

Thus the stored energy will reach half its equilibrium value after 1.2 time constants.

SAMPLE PROBLEM 31-10

A 3.56 H inductor is placed in series with a 12.8 Ω resistor, and an emf of 3.24 V is then suddenly applied across the RL combination.

(a) At 0.278 s (which is one inductive time constant) after the emf is applied, what is the rate P at which energy is being delivered by the battery?

SOLUTION: The current in the circuit is given by Eq. 31-46,

$$i = \frac{\mathcal{E}}{R}(1 - e^{-t/\tau_L}),$$

which, after one time constant, becomes

$$i = \frac{3.24\text{ V}}{12.8\ \Omega}(1 - e^{-1}) = 0.1600\text{ A}.$$

The rate at which the battery delivers energy is then given by Eq. 27-21 with $\mathcal{E}$ replacing V and is

$$P = \mathscr{E}i = (3.24 \text{ V})(0.1600 \text{ A})$$
$$= 0.5184 \text{ W} \approx 518 \text{ mW}. \quad \text{(Answer)}$$

(b) At 0.278 s, at what rate P_R is energy appearing as thermal energy in the resistor?

SOLUTION: This is given by Eq. 27-22:

$$P_R = i^2 R = (0.1600 \text{ A})^2 (12.8 \text{ } \Omega)$$
$$= 0.3277 \text{ W} \approx 328 \text{ mW}. \quad \text{(Answer)}$$

(c) At 0.278 s, at what rate P_B is energy being stored in the magnetic field?

SOLUTION: This is given by Eq. 31-52, which requires that we know di/dt. Differentiating Eq. 31-45 yields

$$\frac{di}{dt} = \frac{\mathscr{E}}{R} \frac{R}{L} (e^{-Rt/L}) = \frac{\mathscr{E}}{L} e^{-t/\tau_L}.$$

After one time constant we have

$$\frac{di}{dt} = \frac{3.24 \text{ V}}{3.56 \text{ H}} e^{-1} = 0.3348 \text{ A/s}.$$

Now from Eq. 31-52 the desired rate is

$$P_B = \frac{dU_B}{dt} = Li \frac{di}{dt}$$
$$= (3.56 \text{ H})(0.1600 \text{ A})(0.3348 \text{ A/s})$$
$$= 0.1907 \text{ W} \approx 191 \text{ mW}. \quad \text{(Answer)}$$

Note that, as required by energy conservation,

$$P = P_R + P_B.$$

31-11 ENERGY DENSITY OF A MAGNETIC FIELD

Consider a length l near the middle of a long solenoid of cross-sectional area A; the volume associated with this length is Al. The energy stored by the length l of the solenoid must lie entirely within this volume because the magnetic field outside such a solenoid is essentially zero. Moreover, the stored energy must be uniformly distributed within the solenoid because the magnetic field is uniform everywhere inside.

Thus the energy per unit volume of the field is

$$u_B = \frac{U_B}{Al}$$

or, since

$$U_B = \tfrac{1}{2} L i^2,$$

we have

$$u_B = \frac{Li^2}{2Al} = \frac{L}{l} \frac{i^2}{2A}.$$

Substituting for L/l from Eq. 31-33, we find

$$u_B = \tfrac{1}{2} \mu_0 n^2 i^2. \quad (31\text{-}55)$$

From Eq. 30-25 ($B = \mu_0 in$) we can write this *energy density* as

$$u_B = \frac{B^2}{2\mu_0} \quad \text{(magnetic energy density).} \quad (31\text{-}56)$$

This equation gives the density of stored energy at any point where the magnetic field is B. Even though we derived it by considering a special case, the solenoid, Eq. 31-56 holds for all magnetic fields, no matter how they are generated. Equation 31-56 is comparable to Eq. 26-23, namely,

$$u_E = \tfrac{1}{2} \epsilon_0 E^2, \quad (31\text{-}57)$$

which gives the energy density (in a vacuum) at any point in an electric field. Note that both u_B and u_E are proportional to the square of the appropriate field quantity, B or E.

CHECKPOINT 7: The table lists the number of turns per unit length, current, and cross-sectional area for three solenoids. Rank the solenoids according to the magnetic energy density within them, greatest first.

SOLENOID	TURNS PER UNIT LENGTH	CURRENT	AREA
a	$2n_1$	i_1	$2A_1$
b	n_1	$2i_1$	A_1
c	n_1	i_1	$6A_1$

SAMPLE PROBLEM 31-11

A long coaxial cable (Fig. 31-23) consists of two thin-walled concentric conducting cylinders with radii a and b. The inner cylinder A carries a steady current i, the outer cylinder B providing the return path.

(a) Calculate the energy stored in the magnetic field between the cylinders for a length l of the cable.

SOLUTION: Consider a volume dV between the two cylinders, consisting of a cylindrical shell whose inner and outer radii are r and $r + dr$ and whose length is l. The energy dU contained within this shell is

$$dU = u_B \, dV,$$

in which u_B (the energy per unit volume) is, from Eq. 31-56, $u_B = B^2/2\mu_0$.

To find B as a function of r, we apply Ampere's law,

$$\oint \mathbf{B} \cdot d\mathbf{s} = \mu_0 i,$$

to the circle of radius r in Fig. 31-23, obtaining

$$(B)(2\pi r) = \mu_0 i,$$

or

$$B = \frac{\mu_0 i}{2\pi r}.$$

The energy density between the cylinders is then

$$u_B = \frac{1}{2\mu_0}\left(\frac{\mu_0 i}{2\pi r}\right)^2 = \frac{\mu_0 i^2}{8\pi^2 r^2}.$$

The volume dV of our shell is $(2\pi r l)(dr)$, so the energy dU contained within the shell is

$$dU = u_B\,dV = \frac{\mu_0 i^2}{8\pi^2 r^2}(2\pi r l)(dr) = \frac{\mu_0 i^2 l}{4\pi}\frac{dr}{r}.$$

The total energy is obtained by integrating this expression over the volume between the cylinders:

$$U = \int dU = \frac{\mu_0 i^2 l}{4\pi}\int_a^b \frac{dr}{r}$$

$$= \frac{\mu_0 i^2 l}{4\pi}\ln\frac{b}{a}. \qquad \text{(Answer)} \qquad (31\text{-}58)$$

No energy is stored outside the outer cylinder or inside the inner cylinder because the magnetic field is zero in both locations, as you can show with Ampere's law.

(b) What is the stored energy per unit length of the cable if $a = 1.2$ mm, $b = 3.5$ mm, and $i = 2.7$ A?

SOLUTION: From Eq. 31-58 we have

$$\frac{U}{l} = \frac{\mu_0 i^2}{4\pi}\ln\frac{b}{a}$$

$$= \frac{(4\pi \times 10^{-7}\ \text{H/m})(2.7\ \text{A})^2}{4\pi}\ln\frac{3.5\ \text{mm}}{1.2\ \text{mm}}$$

$$= 7.8 \times 10^{-7}\ \text{J/m} = 780\ \text{nJ/m}. \qquad \text{(Answer)}$$

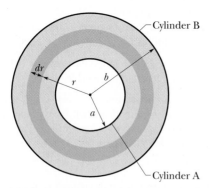

FIGURE 31-23 Sample Problem 31-11. A cross section of a long coaxial cable consisting of two thin-walled conducting cylinders, the inner cylinder of radius a and the outer cylinder of radius b.

31-12 MUTUAL INDUCTION

In this section we return to the case of two interacting coils, which we first discussed in Section 31-2, and we treat it in a somewhat more formal manner. In Fig. 31-2 we saw that if two coils are close together, a steady current i in one coil will set up a magnetic flux Φ through the other coil (*linking* the other coil). If we change i with time, an emf $\mathscr{E}$ given by Faraday's law appears in the second coil; we called this process *induction*. We could better have called it **mutual induction,** to suggest the mutual interaction of the two coils and to distinguish it from **self-induction,** in which only one coil is involved.

Let us look a little more quantitatively at mutual induction. Figure 31-24a shows two circular close-packed coils near each other and sharing a common central axis. There is a steady current i_1 in coil 1, produced by the battery in the external circuit. This current creates a magnetic field represented by the lines of $\mathbf{B}_1$ in the figure. Coil 2 is connected to a sensitive meter but contains no battery; a magnetic flux Φ_{21} (the flux through coil 2 associated with the current in coil 1) links the N_2 turns of coil 2.

We define the mutual inductance M_{21} of coil 2 with respect to coil 1 as

$$M_{21} = \frac{N_2\Phi_{21}}{i_1}. \qquad (31\text{-}59)$$

Compare this with Eq. 31-30 ($L = N\Phi/i$), the definition of (self) inductance. We can recast Eq. 31-59 as

$$M_{21}i_1 = N_2\Phi_{21}.$$

If, by external means, we cause i_1 to vary with time, we have

$$M_{21}\frac{di_1}{dt} = N_2\frac{d\Phi_{21}}{dt}.$$

The right side of this equation, from Faraday's law, is, apart from a difference in sign, just the emf $\mathscr{E}_2$ appearing in coil 2 due to the changing current in coil 1. Thus

$$\mathscr{E}_2 = -M_{21}\frac{di_1}{dt}, \qquad (31\text{-}60)$$

which you should compare with Eq. 31-40 for self-induction ($\mathscr{E} = -L\,di/dt$).

Let us now interchange the roles of coils 1 and 2, as in Fig. 31-24b. That is, we set up a current i_2 in coil 2, by means of a battery, and this produces a magnetic flux Φ_{12} that links coil 1. If we change i_2 with time, we have, by the argument given above,

$$\mathscr{E}_1 = -M_{12}\frac{di_2}{dt}. \qquad (31\text{-}61)$$

Thus we see that the emf induced in either coil is

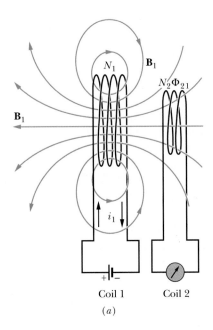

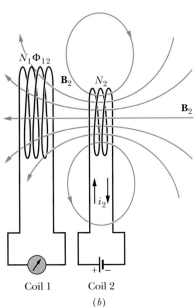

FIGURE 31-24 Mutual induction. (*a*) If the current in coil 1 changes, an emf will be induced in coil 2. (*b*) If the current in coil 2 changes, an emf will be induced in coil 1.

proportional to the rate of change of current in the other coil. The proportionality constants M_{21} and M_{12} seem to be different. We assert, without proof, that they are in fact the same so that no subscripts are needed. (This conclusion is true but is in no way obvious.) Thus we have

$$M_{21} = M_{12} = M, \qquad (31\text{-}62)$$

and we can rewrite Eqs. 31-60 and 31-61 as

$$\mathscr{E}_2 = -M \frac{di_1}{dt} \qquad (31\text{-}63)$$

and

$$\mathscr{E}_1 = -M \frac{di_2}{dt}. \qquad (31\text{-}64)$$

The induction is indeed mutual. The SI unit for M (as for L) is the henry.

SAMPLE PROBLEM 31-12

Figure 31-25 shows two circular close-packed coils, the smaller (radius R_2, with N_2 turns) being coaxial with the larger (radius R_1, with N_1 turns) and in the same plane.

(a) Derive an expression for the mutual inductance M for this arrangement of these two coils, assuming that $R_1 \gg R_2$.

SOLUTION: As Fig. 31-25 suggests, we imagine that we establish a current i_1 in the larger coil and we note the magnetic field B_1 that it sets up. The value of B_1 at the center of this coil is (from Eq. 30-28, with $z = 0$ and after multiplying the right side by N_1)

$$B_1 = \frac{\mu_0 i_1 N_1}{2R_1}.$$

Because we have assumed that $R_1 \gg R_2$, we may take B_1 to be the magnetic field at all points within the boundary of the smaller coil. The flux linkage for the smaller coil is then

$$N_2 \Phi_{21} = N_2(B_1)(\pi R_2^2) = \frac{\pi \mu_0 N_1 N_2 R_2^2 i_1}{2R_1}.$$

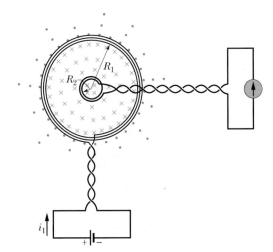

FIGURE 31-25 Sample Problem 31-12. A small coil is located at the center of a large coil. The mutual inductance of the coils can be determined by sending current i_1 through the large coil.

From Eq. 31-59 we then have

$$M = \frac{N_2 \Phi_{21}}{i_1} = \frac{\pi \mu_0 N_1 N_2 R_2^2}{2R_1}. \qquad \text{(Answer)}$$

(b) What is the value of M for $N_1 = N_2 = 1200$ turns, $R_2 = 1.1$ cm, and $R_1 = 15$ cm?

SOLUTION: The equation above yields

$$M = \frac{(\pi)(4\pi \times 10^{-7} \text{ H/m})(1200)(1200)(0.011 \text{ m})^2}{(2)(0.15 \text{ m})}$$

$$= 2.29 \times 10^{-3} \text{ H} \approx 2.3 \text{ mH}. \qquad \text{(Answer)}$$

Consider the situation if we reverse the roles of the two coils in Fig. 31-25, that is, if we produce a current i_2 in the smaller coil and try to calculate M from Eq. 31-59 in the form

$$M = \frac{N_1 \Phi_{12}}{i_2}.$$

The calculation of Φ_{12} (the flux of the smaller coil's magnetic field encompassed by the larger coil) is not simple. If we were to do the calculation numerically using a computer, we would find M to be exactly 2.3 mH, as above! This emphasizes that Eq. 31-62 ($M_{12} = M_{21} = M$) is not obvious.

REVIEW & SUMMARY

Magnetic Flux

The *magnetic flux* Φ_B through an area in a magnetic field **B** is defined as

$$\Phi_B = \int \mathbf{B} \cdot d\mathbf{A}, \qquad (31\text{-}3)$$

where the integral is taken over the area. The SI unit of magnetic flux is the weber, where $1 \text{ Wb} = 1 \text{ T} \cdot \text{m}^2$. If **B** is perpendicular to the area and uniform over it, Eq. 31-3 becomes

$$\Phi_B = BA \qquad (\mathbf{B} \perp \mathbf{A}, \mathbf{B} \text{ uniform}). \qquad (31\text{-}4)$$

Faraday's Law of Induction

If the magnetic flux Φ_B through an area bounded by a closed conducting loop changes with time, a current and an emf are produced in the loop; this process is called *induction*. The induced emf is

$$\mathcal{E} = -\frac{d\Phi_B}{dt}. \qquad \text{(Faraday's law).} \qquad (31\text{-}6)$$

If the loop is replaced by a closely packed coil of N turns, the induced emf is

$$\mathcal{E} = -N \frac{d\Phi_B}{dt}. \qquad (31\text{-}7)$$

Lenz's Law

An induced current has a direction such that the magnetic field *of the current* opposes the change in the magnetic field that produces the current.

Emf and the Induced Electric Field

An emf is induced by a changing magnetic flux even if the loop through which the flux is changing is not a physical conductor but an imaginary line. The changing flux induces an electric field **E** at every point of such a loop; the induced emf is related to **E** by

$$\mathcal{E} = \oint \mathbf{E} \cdot d\mathbf{s}, \qquad (31\text{-}21)$$

where the integration is taken around the loop. From Eq. 31-21 we can write Faraday's law in its most general form,

$$\oint \mathbf{E} \cdot d\mathbf{s} = -\frac{d\Phi_B}{dt} \qquad \text{(Faraday's law).} \qquad (31\text{-}22)$$

The essence of this law is that *a changing magnetic flux $d\Phi_B/dt$ induces an electric field* **E**.

Inductors

An **inductor** is a device that can be used to produce a known magnetic field in a specified region. If a current i is established through each of the N windings of an inductor, a magnetic flux Φ links those windings. The **inductance** L of the inductor is

$$L = \frac{N\Phi}{i} \qquad \text{(inductance defined).} \qquad (31\text{-}30)$$

The SI unit of inductance is the **henry** (H), with

$$1 \text{ henry} = 1 \text{ H} = 1 \text{ T} \cdot \text{m}^2/\text{A}. \qquad (31\text{-}31)$$

The inductance per unit length near the middle of a long solenoid of cross-sectional area A and n turns per unit length is

$$\frac{L}{l} = \mu_0 n^2 A \qquad \text{(solenoid).} \qquad (31\text{-}33)$$

Self-Induction

If a current i in a coil changes with time, an emf is induced in the coil. This self-induced emf is

$$\mathcal{E}_L = -L \frac{di}{dt}. \qquad (31\text{-}40)$$

The direction of $\mathcal{E}_L$ is found from Lenz's law: the self-induced emf acts to oppose the change that produces it.

Series RL Circuits

If a constant emf $\mathcal{E}$ is introduced into a single-loop circuit containing a resistance R and an inductance L, the current rises to an equilibrium value of $\mathcal{E}/R$ according to

$$i = \frac{\mathcal{E}}{R}(1 - e^{-t/\tau_L}) \qquad \text{(rise of current).} \qquad (31\text{-}46)$$

Here $\tau_L \; (= L/R)$ governs the rate of rise of the current and is called the **inductive time constant** of the circuit. When the source of constant emf is removed, the current decays from a value i_0 according to

$$i = i_0 e^{-t/\tau_L} \qquad \text{(decay of current).} \qquad (31\text{-}49)$$

Magnetic Energy

If an inductor L carries a current i, the inductor's magnetic field stores an energy given by

$$U_B = \tfrac{1}{2} L i^2 \qquad \text{(magnetic energy).} \qquad (31\text{-}53)$$

If B is the magnetic field at any point (in an inductor or anywhere else), the density of stored magnetic energy at that point is

$$u_B = \frac{B^2}{2\mu_0} \qquad \text{(magnetic energy density).} \qquad (31\text{-}56)$$

Mutual Induction

If two coils (labeled 1 and 2) are near each other, a changing current in either coil can induce an emf in the other. This mutual induction is described by

$$\mathcal{E}_2 = -M \frac{di_1}{dt} \quad \text{and} \quad \mathcal{E}_1 = -M \frac{di_2}{dt}, \quad (31\text{-}63,\ 31\text{-}64)$$

where M (measured in henries) is the mutual inductance for the coil arrangement.

QUESTIONS

1. Situation 1 in Fig. 31-26 shows a rectangular wire loop of length a and height b in the xy plane. Situations 2 and 3 show wire loops of the same overall dimensions but with sections brought forward parallel to the z axis. In each situation, the same uniform magnetic field pierces the loops and is increasing in magnitude at the same rate. Rank the three situations according to the magnitude of the emf induced in the loops, greatest first, if the magnetic field points in the positive direction of (a) y, (b) z, and (c) x.

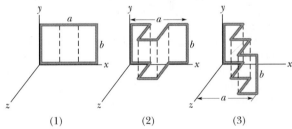

(1)   (2)   (3)

FIGURE 31-26   Question 1.

2. In Fig. 31-27, a long straight wire with current i passes (without touching) three rectangular wire loops with edge lengths L, $1.5L$, and $2L$. The loops are widely spaced (so as to not affect one another). Loops 1 and 3 are symmetric about the long wire. Rank the loops according to the size of the current induced in them if current i is (a) constant and (b) increasing, greatest first.

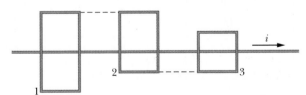

FIGURE 31-27   Question 2.

3. If the circular conductor in Fig. 31-28 undergoes thermal expansion while in a uniform magnetic field, a current will be induced clockwise around it. Is the magnetic field directed into the page or out of it?

FIGURE 31-28   Question 3.

4. In Fig. 31-29, a circular loop moves at constant velocity through regions where uniform magnetic fields of the same magnitude are directed into or out of the page. (The field is zero outside the dashed lines.) At which of the seven indicated loop positions is the emf induced in the loop (a) clockwise, (b) counterclockwise, and (c) zero?

FIGURE 31-29   Question 4.

5. Figure 31-30 shows two circuits in which a conducting bar is slid at the same speed v through the same uniform magnetic field and along a U-shaped wire. The parallel lengths of the wire are separated by $2L$ in circuit 1 and by L in circuit 2. The current induced in circuit 1 is counterclockwise. (a) Is the direction of the magnetic field into or out of the page? (b) Is the direction of the current in circuit 2 clockwise or counterclockwise? (c) Is the current in circuit 1 larger than, smaller than, or the same as that in circuit 2?

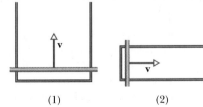

FIGURE 31-30
Question 5.   (1)   (2)

6. In Fig. 31-31, a conducting rod slides at constant velocity v across an incomplete conducting square, with which it makes electrical contact. The square is in a uniform magnetic field that points directly out of the page. During the slide, (a) is the induced current clockwise, counterclockwise, or does it change from one to the other midway, and (b) is the current constant, increasing, or increasing and then decreasing?

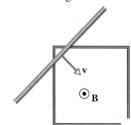

FIGURE 31-31
Question 6.

7. Figure 31-32 shows two coils wrapped around nonconducting rods. Coil X is connected to a battery and a variable resistance. What is the direction of the induced current through the current meter connected to coil Y (a) when coil Y is moved toward coil X and (b) when the current in coil X is decreased without any change in the relative positions of the coils?

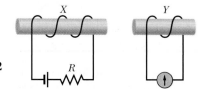

FIGURE 31-32
Question 7.

8. If the variable resistance R in the left-hand circuit of Fig. 31-33 is increased at a steady rate, is the current induced in the right-hand loop clockwise or counterclockwise?

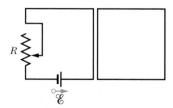

FIGURE 31-33
Question 8.

9. Figure 31-34a shows a circular region in which an increasing uniform magnetic field is directed out of the page, as well as a concentric circular path along which $\oint \mathbf{E} \cdot d\mathbf{s}$ is to be evaluated. The table gives the initial magnitude of the magnetic field, the increase in that magnitude, and the time interval for the increase, in three situations. Rank the situations according to the magnitude of the electric field induced along the path, greatest first.

SITUATION	INITIAL FIELD	INCREASE	TIME
a	B_1	ΔB_1	Δt_1
b	$2B_1$	$\Delta B_1/2$	Δt_1
c	$B_1/4$	ΔB_1	$\Delta t_1/2$

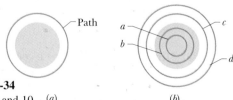

FIGURE 31-34
Questions 9 and 10.

10. Figure 31-34b shows a circular region in which a decreasing uniform magnetic field is directed out of the page, as well as four concentric circular paths. Rank the paths according to the magnitude of $\oint \mathbf{E} \cdot d\mathbf{s}$ evaluated along them, greatest first.

11. The number of turns per unit length, current, and cross-sectional area for three solenoids of the same length are given in the following table. Rank the solenoids according to (a) their inductance and (b) the flux through each turn, greatest first.

SOLENOID	TURNS PER UNIT LENGTH	CURRENT	AREA
1	$2n_1$	i_1	$2A_1$
2	n_1	$2i_1$	A_1
3	n_1	i_1	$4A_1$

12. Figure 31-35 gives the variation with time of the potential difference V_R across a resistor in three circuits wired as in Fig. 31-20. The circuits contain the same resistance R and emf $\mathscr{E}$ but differ in the inductance L. Rank the circuits according to the value of L, greatest first.

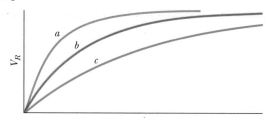

FIGURE 31-35   Question 12.

13. Here are three sets of values for the emf $\mathscr{E}$ of the battery and the potential difference V_R across the resistor in the circuit of Fig. 31-20 at some time after the current begins to increase: (a) 12 V and 3 V; (b) 24 V and 16 V; (c) 18 V and 10 V. Rank the sets according to the potential difference across the inductor at those times, greatest first.

14. Figure 31-36 shows three circuits with identical batteries, inductors, and resistors. Rank the circuits according to the time for the current to reach 50% of its equilibrium value after the switches are closed, greatest first.

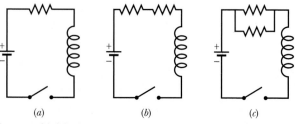

FIGURE 31-36   Question 14.

15. Figure 31-37 shows a circuit with two identical resistors and an inductor. Is the current through the central resistor more than, less than, or the same as that through the other resistor (a) just after the closing of switch S, (b) a long time after the closing of S,

(c) just after the reopening of S, a long time later, and (d) a long time after the reopening of S?

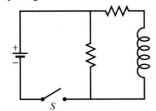

FIGURE 31-37
Question 15.

16. Figure 31-38 shows three circuits with identical batteries, inductors, and resistors. Rank the circuits, greatest first, according to the current through the resistor labeled R (a) long after the switch is closed, (b) just after the switch is reopened a long time later, and (c) long after it is reopened.

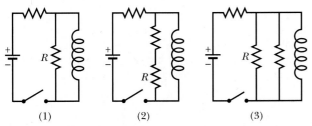

FIGURE 31-38 Question 16.

17. The switch in the circuit of Fig. 31-19 has been closed on a for a long time when it is then thrown to b. The resulting current through the inductor is sketched in Fig. 31-39 for four sets of values for the resistance R and inductance L: (1) R_0 and L_0; (2) $2R_0$ and L_0; (3) R_0 and $2L_0$; (4) $2R_0$ and $2L_0$. Which set goes with which curve?

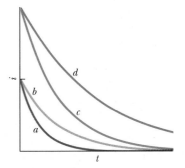

FIGURE 31-39 Question 17.

18. In Fig. 31-24, the current in coil 1 is given, in three situations, by (1) $i_1 = 3\cos(4t)$, (2) $i_1 = 10\cos(t)$, and (3) $i_1 = 5\cos(2t)$, with i_1 in amperes and t in seconds. For the three situations, rank (a) the mutual inductance of the coils and (b) the magnitude of the maximum emf appearing in coil 2 due to i_1, greatest first.

EXERCISES & PROBLEMS

SECTION 31-4 Lenz's Law

1E. At a certain location in the southern hemisphere, Earth's magnetic field has a magnitude of 42 μT and points upward at 57° to the vertical. Calculate the flux through a horizontal surface of area 2.5 m^2; see Fig. 31-40, in which area vector **A** has arbitrarily been chosen to be upward.

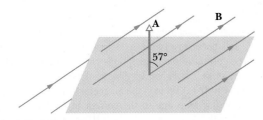

FIGURE 31-40 Exercise 1.

2E. A small loop of area A is inside of, and has its axis in the same direction as, a long solenoid of n turns per unit length and current i. If $i = i_0 \sin \omega t$, find the emf in the loop.

3E. A UHF television loop antenna has a diameter of 11 cm. The magnetic field of a TV signal is normal to the plane of the loop and, at one instant of time, its magnitude is changing at the rate 0.16 T/s. The field is uniform. What is the emf in the antenna?

4E. A uniform magnetic field **B** is perpendicular to the plane of a circular wire loop of radius r. The magnitude of the field varies with time according to $B = B_0 e^{-t/\tau}$, where B_0 and τ are constants. Find the emf in the loop as a function of time.

5E. The magnetic flux through the loop shown in Fig. 31-41 increases according to the relation $\Phi_B = 6.0t^2 + 7.0t$, where Φ_B is in milliwebers and t is in seconds. (a) What is the magnitude of the emf induced in the loop when $t = 2.0$ s? (b) What is the direction of the current through R?

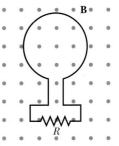

FIGURE 31-41 Exercise 5 and Problem 19.

6E. The magnetic field through a single loop of wire 12 cm in radius and of 8.5 Ω resistance changes with time as shown in Fig. 31-42. Calculate the emf in the loop as a function of time. Consider the time intervals (a) $t = 0$ to $t = 2.0$ s; (b) $t = 2.0$ s to $t = 4.0$ s; (c) $t = 4.0$ s to $t = 6.0$ s. The (uniform) magnetic field is perpendicular to the plane of the loop.

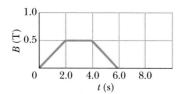

FIGURE 31-42
Exercise 6.

7E. A 20 mΩ square wire loop 20 cm on a side has its plane normal to a uniform magnetic field of magnitude $B = 2.0$ T. If you pull two opposite sides of the loop away from each other, the other two sides automatically draw toward each other, reducing the area enclosed by the loop. If the area is reduced to zero in time $\Delta t = 0.20$ s, what are (a) the average emf and (b) the average current induced in the loop during Δt?

8E. A uniform magnetic field is normal to the plane of a circular loop 10 cm in diameter and made of copper wire (of diameter 2.5 mm). (a) Calculate the resistance of the wire. (See Table 27-1.) (b) At what rate must the magnetic field change with time if an induced current of 10 A is to appear in the loop?

9P. The current in the solenoid of Sample Problem 31-1 changes, not as stated there, but according to $i = 3.0t + 1.0t^2$, where i is in amperes and t is in seconds. (a) Plot the induced emf in the coil from $t = 0$ to $t = 4.0$ s. (b) The resistance of the coil is 0.15 Ω. What is the current in the coil at $t = 2.0$ s?

10P. In Fig. 31-43 a 120-turn coil of radius 1.8 cm and resistance 5.3 Ω is placed *outside* a solenoid like that of Sample Problem 31-1. If the current in the solenoid is changed as in that sample problem, what current appears in the coil while the solenoid current is being changed?

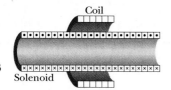

FIGURE 31-43
Problem 10.

11P. A long solenoid with a radius of 25 mm has 100 turns/cm. A single loop of wire of radius 5.0 cm is placed around the solenoid, the central axes of the loop and the solenoid coinciding. In 10 ms the current in the solenoid is reduced from 1.0 A to 0.50 A at a uniform rate. What emf appears in the loop?

12P. Derive an expression for the flux through a toroid of N turns carrying a current i. Assume that the windings have a rectangular cross section of inner radius a, outer radius b, and height h.

13P. A toroid having a 5.00 cm square cross section and an inside radius of 15.0 cm has 500 turns of wire and carries a current of 0.800 A. What is the magnetic flux through the cross section?

14P. An elastic conducting material is stretched into a circular loop of 12.0 cm radius. It is placed with its plane perpendicular to a uniform 0.800 T magnetic field. When released, the radius of the loop starts to shrink at an instantaneous rate of 75.0 cm/s. What emf is induced in the loop at that instant?

15P. A closed loop of wire consists of a pair of equal semicircles, of radius 3.7 cm, lying in mutually perpendicular planes. The loop was formed by folding a plane circular loop along a diameter until the two halves became perpendicular. A uniform magnetic field **B** of magnitude 76 mT is directed perpendicular to the fold diameter and makes equal angles ($= 45°$) with the planes of the semicircles as shown in Fig. 31-44. The magnetic field is reduced to zero at a uniform rate during a time interval of 4.5 ms. Determine the magnitude of the induced emf and the direction of the induced current in the loop during this interval.

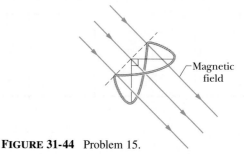

FIGURE 31-44 Problem 15.

16P. In Fig. 31-45, a circular loop of wire 10 cm in diameter (seen edge-on) is placed with its normal **N** making an angle $\theta = 30°$ with the direction of a uniform magnetic field **B** of magnitude 0.50 T. The loop is then rotated such that **N** rotates in a cone about the field direction at the constant rate of 100 rev/min; the angle θ remains unchanged during the process. What is the emf induced in the loop?

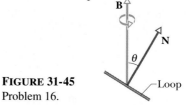

FIGURE 31-45
Problem 16.

17P. A small circular loop of area 2.00 cm^2 is placed in the plane of, and concentric with, a large circular loop of radius 1.00 m. The current in the large loop is changed uniformly from 200 A to -200 A (a change in direction) in a time of 1.00 s, beginning at $t = 0$. (a) What is the magnetic field at the center of the small circular loop due to the current in the large loop at $t = 0$, $t = 0.500$ s, and $t = 1.00$ s? (b) What emf is induced in the small loop at $t = 0.500$ s? (Since the inner loop is small, assume the field **B** due to the outer loop is uniform over the area of the smaller loop.)

18P. Figure 31-46 shows two parallel loops of wire having a common axis. The smaller loop (radius r) is above the larger loop (radius R), by a distance $x \gg R$. Consequently the magnetic field due to the current i in the larger loop is nearly constant throughout the smaller loop. Suppose that x is increasing at the constant rate $dx/dt = v$. (a) Determine the magnetic flux through the area bounded by the smaller loop as a function of *x*. (*Hint:* See Eq. 30-29.) In the smaller loop, find (b) the induced emf and (c) the direction of the induced current.

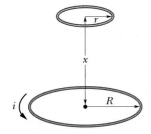

FIGURE 31-46
Problem 18.

19P. In Fig. 31-41 let the flux through the loop be $\Phi_B(0)$ at time $t = 0$. Then let the magnetic field **B** vary in a continuous but unspecified way, in both magnitude and direction, so that at time *t* the flux is represented by $\Phi_B(t)$. (a) Show that the net charge $q(t)$ that has passed through resistor *R* in time *t* is

$$q(t) = \frac{1}{R}[\Phi_B(0) - \Phi_B(t)]$$

and is independent of the way **B** has changed. (b) If $\Phi_B(t) = \Phi_B(0)$ in a particular case, we have $q(t) = 0$. Is the induced current necessarily zero throughout the interval from 0 to *t*?

20P. One hundred turns of insulated copper wire are wrapped around a wooden cylindrical core of cross-sectional area 1.20×10^{-3} m^2. The two terminals are connected to a resistor. The total resistance in the circuit is 13.0 Ω. If an externally applied uniform longitudinal magnetic field in the core changes from 1.60 T in one direction to 1.60 T in the opposite direction, how much charge flows through the circuit? (*Hint:* See Problem 19.)

21P. At a certain place, Earth's magnetic field has magnitude $B = 0.590$ gauss and is inclined downward at an angle of 70.0° to the horizontal. A flat horizontal circular coil of wire with a radius of 10.0 cm has 1000 turns and a total resistance of 85.0 Ω. It is connected to a meter with 140 Ω resistance. The coil is flipped through a half-revolution about a diameter, so that it is again horizontal. How much charge flows through the meter during the flip? (*Hint:* See Problem 19.)

22P. A square wire loop with 2.00 m sides is perpendicular to a uniform magnetic field, with half the area of the loop in the field, as shown in Fig. 31-47. The loop contains a 20.0 V battery with negligible internal resistance. If the magnitude of the field varies with time according to $B = 0.0420 - 0.870t$, with *B* in teslas and *t* in seconds, (a) what is the total emf in the circuit? (b) What is the direction of the current through the battery?

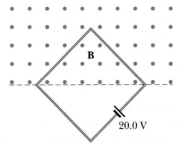

FIGURE 31-47
Problem 22.

23P. A wire is bent into three circular segments, each of radius $r = 10$ cm, as shown in Fig. 31-48. Each segment is a quadrant of a circle, *ab* lying in the *xy* plane, *bc* lying in the *yz* plane, and *ca* lying in the *zx* plane. (a) If a uniform magnetic field **B** points in the positive *x* direction, what is the magnitude of the emf developed in the wire when *B* increases at the rate of 3.0 mT/s? (b) What is the direction of the current in segment *bc*?

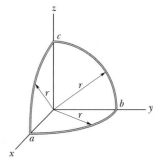

FIGURE 31-48
Problem 23.

24P. A stiff wire bent into a semicircle of radius *a* is rotated with frequency *f* in a uniform magnetic field, as suggested in Fig. 31-49. What are (a) the frequency and (b) the amplitude of the varying emf induced in the loop?

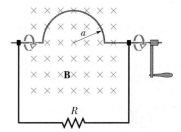

FIGURE 31-49
Problem 24.

25P. A rectangular coil of *N* turns and of length *a* and width *b* is rotated at frequency *f* in a uniform magnetic field **B**, as indicated in Fig. 31-50. The coil is connected to co-rotating cylinders, against which metal brushes slide to make contact. (a) Show that the emf induced in the coil is given (as a function of time *t*) by

$$\mathscr{E} = 2\pi f NabB \sin(2\pi ft) = \mathscr{E}_0 \sin(2\pi ft).$$

This is the principle of the commercial alternating-current generator. (b) Design a loop that will produce an emf with $\mathscr{E}_0 = 150$ V when rotated at 60.0 rev/s in a magnetic field of 0.500 T.

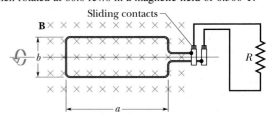

FIGURE 31-50 Problem 25.

26P. An electric generator consists of 100 turns of wire formed into a rectangular loop 50.0 cm by 30.0 cm, placed entirely in a uniform magnetic field with magnitude $B = 3.50$ T. What is the maximum value of the emf produced when the loop is spun at 1000 rev/min about an axis perpendicular to **B**?

27P. For the situation shown in Fig. 31-51, $a = 12.0$ cm and $b = 16.0$ cm. The current in the long straight wire is given by $i = 4.50t^2 - 10.0t$, where i is in amperes and t is in seconds. (a) Find the emf in the square loop at $t = 3.00$ s. (b) What is the direction of the induced current in the loop?

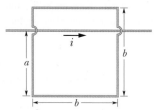

FIGURE 31-51
Problem 27.

28P. In Fig. 31-52, the square loop of wire has sides of length 2.0 cm. A magnetic field points out of the page; its magnitude is given by $B = 4.0t^2y$, where B is in teslas, t is in seconds, and y is in meters. Determine the emf around the square at $t = 2.5$ s and give its direction.

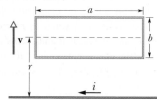

FIGURE 31-52
Problem 28.

29P. A rectangular loop of wire with length a, width b, and resistance R is placed near an infinitely long wire carrying current i, as shown in Fig. 31-53. The distance from the long wire to the center of the loop is r. Find (a) the magnitude of the magnetic flux through the loop and (b) the current in the loop as it moves away from the long wire with speed v.

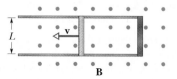

FIGURE 31-53
Problem 29.

30P*. Two long, parallel copper wires (of diameter 2.5 mm) carry currents of 10 A in opposite directions. (a) Assuming that their centers are 20 mm apart, calculate the magnetic flux per meter of wire that exists in the space between the axes of the wires. (b) What fraction of this flux lies inside the wires? (c) Repeat part (a) for parallel currents.

SECTION 31-5 Induction and Energy Transfers

31E. A loop antenna of area A and resistance R is perpendicular to a uniform magnetic field **B.** The field drops linearly to zero in a time interval Δt. Find an expression for the total thermal energy dissipated in the loop.

32E. If 50.0 cm of copper wire (diameter = 1.00 mm) is formed into a circular loop and placed perpendicular to a uniform mag-

netic field that is increasing at the constant rate of 10.0 mT/s, at what rate is thermal energy generated in the loop?

33E. A metal rod is forced to move with constant velocity **v** along two parallel metal rails, connected with a strip of metal at one end, as shown in Fig. 31-54. A magnetic field $B = 0.350$ T points out of the page. (a) If the rails are separated by 25.0 cm and the speed of the rod is 55.0 cm/s, what emf is generated? (b) If the rod has a resistance of 18.0 Ω and the rails and connector have negligible resistance, what is the current in the rod? (c) At what rate is energy being transferred to thermal energy?

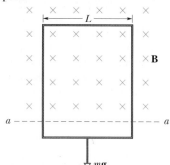

FIGURE 31-54 Exercises 33 and 34.

34E. The conducting rod shown in Fig. 31-54 has a length L and is being pulled along horizontal, frictionless, conducting rails at a constant velocity **v**. The rails are connected at one end with a metal strip. A uniform magnetic field **B**, directed out of the page, fills the region in which the rod moves. Assume that $L = 10$ cm, $v = 5.0$ m/s, and $B = 1.2$ T. (a) What is the induced emf in the rod? (b) What is the current in the conducting loop? Assume that the resistance of the rod is 0.40 Ω and that the resistance of the rails and metal strip is negligibly small. (c) At what rate is thermal energy being generated in the rod? (d) What force must be applied to the rod by an external agent to maintain its motion? (e) At what rate does this external agent do work on the rod? Compare this answer with the answer to (c).

35P. In Fig. 31-55, a long rectangular conducting loop, of width L, resistance R, and mass m, is hung in a horizontal, uniform magnetic field **B** that is directed into the page and that exists only above line aa. The loop is then dropped; during its fall, it accelerates until it reaches a certain terminal speed v_t. Ignoring air drag, find that terminal speed.

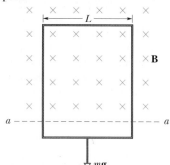

FIGURE 31-55
Problem 35.

36P. Two straight conducting rails form a right angle where their ends are joined. A conducting bar in contact with the rails starts at the vertex at time $t = 0$ and moves with a constant velocity of 5.20 m/s along them, as shown in Fig. 31-56. A 0.350 T magnetic field points out of the page. Calculate (a) the flux through the triangle formed by the rails and bar at $t = 3.00$ s and (b) the emf around the triangle at that time. (c) If we write the emf as $\mathscr{E} = at^n$, where a and n are constants, what is the value of n?

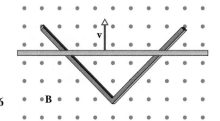

FIGURE 31-56
Problem 36.

37P. Calculate the average power supplied by the generator of Problem 25b if it is connected to a circuit of 42.0 Ω resistance. (*Hint:* The average value of $\sin^2 \theta$ over one cycle is $\frac{1}{2}$.)

38P. In Fig. 31-57 a conducting rod of mass m and length L slides without friction on two long horizontal rails. A uniform vertical magnetic field **B** fills the region in which the rod is free to move. The generator G supplies a constant current i directed as shown. (a) Find the velocity of the rod as a function of time, assuming it to be at rest at $t = 0$. The generator is now replaced by a battery that supplies a constant emf $\mathscr{E}$. (b) Show that the velocity of the rod now approaches a constant terminal value **v** and give its magnitude and direction. (c) What is the current in the rod when this terminal velocity is reached? (d) Analyze this situation and that with the generator from the point of view of energy transfers.

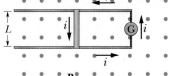

FIGURE 31-57
Problem 38.

39P. Figure 31-58 shows a rod of length L caused to move at constant speed v along horizontal conducting rails. In this case the magnetic field in which the rod moves is not uniform but is provided by a current i in a long wire parallel to the rails. Assume that $v = 5.00$ m/s, $a = 10.0$ mm, $L = 10.0$ cm, and $i = 100$ A. (a) Calculate the emf induced in the rod. (b) What is the current in the conducting loop? Assume that the resistance of the rod is 0.400 Ω and that the resistance of the rails and the strip that connects them at the right is negligible. (c) At what rate is thermal energy being generated in the rod? (d) What force must be applied to the rod by an external agent to maintain its motion? (e) At what rate does this external agent do work on the rod? Compare this answer to that for (c).

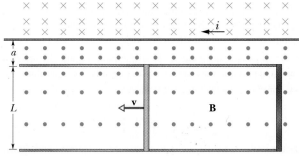

FIGURE 31-58   Problem 39.

SECTION 31-6 Induced Electric Fields

40E. A long solenoid has a diameter of 12.0 cm. When a current i exists in its windings, a uniform magnetic field $B = 30.0$ mT is produced in its interior. By decreasing i, the field is caused to decrease at the rate of 6.50 mT/s. Calculate the magnitude of the induced electric field (a) 2.20 cm and (b) 8.20 cm from the axis of the solenoid.

41E. Figure 31-59 shows two circular regions R_1 and R_2 with radii $r_1 = 20.0$ cm and $r_2 = 30.0$ cm. In R_1 there is a uniform magnetic field $B_1 = 50.0$ mT into the page and in R_2 there is a uniform magnetic field $B_2 = 75.0$ mT out of the page (ignore any fringing of these fields). Both fields are decreasing at the rate of 8.50 mT/s. Calculate the integral $\oint \mathbf{E} \cdot d\mathbf{s}$ for each of the three dashed paths.

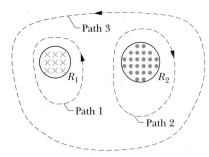

FIGURE 31-59
Exercise 41.

42P. Early in 1981 the Francis Bitter National Magnet Laboratory at M.I.T. commenced operation of a 3.3 cm diameter cylindrical magnet, which produces a 30 T field, then the world's largest steady-state field. The field can be varied sinusoidally between the limits of 29.6 and 30.0 T at a frequency of 15 Hz. When this is done, what is the maximum value of the induced electric field at a radial distance of 1.6 cm from the axis? (*Hint:* See Sample Problem 31-5.)

43P. Figure 31-60 shows a uniform magnetic field **B** confined to a cylindrical volume of radius R. The magnitude of **B** is decreasing at a constant rate of 10 mT/s. What are the instantaneous accelerations (direction and magnitude) experienced by an electron placed at a, at b, and at c? Assume $r = 5.0$ cm.

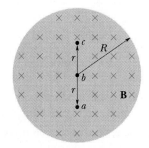

FIGURE 31-60
Problem 43.

44P. Prove that the electric field **E** in a charged parallel-plate capacitor cannot drop abruptly to zero (as is suggested at point a in Fig. 31-61), as one moves perpendicular to the field, say along the horizontal arrow in the figure. Fringing of the field lines always occurs in actual capacitors, which means that **E** approaches zero in a continuous and gradual way (see Problem 45 in

Chapter 30). (*Hint:* Apply Faraday's law to the rectangular path shown by the dashed lines.)

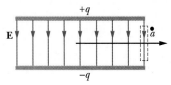

FIGURE 31-61 Problem 44.

SECTION 31-7 Inductors and Inductance

45E. The inductance of a close-packed coil of 400 turns is 8.0 mH. Calculate the magnetic flux through the coil when the current is 5.0 mA.

46E. A circular coil has a 10.0 cm radius and consists of 30.0 closely wound turns of wire. An externally produced magnetic field of 2.60 mT is perpendicular to the coil. (a) If no current is in the coil, what is the flux linkage? (b) When the current in the coil is 3.80 A in a certain direction, the net flux through the coil is found to vanish. What is the inductance of the coil?

47E. A solenoid is wound with a single layer of insulated copper wire (of diameter 2.5 mm) and is 4.0 cm in diameter and 2.0 m long. (a) How many turns are on the solenoid? (b) What is the inductance per meter for the solenoid near its center? Assume that adjacent wires touch and that insulation thickness is negligible.

48P. A long, thin solenoid can be bent into a ring to form a toroid. Show that if the solenoid is long and thin enough, the equation for the inductance of a toroid (Eq. 31-37) is equivalent to that for a solenoid of the appropriate length (Eq. 31-32).

49P. A wide copper strip of width W is bent to form a tube of radius R with two planar extensions, as shown in Fig. 31-62. There is a current i through the strip, distributed uniformly over its width. In this way a ''one-turn solenoid'' is formed. (a) Derive an expression for the magnitude of the magnetic field **B** in the tubular part (far away from the edges). (*Hint:* Assume that the magnetic field outside this one-turn solenoid is negligibly small.) (b) Find the inductance of this one-turn solenoid, neglecting the two planar extensions.

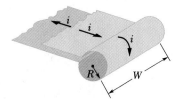

FIGURE 31-62
Problem 49.

50P. Two long parallel wires, each of radius a, whose centers are a distance d apart, carry equal currents in opposite directions. Show that, neglecting the flux within the wires, the inductance of a length l of such a pair of wires is given by

$$L = \frac{\mu_0 l}{\pi} \ln \frac{d-a}{a}.$$

See Sample Problem 30-2. (*Hint:* Calculate the flux through a rectangle of which the wires form two opposite sides.)

SECTION 31-8 Self-Induction

51E. At a given instant the current and self-induced emf in an inductor are as indicated in Fig. 31-63. (a) Is the current increasing or decreasing? (b) The induced emf is 17 V and the rate of change of the current is 25 kA/s; find the inductance.

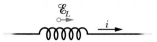

FIGURE 31-63 Exercise 51.

52E. A 12 H inductor carries a steady current of 2.0 A. How can a 60 V self-induced emf be made to appear in the inductor?

53E. A long cylindrical solenoid with 100 turns/cm has a radius of 1.6 cm. Assume that the magnetic field it produces is parallel to its axis and is uniform in its interior. (a) What is its inductance per meter of length? (b) If the current changes at the rate 13 A/s, what emf is induced per meter?

54E. The inductance of a closely wound coil is such that an emf of 3.0 mV is induced when the current changes at the rate of 5.0 A/s. A steady current of 8.0 A produces a magnetic flux of 40 μWb through each turn. (a) Calculate the inductance of the coil. (b) How many turns does the coil have?

55P. The current i through a 4.6 H inductor varies with time t as shown by the graph of Fig. 31-64. The inductor has a resistance of 12 Ω. Find the magnitude of the induced emf $\mathscr{E}$ during the time intervals (a) $t = 0$ to $t = 2$ ms; (b) $t = 2$ ms to $t = 5$ ms; (c) $t = 5$ ms to $t = 6$ ms. (Ignore the behavior at the ends of the intervals.)

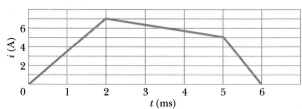

FIGURE 31-64 Problem 55.

56P. *Inductors in series.* Two inductors L_1 and L_2 are connected in series and are separated by a large distance. (a) Show that the equivalent inductance is given by

$$L_{eq} = L_1 + L_2.$$

(*Hint:* Review the derivations for resistors in series and capacitors in series. Which is similar here?) (b) Why must their separation be large for this relationship to hold? (c) What is the generalization of (a) for N inductors in series?

57P. *Inductors in parallel.* Two inductors L_1 and L_2 are connected in parallel and separated by a large distance. (a) Show that the equivalent inductance is given by

$$\frac{1}{L_{eq}} = \frac{1}{L_1} + \frac{1}{L_2}.$$

(*Hint:* Review the derivations for resistors in parallel and capacitors in parallel. Which is similar here?) (b) Why must their separation be large for this relationship to hold? (c) What is the generalization of (a) for N inductors in parallel?

SECTION 31-9 *RL* Circuits

58E. The current in an *RL* circuit builds up to one-third of its steady-state value in 5.00 s. Find the inductive time constant.

59E. In terms of τ_L, how long must we wait for the current in an *RL* circuit to build up to within 0.100% of its equilibrium value?

60E. The current in an *RL* circuit drops from 1.0 A to 10 mA in the first second following removal of the battery from the circuit. If L is 10 H, find the resistance R in the circuit.

61E. How long would it take, following the removal of the battery, for the potential difference across the resistor in an *RL* circuit (with $L = 2.00$ H, $R = 3.00\ \Omega$) to decay to 10.0% of its initial value?

62E. (a) Consider the *RL* circuit of Fig. 31-19. In terms of the battery emf $\mathscr{E}$, what is the self-induced emf $\mathscr{E}_L$ when the switch has just been closed on a? (b) What is $\mathscr{E}_L$ when $t = 2.0\tau_L$? (c) In terms of τ_L, when will $\mathscr{E}_L$ be just one-half the battery emf $\mathscr{E}$?

63E. A solenoid having an inductance of 6.30 μH is connected in series with a 1.20 kΩ resistor. (a) If a 14.0 V battery is switched across the pair, how long will it take for the current through the resistor to reach 80.0% of its final value? (b) What is the current through the resistor at time $t = 1.0\tau_L$?

64E. The flux linkage through a certain coil of 0.75 Ω resistance is 26 mWb when there is a current of 5.5 A in it. (a) Calculate the inductance of the coil. (b) If a 6.0 V battery is suddenly connected across the coil, how long will it take for the current to rise from 0 to 2.5 A?

65P. Suppose the emf of the battery in the circuit of Fig. 31-20 varies with time t so that the current is given by $i(t) = 3.0 + 5.0t$, where i is in amperes and t is in seconds. Take $R = 4.0\ \Omega$, $L = 6.0$ H, and find an expression for the battery emf as a function of time. (*Hint:* Apply the loop rule.)

66P. At $t = 0$ a battery is connected to an inductor and a resistor that are connected in series. The table below gives the measured potential difference across the inductor as a function of time following the connection of the battery. Find (a) the emf of the battery and (b) the time constant of the circuit.

t (ms)	V_L (V)	t (ms)	V_L (V)
1.0	18.2	5.0	5.98
2.0	13.8	6.0	4.53
3.0	10.4	7.0	3.43
4.0	7.90	8.0	2.60

67P. A 45.0 V potential difference is suddenly applied to a coil with $L = 50.0$ mH and $R = 180\ \Omega$. At what rate is the current increasing 1.20 ms later?

68P. A wooden toroidal core with a square cross section has an inner radius of 10 cm and an outer radius of 12 cm. It is wound with one layer of wire (of diameter 1.0 mm and resistance per

meter 0.02 Ω/m). What are (a) the inductance and (b) the inductive time constant of the toroid? Ignore the thickness of the insulation on the wire.

69P. In Fig. 31-65, $\mathscr{E} = 100$ V, $R_1 = 10.0\ \Omega$, $R_2 = 20.0\ \Omega$, $R_3 = 30.0\ \Omega$, and $L = 2.00$ H. Find the values of i_1 and i_2 (a) immediately after the closing of switch S, (b) a long time later, (c) immediately after the reopening of switch S, and (d) a long time after the reopening.

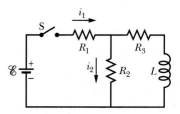

FIGURE 31-65 Problem 69.

70P. In the circuit shown in Fig. 31-66, $\mathscr{E} = 10$ V, $R_1 = 5.0\ \Omega$, $R_2 = 10\ \Omega$, and $L = 5.0$ H. For the two separate conditions (I) switch S just closed and (II) switch S closed for a long time, calculate (a) the current i_1 through R_1, (b) the current i_2 through R_2, (c) the current i through the switch, (d) the potential difference across R_2, (e) the potential difference across L, and (f) the rate of change di_2/dt.

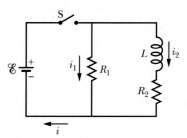

FIGURE 31-66 Problem 70.

71P. Switch S in Fig. 31-67 is closed for $t < 0$ and is opened at $t = 0$. When current i_1 through L_1 and current i_2 through L_2 are *first* equal to each other, what is their common value? (The resistors have the same resistance R.)

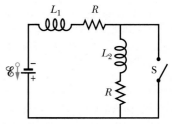

FIGURE 31-67 Problem 71.

72P. In Fig. 31-68, the component in the upper branch is an ideal 3.0 A fuse. It has zero resistance as long as the current through it remains less than 3.0 A. If the current reaches 3.0 A, it "blows" and thereafter has infinite resistance. Switch S is closed at time $t = 0$. (a) When does the fuse blow? (*Hint:* Equation 31-46 does not apply. Rethink Eq. 31-44.) (b) Sketch a graph of the current i

through the inductor as a function of time. Mark the time at which the fuse blows.

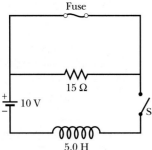

FIGURE 31-68
Problem 72.

73P*. In the circuit shown in Fig. 31-69, switch S is closed at time $t = 0$. Thereafter the constant current source, by varying its emf, maintains a constant current i out of its upper terminal. (a) Derive an expression for the current through the inductor as a function of time. (b) Show that the current through the resistor equals the current through the inductor at time $t = (L/R)\ln 2$.

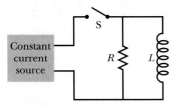

FIGURE 31-69 Problem 73.

SECTION 31-10 Energy Stored in a Magnetic Field

74E. The magnetic energy stored in a certain inductor is 25.0 mJ when the current is 60.0 mA. (a) Calculate the inductance. (b) What current is required for the stored magnetic energy to be four times as much?

75E. Consider the circuit of Fig. 31-20. In terms of the time constant, at what instant after the battery is connected will the energy stored in the magnetic field of the inductor be half its steady-state value?

76E. A coil with an inductance of 2.0 H and a resistance of 10 Ω is suddenly connected to a resistanceless battery with $\mathscr{E} = 100$ V. (a) What is the equilibrium current? (b) How much energy is stored in the magnetic field when this current exists in the coil?

77E. A coil with an inductance of 2.0 H and a resistance of 10 Ω is suddenly connected to a resistanceless battery with $\mathscr{E} = 100$ V. At 0.10 s after the connection is made, what are the rates at which (a) energy is being stored in the magnetic field, (b) thermal energy is appearing in the resistance, and (c) energy is being delivered by the battery?

78P. Suppose that the inductive time constant for the circuit of Fig. 31-20 is 37.0 ms and the current in the circuit is zero at time $t = 0$. At what time does the rate at which energy is dissipated in the resistor equal the rate at which energy is being stored in the inductor?

79P. A coil is connected in series with a 10.0 kΩ resistor. When a 50.0 V battery is applied to the two, the current reaches a value of 2.00 mA after 5.00 ms. (a) Find the inductance of the coil. (b) How much energy is stored in the coil at this same moment?

80P. For the circuit of Fig. 31-20, assume that $\mathscr{E} = 10.0$ V, $R = 6.70$ Ω, and $L = 5.50$ H. The battery is connected at time $t = 0$. (a) How much energy is delivered by the battery during the first 2.00 s? (b) How much of this energy is stored in the magnetic field of the inductor? (c) How much of this energy has been dissipated in the resistor?

81P. A solenoid, with length 80.0 cm and radius 5.00 cm, consists of 3000 turns distributed uniformly over its length. Its total resistance is 10.0 Ω. At 5.00 ms after it is connected to a 12.0 V battery, (a) how much energy is stored in its magnetic field and (b) how much energy has been supplied by the battery up to that time? (Neglect end effects.)

82P. Prove that, after switch S in Fig. 31-19 has been thrown from a to b, all the energy stored in the inductor will ultimately appear as thermal energy in the resistor.

SECTION 31-11 Energy Density of a Magnetic Field

83E. A solenoid 85.0 cm long has a cross-sectional area of 17.0 cm^2. There are 950 turns of wire carrying a current of 6.60 A. (a) Calculate the energy density of the magnetic field inside the solenoid. (b) Find the total energy stored in the magnetic field there (neglect end effects).

84E. A toroidal inductor with an inductance of 90.0 mH encloses a volume of 0.0200 m^3. If the average energy density in the toroid is 70.0 J/m^3, what is the current?

85E. What must be the magnitude of a uniform electric field if it is to have the same energy density as that possessed by a 0.50 T magnetic field?

86E. The magnetic field in the interstellar space of our galaxy has a magnitude of about 10^{-10} T. How much energy is stored in this field in a cube 10 light years on edge? (For scale, note that the nearest star is 4.3 light-years distant and the radius of our galaxy is about 8×10^4 light-years.)

87E. Use the result of Sample Problem 31-11 to obtain an expression for the inductance of a length l of the coaxial cable.

88E. Calculate the energy needed to produce, in a cube that is 10 cm on edge, (a) a uniform electric field of 100 kV/m and (b) a uniform magnetic field of 1.0 T. (Both these large fields are readily available in the laboratory.) (c) From these answers, tell which type of field can store greater amounts of energy.

89E. A circular loop of wire 50 mm in radius carries a current of 100 A. (a) Find the magnetic field strength at the center of the loop. (b) Calculate the energy density at the center of the loop.

90P. (a) For the toroid of Sample Problem 31-6b, find an expression for the energy density as a function of the radial distance r from the center. (b) By integrating the energy density over the volume of the toroid, calculate the total energy stored in the field of the toroid; assume $i = 0.500$ A. (c) Using Eq. 31-53, evaluate the energy stored in the toroid directly from the inductance and compare it with your answer in (b).

91P. A length of copper wire carries a current of 10 A, uniformly distributed. Calculate (a) the energy density of the magnetic field and (b) the energy density of the electric field at the surface of the wire. The wire diameter is 2.5 mm, and its resistance per unit length is 3.3 Ω/km.

92P. (a) What is the energy density of Earth's magnetic field, which has a magnitude of 50 μT? (b) Assuming this density to be relatively constant over distances small compared with Earth's radius and neglecting variations near the magnetic poles, how much energy would be stored between Earth's surface and a spherical shell 16 km above the surface?

SECTION 31-12   Mutual Induction

93E. Two coils are at fixed locations. When coil 1 has no current and the current in coil 2 increases at the rate 15.0 A/s, the emf in coil 1 is 25.0 mV. (a) What is their mutual inductance? (b) When coil 2 has no current and coil 1 has a current of 3.60 A, what is the flux linkage in coil 2?

94E. Coil 1 has $L_1 = 25$ mH and $N_1 = 100$ turns. Coil 2 has $L_2 = 40$ mH and $N_2 = 200$ turns. The coils are rigidly positioned with respect to each other; their mutual inductance M is 3.0 mH. A 6.0 mA current in coil 1 is changing at the rate of 4.0 A/s. (a) What flux Φ_{12} links coil 1, and what self-induced emf appears there? (b) What flux Φ_{21} links coil 2, and what mutually induced emf appears there?

95E. Two solenoids are part of the spark coil of an automobile. When the current in one solenoid falls from 6.0 A to zero in 2.5 ms, an emf of 30 kV is induced in the other solenoid. What is the mutual inductance M of the solenoids?

96P. Two coils, connected as shown in Fig. 31-70, separately have inductances L_1 and L_2. The mutual inductance is M. (a) Show that this combination can be replaced by a single coil of equivalent inductance given by

$$L_{eq} = L_1 + L_2 + 2M.$$

(b) How could the coils in Fig. 31-70 be reconnected to yield an equivalent inductance of

$$L_{eq} = L_1 + L_2 - 2M?$$

(This problem is an extension of Problem 56, but the requirement that the coils be far apart has been removed.)

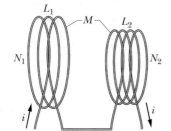

FIGURE 31-70
Problem 96.

97P. A coil C of N turns is placed around a long solenoid S of radius R and n turns per unit length, as in Fig. 31-71. Show that the mutual inductance for the coil–solenoid combination is given

by $M = \mu_0 \pi R^2 nN$. Explain why M does not depend on the shape, size, or possible lack of close-packing of the coil.

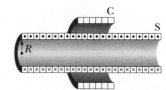

FIGURE 31-71
Problem 97.

98P. Figure 31-72 shows a coil of N_2 turns wound as shown around part of a toroid of N_1 turns. The toroid's inner radius is a, its outer radius is b, and its height is h. Show that the mutual inductance M for the toroid–coil combination is

$$M = \frac{\mu_0 N_1 N_2 h}{2\pi} \ln \frac{b}{a}.$$

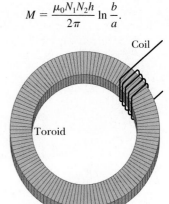

FIGURE 31-72
Problem 98.

99P. Figure 31-73 shows, in cross section, two coaxial solenoids. Show that the mutual inductance M for a length l of this solenoid–solenoid combination is given by $M = \pi R_1^2 l \mu_0 n_1 n_2$, in which n_1 and n_2 are the respective numbers of turns per unit length and R_1 is the radius of the inner solenoid. Why does M depend on R_1 and not on R_2?

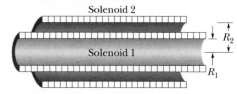

FIGURE 31-73   Problem 99.

100P. A rectangular loop of N close-packed turns is positioned near a long straight wire as in Fig. 31-74. (a) What is the mutual inductance M for the loop–wire combination? (b) Evaluate M for $N = 100$, $a = 1.0$ cm, $b = 8.0$ cm, and $l = 30$ cm.

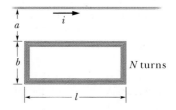

FIGURE 31-74
Problem 100.

32
Magnetism of Matter; Maxwell's Equations

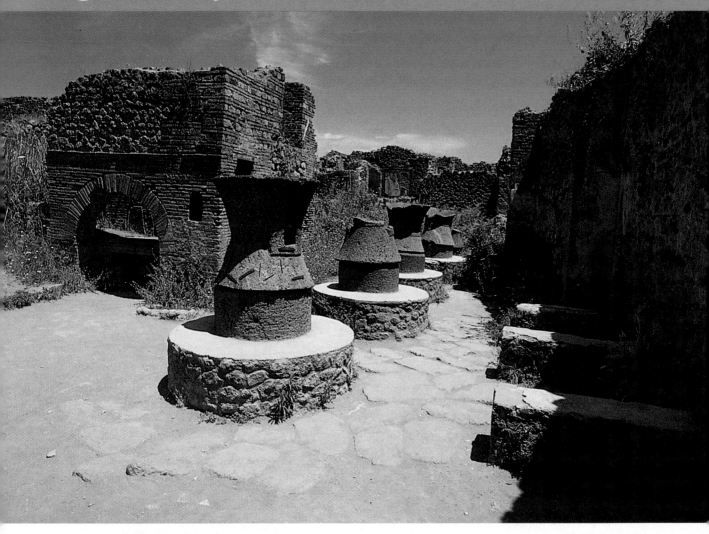

The direction of Earth's magnetic field is not fixed; rather, it gradually wanders. One way we can determine the direction of the field during some period in the past is to examine a clay-walled kiln that was used to bake pottery at that time. But how and why would a clay kiln record Earth's magnetic field?

32-1 MAGNETS

The first known magnets were *lodestones*, which are stones that have been *magnetized* (made magnetic) naturally. When the ancient Greeks and ancient Chinese discovered these rare stones, they were amused by the stones' ability to attract metal over a short distance, as if by magic. Only much later did they learn to use lodestones (and artificially magnetized pieces of iron) in compasses to determine direction.

Today, magnets and magnetic materials are ubiquitous. We find them in VCRs, audio cassettes, ATM and credit cards, audio headsets, and even in the inks for paper money. In fact, some breakfast cereals that are "iron fortified" contain small bits of magnetic materials (you can collect them from a slurry of cereal and water with a magnet). More important, the modern electronics industry as we know it (including the music and information domains) would not exist without magnetic materials.

The magnetic properties of materials can be traced back to their atoms and electrons. We begin here, however, with the bar magnet in Fig. 32-1. As you have seen, iron filings sprinkled around such a magnet tend to align with the magnetic field of the magnet, and their pattern reveals the magnetic field lines. The clustering of the lines at the ends of the magnet suggests that one end is a *source* of the lines (the field diverges from it) and the other end is a *sink* of the lines (the field converges toward it). By convention, we call the source the *north pole* of the magnet and the opposite end the *south pole,* and we say that the magnet, with its two poles, is an example of a **magnetic dipole.**

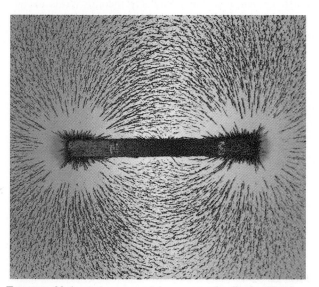

FIGURE 32-1 A bar magnet is a magnetic dipole. The iron filings suggest the magnetic field lines. (The background is illuminated with colored light.)

FIGURE 32-2 If you break a magnet, each fragment becomes a separate magnet, with its own north and south poles.

Suppose we break apart a bar magnet the way we break a piece of chalk (Fig. 32-2). We should, it seems, be able to isolate a single pole, or *monopole.* But surprisingly we cannot, not even if we break the magnet down to its individual atoms and then to its electrons and nuclei. Each fragment of the magnet has a north pole and a south pole. So we must conclude the following:

> The simplest magnetic structure that can exist is a magnetic dipole. Magnetic monopoles do not exist (as far as we know).

32-2 GAUSS' LAW FOR MAGNETIC FIELDS

Gauss' law for magnetic fields is a formal way of saying that magnetic monopoles do not exist. The law asserts that the net magnetic flux Φ_B through any closed Gaussian surface is zero:

$$\Phi_B = \oint \mathbf{B} \cdot d\mathbf{A} = 0 \qquad \text{(Gauss' law for magnetic fields).} \qquad (32\text{-}1)$$

Contrast this with Gauss' law for electric fields,

$$\Phi_E = \oint \mathbf{E} \cdot d\mathbf{A} = \frac{1}{\epsilon_0} q \qquad \text{(Gauss' law for electric fields).}$$

In both equations, the integral is taken over a *closed* Gaussian surface. Gauss' law for electric fields says that this integral (the net electric flux through the surface) is proportional to the net electric charge q enclosed by the surface. Gauss' law for magnetic fields says that there can be no net magnetic flux through the surface because there can be no net "magnetic charge" (individual magnetic poles) enclosed by the surface. The simplest magnetic structure that can exist and thus be enclosed by a Gaussian surface is a dipole, which consists of both a source and a sink for the field lines. Thus, there must always be as much magnetic flux into the surface as out of it, and the net magnetic flux must always be zero.

Gauss' law for magnetic fields holds for more complicated structures than a magnetic dipole, and its holds even

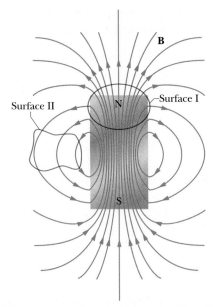

FIGURE 32-3 The field lines for the magnetic field **B** of a short bar magnet. The red curves represent cross sections of closed, three-dimensional Gaussian surfaces.

if the Gaussian surface does not enclose the entire structure. Gaussian surface II near the bar magnet of Fig. 32-3 encloses no poles, and we can easily conclude that the net magnetic flux through it is zero. But Gaussian surface I is more difficult. It may seem to enclose only the north pole of the magnet because it encloses the label N and not the label S. But a south pole must be associated with the lower boundary of the surface, because magnetic field lines enter the surface there. (The enclosed section is like one piece of the broken bar magnet in Fig. 32-2.) Thus, Gaussian surface I encloses a magnetic dipole and the net flux through the surface is zero.

CHECKPOINT **1:** The figure shows four closed surfaces with flat top and bottom faces and curved sides. The table gives the area A of the faces and the magnitudes B of the uniform and perpendicular magnetic fields through those faces; the units of A and B are arbitrary but consistent. Rank the surfaces according to the magnitudes of the magnetic flux through their curved sides, greatest first.

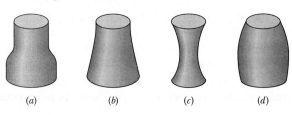

SURFACE	A_{top}	B_{top}	A_{bot}	B_{bot}
a	2	6, outward	4	3, inward
b	2	1, inward	4	2, inward
c	2	6, inward	2	8, outward
d	2	3, outward	3	2, outward

32-3 THE MAGNETISM OF EARTH

Earth is a huge magnet; for points near Earth's surface, its magnetic field can be represented as the field of a huge bar magnet—a magnetic dipole—that straddles the center of the planet. Figure 32-4 is an idealized symmetric depiction of the dipole field, without the distortion caused by passing charged particles from the Sun.

Because Earth's magnetic field is that of a magnetic dipole, a magnetic dipole moment μ is associated with the field. For the idealized field of Fig. 32-4, the magnitude of μ is 8.0×10^{22} J/T and the direction of μ makes an angle of 11.5° with the rotation axis (*RR*) of Earth. The *dipole axis* (*MM* in Fig. 32-4) lies along μ and intersects Earth's surface at the *geomagnetic north pole* in northwest Greenland and the *geomagnetic south pole* in Antarctica. The lines of the magnetic field **B** generally emerge in the southern hemisphere and reenter Earth in the northern hemisphere. Thus, the magnetic pole that is in Earth's northern hemisphere and known as a "north magnetic pole" *is really the south pole of Earth's magnetic dipole.*

The direction of the magnetic field at any location on Earth's surface is commonly specified in terms of two angles. The **field declination** is the angle (left or right)

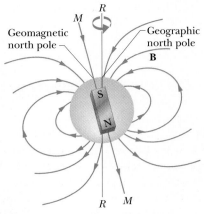

FIGURE 32-4 Earth's magnetic field represented as a dipole field. The dipole axis *MM* makes an angle of 11.5° with Earth's rotational axis *RR*. The south pole of the dipole is in Earth's northern hemisphere.

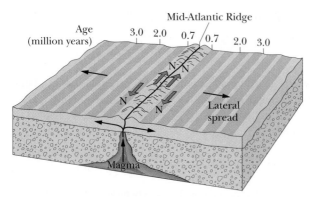

FIGURE 32-5 A magnetic profile of the seafloor on either side of the Mid-Atlantic Ridge. The seafloor, extruded through the ridge and spreading out as part of the tectonic drift system, displays a record of the past magnetic history of Earth's core. The direction of the magnetic field produced by the core reverses about every million years.

between geographic north (which is toward 90° latitude) and the horizontal component of the field. The **field inclination** is the angle (up or down) between a horizontal plane and the field's direction.

Magnetometers measure these angles and determine the field with much precision. However, you can do reasonably well with just a *compass* and a *dip meter*. A compass is simply a needle-shaped magnet that is mounted so that it can rotate freely about a vertical axis. When it is held in a horizontal plane, the north-pole end of the needle points, generally, toward the geomagnetic north pole (really a south magnetic pole, remember). The angle between the needle and geographic north is the field declination. A dip meter is a similar magnet that can rotate freely about a horizontal axis. When its vertical plane of rotation is aligned with the direction of the compass, the angle between the meter's needle and the horizontal is the field inclination.

At any point on Earth's surface, the measured magnetic field may differ appreciably, in both magnitude and direction, from the idealized dipole field of Fig. 32-4. In fact, the point where the field is actually vertical to Earth's surface and inward is not located at the geomagnetic north pole in Greenland as we would expect; instead this so-called *dip north pole* is located in the Queen Elizabeth Islands in northern Canada, far from Greenland.

In addition, the field observed at any location on the surface of Earth varies with time, by measurable amounts over a period of a few years and by substantial amounts over, say, 100 years. For example, between 1580 and 1820 the direction indicated by compass needles in London changed by 35°.

In spite of these local variations, the average dipole

field changes only slowly over such relatively short time periods. Variations over longer periods can be studied by measuring the weak magnetism of the ocean floor on either side of the Mid-Atlantic Ridge (Fig. 32-5). This floor has been formed by molten magma that oozed up through the ridge from Earth's interior, solidified, and was pulled away from the ridge (by the drift of tectonic plates) at the rate of a few centimeters per year. As the magma solidified, it became weakly magnetized with its magnetic field in the direction of Earth's magnetic field at the time of solidification. Study of this solidified magma across the ocean floor reveals that Earth's field has reversed its *polarity* (directions of the north pole and south pole) about every million years. The reason for the reversals is not known. In fact, the mechanism that produces Earth's magnetic field is only vaguely understood.

SAMPLE PROBLEM 32-1

In Tucson, Arizona, in 1964, the north pole of a compass needle pointed 13° east of geographic north, and the north pole of a dip-meter needle pointed downward, 59° below the horizontal. The horizontal component B_h of Earth's magnetic field **B** in Tucson had a magnitude of 26 μT. What was the magnitude B of the field then, in units of gauss? (Earth's field is often measured in gauss.)

SOLUTION: Figure 32-6 shows the given data; it is drawn in a plane that contains the vector **B** and so is angled 13° east of geographic north. From the figure we have

$$B = \frac{B_h}{\cos \theta} = \frac{26 \text{ μT}}{\cos 59°}$$

$$= 50 \text{ μT} = 0.50 \text{ gauss.} \qquad \text{(Answer)}$$

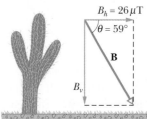

FIGURE 32-6 Sample Problem 32-1. Earth's magnetic field, along with its horizontal and vertical components, at Tucson, Arizona, in 1964.

32-4 MAGNETISM AND ELECTRONS

Magnetic materials, from lodestones to videotapes, are magnetic because of the electrons within them. We have already seen one way in which electrons can generate a magnetic field: send them through a wire as an electric current, and their motion produces a magnetic field around the wire. There are two more ways, each involving a magnetic dipole moment that produces a magnetic field in the surrounding space. However, their explanation requires quantum physics that is beyond the physics presented in this book. So here we shall only outline the results.

Spin Magnetic Dipole Moment

An electron has an intrinsic angular momentum called its **spin angular momentum** (or just **spin**) **S**; associated with this spin is an intrinsic **spin magnetic dipole moment** $\boldsymbol{\mu}_s$. (By *intrinsic*, we mean that **S** and $\boldsymbol{\mu}_s$ are basic characteristics of an electron, like its mass and electric charge.) **S** and $\boldsymbol{\mu}_s$ are related by

$$\boldsymbol{\mu}_s = -\frac{e}{m}\mathbf{S}, \qquad (32\text{-}2)$$

in which e is the elementary charge (1.60×10^{-19} C) and m is the mass of an electron (9.11×10^{-31} kg). The minus sign means that $\boldsymbol{\mu}_s$ and **S** are oppositely directed.

Spin **S** is quite different from the angular momenta of Chapter 12 in two respects:

1. **S** itself cannot be measured. Instead, only its component along an axis can be measured.

2. A measured component of **S** is quantized (restricted to certain values); in fact, it always has the same magnitude (no matter which axis is chosen).

Let us assume that the component of spin **S** is measured along the z axis of a coordinate system. Then the measured component S_z can have only the two values given by

$$S_z = m_s \frac{h}{2\pi}, \qquad \text{for} \quad m_s = \pm\tfrac{1}{2}, \qquad (32\text{-}3)$$

where m_s is the *spin magnetic quantum number* and h ($= 6.63 \times 10^{-34}$ J·s) is the Planck constant, the ubiquitous constant of quantum physics. The signs given in Eq. 32-3 have to do with the direction of S_z along the z axis. When S_z is parallel to the z axis, m_s is $+\tfrac{1}{2}$ and the electron is said to be *spin up*. When S_z is antiparallel to the z axis, m_s is $-\tfrac{1}{2}$ and the electron is said to be *spin down*.

The spin magnetic dipole moment $\boldsymbol{\mu}_s$ of an electron also cannot itself be measured; only a component can be measured, and that component is quantized and always has the same magnitude. We can relate the component $\mu_{s,z}$ measured on the z axis to S_z by rewriting Eq. 32-2 in component form for the z axis as

$$\mu_{s,z} = -\frac{e}{m}S_z.$$

Substituting for S_z from Eq. 32-3 then gives us

$$\mu_{s,z} = \pm\frac{eh}{4\pi m}, \qquad (32\text{-}4)$$

where the plus and minus signs correspond to $\mu_{s,z}$ being parallel and antiparallel to the z axis, respectively.

The quantity on the right side of Eq. 32-4 is called the *Bohr magneton* μ_B:

$$\mu_B = \frac{eh}{4\pi m} = 9.27 \times 10^{-24} \text{ J/T}$$

$$\text{(Bohr magneton).} \qquad (32\text{-}5)$$

Spin magnetic dipole moments of electrons and other elementary particles can be expressed in terms of μ_B. For the electron, the magnitude of the measured component of $\boldsymbol{\mu}_s$ is

$$\mu_{s,z} = 1\mu_B. \qquad (32\text{-}6)$$

(The quantum physics of the electron, called *quantum electrodynamics,* or QED, reveals that $\mu_{s,z}$ is actually slightly greater than $1\mu_B$, but we shall neglect that fact.)

When an electron is placed in an external magnetic field $\mathbf{B}_{\text{ext}}$, a potential energy U can be associated with the orientation of the electron's spin magnetic dipole moment $\boldsymbol{\mu}_s$ just as a potential energy can be associated with the orientation of the magnetic dipole moment $\boldsymbol{\mu}$ of a current loop placed in $\mathbf{B}_{\text{ext}}$. From Eq. 29-36, the potential energy for the electron is

$$U = -\boldsymbol{\mu}_s \cdot \mathbf{B}_{\text{ext}} = -\mu_{s,z}B, \qquad (32\text{-}7)$$

where the z axis is taken to be in the direction of $\mathbf{B}_{\text{ext}}$.

If we imagine an electron to be a microscopic sphere (which it is not), we can represent the spin **S**, the spin magnetic dipole moment $\boldsymbol{\mu}_s$, and the associated magnetic dipole field as in Fig. 32-7. Although we use the word "spin" here, electrons do not spin like tops. How, then, can something have angular momentum without actually rotating? Again, quantum physics provides the answer.

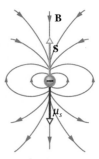

FIGURE 32-7 The spin **S**, spin magnetic dipole moment $\boldsymbol{\mu}_s$, and magnetic field **B** of an electron represented as a microscopic sphere.

Protons and neutrons also have an intrinsic angular momentum called spin and an associated intrinsic spin magnetic dipole moment. For a proton those two vectors have the same direction, and for a neutron they have opposite directions. We shall not examine the contributions of these dipole moments to the magnetic fields of atoms because they are about a thousand times smaller than that due to an electron.

CHECKPOINT 2: The figure shows the spin orientations of two particles in an external magnetic field $\mathbf{B}_{ext}$. (a) If the particles are electrons, which spin orientation is at lower potential energy? (b) If, instead, the particles are protons, which spin orientation is at lower potential energy?

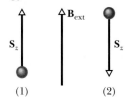

Orbital Magnetic Dipole Moment

When it is in an atom, an electron has an additional angular momentum called its **orbital angular momentum** $\mathbf{L}_{orb}$. Associated with $\mathbf{L}_{orb}$ is an **orbital magnetic dipole moment** $\boldsymbol{\mu}_{orb}$; the two are related by

$$\boldsymbol{\mu}_{orb} = -\frac{e}{2m}\mathbf{L}_{orb}. \qquad (32\text{-}8)$$

The minus sign means that $\boldsymbol{\mu}_{orb}$ and $\mathbf{L}_{orb}$ are in opposite directions.

Orbital angular momentum $\mathbf{L}_{orb}$ cannot be measured; only components along an axis can be measured, and those components are quantized. The components along, say, a z axis can have only the values given by

$$L_{orb,z} = m_l\frac{h}{2\pi}, \qquad \text{for} \quad m_l = 0, \pm 1, \pm 2, \cdots, \pm \text{(limit)}, \qquad (32\text{-}9)$$

in which m_l is the *orbital magnetic quantum number* and "limit" refers to some largest allowed integer value for m_l. The signs in Eq. 32-9 have to do with the direction of $L_{orb,z}$ on the z axis.

The orbital magnetic dipole moment $\boldsymbol{\mu}_{orb}$ of an electron also cannot itself be measured; only a component can be measured, and that component is quantized. By writing Eq. 32-8 in component form for the z axis and then substituting for $L_{orb,z}$ from Eq. 32-9, we can write the z compo-

nent $\mu_{orb,z}$ of the orbital magnetic dipole moment as

$$\mu_{orb,z} = -m_l\frac{eh}{4\pi m} \qquad (32\text{-}10)$$

and, in terms of the Bohr magneton, as

$$\mu_{orb,z} = -m_l\mu_B. \qquad (32\text{-}11)$$

When an atom is placed in an external magnetic field $\mathbf{B}_{ext}$, a potential energy U can be associated with the orientation of the orbital magnetic dipole moment of each electron in the atom. Its value is

$$U = -\boldsymbol{\mu}_{orb}\cdot\mathbf{B}_{ext} = -\mu_{orb,z}B_{ext}, \qquad (32\text{-}12)$$

where the z axis is taken in the direction of $\mathbf{B}_{ext}$.

Although we have used the words "orbit" and "orbital" here, electrons do not orbit the nucleus of an atom like planets orbiting the Sun. How can an electron have an orbital angular momentum without orbiting in the common meaning of the term? Once more the answer requires quantum physics.

Loop Model for Electron Orbits

We can obtain Eq. 32-8 with the nonquantum derivation that follows, in which we assume that an electron moves along a circular path with a radius that is much larger than an atomic radius (hence the name "loop model"). However, the derivation does not apply to an electron within an atom (for which we need quantum physics).

We imagine an electron moving at constant speed v in a circular path of radius r, counterclockwise as shown in Fig. 32-8. The motion of the negative charge of the electron is equivalent to a conventional current i (of positive

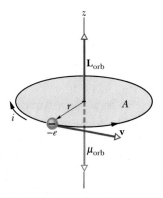

FIGURE 32-8 An electron moving at constant speed v in a circular path of radius r that encloses an area A. The electron has an orbital angular momentum $\mathbf{L}_{orb}$ and an associated orbital magnetic dipole moment $\boldsymbol{\mu}_{orb}$. A clockwise current i (of positive charge) is equivalent to the counterclockwise circulation of the negatively charged electron.

charge) that is clockwise, as also shown in Fig. 32-8. The magnitude of the orbital magnetic dipole moment of such a *current loop* is obtained from Eq. 29-33 with $N = 1$:

$$\mu_{orb} = iA, \tag{32-13}$$

where A is the area enclosed by the loop. The direction of this magnetic dipole moment is, from the right-hand rule of Fig. 30-22, downward in Fig. 32-8.

To evaluate Eq. 32-13, we need the current i. Current is, generally, the rate at which charge passes some point in a circuit. Here, the charge of magnitude e takes a time $T = 2\pi r/v$ to circle from any point back through that point, so

$$i = \frac{\text{charge}}{\text{time}} = \frac{e}{2\pi r/v}. \tag{32-14}$$

Substituting this and the area $A = \pi r^2$ of the loop into Eq. 32-13 gives us

$$\mu_{orb} = \frac{e}{2\pi r/v} \pi r^2 = \frac{evr}{2}. \tag{32-15}$$

To get an expression for the orbital angular momentum $\mathbf{L}_{orb}$ of the electron, we use Eq. 12-25, $\ell = m(\mathbf{r} \times \mathbf{v})$. Because $\mathbf{r}$ and $\mathbf{v}$ are perpendicular, $\mathbf{L}_{orb}$ has the magnitude

$$L_{orb} = mrv \sin 90° = mrv. \tag{32-16}$$

$\mathbf{L}_{orb}$ is directed upward in Fig. 32-8 (see Fig. 12-12). Combining Eqs. 32-15 and 32-16, generalizing to a vector formulation, and indicating the opposite directions of the vectors with a minus sign yield

$$\boldsymbol{\mu}_{orb} = -\frac{e}{2m} \mathbf{L}_{orb},$$

which is Eq. 32-8. Thus by "classical" (nonquantum) analysis we have obtained the same result, in both magnitude and direction, given by quantum physics. You might wonder, since this derivation gives the correct result for an electron within an atom, why the derivation is invalid for that situation. The answer is that this line of reasoning yields other results that are contradicted by experiments.

Loop Model in a Nonuniform Field

We continue to consider an electron orbit as a current loop, as we did in Fig. 32-8. Now, however, we draw the loop in a nonuniform magnetic field $\mathbf{B}_{ext}$ as shown in Fig. 32-9a. (This field is the diverging field near the north pole of the magnet in Fig. 32-3.) We make this change to prepare for the next several sections, in which we shall discuss the forces that act on magnetic materials when the materials are placed in a nonuniform magnetic field. We shall discuss these forces by assuming that the electron orbits in the materials are tiny current loops like that in Fig. 32-9a.

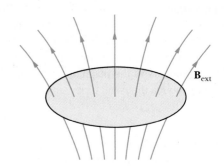

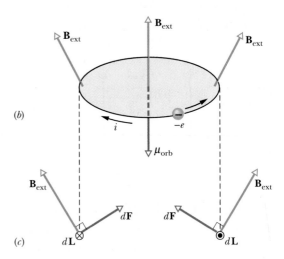

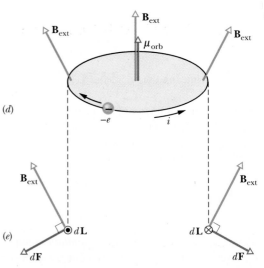

FIGURE 32-9 (*a*) A loop model for an electron orbiting in an atom while in a nonuniform magnetic field $\mathbf{B}_{ext}$. (*b*) Charge $-e$ moves counterclockwise; the associated conventional current i is clockwise. (*c*) The magnetic forces $d\mathbf{F}$ on the left and right sides of the loop, as seen from the plane of the loop. The net force on the loop is upward. (*d*) Charge $-e$ moves clockwise. (*e*) The net force on the loop is now downward.

Here we assume that the magnetic field vectors all around the electron's circular path have the same magnitude and form the same angle with the vertical, as shown in Figs. 32-9b and d. We also assume that all the electrons in an atom move either counterclockwise (Fig. 32-9b) or clockwise (Fig. 32-9d). The associated conventional current i around the current loop and the orbital magnetic dipole moment $\boldsymbol{\mu}_{orb}$ produced by i are shown for each of these directions of motion.

Figures 32-9c and e show diametrically opposite views of an element dL of the loop with the same direction as i, as seen from the plane of the orbit. Also shown are the field $\mathbf{B}_{ext}$ and the resulting magnetic force dF on dL. Recall that a current along an element dL in a magnetic field $\mathbf{B}_{ext}$ experiences a magnetic force dF as given by Eq. 29-27:

$$d\mathbf{F} = i \, d\mathbf{L} \times \mathbf{B}_{ext}. \qquad (32\text{-}17)$$

On the left side of Fig. 32-9c, Eq. 32-17 tells us that the force dF is directed upward and rightward. On the right side, the force dF is just as large and is directed upward and leftward. Because their angles are the same, the horizontal components of these two forces cancel and the vertical components add. The same is true at any other two symmetric points on the loop. So, the net force on the current loop of Fig. 32-9b must be upward. The same reasoning leads to a downward net force on the loop in Fig. 32-9d. We shall use these two results shortly when we examine the behavior of magnetic materials in nonuniform magnetic fields.

32-5 MAGNETIC MATERIALS

Each electron in an atom has an orbital magnetic dipole moment and a spin magnetic dipole moment that combine vectorially. The resultant of these two vectors combines vectorially with similar resultants for all other electrons in the atom. And the resultant for each atom combines with those for all the other atoms in a sample of a material. If the combination of all these magnetic dipole moments produces a magnetic field, then the material is magnetic. There are three general types of magnetism: diamagnetism, paramagnetism, and ferromagnetism.

1. *Diamagnetism* is exhibited by all common materials but is so feeble that it is masked if the material also exhibits magnetism of either of the other two types. In diamagnetism, weak magnetic dipole moments are produced in the atoms of the material when the material is placed in an external magnetic field $\mathbf{B}_{ext}$; the combination of all those induced dipole moments gives the material as a whole only a feeble net magnetic field. The dipole moments and thus their net field disappear when $\mathbf{B}_{ext}$ is removed. The term

diamagnetic material usually refers to materials that exhibit only diamagnetism.

2. *Paramagnetism* is exhibited by materials containing transition elements, rare earth elements, and actinide elements (see Appendix G). Each atom of such a material has a permanent resultant magnetic dipole moment, but the moments are randomly oriented in the material and the material as a whole lacks a net magnetic field. However, an external magnetic field $\mathbf{B}_{ext}$ can partially align the atomic magnetic dipole moments to give the material a net magnetic field. The alignment and thus its field disappear when $\mathbf{B}_{ext}$ is removed. The term *paramagnetic material* usually refers to materials that exhibit primarily paramagnetism.

3. *Ferromagnetism* is a property of iron, nickel, and certain other elements (and of compounds and alloys of these elements). Some of the electrons in these materials align their resultant magnetic dipole moments to produce regions with strong magnetic dipole moments. An external field $\mathbf{B}_{ext}$ can then align the magnetic moments of these regions, producing a strong magnetic field for the material as a whole; the field partially persists when $\mathbf{B}_{ext}$ is removed. We usually use the term *ferromagnetic material,* and even the common term *magnetic material,* to refer to materials that exhibit primarily ferromagnetism.

The next three sections explore these three types of magnetism.

32-6 DIAMAGNETISM

We cannot yet discuss the quantum physical explanation of diamagnetism, but we can provide a classical explanation with the loop model of Figs. 32-8 and 32-9. To begin, we assume that in a diamagnetic material the electrons in an atom orbit only clockwise or counterclockwise as in Fig. 32-9. To account for the lack of magnetism in the absence of an external magnetic field $\mathbf{B}_{ext}$, we assume the atom lacks a net magnetic dipole moment. This implies that before $\mathbf{B}_{ext}$ is applied, as many electrons orbit one way as orbit the other, with the result that the net upward magnetic dipole moment of the atom equals the net downward magnetic dipole moment.

Now let's turn on the nonuniform field $\mathbf{B}_{ext}$ of Fig. 32-9 in which $\mathbf{B}_{ext}$ is directed upward but is diverging (the magnetic field lines are diverging). We could do this by increasing the current through an electromagnet or by moving the north pole of a bar magnet closer to, and below, the orbits. As the magnitude of $\mathbf{B}_{ext}$ increases from zero to its final maximum, steady-state value, a clockwise electric field is induced around the electron's orbital loop according to Faraday's law and Lenz's law. Let us see how this induced electric field affects the orbiting electrons in Figs. 32-9b and d.

In Fig. 32-9b, the counterclockwise electron is accelerated by the clockwise electric field. So as the magnetic field $\mathbf{B}_{ext}$ increases to its maximum value, the electron speed increases to a maximum value. This means that the associated conventional current i and the downward magnetic dipole moment $\mathbf{\mu}$ due to i also *increase*.

In Fig. 32-9d, the clockwise electron is decelerated by the clockwise electric field. So, here, the electron speed, the associated current i, and the downward magnetic dipole moment $\mathbf{\mu}$ due to i all *decrease*. Thus by turning on field $\mathbf{B}_{ext}$, we have given the atom a net magnetic dipole moment that is upward.

The nonuniformity of field $\mathbf{B}_{ext}$ also affects the atom. Because the current i in Fig. 32-9b increases, the upward magnetic forces $d\mathbf{F}$ in Fig. 32-9c also increase, as does the net upward force on the current loop. And because current i in Fig. 32-9d decreases, the downward magnetic forces $d\mathbf{F}$ in Fig. 32-9e also decrease, as does the net downward force on the current loop. Thus by turning on the *nonuniform* field $\mathbf{B}_{ext}$, we have produced a net force on the atom; moreover, that force is directed *away* from the region of greater magnetic field, from which the magnetic field lines diverge or to which they converge.

We have argued with fictitious electron orbits (current loops), but we have ended up with exactly what happens to a diamagnetic material: if we apply the magnetic field of Fig. 32-9, the material develops a downward magnetic dipole moment and experiences an upward force. When the field is removed, both the dipole moment and the force disappear. The external field need not be positioned as shown; similar arguments can be made for other orientations of $\mathbf{B}_{ext}$. In general:

A diamagnetic material placed in an external magnetic field $\mathbf{B}_{ext}$ develops a magnetic dipole moment directed opposite $\mathbf{B}_{ext}$. If the field is nonuniform, the diamagnetic material is repelled from a region of greater magnetic field toward a region of lesser field.

CHECKPOINT **3:** The figure shows two diamagnetic spheres located near the south pole of a bar magnet. Are (a) the magnetic forces on the spheres and (b) the magnetic dipole moments of the spheres directed toward or away from the bar magnet? (c) Is the magnetic force on sphere 1 greater than, less than, or equal to that on sphere 2?

32-7 PARAMAGNETISM

In paramagnetic materials, the spin and orbital magnetic dipole moments of the electrons in each atom do not cancel but add vectorially to give the atom a net (and permanent) magnetic dipole moment $\mathbf{\mu}$. In the absence of an external magnetic field, these atomic dipole moments are randomly oriented, and the net magnetic dipole moment of the material is zero. However, if a sample of the material is placed in an external magnetic field $\mathbf{B}_{ext}$, the dipole moments tend to line up with the field, which gives the sample a net magnetic dipole moment. This alignment with the external field is opposite of what we saw with diamagnetic materials.

A paramagnetic material placed in an external magnetic field $\mathbf{B}_{ext}$ develops a magnetic dipole moment in the direction of $\mathbf{B}_{ext}$. If the field is nonuniform, the paramagnetic material is attracted toward a region of greater magnetic field from a region of lesser field.

A paramagnetic sample with N atoms would have a magnetic dipole moment with a magnitude of $N\mu$ if the alignment of its atomic dipoles were complete. However, random collisions of atoms due to thermal agitation transfer energy among them, disrupting their alignment and thus reducing the sample's magnetic dipole moment.

The importance of thermal agitation may be measured by comparing two energies. One, from Eq. 20-20, is the mean translational kinetic energy $K\ (= \frac{3}{2}kT)$ of an atom at temperature T, where k is the Boltzmann constant (1.38×10^{-23} J/K) and T is in kelvins (not Celsius degrees). The other, from Eq. 29-37, is the difference in energy ΔU_B

Liquid oxygen is suspended between the two pole faces of a magnet because the liquid is paramagnetic and is magnetically attracted to the magnet.

$(= 2\mu B_{ext})$ between parallel alignment and antiparallel alignment of the magnetic dipole moment of an atom and the external field. As we shall show below, $K \gg \Delta U_B$, even for ordinary temperatures and field magnitudes. Thus, energy transfers during collisions among atoms can significantly disrupt the alignment of the atomic dipole moments, keeping the magnetic dipole moment of a sample much less than $N\mu$.

We can express the extent to which a given paramagnetic sample is magnetized by finding the ratio of its magnetic dipole moment to its volume V. This vector quantity, the magnetic dipole moment per unit volume, is called the **magnetization M** of the sample, and its magnitude is

$$M = \frac{\text{measured magnetic moment}}{V}. \quad (32\text{-}18)$$

The unit of **M** is the ampere–square meter per cubic meter, or ampere per meter (A/m). Complete alignment of the atomic dipole moments, called *saturation* of the sample, corresponds to the maximum value $M_{max} = N\mu/V$.

In 1895 Pierre Curie discovered experimentally that the magnetization of a paramagnetic sample is directly proportional to the external magnetic field $\mathbf{B}_{ext}$ and inversely proportional to the temperature T in kelvins; that is,

$$M = C\frac{B_{ext}}{T}. \quad (32\text{-}19)$$

Equation 32-19 is known as *Curie's law,* and C is called the *Curie constant.* Curie's law is reasonable in that increasing B_{ext} tends to align the atomic dipole moments in a sample and thus to increase M, while increasing T tends to disrupt the alignment via thermal agitation and thus to decrease M. However, the law is actually an approximation that is valid only when the ratio B_{ext}/T is not too large.

Figure 32-10 shows the ratio M/M_{max} as a function of B_{ext}/T for a sample of the salt potassium chromium sulfate, in which chromium ions are the paramagnetic substance. The plot is called a *magnetization curve.* The straight line for Curie's law fits the experimental data at the left, for B_{ext}/T below about 0.5 T/K. The curve that fits all the data points is based on quantum physics. The data on the right side, near saturation, are very difficult to obtain because they require very strong magnetic fields (about 100,000 times Earth's field in Sample Problem 32-1), even at the very low temperatures noted in Fig. 32-10.

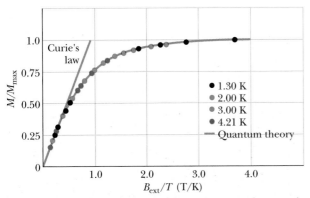

FIGURE 32-10 A *magnetization curve* for potassium chromium sulfate, a paramagnetic salt. The ratio of magnetization M of the salt to the maximum magnetization M_{max} is plotted versus the ratio of the applied magnetic field B_{ext} to the temperature T. Curie's law fits the data at the left; quantum theory fits all the data. After W. E. Henry.

SAMPLE PROBLEM 32-2

A paramagnetic gas at room temperature ($T = 300$ K) is placed in an external uniform magnetic field of magnitude $B = 1.5$ T; the atoms of the gas have magnetic dipole moment $\mu = 1.0\mu_B$. Calculate the mean translational kinetic energy K of an atom of the gas and the energy difference ΔU_B between parallel alignment and antiparallel alignment of the atom's magnetic dipole moment with the external field.

SOLUTION: From Eq. 20-20, we have

$$K = \tfrac{3}{2}kT = \tfrac{3}{2}(1.38 \times 10^{-23} \text{ J/K})(300 \text{ K})$$
$$= 6.2 \times 10^{-21} \text{ J} = 0.039 \text{ eV}. \quad (\text{Answer})$$

From Eqs. 29-37 and 32-5, we have

$$\Delta U_B = 2\mu B = 2(9.27 \times 10^{-24} \text{ J/T})(1.5 \text{ T})$$
$$= 2.8 \times 10^{-23} \text{ J} = 0.00017 \text{ eV}. \quad (\text{Answer})$$

Here K is about 230 times ΔU_B, so energy exchanges among the atoms during their collisions with one another can easily reorient any magnetic dipole moments that might be aligned with the external magnetic field. The magnetic dipole moment exhibited by the gas must then be due to fleeting partial alignments of the atomic dipole moments.

$\mathbb{C}$HECKPOINT 4: The figure shows two paramagnetic spheres located near the south pole of a bar magnet. Are (a) the magnetic forces on the spheres and (b) the magnetic dipole moments of the spheres directed toward or away from the bar magnet? (c) Is the magnetic force on sphere 1 greater than, less than, or equal to that on sphere 2?

32-8 FERROMAGNETISM

When we speak of magnetism in everyday conversation, we almost always have a mental picture of a bar magnet or a disk magnet (probably clinging to a refrigerator door). That is, we picture a ferromagnetic material having strong, permanent magnetism, not a diamagnetic or paramagnetic material having weak, temporary magnetism.

Iron, cobalt, nickel, gadolinium, dysprosium, and alloys of these and other elements exhibit ferromagnetism because of a quantum physical effect called *exchange coupling*. In this process the spins of the electrons in one atom interact with those of neighboring atoms. The result is an alignment of the magnetic dipole moments of the atoms, in spite of the randomizing tendency of the atomic collisions. This persistent alignment is what gives ferromagnetic materials their permanent magnetism.

If the temperature of a ferromagnetic material is raised above a certain critical value, called the *Curie temperature*, the exchange coupling ceases to be effective; most such materials then become simply paramagnetic. That is, the dipoles still tend to align with an external field but much more weakly, and thermal agitation can now more easily disrupt the alignment. The Curie temperature for iron is 1043 K (= 770°C).

The magnetization of a ferromagnetic material such as iron can be studied with an arrangement called a *Rowland ring* (Fig. 32-11). The material is formed into a thin toroidal core of circular cross section. A primary coil P having n turns per unit length is wrapped around the core and carries current i_P. (The coil is essentially a long solenoid bent into a circle.) If the iron core were not present, the magnetic field inside the coil would be, from Eq. 30-25,

$$B_0 = \mu_0 n i_P. \tag{32-20}$$

However, with the iron core present, the magnetic field B inside the coil is greater than B_0, usually by a large amount. We can write this field as

$$B = B_0 + B_M, \tag{32-21}$$

where B_M is the magnetic field contributed by the iron core. This contribution results from the alignment of the atomic dipole moments within the iron, due to exchange coupling and to the applied magnetic field B_0, and is proportional to the magnetization M of the iron. That is, the contribution B_M is proportional to the magnetic dipole moment per unit volume of the iron. To determine B_M we use a secondary coil S to measure B, compute B_0 with Eq. 32-20, and subtract as suggested by Eq. 32-21.

Figure 32-12 shows a magnetization curve for a ferromagnetic material in a Rowland ring: the ratio $B_M/B_{M,\text{max}}$, where $B_{M,\text{max}}$ is the maximum possible value of B_M, corresponding to saturation, is plotted versus B_0. The curve is similar to Fig. 32-10, the magnetization curve for a paramagnetic substance, in that both curves are measures of the extent to which an applied magnetic field can align the atomic dipole moments of a material.

For the ferromagnetic core yielding Fig. 32-12, the alignment of the dipole moments is about 70% complete for $B_0 \approx 1 \times 10^{-3}$ T. If B_0 were increased to 1 T, the alignment would be almost complete ($B_0 = 1$ T, and thus almost complete saturation, is quite difficult to obtain).

Magnetic Domains

Exchange coupling produces strong alignment of adjacent atomic dipoles in a ferromagnetic material at a temperature below the Curie temperature. Why, then, isn't the material naturally at saturation even when there is no applied mag-

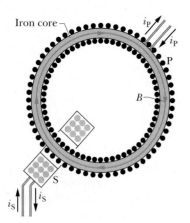

FIGURE 32-11   A Rowland ring. Current i_P is sent through a primary coil P whose core is a ferromagnetic material (here iron) that is magnetized by the current. (The turns of the coil are represented by dots.) The extent of magnetization of the core determines the total magnetic field **B** within coil P. Field **B** can be measured by means of a secondary coil S.

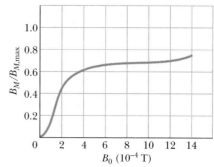

FIGURE 32-12   A magnetization curve for a ferromagnetic core material in the Rowland ring of Fig. 32-11. On the vertical axis, 1.0 corresponds to complete alignment (saturation) of the atomic dipoles within the material.

netic field B_0? That is, why isn't every piece of iron, such as an iron nail, a naturally strong magnet?

To understand this, consider a specimen of a ferromagnetic material such as iron that is in the form of a single crystal. That is, the arrangement of the atoms that make it up—its crystal lattice—extends with unbroken regularity throughout the volume of the specimen. Such a crystal will, in its normal state, be made up of a number of *magnetic domains*. These are regions of the crystal throughout which the alignment of the atomic dipoles is essentially perfect. For the crystal as a whole, however, the domains are so oriented that they largely cancel each other as far as their external magnetic effects are concerned.

Figure 32-13 is a magnified photograph of such an assembly of domains in a single crystal of nickel. It was made by sprinkling a colloidal suspension of finely powdered iron oxide on the surface of the crystal. The domain boundaries, which are thin regions in which the alignment of the elementary dipoles changes from a certain orientation in one domain to a different orientation in the other, are the sites of intense, but highly localized and nonuniform magnetic fields. The suspended colloidal particles are attracted to these boundaries and show up as the white lines. Although the atomic dipoles in each domain are completely aligned as shown by the arrows, the crystal as a whole may have a very small resultant magnetic moment.

Actually, a piece of iron as we ordinarily find it is not a single crystal but an assembly of many tiny crystals, randomly arranged; we call it a *polycrystalline solid*. Each tiny crystal, however, has its array of variously oriented domains, just as in Fig. 32-13. If we magnetize such a specimen by placing it in an external magnetic field of gradually increasing strength, we produce two effects; together they produce a magnetization curve of the shape shown in Fig. 32-12. One effect is a growth in size of the domains that are oriented along the external field at the expense of those that are not. The second effect is a shift of the orientation of the dipoles within a domain, as a unit, to become closer to the field direction.

Exchange coupling and domain shifting give us the following result:

> A ferromagnetic material placed in an external magnetic field $\mathbf{B}_{ext}$ develops a strong magnetic dipole moment in the direction of $\mathbf{B}_{ext}$. If the field is nonuniform, the ferromagnetic material is attracted toward a region of greater magnetic field from a region of lesser field.

You can actually hear sound produced by shifting domains: put an audio cassette player into its play mode

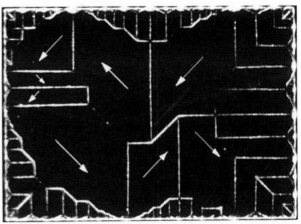

FIGURE 32-13 A photograph of domain patterns within a single crystal of nickel; white lines reveal the boundaries of the domains. The white arrows superimposed on the photograph show the orientations of the magnetic dipoles within the domains and thus the orientations of the net magnetic dipoles of the domains. The crystal as a whole is unmagnetized if the net field (the vector sum over all the domains) is zero.

without a cassette in place (or with a blank cassette) and turn the volume control to maximum. Then bring a strong magnet up to the play head (which is ferromagnetic). The magnetic field causes the domains in the play head to shift abruptly, which abruptly alters the magnetic field through a

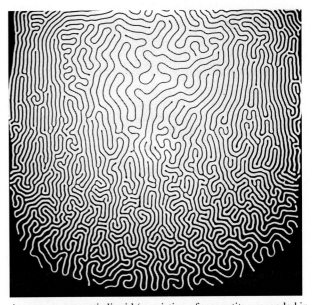

An opaque magnetic liquid (consisting of magnetite suspended in kerosene) and a transparent nonmagnetic liquid have been placed in a thin glass cell. When the cell is upright, the slightly denser magnetic liquid (shown black here) sinks to the bottom of the cell. But when a magnetic field is then applied perpendicular to the cell face, the magnetic liquid quickly snakes its way up into the nonmagnetic liquid, forming the labyrinthine pattern shown.

coil wrapped around the play head. The resulting suddenly induced currents in the coil are amplified and fed to the speaker, producing a fizzing sound.

The Magnetism of Ancient Kilns

The clay in the walls and floor of an ancient kiln behaves similarly to iron because clay contains the iron oxides magnetite and hematite. Grains of magnetite consist of multiple domains that may be as small as 3×10^{-7} m across. Those of hematite consist of single domains that may be as large as 1 mm across. When the clay is heated to several hundred degrees Celsius (as the kiln is used), the domains in grains of both types change. In magnetite the domain walls shift so that domains that are more closely aligned with Earth's magnetic field grow, while others shrink. In hematite, the domains rotate so as to more closely align with Earth's field. For both processes, the result is that the clay then has a magnetic field that is aligned with Earth's field. When the kiln cools after use, the arrangement of the domains, and thus the magnetic field of the clay, is retained, an effect known as *thermoremanent magnetism* (TRM).

To determine the orientation Earth's field had when a kiln was last heated and cooled, an archaeologist outlines a small area of the floor, carefully measures its orientation relative to the horizontal and to geographic north, and then removes that section of the floor. By next determining the direction of the section's magnetic field relative to the section's dimensions, hence its position in the kiln, the archaeologist then knows the direction of Earth's field (relative to the horizontal and to geographic north) when the kiln was last used. If the age of the kiln is found by radiocarbon dating or some other technique, the archaeologist also knows when Earth's field had that direction.

SAMPLE PROBLEM 32-3

A compass needle of pure iron (with density 7900 kg/m^3) has a length L of 3.0 cm, a width of 1.0 mm, and a thickness of 0.50 mm. The magnitude of the magnetic dipole moment associated with an iron atom is $\mu_{Fe} = 2.1 \times 10^{-23}$ J/T.

(a) If the magnetization of the needle is equivalent to the alignment of 10% of the atoms in the needle, what is the magnitude of the needle's magnetic dipole moment μ?

SOLUTION: Alignment of all N atoms in the needle would give the needle a magnetic dipole moment of $N\mu_{Fe}$. We have only 10% alignment, so

$$\mu = 0.10N\mu_{Fe}. \qquad (32\text{-}22)$$

The number of atoms N in the needle is

$$N = \frac{\text{needle's mass}}{\text{atomic mass of iron}}. \qquad (32\text{-}23)$$

The mass m of the needle is the product of its density, 7900 kg/m^3, and its volume (3.0 cm $\times$ 1.0 mm $\times$ 0.50 mm = 1.5×10^{-8} m^3) and works out to be 1.185×10^{-4} kg. The mass of an atom of iron is the ratio of the molar mass M of iron (55.847 g/mol, from Appendix F) to Avogadro's number N_A (6.02×10^{23} atoms/mol). Substituting Eq. 32-23 into Eq. 32-22 and then substituting these quantities and the given data, we find

$$\mu = 0.10 \left(\frac{mN_A}{M} \right) \mu_{Fe}$$

$$= 0.10 \frac{(1.185 \times 10^{-4} \text{ kg})(6.02 \times 10^{23} \text{ atoms/mol})}{(55.847 \text{ g/mol})(10^{-3} \text{ kg/g})}$$

$$\times (2.1 \times 10^{-23} \text{ J/T})$$

$$= 2.682 \times 10^{-3} \text{ J/T} \approx 2.7 \times 10^{-3} \text{ J/T}. \qquad \text{(Answer)}$$

(b) If the compass needle is jarred slightly from its (horizontal) north–south equilibrium position, it oscillates about that position. If the period of oscillation is 2.2 s, what is the horizontal component of the local magnetic field?

SOLUTION: The dipole moment μ of the needle is directed along its length, from its south pole to its north pole. When the needle is jarred from its equilibrium position by an angle θ, so is μ. Earth's magnetic field then exerts a torque about the needle's pivot axis, directed so as to realign μ (and the needle) with the horizontal component B_h of the magnetic field. (Remember, the needle of a compass is free to rotate only horizontally, so we are dealing only with B_h here.) From Eq. 29-34, the magnitude of this torque is

$$\tau = -\mu B_h \sin \theta, \qquad (32\text{-}24)$$

in which the minus sign indicates that τ opposes the angular displacement θ. Because the rotation angle is small, we may write $\sin \theta \approx \theta$ so that

$$\tau = -\mu B_h \theta. \qquad (32\text{-}25)$$

Because μ and B_h are both constant, Eq. 32-25 tells us that the restoring torque is proportional to the negative of the angular displacement. This kind of relation is the hallmark of angular simple harmonic motion, as we saw in Section 16-5. From Eqs. 16-24 and 16-25, the period of oscillation of the motion may then be written as

$$T = 2\pi \sqrt{\frac{I}{\mu B_h}},$$

which yields

$$B_h = \frac{I}{\mu} \left(\frac{2\pi}{T} \right)^2, \qquad (32\text{-}26)$$

where I is the rotational inertia of the needle. Approximating the needle as being a uniform thin rod, we use Table 11-2e to find

$$I = \frac{mL^2}{12} = \frac{(1.185 \times 10^{-4} \text{ kg})(0.030 \text{ m})^2}{12}$$

$$= 8.888 \times 10^{-9} \text{ kg} \cdot \text{m}^2.$$

Substituting this value, the value we obtained for μ, and the given value for T into Eq. 32-26, we find

$$B_h = \frac{8.888 \times 10^{-9} \text{ kg} \cdot \text{m}^2}{2.682 \times 10^{-3} \text{ J/T}} \left(\frac{2\pi}{2.2 \text{ s}} \right)^2$$

$$= 2.7 \times 10^{-5} \text{ T}, \qquad \text{(Answer)}$$

which is approximately the value we used in Sample Problem 32-1 for Tucson. So, even with an inexpensive compass, we can measure a local magnetic field by jarring the needle and timing its oscillations.

Hysteresis

Magnetization curves for ferromagnetic materials do not retrace themselves as we increase and then decrease the external magnetic field B_0. Figure 32-14 is a plot of B_M versus B_0 during the following operations with a Rowland ring: (1) starting with the iron unmagnetized (point a), increase the current in the toroid until B_0 ($= \mu_0 n i$) has the value corresponding to point b; (2) reduce the current in the toroid winding back to zero (point c); (3) reverse the toroid current and increase it in magnitude until B_0 has the value corresponding to point d; (4) reduce the current to zero again (point e); (5) reverse the current once more until point b is reached again.

The lack of retraceability shown in Fig. 32-14 is called **hysteresis,** and the curve $bcdeb$ is called a *hysteresis loop*. Note that at points c and e the iron core is magnetized, even though there is no current in the toroid windings; this is the familiar phenomenon of permanent magnetism.

Hysteresis can be understood through the concept of magnetic domains. Evidently the motions of the domain boundaries and the reorientations of the domain directions are not totally reversible. When the applied magnetic field B_0 is increased and then decreased back to its initial value, the domains do not return completely to their original configuration but retain some ''memory'' of the initial increase. This memory of magnetic materials is essential for the magnetic storage of information, as on cassette tapes and computer disks.

This memory of the alignment of domains can also occur naturally. When lightning sends currents along multiple tortuous paths through the ground, the currents produce intense magnetic fields that can suddenly magnetize any ferromagnetic material in nearby rock. Because of hysteresis, such rock material retains some of that magnetization after the lightning strike (after the currents disappear). Pieces of the rock, later exposed, broken, and loosened by weathering, are then lodestones.

32-9 INDUCED MAGNETIC FIELDS

We have to this point discussed two ways in which a magnetic field can be produced. The first, discussed in Chapter 30, is by means of an electric current; the second, discussed in this chapter, is by means of a magnetic material. There is a third way — by induction.

In Chapter 31 you saw that a changing magnetic flux induces an electric field, and we ended up with Faraday's law of induction in the form

$$\oint \mathbf{E} \cdot d\mathbf{s} = -\frac{d\Phi_B}{dt} \qquad \begin{array}{l}\text{(Faraday's law} \\ \text{of induction).}\end{array} \quad (32\text{-}27)$$

Here $\mathbf{E}$ is the electric field induced along a closed loop by the changing magnetic flux Φ_B through that loop. Because symmetry is often so powerful in physics, we should be tempted to ask whether induction can occur in the opposite sense. That is, can a changing electric flux induce a magnetic field?

The answer is that it can; furthermore, the equation governing the induction of a magnetic field is almost symmetric with Eq. 32-27. We often call it Maxwell's law of induction after James Clerk Maxwell, and we write it as

$$\oint \mathbf{B} \cdot d\mathbf{s} = \mu_0 \epsilon_0 \frac{d\Phi_E}{dt} \qquad \begin{array}{l}\text{(Maxwell's law} \\ \text{of induction).}\end{array} \quad (32\text{-}28)$$

The circle on the integral sign indicates that the integral is taken around a closed loop.

As an example of this sort of induction, we consider the charging of a parallel-plate capacitor with circular plates, as shown in Fig. 32-15a. (Although we shall focus on this particular arrangement, a changing electric flux will always induce a magnetic field whenever it occurs.) We

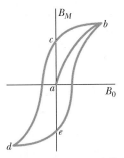

FIGURE 32-14 A magnetization curve (ab) for a ferromagnetic specimen and an associated hysteresis loop ($bcdeb$).

assume that the charge on the capacitor is being increased at a steady rate by a constant current i in the connecting wires. Then the magnitude of the electric field between the plates must also be increasing at a steady rate.

Figure 32-15b is a view of the right-hand plate of Fig. 32-15a from between the plates. The electric field is directed into the page. Let us consider a circular loop through point 1 in Figs. 32-15a and b, concentric with the capacitor plates and with a radius smaller than that of the plates. Because the electric field through the loop is changing, the electric flux through the loop must also be changing. According to Eq. 32-28, this changing electric flux induces a magnetic field around the loop.

Experiment proves that a magnetic field **B** *is* indeed induced around such a loop, directed as shown. This magnetic field has the same magnitude at every point around the loop and thus has circular symmetry about the central axis of the capacitor plates.

If we now consider a larger loop, say through point 2 outside the plates in Figs. 32-15a and b, we find that a magnetic field is induced around that loop as well. Thus, while the electric field is changing, magnetic fields are induced between the plates, both inside and outside the gap. When the electric field stops changing, these induced magnetic fields disappear.

Although Eq. 32-28 is similar to Eq. 32-27, the equations differ in two ways. First, Eq. 32-28 has the two extra symbols, μ_0 and ϵ_0, but they appear only because we employ SI units. Second, Eq. 32-28 lacks the minus sign of Eq. 32-27. That difference in sign means that the induced electric field **E** and the induced magnetic field **B** have opposite directions when they are produced in otherwise similar situations.

To see this opposition of directions, examine Fig. 32-16, in which an increasing magnetic field **B**, directed into the page, induces an electric field **E**. The induced field **E** is counterclockwise, whereas the induced magnetic field **B** in Fig. 32-15b is clockwise.

Now recall that the left side of Eq. 32-28, the integral of the dot product **B** · d**s** around a closed loop, appears in another equation, namely Ampere's law:

$$\oint \mathbf{B} \cdot d\mathbf{s} = \mu_0 i_{\text{enc}} \quad \text{(Ampere's law),} \quad (32\text{-}29)$$

where i_{enc} is the current encircled by the closed loop. Thus, our two equations that specify the magnetic field **B** produced by means other than a magnetic material (that is, by a current and by a changing electric field) give the field in exactly the same form. So we can combine the two equations into the single equation

$$\oint \mathbf{B} \cdot d\mathbf{s} = \mu_0 \epsilon_0 \frac{d\Phi_E}{dt} + \mu_0 i_{\text{enc}}$$
$$\text{(Ampere–Maxwell law).} \quad (32\text{-}30)$$

When there is a current but no change in electric flux (such as with a wire carrying a constant current), the first term on the right side of Eq. 32-30 is zero, and Eq. 32-30 reduces to Eq. 32-29, Ampere's law. When there is a change in electric flux but no current (such as inside or outside the gap of a charging capacitor), the second term on the right side of Eq. 32-30 is zero, and Eq. 32-30 reduces to Eq. 32-28, Maxwell's law of induction.

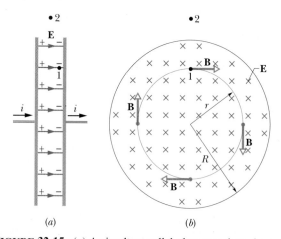

(a)   (b)

FIGURE 32-15 (a) A circular parallel-plate capacitor, shown in side view, is being charged by a constant current i. (b) A view from within the capacitor, toward the plate at the right. The electric field **E** is uniform, is directed into the page (toward the plate), and grows in magnitude as the charge on the capacitor increases. The magnetic field **B** induced by this changing electric field is shown at four points on a circle with a radius r less than the plate radius R.

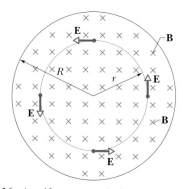

FIGURE 32-16 A uniform magnetic field **B** in a circular region. The field, directed into the page, is increasing in magnitude. The electric field **E** induced by the changing magnetic field is shown at four points on a circle concentric with the circular region. Compare this situation with that of Fig. 32-15b.

SAMPLE PROBLEM 32-4

A parallel-plate capacitor with circular plates of radius R is being charged as in Fig. 32-15a.

(a) Derive an expression for the induced magnetic field at radii r for the case of $r \leq R$.

SOLUTION: There is no current between the plates so that $i_{enc} = 0$ in Eq. 32-30, leaving

$$\oint \mathbf{B} \cdot d\mathbf{s} = \mu_0 \epsilon_0 \frac{d\Phi_E}{dt}. \qquad (32\text{-}31)$$

For an Amperian loop of radius $r \leq R$, the left side of Eq. 32-31 is $(B)(2\pi r)$. The electric flux Φ_E through that loop is $EA = \pi r^2 E$, where E is the magnitude of the electric field between the plates. So we may write Eq. 32-31 as

$$(B)(2\pi r) = \mu_0 \epsilon_0 \frac{d}{dt}(\pi r^2 E) = \mu_0 \epsilon_0 \pi r^2 \frac{dE}{dt}.$$

Solving for B, we find

$$B = \frac{\mu_0 \epsilon_0 r}{2} \frac{dE}{dt} \quad \text{(for } r \leq R\text{)}. \qquad \text{(Answer)}$$

We see that $B = 0$ at the center of the capacitor, where $r = 0$, and that B increases linearly with r out to the edge of the circular capacitor plates.

(b) Evaluate the field magnitude B for $r = R/5 = 11.0$ mm and $dE/dt = 1.50 \times 10^{12}$ V/m·s.

SOLUTION: From the answer to (a), we have

$$\begin{aligned}
B &= \frac{1}{2} \mu_0 \epsilon_0 r \frac{dE}{dt} \\
&= \tfrac{1}{2}(4\pi \times 10^{-7} \text{ T·m/A})(8.85 \times 10^{-12} \text{ C}^2/\text{N·m}^2) \\
&\quad \times (11.0 \times 10^{-3} \text{ m})(1.50 \times 10^{12} \text{ V/m·s}) \\
&= 9.18 \times 10^{-8} \text{ T}. \qquad \text{(Answer)}
\end{aligned}$$

(c) Derive an expression for the induced magnetic field for the case $r \geq R$.

SOLUTION: Outside the plate radius R, the electric field E is equal to zero, so the electric flux through an Amperian loop of radius $r \geq R$ exists only within the area πR^2 and is $\Phi_E = \pi R^2 E$. Then Eq. 32-31 becomes

$$(B)(2\pi r) = \mu_0 \epsilon_0 \frac{d}{dt}(\pi R^2 E) = \mu_0 \epsilon_0 \pi R^2 \frac{dE}{dt}.$$

Solving for B, we find

$$B = \frac{\mu_0 \epsilon_0 R^2}{2r} \frac{dE}{dt} \quad \text{(for } r \geq R\text{)}. \qquad \text{(Answer)}$$

Note that the two expressions for B that we have derived yield the same result (as we expect) for $r = R$. Furthermore, the value of B at $r = R$ is the maximum value of the induced magnetic field.

The value of B calculated in (b) is so small that it can scarcely be measured with simple apparatus. This is in sharp contrast to induced electric fields (Faraday's law), which can be demonstrated easily. This experimental difference exists partly because induced emfs can easily be multiplied by using a coil of many turns. No technique of comparable simplicity exists for multiplying induced magnetic fields. In any case, the experiment suggested by this sample problem has been done, and the presence of the induced magnetic fields has been verified quantitatively.

CHECKPOINT 5: The figure shows graphs of the electric field magnitude E versus time t for four uniform electric fields, all contained within identical circular regions as in Fig. 32-15b. Rank the fields according to the magnitude of the induced magnetic field at the edge of the region, greatest first.

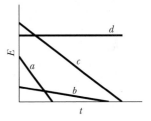

32-10 DISPLACEMENT CURRENT

If you compare the two terms on the right side of Eq. 32-30, you will see that the portion $\epsilon_0(d\Phi_E/dt)$ of the first term must have the dimension of a current. In fact, historically, that portion has been treated as being a fictitious current called the **displacement current** i_d:

$$i_d = \epsilon_0 \frac{d\Phi_E}{dt} \quad \text{(displacement current)}. \qquad (32\text{-}32)$$

''Displacement'' is poorly chosen in that nothing is being displaced, but we are stuck with the word. Nevertheless, we can now rewrite Eq. 32-30 as

$$\oint \mathbf{B} \cdot d\mathbf{s} = \mu_0 i_{d,enc} + \mu_0 i_{enc}$$

$$\text{(Ampere–Maxwell law)}, \qquad (32\text{-}33)$$

in which $i_{d,enc}$ is the displacement current that is encircled by the integration loop.

Let us again focus on a charging capacitor with circular plates, as in Fig. 32-17a. The real current i that is charging the plates changes the electric field $\mathbf{E}$ between the plates. The fictitious displacement current i_d between the plates is associated with that changing field $\mathbf{E}$. Let us relate these two currents.

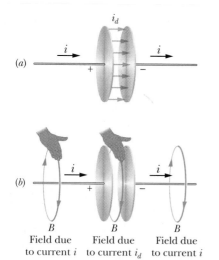

FIGURE 32-17 (a) The displacement current i_d between the plates of a capacitor that is being charged by a current i. (b) The right-hand rule for finding the direction of the magnetic field around a wire with a real current (as at the left) also gives the direction of the magnetic field around a displacement current (as in the center).

The charge q on the plates at any time is related to the magnitude E of the field between the plates at that time by Eq. 26-4:

$$q = \epsilon_0 A E, \tag{32-34}$$

in which A is the plate area. To get the real current i, we differentiate Eq. 32-34 with respect to time, finding

$$\frac{dq}{dt} = i = \epsilon_0 A \frac{dE}{dt}. \tag{32-35}$$

To get the displacement current i_d, we can use Eq. 32-32. Assuming that the electric field **E** between the two plates is uniform (we neglect any fringing), we can replace the electric flux Φ_E in that equation with EA. Then Eq. 32-32 becomes

$$i_d = \epsilon_0 \frac{d\Phi_E}{dt} = \epsilon_0 \frac{d(EA)}{dt} = \epsilon_0 A \frac{dE}{dt}. \tag{32-36}$$

Comparing Eqs. 32-35 and 32-36, we see that the real current i charging the capacitor and the fictitious displacement current i_d between the plates have the same magnitude:

$$i_d = i \quad \begin{array}{l}\text{(displacement current} \\ \text{in a capacitor).}\end{array} \tag{32-37}$$

Thus, we can consider the fictitious displacement current i_d to be simply a continuation of the real current i from one plate, across the capacitor gap, to the other plate. Because

the electric field is uniformly spread over the plates, the same is true of this fictitious displacement current i_d, as suggested by the spread of current arrows in Fig. 32-17a. Although no charge actually moves across the gap between the plates, the idea of the fictitious current i_d can help us to quickly find the direction and magnitude of an induced magnetic field, as follows.

Finding the Induced Magnetic Field

In Chapter 30 we found the direction of the magnetic field produced by a real current i by using the right-hand rule of Fig. 30-4. We can apply the same rule to find the direction of an induced magnetic field produced by a fictitious displacement current i_d, as is shown in the center of Fig. 32-17b for a capacitor.

We can also use i_d to find the magnitude of the induced magnetic field for a charging capacitor with parallel circular plates of radius R. We simply consider the space between the plates to be an imaginary circular wire of radius R carrying the imaginary current i_d. Then, from Eq. 30-22, the magnitude of the magnetic field at a point inside the capacitor at radius r from the center is

$$B = \left(\frac{\mu_0 i_d}{2\pi R^2}\right) r \quad \text{(inside a circular capacitor).} \tag{32-38}$$

Similarly, from Eq. 30-19, the magnitude of the magnetic field at a point outside the capacitor at radius r is

$$B = \frac{\mu_0 i_d}{2\pi r} \quad \text{(outside a circular capacitor).} \tag{32-39}$$

SAMPLE PROBLEM 32-5

The circular parallel-plate capacitor in Sample Problem 32-4 is being charged with a current i.

(a) Between the plates, what is the magnitude of $\oint \mathbf{B} \cdot d\mathbf{s}$, in terms of μ_0 and i, at a radius $r = R/5$?

SOLUTION: The magnetic field within the capacitor is produced by the displacement current i_d between the plates. So, we write Ampere's law for a circular loop of radius r concentric with the central axis of the capacitor as

$$\oint \mathbf{B} \cdot d\mathbf{s} = \mu_0 i_{d,\text{enc}}. \tag{32-40}$$

We assume that the displacement current is uniformly spread over the full plate area. Thus the current $i_{d,\text{enc}}$ that is encircled by the loop is proportional to the area encircled by the loop;

that is,

$$\frac{i_{d,\text{enc}}}{i_d} = \frac{\text{encircled area}}{\text{full plate area}},$$

from which we find that

$$i_{d,\text{enc}} = i_d \frac{\pi r^2}{\pi R^2}.$$

Substituting this into Eq. 32-40, we obtain

$$\oint \mathbf{B} \cdot d\mathbf{s} = \mu_0 i_d \frac{\pi r^2}{\pi R^2}. \qquad (32\text{-}41)$$

Now substituting $i_d = i$ (from Eq. 32-37) and $r = R/5$ into Eq. 32-41 gives us

$$\oint \mathbf{B} \cdot d\mathbf{s} = \mu_0 i \frac{\pi (R/5)^2}{\pi R^2} = \frac{\mu_0 i}{25}. \qquad \text{(Answer)}$$

(b) In terms of the maximum field, what is the magnitude of the magnetic field at $r = R/5$ inside the capacitor?

SOLUTION: Because the capacitor has parallel circular plates, we can use Eq. 32-38 to find B. At $r = R/5$, that equation yields

$$B = \left(\frac{\mu_0 i_d}{2\pi R^2}\right) r = \frac{\mu_0 i_d (R/5)}{2\pi R^2} = \frac{\mu_0 i_d}{10\pi R}. \qquad (32\text{-}42)$$

The maximum field B_{max} within the capacitor occurs at $r = R$. It is

$$B_{\text{max}} = \left(\frac{\mu_0 i_d}{2\pi R^2}\right) r = \frac{\mu_0 i_d R}{2\pi R^2} = \frac{\mu_0 i_d}{2\pi R}. \qquad (32\text{-}43)$$

Dividing Eq. 32-42 by Eq. 32-43 and rearranging the result, we find

$$B = \frac{B_{\text{max}}}{5}. \qquad \text{(Answer)}$$

We should be able to obtain this result with a little reasoning and less work. Equation 32-38 tells us that inside the capacitor, B increases linearly with r. So a point $\frac{1}{5}$ the distance out to the full radius R of the plates, where B_{max} occurs, should have a field B that is $\frac{1}{5} B_{\text{max}}$.

CHECKPOINT 6: The figure is a view of one plate of a parallel-plate capacitor from within the capacitor. The dashed lines show four integration paths (path b follows the edge of the plate). Rank the paths according to the magnitude of $\oint \mathbf{B} \cdot d\mathbf{s}$ along the paths during the discharging of the capacitor, greatest first.

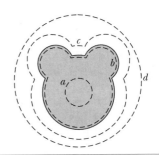

32-11 MAXWELL'S EQUATIONS

Equation 32-30 is the last of the four fundamental equations of electromagnetism, called *Maxwell's equations* and displayed in Table 32-1. These four equations explain a diverse range of phenomena, from why a compass needle points north to why a car starts when you turn the ignition key. They are the basis for the functioning of such electromagnetic devices as electric motors, cyclotrons, television transmitters and receivers, telephones, fax machines, radar, and microwave ovens.

Maxwell's equations are the basis from which many of the equations you have seen since Chapter 22 can be derived. They are also the basis of many of the equations you will see in Chapters 34 through 37, which introduce you to optics, as well as such optical devices as telescopes and eyeglasses.

TABLE 32-1   MAXWELL'S EQUATIONS[a]

NAME	EQUATION
Gauss' law for electricity	$\oint \mathbf{E} \cdot d\mathbf{A} = q/\epsilon_0$
Relates net electric flux to net enclosed electric charge	
Gauss' law for magnetism	$\oint \mathbf{B} \cdot d\mathbf{A} = 0$
Relates net magnetic flux to net enclosed magnetic charge	
Faraday's law	$\oint \mathbf{E} \cdot d\mathbf{s} = -\dfrac{d\Phi_B}{dt}$
Relates induced electric field to changing magnetic flux	
Ampere–Maxwell law	$\oint \mathbf{B} \cdot d\mathbf{s} = \mu_0 \epsilon_0 \dfrac{d\Phi_E}{dt} + \mu_0 i_{\text{enc}}$
Relates induced magnetic field to changing electric flux and to current	

[a]Written on the assumption that no dielectric or magnetic materials are present.

REVIEW & SUMMARY

Gauss' Law for Magnetic Fields

The simplest magnetic structures are magnetic dipoles. Magnetic monopoles do not exist (as far as we know). **Gauss' law** for magnetic fields,

$$\Phi_B = \oint \mathbf{B} \cdot d\mathbf{A} = 0, \qquad (32\text{-}1)$$

states that the net magnetic flux through any (closed) Gaussian surface is zero. It implies that magnetic monopoles don't exist.

Earth's Magnetic Field

Earth's magnetic field can be approximated as being that of a magnetic dipole whose dipole moment makes an angle of $11.5°$ with Earth's rotation axis, and with the south pole of the dipole in the northern hemisphere. The direction of the local magnetic field at any point on Earth's surface is given by the *field declination* (the angle left or right from geographic north) and the *field inclination* (the angle up or down from the horizontal).

Spin Magnetic Dipole Moment

An electron has an intrinsic angular momentum called *spin angular momentum* (or *spin*) $\mathbf{S}$, with which an intrinsic *spin magnetic dipole moment* $\boldsymbol{\mu}_s$ is associated:

$$\boldsymbol{\mu}_s = -\frac{e}{m}\mathbf{S}. \qquad (32\text{-}2)$$

Spin $\mathbf{S}$ cannot itself be measured, but a component can be measured. Assuming that the measurement is along a z axis of a coordinate system, the component S_z can have only the values given by

$$S_z = m_s \frac{h}{2\pi}, \qquad \text{for} \qquad m_s = \pm\tfrac{1}{2}, \qquad (32\text{-}3)$$

where h $(= 6.63 \times 10^{-34}\ \text{J} \cdot \text{s})$ is the Planck constant. Similarly, the electron's spin magnetic dipole moment $\boldsymbol{\mu}_s$ cannot itself be measured but its component can be measured. Along a z axis, the component is

$$\mu_{s,z} = \pm\frac{eh}{4\pi m} = \pm\mu_B, \qquad (32\text{-}4, 32\text{-}6)$$

where μ_B is the *Bohr magneton:*

$$\mu_B = \frac{eh}{4\pi m} = 9.27 \times 10^{-24}\ \text{J/T}. \qquad (32\text{-}5)$$

The potential energy U associated with the orientation of the spin magnetic dipole moment in an external magnetic field $\mathbf{B}_{\text{ext}}$ is

$$U = -\boldsymbol{\mu}_s \cdot \mathbf{B}_{\text{ext}} = -\mu_{s,z}B. \qquad (32\text{-}7)$$

Orbital Magnetic Dipole Moment

An electron in an atom has an additional angular momentum called its *orbital angular momentum* $\mathbf{L}_{\text{orb}}$, with which an *orbital magnetic dipole moment* $\boldsymbol{\mu}_{\text{orb}}$ is associated:

$$\boldsymbol{\mu}_{\text{orb}} = -\frac{e}{2m}\mathbf{L}_{\text{orb}}. \qquad (32\text{-}8)$$

Orbital angular momentum is quantized and can have only values given by

$$L_{\text{orb},z} = m_l \frac{h}{2\pi}, \quad \text{for}\ m_l = 0, \pm 1, \pm 2, \ldots, \pm(\text{limit}). \quad (32\text{-}9)$$

So, the magnitude of the orbital angular momentum is

$$\mu_{\text{orb},z} = -m_l \frac{eh}{4\pi m} = -m_l \mu_B. \qquad (32\text{-}10, 32\text{-}11)$$

The potential energy U associated with the orientation of the orbital magnetic dipole moment in an external magnetic field $\mathbf{B}_{\text{ext}}$ is

$$U = -\boldsymbol{\mu}_{\text{orb}} \cdot \mathbf{B}_{\text{ext}} = -\mu_{\text{orb},z}B_{\text{ext}}. \qquad (32\text{-}12)$$

Diamagnetism

Diamagnetic materials do not exhibit magnetism until they are placed in an external magnetic field $\mathbf{B}_{\text{ext}}$. They then develop a magnetic dipole moment directed opposite $\mathbf{B}_{\text{ext}}$. If the field is nonuniform, the diamagnetic material is repelled from regions of greater magnetic field. This property is called *diamagnetism.*

Paramagnetism

In a *paramagnetic material,* each atom has a permanent magnetic dipole moment $\boldsymbol{\mu}$, but the dipole moments are randomly oriented and the material as a whole lacks a magnetic field. However, an external magnetic field $\mathbf{B}_{\text{ext}}$ can partially align the atomic dipole moments to give the material a net magnetic dipole moment in the direction of $\mathbf{B}_{\text{ext}}$. If $\mathbf{B}_{\text{ext}}$ is nonuniform, the material is attracted to regions of greater magnetic field. These properties are called *paramagnetism.*

The alignment of the atomic dipole moments increases with an increase in $\mathbf{B}_{\text{ext}}$ and decreases with an increase in temperature T. The extent to which a sample of volume V is magnetized is given by its *magnetization* $\mathbf{M}$, whose magnitude is

$$M = \frac{\text{measured magnetic moment}}{V}. \qquad (32\text{-}18)$$

Complete alignment of all N atomic magnetic dipoles in a sample, called *saturation* of the sample, corresponds to the maximum magnetization value $M_{\text{max}} = N\mu/V$. For low values of the ratio B_{ext}/T, we have the approximation

$$M = C\frac{B_{\text{ext}}}{T} \qquad \text{(Curie's law)}, \qquad (32\text{-}19)$$

where C is called the *Curie constant.*

Ferromagnetism

In the absence of an external magnetic field, some of the electrons in a ferromagnetic material have their magnetic dipole moments aligned by means of a quantum physical interaction called *exchange coupling*, producing regions (domains) within the material with strong magnetic dipole moments. An external field $\mathbf{B}_{\text{ext}}$ can align the magnetic dipole moments of those regions, producing a strong net magnetic dipole moment for the material as a whole, in the direction of $\mathbf{B}_{\text{ext}}$. This net magnetic dipole moment can partially persist when $\mathbf{B}_{ext}$ is removed. If $\mathbf{B}_{ext}$ is nonuniform, the ferromagnetic material is attracted to regions of greater magnetic field. These properties are called *ferromagnetism*. Exchange coupling disappears when a sample's temperature exceeds its *Curie temperature*, and then the sample has only paramagnetism.

Maxwell's Extension of Ampere's Law

A changing electric flux induces a magnetic field $\mathbf{B}$. Maxwell's law,

$$\oint \mathbf{B} \cdot d\mathbf{s} = \mu_0 \epsilon_0 \frac{d\Phi_E}{dt} \quad \text{(Maxwell's law of induction),} \quad (32\text{-}28)$$

relates the magnetic field induced along a closed loop to the changing electric flux Φ_E through the loop. Ampere's law, $\oint \mathbf{B} \cdot d\mathbf{s} = \mu_0 i_{enc}$ (Eq. 32-29), gives the magnetic field generated by a current i_{enc} encircled by a loop. Maxwell's law and Ampere's law can be written as the single equation:

$$\oint \mathbf{B} \cdot d\mathbf{s} = \mu_0 \epsilon_0 \frac{d\Phi_E}{dt} + \mu_0 i_{enc}$$

$$\text{(Ampere–Maxwell law).} \quad (32\text{-}30)$$

Displacement Current

We define the fictitious *displacement current* due to a changing electric field as

$$i_d = \epsilon_0 \frac{d\Phi_E}{dt}. \quad (32\text{-}32)$$

Equation 32-30 then becomes

$$\oint \mathbf{B} \cdot d\mathbf{s} = \mu_0 i_{d,enc} + \mu_0 i_{enc}$$

$$\text{(Ampere–Maxwell law),} \quad (32\text{-}33)$$

where $i_{d,enc}$ is the displacement current encircled by the integration loop. The idea of a displacement current allows us to retain the notion of continuity of current through a capacitor. However, displacement current is *not* a transfer of charge.

Maxwell's Equations

Maxwell's equations, displayed in Table 32-1, summarize electromagnetism and form its foundation.

QUESTIONS

1. Figure 32-18 shows four steel bars; three are permanent magnets. One of the poles is indicated. Through experiment we find that ends a and d attract each other, ends c and f repel, ends e and h attract, and ends a and h attract. (a) Which ends are north poles? (b) Which bar is not a magnet?

FIGURE 32-18 Question 1.

2. Figure 32-19 shows four arrangements of a pair of small compass needles in a region otherwise free of magnetic field. The arrows indicate the directions of the needles and thus also the directions of the magnetic dipole moments. Which pairs are in stable equilibrium?

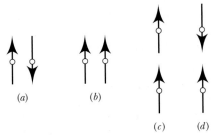

FIGURE 32-19 Question 2.

3. An electron is in an external magnetic field $\mathbf{B}_{ext}$ with spin angular momentum S_z antiparallel to $\mathbf{B}_{ext}$. If the electron under-

goes a *spin-flip* so that S_z is then parallel with $\mathbf{B}_{ext}$, must energy be supplied to or lost by the electron?

4. Figure 32-20a shows the opposite spin orientations of an electron in an external magnetic field $\mathbf{B}_{ext}$. Figure 32-20b gives three choices for the graph of the potential energies associated with those orientations as a function of the magnitude of $\mathbf{B}_{ext}$. Choices b and c consist of intersecting lines, choice a of parallel lines. Which is the correct choice?

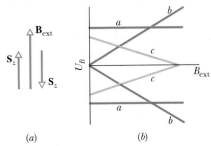

FIGURE 32-20 Question 4.

5. Figure 32-21 shows three loop models of an electron orbiting counterclockwise within a magnetic field. The fields are nonuniform for models 1 and 2 and uniform for model 3. For each model, are (a) the magnetic dipole moment of the loop and (b) the magnetic force on the loop directed up, directed down, or zero?

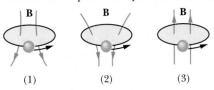

FIGURE 32-21 Questions 5, 7, and 8.

6. Does the magnitude of the net force on the loop of Figs. 32-9a and b increase, decrease, or remain the same if we increase (a) the magnitude of $\mathbf{B}_{\text{ext}}$ and (b) the divergence of $\mathbf{B}_{\text{ext}}$?

7. Replace the current loops of Question 5 and Fig. 32-21 with diamagnetic spheres. For each field, are (a) the magnetic dipole moment of the sphere and (b) the magnetic force on the sphere directed up, directed down, or zero?

8. Replace the current loops of Question 5 and Fig. 32-21 with paramagnetic spheres. For each field, are (a) the magnetic dipole moment of the sphere and (b) the magnetic force on the sphere up, down, or zero?

9. In Fig. 32-22 a nonuniform magnetic field extends between the two pole faces shown. An electron is sent through the field along a path perpendicular to the page (at the dot). Is the force on the electron due to the interaction of its intrinsic spin magnetic dipole moment with the field leftward, rightward, or zero if the electron's spin S_z is directed (a) leftward and (b) rightward? (*Hint:* Assume that the electron is a spinning ball with negative charge on its surface, resulting in a current loop like those in Fig. 32-9.)

FIGURE 32-22 Question 9.

10. Figure 32-23 shows a ferromagnetic sphere that initially has no net magnetic dipole moment; it is held in place by two very fine, taut wires. When a uniform upward magnetic field $\mathbf{B}$ is switched on, the sphere acquires an upward magnetic dipole moment. As the field turns on, does the sphere rotate about the wire clockwise (as shown) or counterclockwise? (*Hint:* Consider the orientations of the spin angular momenta of the electrons and the conservation of angular momentum.)

FIGURE 32-23 Question 10.

11. Figure 32-24 shows, in two situations, an electric field vector $\mathbf{E}$ and an induced magnetic field line. In each, is the magnitude of $\mathbf{E}$ increasing or decreasing?

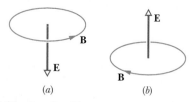

FIGURE 32-24 Question 11.

12. In Figure 32-15b, $\mathbf{E}$ is directed into the page and is increasing in magnitude. Is the direction of the magnetic field $\mathbf{B}$ clockwise or counterclockwise if, instead, $\mathbf{E}$ is out of the page and (a) increasing and (b) decreasing? (c) What is the direction of $\mathbf{B}$ if $\mathbf{E}$ is out of the page and not changing?

13. Figure 32-25 shows a face-on view of one of the two square plates of a parallel-plate capacitor, as well as four loops that are located between the plates. The capacitor is being discharged. (a) Neglecting fringing of the magnetic field, rank the loops according to the magnitude of $\oint \mathbf{B} \cdot d\mathbf{s}$ along them, greatest first. (b) Along which loop, if any, is the angle between $\mathbf{B}$ and $d\mathbf{s}$ constant (so that their dot product can easily be evaluated)? (c) Along which loop, if any, is B constant (so that B can be brought in front of the integral sign in Eq. 32-28)?

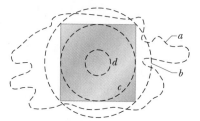

FIGURE 32-25 Question 13.

14. Figure 32-26 shows a parallel-plate capacitor and the current in the connecting wires that is discharging the capacitor. Are the directions of (a) $\mathbf{E}$ and (b) i_d leftward or rightward? (c) Is the magnetic field at point P into the page or out of the page?

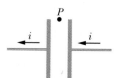

FIGURE 32-26 Question 14.

15. A parallel-plate capacitor with rectangular plates is being discharged. Consider a rectangular loop centered on the plates and between them. The loop measures L by $2L$; the plate measures $2L$ by $4L$. What fraction of the displacement current is encircled by the loop?

16. Figure 32-27a shows a capacitor that is being charged. Point a (near one of the connecting wires) and point b (inside the capacitor gap) are equidistant from the central axis, as are point c (near the wire) and point d (between the plates but outside the gap). In Fig. 32-27b, one curve gives the variation with distance r of the magnitude of the magnetic field inside and outside the wire. The other curve gives the variation with distance r of the magnitude of the magnetic field inside and outside the gap. The two curves partially overlap. Which of the three points on the curves correspond to which of the four points of Fig. 32-27a?

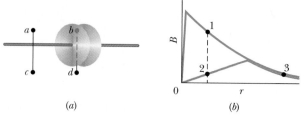

FIGURE 32-27 Question 16.

EXERCISES & PROBLEMS

SECTION 32-2 Gauss' Law for Magnetic Fields

1E. Imagine rolling a sheet of paper into a cylinder and placing a bar magnet near its end as shown in Fig. 32-28. (a) Sketch the magnetic field lines that pass through the surface of the cylinder. (b) What can you say about the sign of $\mathbf{B} \cdot d\mathbf{A}$ for every area $d\mathbf{A}$ on the surface? (c) Does this contradict Gauss' law for magnetism? Explain.

FIGURE 32-28 Exercise 1.

2E. The magnetic flux through each of five faces of a die (singular of "dice") is given by $\Phi_B = \pm N$ Wb, where $N (= 1$ to $5)$ is the number of spots on the face. The flux is positive (outward) for N even and negative (inward) for N odd. What is the flux through the sixth face of the die?

3P. A Gaussian surface in the shape of a right-circular cylinder with end caps has a radius of 12.0 cm and a length of 80.0 cm. Through one end there is an inward magnetic flux of 25.0 μWb. At the other end there is a uniform magnetic field of 1.60 mT, normal to the surface and directed outward. What is the net magnetic flux through the curved surface?

4P*. Two wires, parallel to the z axis and a distance $4r$ apart, carry equal currents i in opposite directions, as shown in Fig. 32-29. A circular cylinder of radius r and length L has its axis on the z axis, midway between the wires. Use Gauss' law for magnetism to calculate the net outward magnetic flux through the half of the cylindrical surface above the x axis. (*Hint:* Find the flux through that portion of the xz plane that is within the cylinder.)

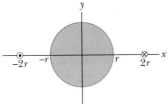

FIGURE 32-29 Problem 4.

SECTION 32-3 The Magnetism of Earth

5E. In New Hampshire the average horizontal component of Earth's magnetic field in 1912 was 16 μT and the average inclination or "dip" was 73°. What was the corresponding magnitude of Earth's magnetic field?

6E. In Sample Problem 32-1 the vertical component of Earth's magnetic field in Tucson, Arizona, was found to be 43 μT. Assume this is the average value for all of Arizona, which has an area of 295,000 km², and calculate the net magnetic flux through the rest of Earth's surface (the entire surface excluding Arizona). Is the flux outward or inward?

7E. Earth has a magnetic dipole moment of 8.0×10^{22} J/T. (a) What current would have to be produced in a single turn of wire extending around Earth at its geomagnetic equator if we wished to set up such a dipole? Could such an arrangement be used to cancel out Earth's magnetism (b) at points in space well above Earth's surface or (c) on Earth's surface?

8P. The magnetic field of Earth can be approximated as the magnetic field of a dipole, with horizontal and vertical components, at a point a distance r from Earth's center, given by

$$B_h = \frac{\mu_0 \mu}{4\pi r^3} \cos \lambda_m, \qquad B_v = \frac{\mu_0 \mu}{2\pi r^3} \sin \lambda_m,$$

where λ_m is the *magnetic latitude* (latitude measured from the geomagnetic equator toward the north or south geomagnetic pole). Assume that Earth's magnetic dipole moment is $\mu = 8.00 \times 10^{22}$ A·m². (a) Show that the magnitude of Earth's field at latitude λ_m is given by

$$B = \frac{\mu_0 \mu}{4\pi r^3} \sqrt{1 + 3 \sin^2 \lambda_m}.$$

(b) Show that the inclination ϕ_i of the magnetic field is related to the magnetic latitude λ_m by

$$\tan \phi_i = 2 \tan \lambda_m.$$

9P. Use the results displayed in Problem 8 to predict Earth's magnetic field (both magnitude and inclination) at (a) the geomagnetic equator; (b) a point at geomagnetic latitude 60°; (c) the north geomagnetic pole.

10P. Using the approximations given in Problem 8, find (a) the altitude above Earth's surface where the magnitude of its magnetic field is 50% of the surface value at the same latitude, (b) the maximum magnitude of the magnetic field at the core–mantle boundary, 2900 km below Earth's surface, and (c) the magnitude and inclination of Earth's magnetic field at the north geographic pole. Suggest why the values you calculated for (c) differ from measured values.

SECTION 32-4 Magnetism and Electrons

11E. What is the energy difference between parallel and antiparallel alignment of the z component of an electron's spin magnetic dipole moment with an external magnetic field of magnitude 0.25 T, directed parallel to the z axis?

12E. What is the measured component of the orbital magnetic dipole moment of an electron with (a) $m_l = 1$ and (b) $m_l = -2$?

13E. In the lowest energy state of the hydrogen atom, the most probable distance of the single electron from the central proton (the nucleus) is 5.2×10^{-11} m. (a) Compute the magnitude of the proton's electric field at that distance. The component $\mu_{s,z}$ of the proton's spin magnetic dipole moment measured on a z axis is 1.4×10^{-26} J/T. (b) Compute the magnitude of the proton's magnetic field at the distance 5.2×10^{-11} m on the z axis. (*Hint:*

Use Eq. 30-29.) (c) What is the ratio of the spin magnetic dipole moment of the electron to that of the proton?

14E. If an electron in an atom has an orbital angular momentum with $m_l = 0$, what are the components (a) $L_{orb,z}$ and (b) $\mu_{orb,z}$? If the atom is in an external magnetic field **B** of magnitude 35 mT and directed along the z axis, what are the potential energies associated with the orientations of (c) the electron's orbital magnetic dipole moment and (d) the electron's spin magnetic dipole moment? (e) Repeat (a) through (d) for $m_l = -3$.

15E. If an electron in an atom has orbital angular momentum with m_l values limited by ±3, how many values of (a) $L_{orb,z}$ and (b) $\mu_{orb,z}$ can it have? In terms of h, m, and e, what are the greatest and least allowed magnitudes for (c) $L_{orb,z}$ and (d) $\mu_{orb,z}$? (e) What is the greatest allowed magnitude for the z component of its *net* angular momentum (orbital plus spin)? (f) How many values (signs included) are allowed for the z component of its net angular momentum?

16P. Figure 32-30a is a one-axis graph along which two of the allowed energy values (*levels*) of an atom are plotted (as in Fig. 8-17). When the atom is placed in a magnetic field of 0.50 T, the graph changes to that of Fig. 32-30b because of the energy associated with $\boldsymbol{\mu}_{orb} \cdot \mathbf{B}$. (We neglect $\boldsymbol{\mu}_s$.) Level E_1 is unchanged, but level E_2 splits into a (closely spaced) triplet of levels. What are the allowed values of m_l associated with (a) energy level E_1 and (b) energy level E_2? (c) On the graph, what is the spacing between the triplet levels?

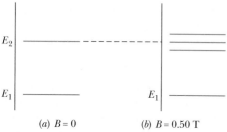

(a) $B = 0$ (b) $B = 0.50$ T

FIGURE 32-30   Problem 16.

17P. A charge q is distributed uniformly around a thin ring of radius r. The ring is rotating about an axis through its center and perpendicular to its plane, at an angular speed ω. (a) Show that the magnetic moment due to the rotating charge is

$$\mu = \tfrac{1}{2}q\omega r^2.$$

(b) What is the direction of this magnetic moment if the charge is positive?

SECTION 32-6   Diamagnetism

18E. Figure 32-31 shows a loop model (loop L) for a diamagnetic material. (a) Sketch the magnetic field lines through and about the material due to the bar magnet. (b) What are the direction of the loop's net magnetic dipole moment $\boldsymbol{\mu}$ and the direction

FIGURE 32-31   Exercises 18 and 22.

of the conventional current i in the loop? (c) What is the direction of the magnetic force on the loop?

19P*. Assume that an electron of mass m and charge magnitude e moves in a circular orbit of radius r about a nucleus. A uniform magnetic field **B** is then established perpendicular to the plane of the orbit. Assuming also that the radius of the orbit does not change and that the change in the speed of the electron due to field **B** is small, find an expression for the change in the orbital magnetic dipole moment of the electron.

SECTION 32-7   Paramagnetism

20E. A 0.50 T magnetic field is applied to a paramagnetic gas whose atoms have an intrinsic magnetic dipole moment of 1.0×10^{-23} J/T. At what temperature will the mean kinetic energy of translation of the gas atoms be equal to the energy required to reverse such a dipole end for end in this magnetic field?

21E. A magnet in the form of a cylindrical rod has a length of 5.00 cm and a diameter of 1.00 cm. It has a uniform magnetization of 5.30×10^3 A/m. What is its magnetic dipole moment?

22E. Repeat Exercise 18 for the case in which loop L is the model for a paramagnetic material.

23E. A sample of the paramagnetic salt to which the magnetization curve of Fig. 32-10 applies is to be tested to see whether it obeys Curie's law. The sample is placed in a uniform 0.50 T magnetic field that remains constant throughout the experiment. The magnetization M is then measured at temperatures ranging from 10 to 300 K. Will it be found that Curie's law is valid under these conditions?

24E. A sample of the paramagnetic salt to which the magnetization curve of Fig. 32-10 applies is held at room temperature (300 K). At what applied magnetic field will the degree of magnetic saturation of the sample be (a) 50% and (b) 90%? (c) Are these fields attainable in the laboratory?

25E. A sample of the paramagnetic salt to which the magnetization curve of Fig. 32-10 applies is immersed in a uniform magnetic field of 2.0 T. At what temperature will the degree of magnetic saturation of the sample be (a) 50% and (b) 90%?

26P. An electron with kinetic energy K_e travels in a circular path that is perpendicular to a uniform magnetic field, the electron's motion subject only to the force of the field. (a) Show that the magnetic dipole moment of the electron due to its orbital motion has magnitude $\mu = K_e/B$ and that it is in the direction opposite that of **B**. (b) What are the magnitude and direction of the magnetic dipole moment of a positive ion with kinetic energy K_i under the same circumstances? (c) An ionized gas consists of 5.3×10^{21} electrons/m^3 and the same number density of ions. Take the average electron kinetic energy to be 6.2×10^{-20} J and the average ion kinetic energy to be 7.6×10^{-21} J. Calculate the magnetization of the gas when it is in a magnetic field of 1.2 T.

27P. Consider a solid containing N atoms per unit volume, each atom having a magnetic dipole moment $\boldsymbol{\mu}$. Suppose the direction of $\boldsymbol{\mu}$ can be only parallel or antiparallel to an externally applied magnetic field **B** (this will be the case if $\boldsymbol{\mu}$ is due to the spin of a single electron). According to statistical mechanics, it can be shown that the probability of an atom being in a state with energy U is proportional to $e^{-U/kT}$, where T is the temperature and k is Boltzmann's constant. Thus, since $U = -\boldsymbol{\mu} \cdot \mathbf{B}$, the fraction of atoms whose dipole moment is parallel to $\mathbf{B}$ is proportional to $e^{\mu B/kT}$ and the fraction of atoms whose dipole moment is antiparallel to $\mathbf{B}$ is proportional to $e^{-\mu B/kT}$. (a) Show that the magnetization of this solid is $M = N\mu \tanh(\mu B/kT)$. Here tanh is the hyperbolic tangent function: $\tanh(x) = (e^x - e^{-x})/(e^x + e^{-x})$. (b) Show that the result given in (a) reduces to $M = N\mu^2 B/kT$ for $\mu B \ll kT$. (c) Show that the result of (a) reduces to $M = N\mu$ for $\mu B \gg kT$. (d) Show that both (b) and (c) agree qualitatively with Fig. 32-10.

SECTION 32-8 Ferromagnetism

28E. Measurements in mines and boreholes indicate that Earth's interior temperature increases with depth at the average rate of 30 C°/km. Assuming a surface temperature of 10°C, at what depth does iron cease to be ferromagnetic? (The Curie temperature of iron varies very little with pressure.)

29E. The exchange coupling mentioned in Section 32-8 as being responsible for ferromagnetism is *not* the mutual magnetic interaction between two elementary magnetic dipoles. To show this, calculate (a) the magnetic field a distance of 10 nm away, along the dipole axis, from an atom with magnetic dipole moment 1.5×10^{-23} J/T (cobalt), and (b) the minimum energy required to turn a second identical dipole end for end in this field. By comparing the latter with the results of Sample Problem 32-2, what can you conclude?

30E. The saturation magnetization M_{max} of the ferromagnetic metal nickel is 4.70×10^5 A/m. Calculate the magnetic moment of a single nickel atom. (The density of nickel is 8.90 g/cm³ and its molar mass is 58.71 g/mol.)

31E. The dipole moment associated with an atom of iron in an iron bar is 2.1×10^{-23} J/T. Assume that all the atoms in the bar, which is 5.0 cm long and has a cross-sectional area of 1.0 cm², have their dipole moments aligned. (a) What is the dipole moment of the bar? (b) What torque must be exerted to hold this magnet perpendicular to an external field of 1.5 T? The density of iron is 7.9 g/cm³.

32P. The magnetic dipole moment of Earth is 8.0×10^{22} J/T. (a) If the origin of this magnetism were a magnetized iron sphere at the center of Earth, what would be its radius? (b) What fraction of the volume of Earth would such a sphere occupy? Assume complete alignment of the dipoles. The density of Earth's inner core is 14 g/cm³. The magnetic dipole moment of an iron atom is 2.1×10^{-23} J/T. (*Note:* Earth's inner core is in fact thought to be in both liquid and solid forms and partly iron, but a permanent magnet as the source of Earth's magnetism has been ruled out by several considerations. For one, the temperature is certainly above the Curie point.)

33P. Figure 32-32 shows the apparatus used in a lecture demonstration of para- and diamagnetism. A sample of the magnetic material is suspended by a long string in the nonuniform field ($d = 2$ cm) between the poles of a powerful electromagnet. Pole

P_1 is sharply pointed and pole P_2 is rounded as indicated. Any deflection of the string from the vertical is visible to the audience by means of an optical projection system (not shown). (a) First a bismuth (highly diamagnetic) sample is used. When the electromagnet is turned on, the sample is observed to deflect slightly (about 1 mm) toward one of the poles. What is the direction of this deflection? (b) Next an aluminum (paramagnetic, conducting) sample is used. When the electromagnet is turned on, the sample is observed to deflect strongly (about 1 cm) toward one pole for about a second and then deflect moderately (a few millimeters) toward the other pole. Explain and indicate the direction of these deflections. (*Hint:* The aluminum sample is a conductor, for which Lenz's law applies.) (c) What would happen if a ferromagnetic sample were used?

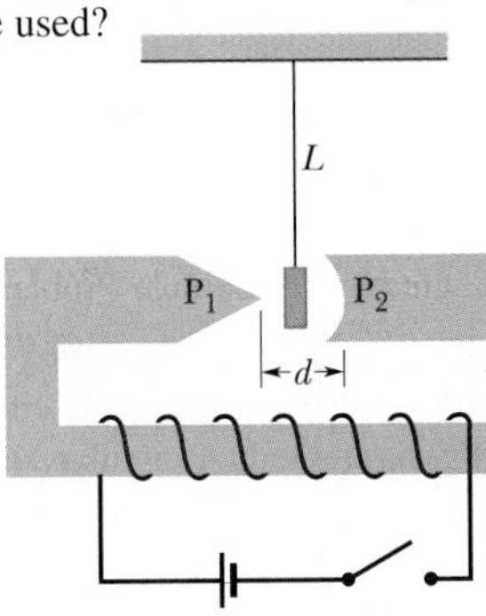

FIGURE 32-32
Problem 33.

34P. A magnetic compass has its needle of mass 0.050 kg and length 4.0 cm aligned with the horizontal component of Earth's magnetic field at a place where that component has the value $B_h = 16 \ \mu$T. After the compass is given a momentary gentle shake, the needle oscillates with angular frequency $\omega = 45$ rad/s. Assuming that the needle is a uniform thin rod mounted at its center, find its magnetic dipole moment.

35P. A Rowland ring is formed of ferromagnetic material. It is circular in cross section, with an inner radius of 5.0 cm and an outer radius of 6.0 cm, and is wound with 400 turns of wire. (a) What current must be set up in the windings to attain a toroidal field $B_0 = 0.20$ mT? (b) A secondary coil wound around the toroid has 50 turns and resistance 8.0 Ω. If, for this value of B_0, we have $B_M = 800B_0$, how much charge moves through the secondary coil when the current in the toroid windings is turned on?

SECTION 32-9 Induced Magnetic Fields

36E. For the situation of Sample Problem 32-4, at what radius r is the induced magnetic field equal to 50% of its maximum value?

37E. The induced magnetic field 6.0 mm from the central axis of a circular parallel-plate capacitor and between the plates is 2.0×10^{-7} T. The plates have radius 3.0 mm. At what rate dE/dt is the electric field between the plates changing?

38P. A parallel-plate capacitor, with circular plates of radius $R = 16$ mm and gap width $d = 5.0$ mm, has a uniform electric field between the plates. Starting at time $t = 0$, the potential difference between the plates is $V = (100 \text{ V})e^{-t/\tau}$, where the time constant $\tau = 12$ ms. At the radial distance $r = 0.80R$ from the capacitor axis within the gap, what is the magnetic field (a) as a function of time for $t \ge 0$ and (b) at time $t = 3\tau$?

39P. Suppose that a parallel-plate capacitor has circular plates with radius $R = 30$ mm and a plate separation of 5.0 mm. Suppose also that a sinusoidal potential difference with a maximum value of 150 V and a frequency of 60 Hz is applied across the plates; that is,

$$V = (150 \text{ V}) \sin[2\pi(60 \text{ Hz})t].$$

(a) Find $B_{max}(R)$, the maximum value of the induced magnetic field which occurs at $r = R$. (b) Plot $B_{max}(r)$ for $0 < r < 10$ cm.

SECTION 32-10   Displacement Current

40E. Prove that the displacement current in a parallel-plate capacitor of capacitance C can be written as $i_d = C(dV/dt)$, where V is the potential difference between the plates.

41E. At what rate must the potential difference between the plates of a parallel-plate capacitor with a 2.0 μF capacitance be changed to produce a displacement current of 1.5 A?

42E. For the situation of Sample Problem 32-4, show that the *displacement current density* is $J_d = \epsilon_0(dE/dt)$ for $r \le R$.

43E. A parallel-plate capacitor with circular plates of radius 0.10 m is being discharged. A circular loop of radius 0.20 m is concentric with the capacitor and halfway between the plates. The displacement current through the loop is 2.0 A. At what rate is the electric field between the plates changing?

44E. A parallel-plate capacitor with circular plates of radius R is being discharged. The displacement current through a central circular area, parallel to the plates and with radius $R/2$, is 2.0 A. What is the discharge current?

45P. The magnitude of the electric field between the two circular plates of the parallel-plate capacitor in Fig. 32-33 is $E = (4.0 \times 10^5) - (6.0 \times 10^4 t)$, with E in volts per meter and t in seconds. At $t = 0$, the field is upward as shown. The plate area is 4.0×10^{-2} m^2. For $t \ge 0$, (a) what are the magnitude and direction of the displacement current between the plates and (b) is the direction of the induced magnetic field clockwise or counterclockwise around the plates?

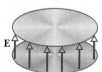

FIGURE 32-33   Problem 45.

46P. A parallel-plate capacitor with circular plates of radius R is being discharged by a current of 6.0 A. (a) At what distances from the central axis is the induced magnetic field equal to 75% of the maximum value of that field? (b) What is the maximum value of that field if $R = 0.040$ m?

47P. As a parallel-plate capacitor with circular plates 20 cm in diameter is being charged, the displacement current density throughout the region between the plates is uniform and has a magnitude of 20 A/m^2. (a) Calculate the magnitude B of the magnetic field at a distance $r = 50$ mm from the axis of symmetry of the region. (b) Calculate dE/dt in this region.

48P. A uniform electric field collapses to zero from an initial strength of 6.0×10^5 N/C in a time of 15 μs in the manner

shown in Fig. 32-34. Calculate the magnitude of the displacement current, through a 1.6 m^2 region perpendicular to the field, during each of the time intervals a, b, and c shown on the graph. (Ignore the behavior at the ends of the intervals.)

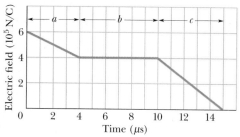

FIGURE 32-34   Problem 48.

49P. A parallel-plate capacitor has square plates 1.0 m on a side as in Fig. 32-35. A current of 2.0 A charges the capacitor, producing a uniform electric field **E** between the plates, with **E** perpendicular to the plates. (a) What is the displacement current through the region between the plates? (b) What is dE/dt in this region? (c) What is the displacement current through the square dashed path between the plates? (d) What is $\oint \mathbf{B} \cdot d\mathbf{s}$ around this square dashed path?

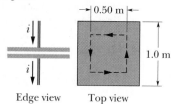

FIGURE 32-35   Problem 49.

50P. A silver wire has resistivity $\rho = 1.62 \times 10^{-8}$ $\Omega \cdot$m and a cross-sectional area of 5.00 mm^2. The current in the wire is uniform and changing at the rate of 2000 A/s when the current is 100 A. (a) What is the (uniform) electric field in the wire when the current in the wire is 100 A? (b) What is the displacement current in the wire at that time? (c) What is the ratio of the magnetic field due to the displacement current to that due to the current at a distance r from the wire?

51P. The capacitor in Fig. 32-36 with circular plates of radius $R = 18.0$ cm is connected to a source of emf $\mathscr{E} = \mathscr{E}_m \sin \omega t$, where $\mathscr{E}_m = 220$ V and $\omega = 130$ rad/s. The maximum value of the displacement current is $i_d = 7.60$ μA. Neglect fringing of the electric field at the edges of the plates. (a) What is the maximum value of the current i? (b) What is the maximum value of $d\Phi_E/dt$, where Φ_E is the electric flux through the region between the plates? (c) What is the separation d between the plates? (d) Find the maximum value of the magnitude of **B** between the plates at a distance $r = 11.0$ cm from the center.

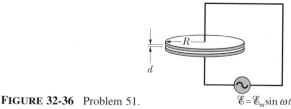

FIGURE 32-36   Problem 51.

33
Electromagnetic Oscillations and Alternating Current

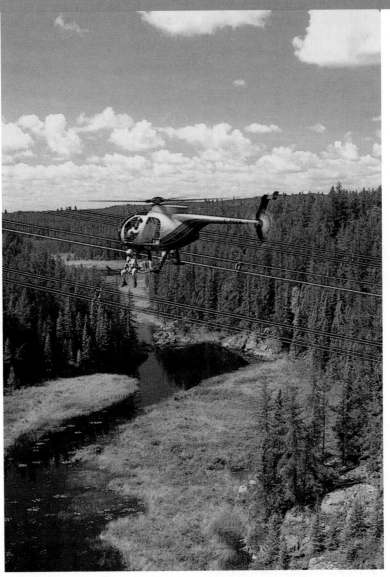

When a high-voltage power transmission line requires repair, a utility company cannot just shut it down, blacking out, perhaps, an entire city. So repairs must be made while the lines are electrically "hot." The man outside the helicopter has just replaced a spacer between 500 kV lines <u>by hand</u>, a procedure that requires considerable expertise. Why, exactly, is the potential of such power transmission lines kept so high? Surprisingly, the current through them, although highly lethal, is not very large. Shouldn't it be large?

33-1 NEW PHYSICS— OLD MATHEMATICS

In this chapter you will see how the electric charge q varies with time in a circuit made up of an inductor L, a capacitor C, and a resistor R. From another point of view, we shall discuss how energy shuttles back and forth between the magnetic field of the inductor and the electric field of the capacitor, being gradually dissipated—while these oscillations continue—as thermal energy in the resistor.

We have discussed oscillations before, in another context. In Chapter 16 we saw how displacement x varies with time in a mechanical oscillating system made up of a block of mass m, a spring of spring constant k, and a viscous or frictional element such as oil; Fig. 16-17 shows such a system. You also saw how energy shuttles back and forth between the kinetic energy of the oscillating mass and the potential energy of the spring, being gradually dissipated—while the oscillations continue—as thermal energy.

The parallel between these two idealized systems is exact, and the controlling differential equations are identical. Thus there is no new mathematics to be learned; we can simply change the symbols and give our full attention to the physics of the situation.

33-2 *LC* OSCILLATIONS, QUALITATIVELY

Of the three circuit elements, resistance R, capacitance C, and inductance L, we have so far discussed the series combinations RC (in Section 28-8) and RL (in Section 31-9). In these two kinds of circuit we found that the charge, current, and potential difference grow and decay exponentially. The time scale of the growth or decay is given by a *time constant* τ, which is either capacitive or inductive.

We now examine the remaining two-element circuit combination LC. You will see that in this case the charge, current, and potential difference do not decay exponentially with time but vary sinusoidally (with period T and angular frequency ω). The circuit is said to *oscillate,* and the resulting oscillations of the capacitor's electric field

and the inductor's magnetic field are said to be **electromagnetic oscillations.**

Parts a through h of Fig. 33-1 show succeeding stages of the oscillations in a simple LC circuit. From Eq. 26-21, the energy stored in the electric field of the capacitor at any time is

$$U_E = \frac{q^2}{2C}, \qquad (33\text{-}1)$$

where q is the charge on the capacitor. From Eq. 31-53, the energy stored in the magnetic field of the inductor at any time is

$$U_B = \frac{Li^2}{2}, \qquad (33\text{-}2)$$

where i is the current through the inductor.

We now adopt the convention of representing *instantaneous values* of the electrical quantities of a sinusoidally oscillating circuit with small letters, such as q, and the *amplitudes* of those quantities with capital letters, such as Q. With this convention in mind, let us assume that initially the charge q on the capacitor in Fig. 33-1 is at its maximum value Q and that the current i through the inductor is zero. This initial state of the circuit is shown in Fig. 33-1a. The bar graphs for energy included there indicate that at this instant, with zero current through the inductor and maximum charge on the capacitor, the energy U_B of the magnetic field is zero and the energy U_E of the electric field is a maximum.

The capacitor now starts to discharge through the inductor, positive charge carriers moving counterclockwise, as shown in Fig. 33-1b. This means that a current i, given by dq/dt and pointing down in the inductor, is established. As the capacitor's charge decreases, the energy stored in the electric field within the capacitor also decreases. This energy is transferred to the magnetic field that appears around the inductor because of the current i that is building up there. Thus the electric field decreases and the magnetic field builds up as energy is transferred from the electric field to the magnetic field.

The capacitor eventually loses all its charge (Fig. 33-1c) and thus also loses its electric field and the energy stored in that field. The energy has then been fully transferred to the magnetic field of the inductor. Because the magnetic field is then at its maximum magnitude, the current through the inductor is then at its maximum value I.

Although the charge on the capacitor is now zero, the counterclockwise current must continue because the inductor does not allow it to change suddenly to zero. So, the current continues to transfer positive charge from the top

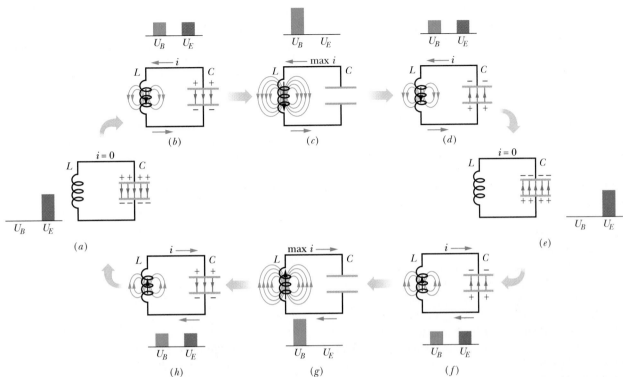

FIGURE 33-1 Eight stages in a single cycle of oscillation of a resistanceless *LC* circuit. The bar graphs by each figure show the stored magnetic and electric energies. The magnetic field lines of the inductor and the electric field lines of the capacitor are shown. (*a*) Capacitor with maximum charge, no current. (*b*) Capacitor discharging, current increasing. (*c*) Capacitor fully discharged, current maximum. (*d*) Capacitor charging but with polarity opposite that in (*a*), current decreasing. (*e*) Capacitor with maximum charge having polarity opposite that in (*a*), no current. (*f*) Capacitor discharging, current increasing with direction opposite that in (*b*). (*g*) Capacitor fully discharged, current maximum. (*h*) Capacitor charging, current decreasing.

plate to the bottom plate through the circuit (Fig. 33-1*d*). Energy now flows from the inductor back to the capacitor as the electric field within the capacitor builds up again. The current gradually decreases during this energy transfer. When, eventually, the energy has been transferred completely back to the capacitor (Fig. 33-1*e*), the current has decreased to zero (momentarily). The situation of Fig. 33-1*e* is like the initial situation, except that the capacitor is now charged oppositely.

The capacitor then starts to discharge again but now with a clockwise current (Fig. 33-1*f*). Reasoning as before, we see that the clockwise current builds to a maximum (Fig. 33-1*g*) and then decreases (Fig. 33-1*h*), until the circuit eventually returns to its initial situation (Fig. 33-1*a*). The process then repeats at a frequency *f* and thus at an angular frequency $\omega = 2\pi f$. In the ideal circuit with no resistance, there are no energy transfers other than that between the electric field of the capacitor and the magnetic field of the inductor. Owing to the conservation of energy, the oscillations continue indefinitely. The oscillations need

not begin with the energy all in the electric field; the initial situation could be any other stage of the oscillation.

To find the charge *q* on the capacitor as a function of time, we can use a voltmeter to measure the time-varying potential difference (or *voltage*) v_C that exists across the capacitor *C*. From Eq. 26-1 we can write

$$v_C = \left(\frac{1}{C}\right) q,$$

which allows us to find *q*. To measure the current, we can connect a small resistance *R* in series with the capacitor and inductor and measure the time-varying potential difference v_R across it; v_R is proportional to *i* through the relation

$$v_R = iR.$$

We assume here that *R* is so small that its effect on the behavior of the circuit is negligible. The variations in time of v_C and v_R, and thus of *q* and *i*, are shown in Fig. 33-2. All four quantities vary sinusoidally.

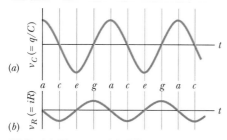

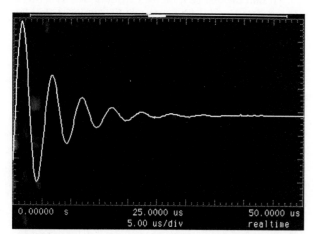

FIGURE 33-2 (*a*) The potential difference across the capacitor of the circuit of Fig. 33-1 as a function of time. This quantity is proportional to the charge on the capacitor. (*b*) A potential proportional to the current in the circuit of Fig. 33-1. The letters refer to the correspondingly labeled oscillation stages in Fig. 33-1.

In an actual *LC* circuit, the oscillations will not continue indefinitely because there is always some resistance present that will drain energy from the electric and magnetic fields and dissipate it as thermal energy; the circuit may become warmer. The oscillations, once started, will die away as Fig. 33-3 suggests. Compare this figure with Fig. 16-18, which shows the decay of mechanical oscillations caused by frictional damping in a block–spring system.

C̲HECKPOINT **1:** A charged capacitor and an inductor are connected in series at time $t = 0$. In terms of the period T of the resulting oscillations, determine how much later the following reach their maximums: (a) the charge on the capacitor, (b) the voltage across the capacitor, with its original polarity, (c) the energy stored in the electric field, and (d) the current.

FIGURE 33-3 An oscilloscope trace showing how the oscillations in an *RLC* circuit actually die away because energy is dissipated in the resistor in thermal form.

SAMPLE PROBLEM 33-1

A 1.5 μF capacitor is charged to 57 V. The charging battery is then disconnected, and a 12 mH coil is connected in series with the capacitor so that *LC* oscillations occur. What is the maximum current in the coil? Assume that the circuit contains no resistance.

SOLUTION: From the principle of conservation of energy, the maximum stored energy in the capacitor must equal the maximum stored energy in the inductor. This leads, from Eqs. 33-1 and 33-2 to

$$\frac{Q^2}{2C} = \frac{LI^2}{2},$$

where I is the maximum current and Q is the maximum charge. (The maximum current and the maximum charge occur not at the same time but one-fourth of a cycle apart, as is evident from Figs. 33-1 and 33-2.) Solving for I and substituting CV for Q, we find

$$I = V\sqrt{\frac{C}{L}} = (57 \text{ V})\sqrt{\frac{1.5 \times 10^{-6} \text{ F}}{12 \times 10^{-3} \text{ H}}}$$

$$= 0.637 \text{ A} \approx 640 \text{ mA}. \qquad \text{(Answer)}$$

33-3 THE ELECTRICAL– MECHANICAL ANALOGY

Let us look a little closer at the analogy between the oscillating *LC* system of Fig. 33-1 and an oscillating block–spring system. Two kinds of energy are involved in the block–spring system. One is potential energy of the compressed or extended spring; the other is kinetic energy of the moving block. These two energies are given by the familiar formulas at the left in Table 33-1.

The table also shows the two kinds of energy involved in *LC* oscillations. We can see an analogy between the two pairs of energies—the mechanical energies of the block–spring system and the electromagnetic energies of the *LC* oscillator. The equations for v and i at the bottom of the table help us to improve the analogy. They tell us that q

TABLE 33-1 THE ENERGY IN TWO OSCILLATING SYSTEMS COMPARED

BLOCK–SPRING SYSTEM		LC OSCILLATOR	
ELEMENT	ENERGY	ELEMENT	ENERGY
Spring	Potential, $\frac{1}{2}kx^2$	Capacitor	Electric, $\frac{1}{2}(1/C)q^2$
Block	Kinetic, $\frac{1}{2}mv^2$	Inductor	Magnetic, $\frac{1}{2}Li^2$
	$v = dx/dt$		$i = dq/dt$

corresponds to *x*, and *i* corresponds to *v* (in both equations, the former is differentiated to obtain the latter). These correspondences lead us to pair the individual energies horizontally as shown in the table, hence suggest that $1/C$ corresponds to *k* and *L* corresponds to *m*. Thus

q corresponds to x, $1/C$ corresponds to k,
i corresponds to v, and L corresponds to m.

These correspondences suggest that in an *LC* oscillator, the capacitor is mathematically like the spring in a block–spring system, and the inductor is like the block.

In Section 16-3 we saw that the angular frequency of oscillation of a (frictionless) block–spring system is

$$\omega = \sqrt{\frac{k}{m}} \quad \text{(block–spring system).} \quad (33\text{-}3)$$

The correspondences listed above suggest that to find the angular frequency of oscillation for a (resistanceless) *LC* circuit, *k* should be replaced by $1/C$ and *m* by *L*, yielding

$$\omega = \frac{1}{\sqrt{LC}} \quad \text{(LC circuit).} \quad (33\text{-}4)$$

We derive this result in the next section.

33-4 *LC* OSCILLATIONS, QUANTITATIVELY

Here we want to show explicitly that Eq. 33-4 for the angular frequency of *LC* oscillations is correct. At the same time, we want to examine even more closely the analogy between *LC* oscillations and block–spring oscillations. We start by extending somewhat our earlier treatment of the mechanical block–spring oscillator.

The Block–Spring Oscillator

We analyzed block–spring oscillations in Chapter 16 in terms of energy transfers and did not—at that early stage —derive the fundamental differential equation that governs those oscillations. We do so now.

We can write, for the total energy *U* of a block–spring oscillator at any instant,

$$U = U_b + U_s = \tfrac{1}{2}mv^2 + \tfrac{1}{2}kx^2, \quad (33\text{-}5)$$

where U_b and U_s are, respectively, the kinetic energy of the moving block and the potential energy of the stretched or compressed spring. If there is no friction—which we assume—the total energy *U* remains constant with time, even though *v* and *x* vary. In more formal language,

$dU/dt = 0$. This leads to

$$\frac{dU}{dt} = \frac{d}{dt}(\tfrac{1}{2}mv^2 + \tfrac{1}{2}kx^2)$$

$$= mv\frac{dv}{dt} + kx\frac{dx}{dt} = 0. \quad (33\text{-}6)$$

But $v = dx/dt$ and $dv/dt = d^2x/dt^2$. With these substitutions, Eq. 33-6 becomes

$$m\frac{d^2x}{dt^2} + kx = 0 \quad \begin{array}{l}\text{(block–spring}\\ \text{oscillations).}\end{array} \quad (33\text{-}7)$$

Equation 33-7 is the fundamental *differential equation* that governs the frictionless block–spring oscillations. It involves the displacement *x* and its second derivative with respect to time.

The general solution to Eq. 33-7, that is, the function $x(t)$ that describes the block–spring oscillations, is (as we saw in Eq. 16-3)

$$x = X\cos(\omega t + \phi) \quad \text{(displacement),} \quad (33\text{-}8)$$

in which *X* is the amplitude of the mechanical oscillations (represented by x_m in Chapter 16), ω is the angular frequency of the oscillations, and ϕ is a phase constant.

The *LC* Oscillator

Now let us analyze the oscillations of a resistanceless *LC* circuit, proceeding exactly as we just did for the block–spring oscillator. The total energy *U* present at any instant in an oscillating *LC* circuit is given by

$$U = U_B + U_E = \frac{Li^2}{2} + \frac{q^2}{2C}, \quad (33\text{-}9)$$

in which U_B is the energy stored in the magnetic field of the inductor and U_E is the energy stored in the electric field of the capacitor. Since we have assumed the circuit resistance to be zero, no energy is transferred to thermal energy and *U* remains constant with time. In more formal language, dU/dt must be zero. This leads to

$$\frac{dU}{dt} = \frac{d}{dt}\left(\frac{Li^2}{2} + \frac{q^2}{2C}\right)$$

$$= Li\frac{di}{dt} + \frac{q}{C}\frac{dq}{dt} = 0. \quad (33\text{-}10)$$

But $i = dq/dt$ and $di/dt = d^2q/dt^2$. With these substitutions, Eq. 33-10 becomes

$$L\frac{d^2q}{dt^2} + \frac{1}{C}q = 0 \quad \text{(LC oscillations).} \quad (33\text{-}11)$$

This is the *differential equation* that describes the oscillations of a resistanceless LC circuit. Careful comparison shows that Eqs. 33-11 and 33-7 are exactly of the same mathematical form, differing only in the symbols used.

Since the differential equations are mathematically identical, their solutions must also be mathematically identical. Because q corresponds to x, we can write the general solution of Eq. 33-11, giving q as a function of time, by analogy to Eq. 33-8 as

$$q = Q \cos(\omega t + \phi) \quad \text{(charge)}, \quad (33\text{-}12)$$

where Q is the amplitude of the charge variations, ω is the angular frequency of the electromagnetic oscillations, and ϕ is the phase constant.

Taking the first derivative of Eq. 33-12 with respect to time gives us the current of the LC oscillator:

$$i = \frac{dq}{dt} = -\omega Q \sin(\omega t + \phi) \quad \text{(current)}. \quad (33\text{-}13)$$

The amplitude I of this sinusoidally varying current is

$$I = \omega Q, \quad (33\text{-}14)$$

so we can rewrite Eq. 33-13 as

$$i = -I \sin(\omega t + \phi). \quad (33\text{-}15)$$

We can test whether Eq. 33-12 is a solution of Eq. 33-11 by substituting it and its second derivative with respect to time into Eq. 33-11. The first derivative of Eq. 33-12 is Eq. 33-13. The second derivative is then

$$\frac{d^2q}{dt^2} = -\omega^2 Q \cos(\omega t + \phi).$$

Substituting for q and d^2q/dt^2 into Eq. 33-11, we obtain

$$-L\omega^2 Q \cos(\omega t + \phi) + \frac{1}{C} Q \cos(\omega t + \phi) = 0.$$

Canceling $Q \cos(\omega t + \phi)$ and rearranging lead to

$$\omega = \frac{1}{\sqrt{LC}}.$$

Thus Eq. 33-12 is indeed a solution of Eq. 33-11 if ω has the constant value $1/\sqrt{LC}$. Note that this expression for ω is exactly that given by Eq. 33-4, which we arrived at by examining correspondences.

The phase constant ϕ in Eq. 33-12 is determined by the conditions that prevail at $t = 0$. If those conditions yield $\phi = 0$, for example, then at $t = 0$, Eq. 33-12 requires that $q = Q$ and Eq. 33-13 requires that $i = 0$; these are the initial conditions represented by Fig. 33-1a.

The electric energy stored in the LC circuit at any time t is, from Eqs. 33-1 and 33-12,

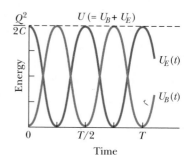

FIGURE 33-4 The stored magnetic energy and electric energy in the circuit of Fig. 33-1 as a function of time. Note that their sum remains constant. T is the period of oscillation.

$$U_E = \frac{q^2}{2C} = \frac{Q^2}{2C} \cos^2(\omega t + \phi), \quad (33\text{-}16)$$

and the magnetic energy is, from Eqs. 33-2 and 33-13,

$$U_B = \tfrac{1}{2}Li^2 = \tfrac{1}{2}L\omega^2 Q^2 \sin^2(\omega t + \phi).$$

Substituting for ω from Eq. 33-4 then gives us

$$U_B = \frac{Q^2}{2C} \sin^2(\omega t + \phi). \quad (33\text{-}17)$$

Figure 33-4 shows plots of $U_E(t)$ and $U_B(t)$ for the case of $\phi = 0$. Note that:

1. The maximum values of U_E and U_B are both $Q^2/2C$.

2. At any instant the sum of U_E and U_B is equal to $Q^2/2C$, a constant.

3. When U_E is maximum, U_B is zero, and conversely.

CHECKPOINT 2: A capacitor in an LC oscillator has a maximum potential difference of 17 V and a maximum energy of 160 μJ. When the capacitor has a potential difference of 5 V and an energy of 10 μJ, what are (a) the emf across the inductor and (b) the energy stored in the magnetic field?

SAMPLE PROBLEM 33-2

(a) In an oscillating LC circuit, what charge q, expressed in terms of the maximum charge Q, is present on the capacitor when the energy is shared equally between the electric and magnetic fields? Assume that $L = 12$ mH and $C = 1.7 \ \mu$F.

SOLUTION: The problem requires that $U_E = \tfrac{1}{2}U_{E,\text{max}}$. The instantaneous and maximum stored energy in the capacitor

are, respectively,

$$U_E = \frac{q^2}{2C} \quad \text{and} \quad U_{E,\text{max}} = \frac{Q^2}{2C},$$

so the problem requires that

$$\frac{q^2}{2C} = \frac{1}{2}\frac{Q^2}{2C},$$

or

$$q = \frac{1}{\sqrt{2}}Q = 0.707Q. \qquad \text{(Answer)}$$

(b) When does this condition occur if the capacitor has its maximum charge at time $t = 0$?

SOLUTION: Equation 33-12 tells us how q varies with time. Because $q = Q$ at time $t = 0$, the phase constant ϕ is zero. Substituting $\phi = 0$ and $q = 0.707Q$ into Eq. 33-12 gives

$$0.707Q = Q \cos \omega t$$

from which

$$\omega t = 45° = \frac{\pi}{4} \text{ rad.}$$

This corresponds to one-eighth of a full oscillation of 2π rad. Substituting for ω from Eq. 33-4 and solving for t yield

$$t = \frac{\pi}{4\omega} = \frac{\pi \sqrt{LC}}{4} = \frac{\pi\sqrt{(12 \times 10^{-3} \text{ H})(1.7 \times 10^{-6} \text{ F})}}{4}$$

$$= 1.12 \times 10^{-4} \text{ s} \approx 110 \ \mu\text{s}. \qquad \text{(Answer)}$$

33-5 DAMPED OSCILLATIONS IN AN *RLC* CIRCUIT

A circuit containing resistance, inductance, and capacitance is called an *RLC circuit*. We shall here discuss only *series RLC circuits* like that shown in Fig. 33-5. With a resistance R present, the total *electromagnetic energy U* of the circuit (the sum of the electric energy and magnetic energy) is no longer constant; instead, it decreases with time as energy is transferred to thermal energy in the resistance. Because of this loss of energy, the oscillations of charge, current, and potential difference continuously de-

FIGURE 33-5 A series *RLC* circuit. As the charge contained in the circuit oscillates back and forth through the resistance, electromagnetic energy is dissipated as thermal energy, damping (decreasing the amplitude of) the oscillations.

crease in amplitude, and the oscillations are said to be *damped*. As you will see, they are damped in exactly the same way as those of the damped block–spring oscillator of Section 16-8.

To analyze the oscillations of this circuit, we write an equation for the total electromagnetic energy U in the circuit at any instant. Because the resistance does not store electromagnetic energy, we can use Eq. 33-9:

$$U = U_B + U_E = \frac{Li^2}{2} + \frac{q^2}{2C}. \qquad (33\text{-}18)$$

Now, however, this total energy decreases as energy is transferred to thermal energy. The rate of that transfer is, from Eq. 27-22,

$$\frac{dU}{dt} = -i^2R, \qquad (33\text{-}19)$$

where the minus sign indicates that U decreases. By differentiating Eq. 33-18 with respect to time and then substituting the result in Eq. 33-19, we obtain

$$\frac{dU}{dt} = Li\frac{di}{dt} + \frac{q}{C}\frac{dq}{dt} = -i^2R.$$

Substituting dq/dt for i and d^2q/dt^2 for di/dt, we obtain

$$L\frac{d^2q}{dt^2} + R\frac{dq}{dt} + \frac{1}{C}q = 0 \quad (RLC \text{ circuit}), \quad (33\text{-}20)$$

which is the differential equation that describes damped charge oscillations in an *RLC* circuit.

The solution to Eq. 33-20 is

$$q = Qe^{-Rt/2L}\cos(\omega' t + \phi), \qquad (33\text{-}21)$$

in which

$$\omega' = \sqrt{\omega^2 - (R/2L)^2}, \qquad (33\text{-}22)$$

where $\omega = 1/\sqrt{LC}$, as with an undamped oscillator. Equation 33-21 tells us how the charge on the capacitor oscillates in a damped *RLC* circuit; the equation is the electromagnetic counterpart of Eq. 16-40, which gives the displacement of a damped block–spring oscillator.

Equation 33-21 describes a sinusoidal oscillation (the cosine term) with an *exponentially decaying amplitude* $Qe^{-Rt/2L}$ (the terms that multiply the cosine). The angular frequency ω' of the damped oscillations is always less than the angular frequency ω of the undamped oscillations; however, we shall here consider only situations in which R is small enough for us to replace ω' with ω.

Let us next find an expression for the total electromagnetic energy U of the circuit as a function of time. One way to do so is to monitor the energy of the electric field in the capacitor, which is given by Eq. 33-1 ($U_E = q^2/2C$).

By substituting Eq. 33-21 into Eq. 33-1, we obtain

$$U_E = \frac{q^2}{2C} = \frac{[Qe^{-Rt/2L}\cos(\omega't + \phi)]^2}{2C}$$

$$= \frac{Q^2}{2C} e^{-Rt/L}\cos^2(\omega't + \phi). \qquad (33\text{-}23)$$

Equation 33-23 shows that the energy of the electric field oscillates according to a cosine-squared term and that the amplitude of that oscillation decreases exponentially with time. Thus, the total electromagnetic energy U must also be decreasing exponentially with time. That total energy is stored in the capacitor whenever U_E reaches a maximum, which occurs whenever $\cos^2(\omega't + \phi)$ is 1. So we can find an expression for the total energy simply by setting the cosine-squared term in Eq. 33-23 equal to 1. We get

$$U = \frac{Q^2}{2C} e^{-Rt/L}. \qquad (33\text{-}24)$$

CHECKPOINT 3: (a) The figure shows the graphs of the total electromagnetic energy U in two RLC circuits with identical capacitors and inductors. Which curve corresponds to the circuit with greater R? (b) If, instead, the circuits have the same R and C, which curve corresponds to the circuit with greater L?

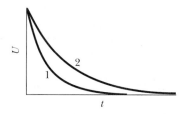

SAMPLE PROBLEM 33-3

A series RLC circuit has inductance $L = 12$ mH, capacitance $C = 1.6$ μF, and resistance $R = 1.5$ Ω.

(a) At what time t will the amplitude of the charge oscillations in the circuit be 50% of its initial value?

SOLUTION: Equation 33-21 gives the exponential decay of the charge oscillations. The amplitude of the charge oscillations has decayed to 50% of its initial value when the amplitude $Qe^{-Rt/2L}$ in Eq. 33-21 has decayed to $0.50Q$, that is, when

$$Qe^{-Rt/2L} = 0.50Q.$$

Canceling Q and taking the natural logarithms of both sides, we have

$$-\frac{Rt}{2L} = \ln 0.50.$$

Solving for t and then substituting given data yield

$$t = -\frac{2L}{R} \ln 0.50 = -\frac{(2)(12 \times 10^{-3} \text{ H})(\ln 0.50)}{1.5 \ \Omega}$$

$$= 0.0111 \text{ s} \approx 11 \text{ ms.} \qquad \text{(Answer)}$$

(b) How many oscillations are completed within this time?

SOLUTION: Since the period of oscillation is $T = 2\pi/\omega$ and the angular frequency is $\omega = 1/\sqrt{LC}$, we have $T = 2\pi\sqrt{LC}$. Each oscillation takes one period, so, in a time interval $\Delta t = 0.0111$ s, the number of complete oscillations is

$$\frac{\Delta t}{T} = \frac{\Delta t}{2\pi\sqrt{LC}}$$

$$= \frac{0.0111 \text{ s}}{2\pi[(12 \times 10^{-3} \text{ H})(1.6 \times 10^{-6} \text{ F})]^{1/2}} \approx 13.$$

$$\text{(Answer)}$$

Thus the amplitude decays by 50% in about 13 complete oscillations. This damping is less severe than that shown in Fig. 33-3, where the amplitude decays by a little more than 50% in one oscillation.

33-6 ALTERNATING CURRENT

The oscillations in an RLC circuit will not damp out if an external emf device supplies enough energy to make up for the energy dissipated as thermal energy in the resistance R. Circuits in homes, offices, and factories, including countless RLC circuits, receive such energy from local power companies. In most countries the energy is supplied via oscillating emfs and currents—the current is said to be an **alternating current,** or **ac** for short. (The nonoscillating current from a battery is said to be a **direct current,** or **dc**.) These oscillating emfs and currents vary sinusoidally with time, reversing direction (in North America) 120 times per second and thus having frequency $f = 60$ Hz.

At first sight this may seem to be a strange arrangement. We have seen that the drift speed of the conduction electrons in household wiring may typically be 4×10^{-5} m/s. If we now reverse their direction every $\frac{1}{120}$ s, such electrons can move only about 3×10^{-7} m in a half-cycle. At this rate, a typical electron can drift past no more than about 10 atoms in the wiring before it is required to reverse its direction. How, you may wonder, can the electron ever get anywhere?

Although this question may be worrisome, it is a needless concern. The conduction electrons do not have to "get anywhere." When we say that the current in a wire is one ampere, we mean that charge carriers pass through any plane cutting across that wire at the rate of one coulomb per second. The speed at which the carriers cross that plane does not matter directly; one ampere may correspond to many charge carriers moving very slowly or to a few moving very rapidly. Furthermore, the signal to the electrons to reverse directions—which originates in the alternating emf provided by the power company's generator—is propagated along the conductor at a speed close to that of light. All electrons, no matter where they are located, get their reversal instructions at about the same instant. Finally, we note that for many devices, such as lightbulbs and toasters, the direction of motion is unimportant as long as the electrons do move so as to transfer energy to the device via collisions with atoms in the device.

The basic advantage of alternating currents is this: *as the current alternates, so does the magnetic field that surrounds the conductor.* This makes possible the use of Faraday's law of induction, which, among other things, means that we can step up (increase) or step down (decrease) the magnitude of an alternating potential difference at will, using a device called a transformer, as you will see later in this chapter. Moreover, alternating current is more readily adaptable to rotating machinery such as generators and motors than is (nonalternating) direct current.

Figure 33-6 shows a simple model of an ac generator. As the conducting loop is forced to rotate through the external magnetic field **B**, a sinusoidally oscillating emf $\mathscr{E}$ is induced in the loop:

$$\mathscr{E} = \mathscr{E}_m \sin \omega_d t. \tag{33-25}$$

The *angular frequency* ω_d of the emf is equal to the angular speed with which the loop rotates in the magnetic field; the *phase* of the emf is $\omega_d t$; and the *amplitude* of the emf is $\mathscr{E}_m$ (where the subscript stands for maximum). When the rotating loop is part of a closed conducting path, this emf

produces (*drives*) a sinusoidal (alternating) current along the path with the same angular frequency ω_d, which then is called the **driving angular frequency.** We can write the current as

$$i = I \sin(\omega_d t - \phi), \tag{33-26}$$

in which I is the amplitude of the driven current. (The phase $\omega_d t - \phi$ of the current is traditionally written with a minus sign instead of as $\omega_d t + \phi$.) We include a phase constant ϕ in Eq. 33-26 because the current i may not be in phase with the emf $\mathscr{E}$. (As you will see, the phase constant depends on the circuit to which the generator is connected.)

33-7 FORCED OSCILLATIONS

We have seen that once started, the charge, potential difference, and current in both undamped LC circuits and damped RLC circuits (with small enough R) oscillate at angular frequency $\omega = 1/\sqrt{LC}$. Such oscillations are said to be *free oscillations* (free of any external emf), and the angular frequency ω is said to be the circuit's **natural angular frequency.**

When the external alternating emf of Eq. 33-25 is connected to an RLC circuit, the oscillations of charge, potential difference, and current are said to be *driven oscillations* or *forced oscillations*. These oscillations always occur at the driving angular frequency ω_d:

> Whatever the natural angular frequency ω of a circuit may be, forced oscillations of charge, current, and potential difference in the circuit always occur at the driving angular frequency ω_d.

However, as you will see in Section 33-9, the amplitudes of the oscillations very much depend on how close ω_d is to ω. When the two angular frequencies match—a condition known as **resonance**—the amplitude I of the current in the circuit is maximum.

33-8 THREE SIMPLE CIRCUITS

Later, we shall connect an external alternating emf device to a series RLC circuit as in Fig. 33-7. We shall then find expressions for the amplitude I and phase constant ϕ of the sinusoidally oscillating current in terms of the amplitude $\mathscr{E}_m$ and angular frequency ω_d of the external emf. But first let us consider three simpler circuits, each having an external emf and only one other circuit element: R, C, or L. We start with a resistive element (a purely *resistive load*).

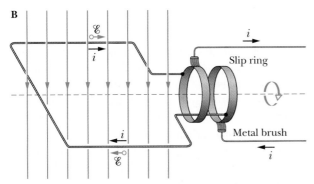

FIGURE 33-6 The basic mechanism of an alternating-current generator is a conducting loop rotated in an external magnetic field. In practice, the alternating emf induced in a coil of many turns of wire is made accessible by means of slip rings attached to the rotating loop, each ring connected to one end of the loop wire and electrically connected by a conducting brush (against which it slips) to the rest of the generator circuit.

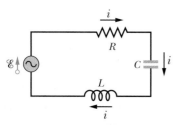

FIGURE 33-7 A single-loop circuit containing a resistor, a capacitor, and an inductor. A generator, represented by a sine wave in a circle, produces an alternating emf that establishes an alternating current; the directions of the emf and current are indicated here at only one instant.

A Resistive Load

Figure 33-8a shows a circuit containing a resistance element of value R and an ac generator with the alternating emf of Eq. 33-25. By the loop rule we have

$$\mathscr{E} - v_R = 0.$$

With Eq. 33-25, this gives us

$$v_R = \mathscr{E}_m \sin \omega_d t.$$

Because the amplitude V_R of the alternating potential dif-

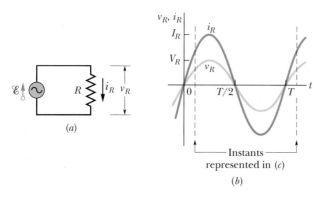

(a)

(b)

— Instants represented in (c) —

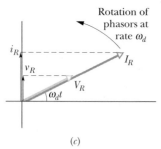

(c)

FIGURE 33-8 (a) A resistor is connected across an alternating-current generator. (b) The current and the potential difference across the resistor are in phase and complete one cycle in one period T. (c) A phasor diagram shows the same thing as (b).

ference (or voltage) across the resistance is equal to the amplitude $\mathscr{E}_m$ of the alternating emf, we can write this as

$$v_R = V_R \sin \omega_d t. \qquad (33\text{-}27)$$

From the definition of resistance ($R = V/i$), we can now write the current i_R in the resistance as

$$i_R = \frac{v_R}{R} = \frac{V_R}{R} \sin \omega_d t. \qquad (33\text{-}28)$$

From Eq. 33-26, we can also write this current as

$$i_R = I_R \sin(\omega_d t - \phi), \qquad (33\text{-}29)$$

where I_R is the amplitude of the current i_R in the resistance. Comparing Eqs. 33-28 and 33-29, we see that for a purely resistive load the phase constant $\phi = 0°$. We also see that the voltage amplitude and current amplitude are related by

$$V_R = I_R R \quad \text{(resistor)}. \qquad (33\text{-}30)$$

Although we found this relation for the circuit of Fig. 33-8a, it applies to any resistance in any ac circuit.

By comparing Eqs. 33-27 and 33-28, we see that the time-varying quantities v_R and i_R are both functions of sin $\omega_d t$ with $\phi = 0°$. Thus these two quantities are *in phase*, which means that their corresponding maxima (and minima) occur at the same times. Figure 33-8b, which is a plot of $v_R(t)$ and $i_R(t)$, illustrates this fact. Note that v_R and i_R do not decay here, because the generator supplies energy to the circuit to make up for the energy dissipated in R.

The time-varying quantities v_R and i_R can also be represented geometrically by *phasors*. Recall from Section 17-10 that phasors are vectors that rotate around an origin. Those that represent the voltage across and current in the resistor of Fig. 33-8a are shown in Fig. 33-8c at an arbitrary time t. Such phasors have the following properties:

Angular speed: Both phasors rotate counterclockwise about the origin with an angular speed equal to the angular frequency ω_d of v_R and i_R.

Length: The length of each phasor represents the amplitude of the alternating quantity: V_R for the voltage and I_R for the current.

Projection: The projection of each phasor on the *vertical* axis represents the value of the alternating quantity at time t: v_R for the voltage and i_R for the current.

Rotation angle: The rotation angle of each phasor is equal to the phase of the alternating quantity at time t. In Fig. 33-8c, the voltage and current are in phase. So their phasors have the same phase $\omega_d t$ and the same rotation angle, and thus they rotate together.

Mentally follow the rotation. Can you see that when the phasors have rotated so that $\omega_d t = 90°$ (they point vertically upward), they indicate that just then $v_R = V_R$ and $i_R = I_R$? Equations 33-27 and 33-29 give the same results.

A Capacitive Load

Figure 33-9a shows a circuit containing a capacitance and a generator with the alternating emf of Eq. 33-25. Using the loop rule and proceeding as we did when we obtained Eq. 33-27, we find that the potential difference across the capacitor is

$$v_C = V_C \sin \omega_d t, \qquad (33\text{-}31)$$

where V_C is the voltage amplitude across the capacitor. From the definition of capacitance we can also write

$$q_C = Cv_C = CV_C \sin \omega_d t. \qquad (33\text{-}32)$$

Our concern, however, is with the current rather than the charge. Thus we differentiate Eq. 33-32 to find

$$i_C = \frac{dq_C}{dt} = \omega_d CV_C \cos \omega_d t. \qquad (33\text{-}33)$$

We now recast Eq. 33-33 in two ways. First, for reasons of symmetry of notation, we introduce the quantity X_C, called the **capacitive reactance** of the capacitor, defined as

$$X_C = \frac{1}{\omega_d C} \qquad \text{(capacitive reactance).} \qquad (33\text{-}34)$$

Its value depends not only on the capacitance but also on the driving angular frequency ω_d. We know from the definition of the capacitive time constant ($\tau = RC$) that the SI unit for C can be expressed as seconds per ohm. Applying this to Eq. 33-34 shows that the SI unit of X_C is the *ohm*, just as for resistance R.

As our second modification of Eq. 33-33, we replace $\cos \omega_d t$ with a phase-shifted sine, namely,

$$\cos \omega_d t = \sin(\omega_d t + 90°).$$

You can verify this identity by expanding the right-hand side with the formula for $\sin(\alpha + \beta)$ listed in Appendix E.

With these two modifications, Eq. 33-33 becomes

$$i_C = \left(\frac{V_C}{X_C}\right) \sin(\omega_d t + 90°). \qquad (33\text{-}35)$$

From Eq. 33-26, we can also write the current i_C in C as

$$i_C = I_C \sin(\omega_d t - \phi), \qquad (33\text{-}36)$$

where I_C is the amplitude of i_C. Comparing Eqs. 33-35 and 33-36, we see that for a purely capacitive load the phase constant $\phi = -90°$. We also see that the voltage amplitude and current amplitude are related by

$$V_C = I_C X_C \qquad \text{(capacitor).} \qquad (33\text{-}37)$$

Although we found this relation for the circuit of Fig. 33-9a, it applies to any capacitance in any ac circuit.

Comparison of Eqs. 33-31 and 33-35, or inspection of Fig. 33-9b, shows that the quantities v_C and i_C are 90°, or one-quarter-cycle, out of phase. Furthermore, we see that i_C *leads* v_C, which means that, if you monitored the current i_C and the potential difference v_C in the circuit of Fig. 33-9a, you would find that i_C reaches its maximum *before* v_C does, by one-quarter cycle.

This relation between i_C and v_C is illustrated by the phasor diagram of Fig. 33-9c. As the phasors representing these two quantities rotate counterclockwise together, the phasor labeled I_C does indeed lead that labeled V_C, and by an angle of 90°. That is, the phasor I_C coincides with the vertical axis one-quarter cycle before the phasor V_C does. Be sure to convince yourself that the phasor diagram of Fig. 33-9c is consistent with Eqs. 33-31 and 33-35.

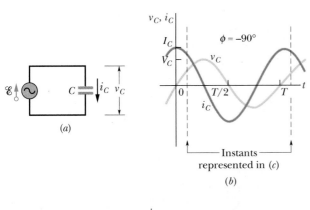

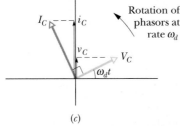

(c)

FIGURE 33-9 (a) A capacitor is connected across an alternating-current generator. (b) The current in the capacitor leads the voltage by 90°. (c) A phasor diagram shows the same thing.

CHECKPOINT 4: The figure shows, in (a), a sine curve $S(t) = \sin(\omega_d t)$ and three other sinusoidal curves $A(t)$, $B(t)$, and $C(t)$, each of the form $\sin(\omega_d t - \phi)$. (a) Rank the three other curves according to the value of ϕ, most positive first and most negative last. (b) Which curve corresponds to which phasor in (b) of the figure? (c) Which curve leads the others?

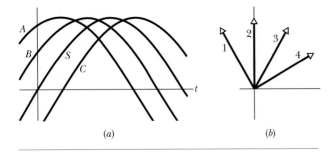

(a)   (b)

An Inductive Load

Figure 33-10a shows a circuit containing an inductance and a generator with the alternating emf of Eq. 33-25. Using the loop rule and proceeding as we did to obtain Eq. 33-27, we find that the potential difference across the inductance is

$$v_L = V_L \sin \omega_d t, \qquad (33\text{-}38)$$

where V_L is the amplitude of v_L. From Eq. 31-40, we can write the potential difference across an inductance L, in which the current is changing at the rate di_L/dt, as

$$v_L = L \frac{di_L}{dt}. \qquad (33\text{-}39)$$

If we combine Eqs. 33-38 and 33-39 we have

$$\frac{di_L}{dt} = \frac{V_L}{L} \sin \omega_d t. \qquad (33\text{-}40)$$

Our concern, however, is with the current rather than with its time derivative. We find the former by integrating Eq. 33-40, obtaining

$$i_L = \int di_L = \frac{V_L}{L} \int \sin \omega_d t \, dt$$

$$= -\left(\frac{V_L}{\omega_d L}\right) \cos \omega_d t. \qquad (33\text{-}41)$$

We now recast this equation in two ways. First, for reasons of symmetry of notation, we introduce the quantity X_L, called the **inductive reactance** of the inductor, which is defined as

$$X_L = \omega_d L \quad \text{(inductive reactance).} \qquad (33\text{-}42)$$

The value of X_L depends on the driving angular frequency ω_d. The unit of the inductive time constant τ_L indicates that the SI unit of X_L is the *ohm*, just as it is for X_C and for R.

Second, we replace the function $-\cos \omega_d t$ in Eq. 33-

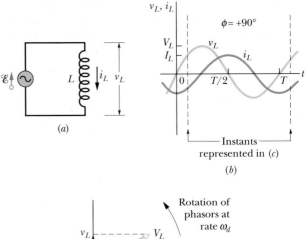

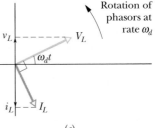

FIGURE 33-10   (a) An inductor is connected across an alternating-current generator. (b) The current in the inductor lags the voltage by 90°. (c) A phasor diagram shows the same thing.

41 with a phase-shifted sine, namely,

$$-\cos \omega_d t = \sin(\omega_d t - 90°).$$

You can verify this identity by expanding its right-hand side with the formula for $\sin(\alpha - \beta)$ given in Appendix E.

With these two changes, Eq. 33-41 becomes

$$i_L = \left(\frac{V_L}{X_L}\right) \sin(\omega_d t - 90°). \qquad (33\text{-}43)$$

From Eq. 33-26, we can also write this current in the inductance as

$$i_L = I_L \sin(\omega_d t - \phi), \qquad (33\text{-}44)$$

where I_L is the amplitude of the current i_L. Comparing Eqs. 33-43 and 33-44, we see that for a purely inductive load the phase constant $\phi = +90°$. We also see that the voltage amplitude and the current amplitude are related by

$$V_L = I_L X_L \quad \text{(inductor).} \qquad (33\text{-}45)$$

Although we found this relation for the circuit of Fig. 33-10a, it applies to any inductance in any ac circuit.

Comparison of Eqs. 33-38 and 33-43, or inspection of Fig. 33-10b, shows that the quantities i_L and v_L are 90° out of phase. In this case, however, i_L *lags* v_L. That is, if you monitored the current i_L and the potential difference v_L in the circuit of Fig. 33-10a, you would find that i_L reaches its maximum value *after* v_L does, by one-quarter cycle.

The phasor diagram of Fig. 33-10*c* also contains this information. As the phasors rotate counterclockwise in the figure, the phasor labeled I_L does indeed lag that labeled V_L, and by an angle of 90°. Be sure to convince yourself that Fig. 33-10*c* represents Eqs. 33-38 and 33-43.

PROBLEM SOLVING TACTICS

TACTIC 1: *Leading and Lagging in ac Circuits*

Table 33-2 summarizes the relations between the current *i* and the voltage *v* for each of the three kinds of circuit elements we have considered. When an applied alternating voltage produces an alternating current in them, the current is in phase with the voltage across a resistor, leads the voltage across a capacitor, and lags the voltage across an inductor.

Many students remember these results with the mnemonic "*ELI* the *ICE* man." *ELI* contains the letter *L* (for inductor), and in it the letter *I* (for current) comes after the letter *E* (for emf or voltage). Thus, for an inductor, the current *lags* the voltage. Similarly *ICE* (which contains a *C* for capacitor) means that the current leads the voltage.

You might also use the modified mnemonic "*ELI positively* is the *ICE* man" to remember that the phase constant ϕ is positive for an inductor.

If you have difficulty in remembering whether X_C is equal to $\omega_d C$ (wrong) or $1/\omega_d C$ (right), try remembering that *C* is in the "cellar," that is, in the denominator.

SAMPLE PROBLEM 33-4

(a) In Fig. 33-9*a*, let $C = 15.0 \ \mu F$ and $\mathcal{E}_m = V_C = 36.0$ V, and let the *driving frequency f_d* of the alternating emf device be 60.0 Hz. What is the current amplitude I_C?

SOLUTION: We can get this amplitude from Eq. 33-37 ($V_C = I_C X_C$) if we first find the capacitive reactance X_C. From Eq. 33-34 with $\omega_d = 2 \pi f_d$, we have

$$X_C = \frac{1}{2 \pi f_d C} = \frac{1}{(2\pi)(60.0 \text{ Hz})(15.0 \times 10^{-6} \text{ F})}$$

$$= 177 \ \Omega.$$

Then from Eq. 33-37,

$$I_C = \frac{V_C}{X_C} = \frac{36.0 \text{ V}}{177 \ \Omega} = 0.203 \text{ A}. \qquad \text{(Answer)}$$

(b) In Fig. 33-10*a*, let $L = 230$ mH, $\mathcal{E}_m = V_L = 36.0$ V, and $f_d = 60.0$ Hz. What is the current amplitude I_L?

SOLUTION: We can get this amplitude from Eq. 33-45 ($I_L = V_L/X_L$) if we first find the inductive reactance X_L. From Eq. 33-42 with $\omega_d = 2 \pi f_d$, we have

$$X_L = 2 \pi f_d L = (2\pi)(60.0 \text{ Hz})(230 \times 10^{-3} \text{ H}) = 86.7 \ \Omega.$$

Then, from Eq. 33-45,

$$I_L = \frac{V_L}{X_L} = \frac{36.0 \text{ V}}{86.7 \ \Omega} = 0.415 \text{ A}. \qquad \text{(Answer)}$$

(c) Write an expression for the time-varying current i_L in the circuit of (b).

SOLUTION: Equation 33-44 is the general equation for i_L. With I_L computed above as 0.415 A, with

$$\omega_d = 2 \pi f_d = 120 \pi \text{ Hz},$$

and with $\phi = 90° = \pi/2$ rad for this purely inductive circuit, we have

$$i_L = I_L \sin(\omega_d t - \phi)$$

$$= (0.415 \text{ A}) \sin\left(120 \pi t - \frac{\pi}{2} \right). \qquad \text{(Answer)}$$

$\mathbf{C}$HECKPOINT 5: If we increase the driving frequency f_d in the circuits of (a) Fig. 33-9*a* and (b) Fig. 33-10*a*, does the current amplitude *I* increase, decrease, or stay the same?

33-9 THE SERIES *RLC* CIRCUIT

We are now ready to apply the alternating emf of Eq. 33-25,

$$\mathcal{E} = \mathcal{E}_m \sin \omega_d t \quad \text{(applied emf)}, \qquad (33\text{-}46)$$

to the full *RLC* circuit of Fig. 33-7. Because *R, L,* and *C* are in series, the same current

$$i = I \sin(\omega_d t - \phi) \qquad (33\text{-}47)$$

TABLE 33-2 PHASE AND AMPLITUDE RELATIONS FOR ALTERNATING CURRENTS AND VOLTAGES

CIRCUIT ELEMENT	SYMBOL	RESISTANCE OR REACTANCE	PHASE OF THE CURRENT	PHASE ANGLE ϕ	AMPLITUDE RELATION
Resistor	R	R	In phase with v_R	0°	$V_R = I_R R$
Capacitor	C	$X_C = 1/\omega_d C$	Leads v_C by 90°	−90°	$V_C = I_C X_C$
Inductor	L	$X_L = \omega_d L$	Lags v_L by 90°	+90°	$V_L = I_L X_L$

is driven in all three of them. We wish to find the current amplitude I and the phase constant ϕ.

The solution is simplified by the use of phasor diagrams. We start with Fig. 33-11a, which shows the phasor representing the current of Eq. 33-47 at an arbitrary time t. The length of the phasor is the amplitude I, the projection of the phasor on the vertical axis is the current i at time t, and the angle of rotation of the phasor is the phase $\omega_d t - \phi$ of the current at time t.

Figure 33-11b shows the phasors representing the voltages across R, L, and C at the same time t. Each phasor is oriented relative to the angle of rotation of current phasor I in Fig. 33-11a, based on the information in Table 33-2:

Resistor: Here current and voltage are in phase; so the angle of rotation of voltage phasor V_R is the same as that of phasor I.

Capacitor: Here current leads the voltage by 90°; so the angle of rotation of voltage phasor V_C is 90° less than that of phasor I.

Inductor: Here current lags the voltage by 90°; so the angle of rotation of voltage phasor v_L is 90° greater than that of phasor I.

Figure 33-11b also shows the instantaneous voltages v_R, v_C, and v_L across R, C, and L at time t; those voltages are the projections of the corresponding phasors on the vertical axis of the figure.

Figure 33-11c shows the phasor representing the applied emf of Eq. 33-46. The length of the phasor is the amplitude $\mathscr{E}_m$, the projection of the phasor on the vertical axis is the emf $\mathscr{E}$ at time t, and the angle of rotation of the phasor is the phase $\omega_d t$ of the emf at time t.

Now, from the loop rule we know that at any instant the sum of the voltages v_R, v_C, and v_L is equal to the applied emf $\mathscr{E}$:

$$\mathscr{E} = v_R + v_C + v_L. \tag{33-48}$$

Thus at time t the projection $\mathscr{E}$ in Fig. 33-11c is equal to the algebraic sum of the projections v_R, v_C, and v_L in Fig. 33-11b. In fact, as the phasors rotate together, this equality always holds. This means that phasor $\mathscr{E}_m$ in Fig. 33-11c must be equal to the vector sum of the three voltage phasors V_R, V_C, and V_L in Fig. 33-11b.

We can simplify this vector sum by first noting that phasors V_C and V_L have opposite directions. We can combine them into the single phasor $V_L - V_C$ as shown in Fig. 3-11d. And we can find the vector sum of the three voltage phasors in Fig. 3-11b by finding the resultant of the two phasors V_R and $(V_L - V_C)$ in Fig. 3-11d. That resultant must coincide with phasor $\mathscr{E}_m$ as shown.

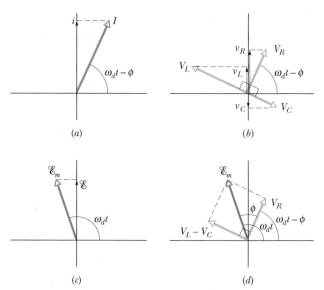

(a) (b)

(c) (d)

FIGURE 33-11 (a) A phasor representing the alternating current in the driven RLC circuit of Fig. 33-7 at time t. The amplitude I, the instantaneous value i, and the phase $(\omega_d t - \phi)$ are shown. (b) Phasors representing the voltages across the inductor, resistor, and capacitor, oriented with respect to the current phasor in (a). (c) A phasor representing the alternating emf that drives the current of (a). (d) The emf phasor is equal to the vector sum of the three voltage phasors of (b). Here, voltage phasors V_L and V_C have been added to yield their net phasor $(V_L - V_C)$.

Both triangles in Fig. 3-11d are right triangles. Applying the Pythagorean theorem to either one yields

$$\mathscr{E}_m^2 = V_R^2 + (V_L - V_C)^2. \tag{33-49}$$

From the amplitude information displayed in Table 33-2 we can rewrite this as

$$\mathscr{E}_m^2 = (IR)^2 + (IX_L - IX_C)^2, \tag{33-50}$$

and then rearrange it to the form

$$I = \frac{\mathscr{E}_m}{\sqrt{R^2 + (X_L - X_C)^2}}. \tag{33-51}$$

The denominator in Eq. 33-51 is called the **impedance** Z of the circuit for the driving angular frequency ω_d:

$$Z = \sqrt{R^2 + (X_L - X_C)^2} \quad \text{(impedance defined).} \tag{33-52}$$

We can then write Eq. 33-51 as

$$I = \frac{\mathscr{E}_m}{Z}. \tag{33-53}$$

If we substitute for X_C and X_L from Eqs. 33-34 and 33-42, we can write Eq. 33-51 more explicitly as

$$I = \frac{\mathscr{E}_m}{\sqrt{R^2 + (\omega_d L - 1/\omega_d C)^2}} \qquad \text{(current amplitude).} \qquad (33\text{-}54)$$

We have now accomplished half our goal: we have obtained an expression for the current amplitude I in terms of the sinusoidal driving emf and the circuit elements in a series *RLC* circuit.

The value of I depends on the difference between $\omega_d L$ and $1/\omega_d C$ in Eq. 33-54 or, equivalently, on the difference between X_L and X_C in Eq. 33-51. In either equation, it does not matter which of the two quantities is greater because the difference is always squared.

The current that we have been describing in this section is the *steady-state current* that occurs after the alternating emf has been applied for some time. When the emf is first applied to a circuit, a brief *transient current* occurs. Its duration (before settling down into the steady-state current) is determined by the time constants $\tau_L = L/R$ and $\tau_C = RC$ as the inductive and capacitive elements ''turn on.'' This transient current can be large and can, for example, destroy a motor on start-up if it is not properly taken into account in the motor's circuit design.

The Phase Constant

We still need to find an expression for the phase constant ϕ. From the right-hand phasor triangle in Fig. 33-11d and from Table 33-2 we can write

$$\tan \phi = \frac{V_L - V_C}{V_R} = \frac{IX_L - IX_C}{IR}, \qquad (33\text{-}55)$$

which gives us

$$\tan \phi = \frac{X_L - X_C}{R} \qquad \text{(phase constant).} \qquad (33\text{-}56)$$

This is the other half of our goal: an expression for the phase constant ϕ in a sinusoidally driven series *RLC* circuit. The expression gives us three different results for the phase constant, depending on the relative values of X_L and X_C:

$X_L > X_C$: The circuit is said to be *more inductive than capacitive*. Equation 33-56 tells us that ϕ is positive for such a circuit, which means that phasor $\mathscr{E}_m$ rotates ahead of phasor I (Fig. 33-12a). A plot of $\mathscr{E}$ and i versus time is like that in Fig. 33-12b. (The phasors in Figs. 33-11c and d were drawn assuming $X_L > X_C$.)

$X_C > X_L$: The circuit is said to be *more capacitive than inductive*. Equation 33-56 tells us that ϕ is nega-

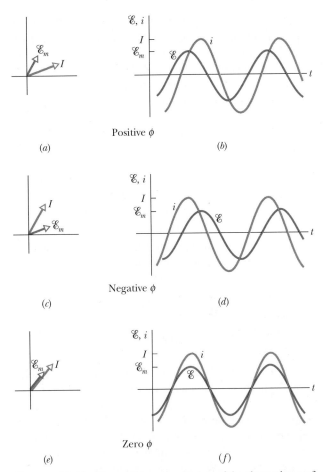

FIGURE 33-12 Phasor diagrams and plots of the alternating emf and current for the driven *RLC* circuit of Fig. 33-7, for (*a, b*) positive ϕ, (*c, d*) negative ϕ, and (*e, f*) zero ϕ.

tive for such a circuit, which means that phasor $\mathscr{E}_m$ rotates behind phasor I (Fig. 33-12c). A plot of $\mathscr{E}$ and i versus time is like that in Fig. 33-12d.

$X_C = X_L$: The circuit is said to be in *resonance*, a term that is explained below. Equation 33-56 tells us that $\phi = 0°$ for such a circuit, which means that phasors $\mathscr{E}_m$ and I rotate together (Fig. 33-12e). A plot of $\mathscr{E}$ and i versus time is like that in Fig. 33-12f.

As illustration, let us reconsider two extreme circuits: In the *purely inductive circuit* of Fig. 33-10a, where X_L is nonzero and $X_C = R = 0$, Eq. 33-56 tells us that $\phi = +90°$ (the greatest value of ϕ), consistent with Fig. 33-10c. In the *purely capacitive circuit* of Fig. 33-9a, where X_C is nonzero and $X_L = R = 0$, Eq. 33-56 tells us that $\phi = -90°$ (the least value of ϕ), consistent with Fig. 33-9c.

Resonance

Equation 33-54 gives the current amplitude I in an RLC circuit as a function of the driving angular frequency ω_d of the external alternating emf. For a given resistance R, that amplitude is a maximum when the quantity $\omega_d L - 1/\omega_d C$ in the denominator is zero, that is, when

$$\omega_d L = \frac{1}{\omega_d C}$$

or

$$\omega_d = \frac{1}{\sqrt{LC}} \quad \text{(maximum } I\text{).} \quad (33\text{-}57)$$

Because the natural angular frequency ω of the RLC circuit is also equal to $1/\sqrt{LC}$, the maximum value of I occurs when the driving angular frequency matches the natural angular frequency, that is, at resonance. So in an RLC circuit, resonance and maximum current amplitude I occur when

$$\omega_d = \omega = \frac{1}{\sqrt{LC}} \quad \text{(resonance).} \quad (33\text{-}58)$$

Figure 33-13 shows three *resonance curves* for sinusoidally driven oscillations in three series RLC circuits differing only in R. Each curve peaks at its maximum current amplitude when the ratio $\omega_d/\omega = 1.00$, but the maximum value of I decreases with increasing R. (The maximum I is always $\mathcal{E}_m/R$; to see why, combine Eqs. 33-52 and 33-53.) In addition, the curves increase in width (measured in Fig. 33-13 at half the maximum value of I) with increasing R.

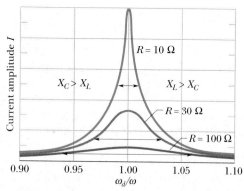

FIGURE 33-13   *Resonance curves* for the driven RLC circuit of Fig. 33-7 with $L = 100\ \mu H$, $C = 100$ pF, and three values of R. The current amplitude I of the alternating current depends on how close the driving angular frequency ω_d is to the natural angular frequency ω. The horizontal arrow on each curve measures the curve's width at the half-maximum level, a measure of the sharpness of the resonance. To the left of $\omega_d/\omega = 1.00$, the circuit is mainly capacitive, with $X_C > X_L$; to the right, it is mainly inductive, with $X_L > X_C$.

We can make physical sense of the resonance curves in Fig. 33-13 by considering how the reactances X_L and X_C change as we increase the driving angular frequency ω_d, starting with a value much less than the natural frequency ω. For small ω_d, reactance X_L ($= \omega_d L$) is small and reactance X_C ($= 1/\omega_d C$) is large. So, the circuit is mainly capacitive and the impedance is dominated by the large X_C, which keeps the current low.

As we increase ω_d, reactance X_C remains dominant but decreases while reactance X_L increases. The decrease in X_C decreases the impedance, allowing the current to increase, as we see on the left side of any resonance curve in Fig. 33-13. When the increasing X_L and the decreasing X_C reach equal values, the current is greatest and the circuit is in resonance, with $\omega_d = \omega$.

As we continue to increase ω_d, the increasing reactance X_L becomes progressively more dominant over the decreasing reactance X_C. So, the impedance increases because of X_L and the current decreases, as on the right side of any resonance curve in Fig. 33-13. In summary, then: the low-angular-frequency side of a resonance curve is dominated by the capacitor's reactance; the high-angular-frequency side is dominated by the inductor's reactance; and resonance occurs between the two regions.

SAMPLE PROBLEM 33-5

In Fig. 33-7 let $R = 160\ \Omega$, $C = 15.0\ \mu F$, $L = 230$ mH, $f_d = 60.0$ Hz, and $\mathcal{E}_m = 36.0$ V. (Except for R, these parameters are the same as in Sample Problem 33-4.)

(a) What is the current amplitude I?

SOLUTION: We can get the current amplitude with Eq. 33-53 ($I = \mathcal{E}_m/Z$) if we first find the impedance Z of the circuit with Eq. 33-52. From Sample Problem 33-4a we know that the capacitive reactance X_C for the capacitor (and thus for the circuit) is 177 Ω; from Sample Problem 33-4b we know that the inductive reactance X_L for the inductor is 86.7 Ω. So, Eq. 33-52 tells us that

$$Z = \sqrt{R^2 + (X_L - X_C)^2}$$
$$= \sqrt{(160\ \Omega)^2 + (86.7\ \Omega - 177\ \Omega)^2}$$
$$= 184\ \Omega.$$

We then find

$$I = \frac{\mathcal{E}_m}{Z} = \frac{36.0\ \text{V}}{184\ \Omega} = 0.196\ \text{A.} \quad \text{(Answer)}$$

(b) What is the phase constant ϕ?

SOLUTION: From Eq. 33-56,

$$\tan \phi = \frac{X_L - X_C}{R} = \frac{86.7\ \Omega - 177\ \Omega}{160\ \Omega} = -0.564.$$

Thus we have

$$\phi = \tan^{-1}(-0.564) = -29.4° = -0.513 \text{ rad.} \quad \text{(Answer)}$$

The negative phase angle is consistent with the fact that the load is mainly capacitive; that is, $X_C > X_L$.

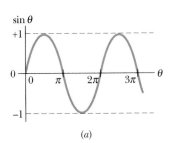

(a)

CHECKPOINT **6:** Here are the capacitive reactance and inductive reactance, respectively, for three sinusoidally driven series *RLC* circuits: (1) 50 Ω, 100 Ω; (2) 100 Ω, 50 Ω; (3) 50 Ω, 50 Ω. (a) For each, does the current lead or lag the applied emf, or are the two in phase? (b) Which circuit is in resonance?

33-10 POWER IN ALTERNATING-CURRENT CIRCUITS

In the *RLC* circuit of Fig. 33-7 the source of energy is the alternating-current generator. Some of the energy that it provides is stored in the electric field in the capacitor, some is stored in the magnetic field in the inductor, and some is dissipated as thermal energy in the resistor. In steady-state operation — which we assume — the average energy stored in the capacitor and in the inductor remains constant. The net transfer of energy is thus from the generator to the resistor, where the electromagnetic energy is dissipated as thermal energy.

The instantaneous rate at which energy is dissipated in the resistor can be written, with the help of Eqs. 27-22 and

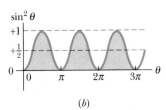

FIGURE 33-14 (a) A plot of $\sin \theta$ versus θ. The average value over one cycle is zero. (b) A plot of $\sin^2 \theta$ versus θ. The average value over one cycle is $\frac{1}{2}$.

(b)

33-26, as

$$P = i^2R = [I \sin(\omega_d t - \phi)]^2R$$
$$= I^2R \sin^2(\omega_d t - \phi). \quad (33\text{-}59)$$

The *average* rate at which energy is dissipated in the resistor, however, is the average of Eq. 33-59 over time. Although the average value of $\sin \theta$, where θ is any variable, is zero (Fig. 33-14a), the average value of $\sin^2 \theta$ over one complete cycle is $\frac{1}{2}$ (Fig 33-14b). (Note in Fig. 33-14b how the shaded parts of the curve that lie above the horizontal line marked $+\frac{1}{2}$ exactly fill in the empty spaces below that

In August 1988, after playing only day games for 72 years, the Chicago Cubs got lamps for night games: 540 metal halide lamps, rated at 1500 W each, lit up the playing field. However, the first night game (against the Phillies) was stopped because of a thunderstorm, which some fans took as a sign that the Cubs should have stayed with day games.

At 5:17 P.M. on November 9, 1965, a faulty relay in the power system near Niagara Falls opened a circuit breaker on a transmission line, automatically causing the current to switch to other lines, which overloaded those lines and made other circuit breakers open. Within minutes the runaway shutdown blacked out much of New York, New England, and Ontario.

line.) Thus we can write, from Eq. 33-59,

$$P_{av} = \frac{I^2R}{2} = \left(\frac{I}{\sqrt{2}}\right)^2 R. \qquad (33\text{-}60)$$

The quantity $I/\sqrt{2}$ is called the **root-mean-square**, or **rms,** value of the current i:

$$I_{rms} = \frac{I}{\sqrt{2}} \quad \text{(rms current).} \qquad (33\text{-}61)$$

We can now rewrite Eq. 33-60 as

$$P_{av} = I_{rms}^2 R \quad \text{(average power).} \qquad (33\text{-}62)$$

Equation 33-62 looks much like Eq. 27-22 ($P = i^2R$); the message is that if we switch to the rms current, we can compute the average rate of energy dissipation for alternating-current circuits just as for direct-current circuits.

We can also define rms values of voltage and emf for alternating-current circuits:

$$V_{rms} = \frac{V}{\sqrt{2}} \quad \text{and} \quad \mathscr{E}_{rms} = \frac{\mathscr{E}_m}{\sqrt{2}} \qquad (33\text{-}63)$$
$$\text{(rms voltage; rms emf).}$$

Alternating-current instruments, such as ammeters and voltmeters, are usually calibrated to read I_{rms}, V_{rms}, and $\mathscr{E}_{rms}$. Thus if you plug an alternating-current voltmeter into a household electric outlet and it reads 120 V, that is an rms voltage. The *maximum* value of the potential difference at the outlet is $\sqrt{2} \times$ (120 V), or 170 V.

Because the proportionality factor $1/\sqrt{2}$ in Eqs. 33-61 and 33-63 is the same for all three variables, we can write Eqs. 33-53 and 33-51 as

$$I_{rms} = \frac{\mathscr{E}_{rms}}{Z} = \frac{\mathscr{E}_{rms}}{\sqrt{R^2 + (X_L - X_C)^2}}, \qquad (33\text{-}64)$$

and, indeed, this is the form that we almost always use.

We can use the relationship $I_{rms} = \mathscr{E}_{rms}/Z$ to recast Eq. 33-62 in a useful equivalent way. We write

$$P_{av} = \frac{\mathscr{E}_{rms}}{Z} I_{rms} R = \mathscr{E}_{rms} I_{rms} \frac{R}{Z}. \qquad (33\text{-}65)$$

From Fig. 33-11d, Table 33-2, and Eq. 33-53, however, we see that R/Z is just the cosine of the phase constant ϕ:

$$\cos \phi = \frac{V_R}{\mathscr{E}_m} = \frac{IR}{IZ} = \frac{R}{Z}. \qquad (33\text{-}66)$$

Equation 33-65 then becomes

$$P_{av} = \mathscr{E}_{rms} I_{rms} \cos \phi \quad \text{(average power),} \qquad (33\text{-}67)$$

in which the term $\cos \phi$ is called the **power factor.** Because $\cos \phi = \cos(-\phi)$, Eq. 33-67 is independent of the sign of the phase constant ϕ.

To maximize the rate at which energy is supplied to a resistive load in an *RLC* circuit, we should keep the power factor $\cos \phi$ as close to unity as possible. This is equivalent to keeping the phase constant ϕ in Eq. 33-26 as close to zero as possible. If, for example, the circuit is highly inductive, it can be made less so by adding capacitance to the circuit, thus reducing the phase constant and increasing the power factor in Eq. 33-67. Power companies place capacitors throughout their transmission systems to do this.

CHECKPOINT 7: (a) If the current in a sinusoidally driven series *RLC* circuit leads the emf, would we increase or decrease the capacitance to increase the rate at which energy is supplied to the resistance? (b) Will this change bring the resonant angular frequency of the circuit closer to the angular frequency of the emf or put it further away?

SAMPLE PROBLEM 33-6

A series *RLC* circuit, driven with $\mathscr{E}_{rms} = 120$ V at frequency $f_d = 60.0$ Hz, contains a resistance $R = 200 \ \Omega$, an inductance with $X_L = 80.0 \ \Omega$, and a capacitance with $X_C = 150 \ \Omega$.

(a) What are the power factor $\cos \phi$ and phase constant ϕ of the circuit?

SOLUTION: We can obtain the power factor from Eq. 33-66 ($\cos \phi = R/Z$) if we first find the impedance Z. From Eq. 33-52, we have

$$Z = \sqrt{R^2 + (X_L - X_C)^2}$$
$$= \sqrt{(200 \ \Omega)^2 + (80.0 \ \Omega - 150 \ \Omega)^2} = 211.90 \ \Omega.$$

Equation 33-66 then gives us

$$\cos \phi = \frac{R}{Z} = \frac{200 \ \Omega}{211.90 \ \Omega} = 0.944. \quad \text{(Answer)}$$

Taking inverse cosine then yields

$$\phi = \cos^{-1} 0.944 = 19.3°.$$

The implied plus sign here is wrong, however: with $X_C > X_L$, the circuit is mainly capacitive and ϕ must be negative. The problem is that the sign of the difference $X_L - X_C$ is lost in the calculation of Z. Inserting a minus sign, we then have

$$\phi = -19.3°. \quad \text{(Answer)}$$

(We could, instead, insert the known data into Eq. 33-56; we arrive at the same answer, complete with the minus sign.)

(b) What is the average rate P_{av} at which energy is dissipated in the resistance?

SOLUTION: We can calculate P_{av} with Eq. 33-67 if we first calculate I_{rms}. From Eq. 33-64 we have

$$I_{rms} = \frac{\mathscr{E}_{rms}}{Z} = \frac{120 \text{ V}}{211.90 \text{ } \Omega} = 0.5663 \text{ A}.$$

Substituting this and known data into Eq. 33-67, we find

$$P_{av} = \mathscr{E}_{rms}I_{rms} \cos \phi = (120 \text{ V})(0.5663 \text{ A})(0.944)$$
$$= 64.2 \text{ W}. \hspace{2cm} \text{(Answer)}$$

(c) What change ΔC in the capacitance is needed to maximize P_{av} if the other parameters of the circuit are not changed?

SOLUTION: From Eq. 33-34 ($X_C = 1/\omega_d C$), the original capacitance is

$$C = \frac{1}{2\pi f_d X_C} = \frac{1}{(2\pi)(60.0 \text{ Hz})(150 \text{ } \Omega)} = 17.7 \text{ } \mu\text{F}.$$

The energy dissipation rate P_{av} is maximized when the circuit is in resonance, that is, when X_C is equal to X_L. Again from Eq. 33-34, now with $X_C = X_L = 80.0 \text{ } \Omega$, the capacitance would then be

$$C' = \frac{1}{2\pi f_d X_C} = \frac{1}{(2\pi)(60 \text{ Hz})(80.0 \text{ } \Omega)} = 33.2 \text{ } \mu\text{F}.$$

Thus, the required change in capacitance would be

$$\Delta C = C' - C = 33.2 \text{ } \mu\text{F} - 17.7 \text{ } \mu\text{F}$$
$$= +15.5 \text{ } \mu\text{F}. \hspace{2cm} \text{(Answer)}$$

(d) With that change in capacitance, what would P_{av} be?

SOLUTION: The change would give $X_C = X_L$. Then from Eqs. 33-52 and 33-66, we would have $Z = R$ and $\cos \phi = 1$. The rms current would be, from Eq. 33-64,

$$I_{rms} = \frac{\mathscr{E}_{rms}}{Z} = \frac{120 \text{ V}}{200 \text{ } \Omega} = 0.600 \text{ A},$$

and the average power should be

$$P_{av} = \mathscr{E}_{rms}I_{rms} \cos \phi = (120 \text{V})(0.600 \text{ A})(1.0)$$
$$= 72.0 \text{ W}. \hspace{2cm} \text{(Answer)}$$

33-11 TRANSFORMERS

Energy Transmission Requirements

When an ac circuit has only a resistive load, the power factor in Eq. 33-67 is $\cos 0° = 1$ and the applied rms emf $\mathscr{E}$ is equal to the rms voltage V across the load. So with an rms current I in the load, energy is supplied and dissipated at the average rate of

$$P_{av} = \mathscr{E}I = IV. \hspace{2cm} (33\text{-}68)$$

(In this section, we follow conventional practice and drop the subscripts identifying rms quantities. Engineers and scientists assume that all time-varying currents and voltages are reported as rms values; that is what the meters read.) Equation 33-68 tells us that, to satisfy a given power requirement, we have a range of choices, from a relatively large current I and a relatively small voltage V to just the reverse, provided only that the product IV is as required.

In electric power distribution systems it is desirable for reasons of safety and for efficient equipment design to deal with relatively low voltages at both the generating end (the electric power plant) and the receiving end (the home or factory). Nobody wants an electric toaster or a child's electric train to operate at, say, 10 kV. On the other hand, in the transmission of electric energy from the generating plant to the consumer, we want the lowest practical current (hence the largest practical voltage) to minimize I^2R losses (often called *ohmic losses*) in the transmission line.

As an example, consider the 735 kV line used to transmit electric energy from the La Grande 2 hydroelectric plant in Quebec to Montreal, 1000 km away. Suppose that the current is 500 A and the power factor is close to unity. Then from Eq. 33-67, energy is supplied at the average rate

$$P_{av} = \mathscr{E}I = (7.35 \times 10^5 \text{ V})(500 \text{ A}) = 368 \text{ MW}.$$

The resistance per kilometer is about 0.220 Ω/km for the line; thus there is a total resistance of about 220 Ω for the 1000 km stretch. Energy is dissipated owing to that resistance at a rate of about

$$P_{av} = I^2R = (500 \text{ A})^2(220 \text{ } \Omega) = 55.0 \text{ MW},$$

which is nearly 15% of the supply rate.

Imagine what would happen if we doubled the current and halved the voltage. Energy would be supplied by the plant at the same average rate of 368 MW as previously, but now energy would be dissipated at the rate of about

$$P_{av} = I^2R = (1000 \text{ A})^2(220 \text{ } \Omega) = 220 \text{ MW},$$

which is *almost 60% of the supply rate.* Hence the general energy transmission rule: transmit at the highest possible voltage and the lowest possible current.

The Ideal Transformer

The transmission rule leads to a fundamental mismatch between the requirement for efficient high-voltage transmission and the need for safe low-voltage generation and consumption. We need a device with which we can raise (for transmission) and lower (for use) the voltage in a circuit, keeping the product current × voltage essentially constant. The **transformer** is such a device. It has no moving parts, operates by Faraday's law of induction, and has no simple direct-current counterpart.

The *ideal transformer* in Fig. 33-15 consists of two coils, with different numbers of turns, wound around an iron core. (The coils are insulated from the core.) In use, the primary winding, of N_p turns, is connected to an alternating-current generator whose emf $\mathscr{E}$ is given by

$$\mathscr{E} = \mathscr{E}_m \sin \omega t. \qquad (33\text{-}69)$$

The secondary winding, of N_s turns, is connected to load resistance R, but its circuit is an open circuit as long as switch S is open (which we assume for the present). Thus there can be no current through the secondary coil. We assume further for this ideal transformer that the resistances of the primary and secondary windings are negligible, as are energy losses due to magnetic hysteresis in the iron core. Well-designed, high-capacity transformers can have energy losses as low as 1%, so our assumptions are reasonable.

For the assumed conditions, the primary winding (or *primary*) is a pure inductance, and the primary circuit is like that in Fig. 33-10a. Thus the (very small) primary current, also called the *magnetizing current* I_{mag}, lags the primary voltage V_p by 90°; the primary's power factor ($= \cos \phi$ in Eq. 33-67) is zero, and thus no power is delivered from the generator to the transformer.

However, the small alternating primary current I_{mag} induces an alternating magnetic flux Φ_B in the iron core. Because the core extends through the secondary winding (or *secondary*), this induced flux also extends through the turns of the secondary. From Faraday's law of induction (Eq. 31-6), the induced emf per turn $\mathscr{E}_{turn}$ is the same for both the primary and the secondary. Also, the voltage V_p across the primary is equal to the emf induced in the primary, and the voltage V_s across the secondary is equal to the emf induced in the secondary. Thus we can write

$$\mathscr{E}_{turn} = \frac{d\Phi_B}{dt} = \frac{V_p}{N_p} = \frac{V_s}{N_s}$$

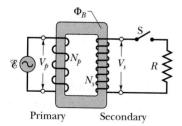

Primary   Secondary

FIGURE 33-15 An ideal transformer, two coils wound on an iron core, in a basic transformer circuit. An ac generator produces current in the coil at the left (the *primary*). The coil at the right (the *secondary*) is connected to the resistive load R when switch S is closed.

and thus

$$V_s = V_p \frac{N_s}{N_p} \qquad \begin{array}{c}\text{(transformation} \\ \text{of voltage).}\end{array} \qquad (33\text{-}70)$$

If $N_s > N_p$, the transformer is called a *step-up transformer* because it steps the primary's voltage V_p up to a higher voltage V_s. Similarly, if $N_s < N_p$, the device is a *step-down transformer*.

So far, with switch S open, no energy is transferred from the generator to the rest of the circuit. Now let us close S to connect the secondary to the resistive load R. (In general, the load would also contain inductive and capacitive elements, but here we consider just resistance R.) We find that now energy *is* transferred from the generator. Let us see why.

Several things happen when we close switch S. (1) An alternating current I_s appears in the secondary circuit, with corresponding energy dissipation rate $I_s^2 R$ ($= V_s^2/R$) in the resistive load. (2) This current produces its own alternating magnetic flux in the iron core, and this flux induces (from Faraday's law and Lenz's law) an opposing emf in the primary windings. (3) The voltage V_p of the primary, however, cannot change in response to this opposing emf because it must always be equal to the emf $\mathscr{E}$ that is provided by the generator; closing switch S cannot change this fact. (4) To maintain V_p, the generator now produces (in addition to I_{mag}) an alternating current I_p in the primary circuit; the magnitude and phase constant of I_p are just those required for the emf induced by I_p in the primary to exactly cancel the emf induced there by I_s. Because the phase constant of I_p is not 90° like that of I_{mag}, this current I_p can transfer energy to the primary.

We want to relate I_s to I_p. However, rather than analyze the foregoing complex process in detail, let us just apply the principle of conservation of energy. The rate at which the generator transfers energy to the primary is equal to $I_p V_p$. The rate at which the primary then transfers energy to the secondary (via the alternating magnetic field linking the two coils) is $I_s V_s$. Because we assume that no energy is lost along the way, conservation of energy requires that

$$I_p V_p = I_s V_s.$$

Substituting for V_s from Eq. 33-70, we find that

$$I_s = I_p \frac{N_p}{N_s} \qquad \text{(transformation of currents).} \qquad (33\text{-}71)$$

This equation tells us that the current I_s in the secondary can be greater than, less than, or the same as the current I_p in the primary, depending on the *turns ratio* N_p/N_s.

Current I_p appears in the primary circuit because of the resistive load R in the secondary circuit. To find I_p, we substitute $I_s = V_s/R$ into Eq. 33-71 and then we substitute for V_s from Eq. 33-70. We find

$$I_p = \frac{1}{R} \left(\frac{N_s}{N_p}\right)^2 V_p. \qquad (33\text{-}72)$$

This equation has the form $I_p = V_p/R_{eq}$, where equivalent resistance R_{eq} is

$$R_{eq} = \left(\frac{N_p}{N_s}\right)^2 R. \qquad (33\text{-}73)$$

This R_{eq} is the value of the load resistance as "seen" by the generator; the generator produces the current I_p and voltage V_p as if it were connected to a resistance R_{eq}.

Impedance Matching

Equation 33-73 suggests still another function for the transformer. For maximum transfer of energy from an emf device to a resistive load, the resistance of the emf device and the resistance of the load must be equal. The same relation holds for ac circuits except that the *impedance* (rather than just the resistance) of the generator must be matched to that of the load. It often happens—as when we wish to connect a speaker set to an amplifier—that this condition is far from met, the amplifier and the speaker set being of high and low impedance, respectively. We can match the impedances of the two devices by coupling them through a transformer with a suitable turns ratio N_p/N_s.

SAMPLE PROBLEM 33-7

A transformer on a utility pole operates at $V_p = 8.5$ kV on the primary side and supplies electric energy to a number of nearby houses at $V_s = 120$ V, both quantities being rms values. Assume an ideal step-down transformer, a purely resistive load, and a power factor of unity.

(a) What is the turns ratio N_p/N_s of the transformer?

SOLUTION: From Eq. 33-70 we have

$$\frac{N_p}{N_s} = \frac{V_p}{V_s} = \frac{8.5 \times 10^3 \text{ V}}{120 \text{ V}} = 70.83 \approx 71. \quad \text{(Answer)}$$

(b) The average rate of energy consumption in the houses served by the transformer is 78 kW. What are the rms currents in the primary and secondary of the transformer?

SOLUTION: From Eq. 33-67 we have (with $\cos \phi = 1$ and $V_p = \mathcal{E}$ for a transformer connected across an emf device)

$$I_p = \frac{P_{av}}{V_p} = \frac{78 \times 10^3 \text{ W}}{8.5 \times 10^3 \text{ V}} = 9.176 \text{ A} \approx 9.2 \text{ A} \quad \text{(Answer)}$$

and

$$I_s = \frac{P_{av}}{V_s} = \frac{78 \times 10^3 \text{ W}}{120 \text{ V}} = 650 \text{ A}. \quad \text{(Answer)}$$

(c) What is the resistive load in the secondary circuit?

SOLUTION: In the secondary circuit we have

$$R_s = \frac{V_s}{I_s} = \frac{120 \text{ V}}{650 \text{ A}} = 0.1846 \ \Omega \approx 0.18 \ \Omega. \quad \text{(Answer)}$$

(d) What is the resistive load in the primary circuit?

SOLUTION: In the primary circuit we find

$$R_p = \frac{V_p}{I_p} = \frac{8.5 \times 10^3 \text{ V}}{9.176 \text{ A}} = 926 \ \Omega \approx 930 \ \Omega \quad \text{(Answer)}$$

or, using Eq. 33-73 with $R_p = R_{eq}$ and $R_s = R$,

$$R_p = \left(\frac{N_p}{N_s}\right)^2 R_s = (70.83)^2 (0.1846 \ \Omega)$$

$$= 926 \ \Omega \approx 930 \ \Omega. \quad \text{(Answer)}$$

$\mathbb{C}$HECKPOINT **8:** An alternating-current emf device has a smaller resistance than that of the resistive load; to increase the transfer of energy from the device to the load, a transformer will be connected between the two. Should it be a step-up or step-down transformer?

REVIEW & SUMMARY

LC Energy Transfers

In an oscillating *LC* circuit, energy is shuttled periodically between the electric field of the capacitor and the magnetic field of the inductor; instantaneous values of the two forms of energy are

$$U_E = \frac{q^2}{2C} \quad \text{and} \quad U_B = \frac{Li^2}{2}. \qquad (33\text{-}1, 33\text{-}2)$$

The total energy $U (= U_E + U_B)$ remains constant.

LC Charge and Current Oscillations

The principle of conservation of energy leads to

$$L\frac{d^2q}{dt^2} + \frac{q}{C} = 0 \quad (LC \text{ oscillations}) \qquad (33\text{-}11)$$

as the differential equation of *LC* oscillations (with no resistance).

The solution of Eq. 33-11 is

$$q = Q \cos(\omega t + \phi) \quad \text{(charge)}, \qquad (33\text{-}12)$$

in which Q is the *charge amplitude* (maximum charge on the capacitor) and the angular frequency ω of the oscillations is

$$\omega = \frac{1}{\sqrt{LC}}. \qquad (33\text{-}4)$$

The phase constant ϕ in Eq. 33-12 is determined by the initial conditions (at $t = 0$) of the system.

The current i in the system is

$$i = -\omega Q \sin(\omega t + \phi) \quad \text{(current)}, \qquad (33\text{-}13)$$

in which ωQ is the *current amplitude I*.

Damped Oscillations

Oscillations in an *LC* circuit are damped when a dissipative element R is also present in the circuit, and then the differential equation of oscillation is

$$L \frac{d^2q}{dt^2} + R \frac{dq}{dt} + \frac{1}{C} q = 0 \quad \text{(RLC circuit)}. \qquad (33\text{-}20)$$

Its solution is

$$q = Qe^{-Rt/2L} \cos(\omega' t + \phi), \qquad (33\text{-}21)$$

where
$$\omega' = \sqrt{\omega^2 - (R/2L)^2}. \qquad (33\text{-}22)$$

We consider only situations with small R and thus small damping; then $\omega' \approx \omega$ and the maximum energy of the electric field in the capacitor is

$$U = \frac{Q^2}{2C} e^{-Rt/L}. \qquad (33\text{-}24)$$

Alternating Currents; Forced Oscillations

A series *RLC* circuit may be set into *forced oscillation* at a *driving angular frequency* ω_d by an external alternating emf

$$\mathcal{E} = \mathcal{E}_m \sin \omega_d t. \qquad (33\text{-}25)$$

The current driven in the circuit by the emf is

$$i = I \sin(\omega_d t - \phi), \qquad (33\text{-}26)$$

where ϕ is a phase constant.

Resonance

The current amplitude I in a series *RLC* circuit driven by a sinusoidal external emf is a maximum ($I = \mathcal{E}_m/R$) when the driving angular frequency ω_d equals the natural angular frequency ω of the circuit (at *resonance*). Then $X_C = X_L$, $\phi = 0$, and the current is in phase with the emf.

Single Circuit Elements

The alternating potential difference across a resistor has amplitude $V_R = IR$; the current is in phase with the potential differ-

ence. For a *capacitor*, $V_C = IX_C$, in which $X_C = 1/\omega_d C$ is the **capacitive reactance;** the current here leads the potential difference by 90°. For an *inductor*, $V_L = IX_L$, in which $X_L = \omega_d L$ is the **inductive reactance;** the current here lags the potential difference by 90°.

Series RLC Circuits

For a series *RLC* circuit with external emf given by Eq. 33-25 and current given by Eq. 33-26,

$$I = \frac{\mathcal{E}_m}{\sqrt{R^2 + (X_L - X_C)^2}} = \frac{\mathcal{E}_m}{\sqrt{R^2 + (\omega_d L - 1/\omega_d C)^2}}$$
$$\text{(current amplitude)} \qquad (33\text{-}51, 33\text{-}54)$$

and
$$\tan \phi = \frac{X_L - X_C}{R} \quad \text{(phase constant)}. \qquad (33\text{-}56)$$

Defining the impedance Z of the circuit as

$$Z = \sqrt{R^2 + (X_L - X_C)^2} \quad \text{(impedance)} \qquad (33\text{-}52)$$

allows us to write Eq. 33-51 as $I = \mathcal{E}_m/Z$.

Power

In a series *RLC* circuit, the **average power** P_{av} of the generator is equal to the production rate of thermal energy in the resistor:

$$P_{av} = I_{rms}^2 R = \mathcal{E}_{rms} I_{rms} \cos \phi$$
$$\text{(average power)}. \qquad (33\text{-}62, 33\text{-}67)$$

Here "rms" stands for **root-mean-square;** rms quantities are related to maximum quantities by $I_{rms} = I/\sqrt{2}$, $V_{rms} = V_m/\sqrt{2}$, and $\mathcal{E}_{rms} = \mathcal{E}_m/\sqrt{2}$. The term $\cos \phi$ is called the **power factor.**

Transformers

A *transformer* (assumed to be ideal) is an iron core on which are wound a primary coil of N_p turns and a secondary coil of N_s turns. If the primary coil is connected across an alternating-current generator, the primary and secondary voltages are related by

$$V_s = V_p \frac{N_s}{N_p} \quad \begin{array}{l}\text{(transformation}\\\text{of voltage)}.\end{array} \qquad (33\text{-}70)$$

The currents are related by

$$I_s = I_p \frac{N_p}{N_s} \quad \begin{array}{l}\text{(transformation}\\\text{of currents)},\end{array} \qquad (33\text{-}71)$$

and the equivalent resistance of the secondary circuit, as seen by the generator, is

$$R_{eq} = \left(\frac{N_p}{N_s}\right)^2 R, \qquad (33\text{-}73)$$

where R is the resistive load in the secondary circuit.

QUESTIONS

1. A charged capacitor and an inductor are connected at time $t = 0$. In terms of the period T of the resulting oscillations, what is the first later time at which the following reach a maximum: (a) U_B, (b) the magnetic flux through the inductor, (c) di/dt, and (d) the emf of the inductor?

2. What values of the phase constant ϕ in Eq. 33-12 would permit situations (a), (c), (e), and (g) of Fig. 33-1 to be the situation at $t = 0$?

3. Figure 33-16 shows three oscillating LC circuits with identical inductors and capacitors. Rank the circuits according to the time taken to fully discharge the capacitors during the oscillations, greatest first.

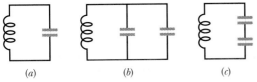

(a)   (b)   (c)

FIGURE 33-16 Question 3.

4. Figure 33-17 shows graphs of capacitor voltage v_C for LC circuits 1 and 2, which contain identical capacitances and have the same maximum charge Q. Are (a) the inductance L and (b) the maximum current I in circuit 1 greater than, less than, or the same as those in circuit 2?

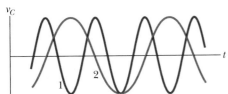

FIGURE 33-17 Question 4.

5. Charges on the capacitors in three oscillating LC circuits vary as follows: (1) $q = 2 \cos 4t$; (2) $q = 4 \cos t$; (3) $q = 3 \cos 4t$ (with q in coulombs and t in seconds). Rank the circuits according to (a) the current amplitude and (b) the period, greatest first.

6. If you increase the inductance L in an oscillating LC circuit having a given maximum charge Q, do (a) the current magnitude I and (b) the maximum magnetic energy U_B increase, decrease, or stay the same?

7. In a damped oscillating RLC circuit, does the charge decay faster than, slower than, or at the same rate as the energy?

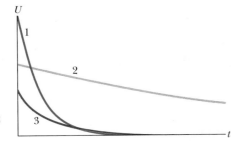

FIGURE 33-18
Question 8.

8. Figure 33-18 shows the decrease in energy with time of three damped oscillating RLC circuits with the same initial charge Q. Rank the circuits according to their (a) capacitances C and (b) values of L/R, greatest first.

9. The values of the phase constant ϕ for four sinusoidally driven series RLC circuits are (1) $-15°$, (2) $+35°$, (3) $\pi/3$ rad, and (4) $-\pi/6$ rad. (a) In which is the load primarily capacitive? (b) In which does the alternating emf lead the current?

10. Figure 33-19 shows the current i and driving emf $\mathscr{E}$ for a series RLC circuit. (a) Does the current lead or lag the emf? (b) Is the circuit's load mainly capacitive or mainly inductive? (c) Is the angular frequency of the emf greater than or less than the natural angular frequency of the circuit?

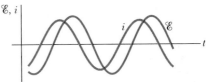

FIGURE 33-19 Questions 10 and 15.

11. The following table gives, for three series RLC circuits, the amplitude $\mathscr{E}_m$ of the driving emf and the values of R, L, and C. Without written calculation, rank the circuits according to (a) the amplitude I of the current at resonance and (b) the angular frequency at resonance, greatest first.

CIRCUIT	$\mathscr{E}_m$ (V)	R (Ω)	L (mH)	C (μF)
1	25	5.0	200	10
2	60	12	100	5.0
3	80	10	300	10

12. Suppose that for a particular driving angular frequency, the emf leads the current in a series RLC circuit. If you decrease the driving angular frequency slightly, do (a) the phase constant and (b) the current amplitude increase, decrease, or remain the same?

13. The driving angular frequency in a certain series RLC circuit is less than the natural angular frequency of the circuit. (a) Is the phase constant ϕ positive, negative, or zero? (b) Does the current lead or lag the emf?

14. Figure 33-20 shows three situations like those of Fig. 33-12. For each situation, is the driving angular frequency greater than, less than, or equal to the resonant angular frequency?

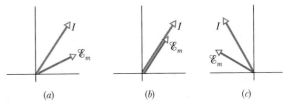

(a)   (b)   (c)

FIGURE 33-20 Question 14.

15. Figure 33-19 shows the current i and driving emf $\mathcal{E}$ for a series RLC circuit. Relative to the emf curve, does the current curve shift leftward or rightward and does the amplitude of that curve increase or decrease if we slightly increase (a) L, (b) C, and (c) the driving angular frequency of the emf?

16. Figure 33-21 shows the current i and driving emf $\mathcal{E}$ for a series RLC circuit. (a) Is the phase constant positive or negative? (b) To increase the rate at which energy is transferred to the resistive load, should L be increased or decreased? (c) Should, instead, C be increased or decreased?

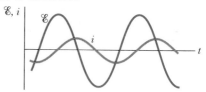

FIGURE 33-21
Question 16.

EXERCISES & PROBLEMS

SECTION 33-2 LC Oscillations, Qualitatively

1E. What is the capacitance of an oscillating LC circuit if the maximum charge on the capacitor is 1.60 μC and the total energy is 140 μJ?

2E. A 1.50 mH inductor in an oscillating LC circuit stores a maximum energy of 10.0 μJ. What is the maximum current?

3E. In an oscillating LC circuit $L = 1.10$ mH and $C = 4.00$ μF. The maximum charge on the capacitor is 3.00 μC. Find the maximum current.

4E. An oscillating LC circuit consists of a 75.0 mH inductor and a 3.60 μF capacitor. If the maximum charge on the capacitor is 2.90 μC, (a) what is the total energy in the circuit and (b) what is the maximum current?

5E. In a certain oscillating LC circuit the total energy is converted from electric energy in the capacitor to magnetic energy in the inductor in 1.50 μs. (a) What is the period of oscillation? (b) What is the frequency of oscillation? (c) How long after the magnetic energy is a maximum will it be a maximum again?

6P. The frequency of oscillation of a certain LC circuit is 200 kHz. At time $t = 0$, plate A of the capacitor has maximum positive charge. At what times $t > 0$ will (a) plate A again have maximum positive charge, (b) the other plate of the capacitor have maximum positive charge, and (c) the inductor have maximum magnetic field?

SECTION 33-3 The Electrical–Mechanical Analogy

7E. A 0.50 kg body oscillates on a spring that, when extended 2.0 mm from equilibrium, has a restoring force of 8.0 N. (a) What is the angular frequency of oscillation? (b) What is the period of oscillation? (c) What is the capacitance of the corresponding LC circuit if L is chosen to be 5.0 H?

8P. The energy in an oscillating LC circuit containing a 1.25 H inductor is 5.70 μJ. The maximum charge on the capacitor is 175 μC. Find (a) the mass, (b) the spring constant, (c) the maximum displacement, and (d) the maximum speed for the corresponding mechanical system.

SECTION 33-4 LC Oscillations, Quantitatively

9E. LC oscillators have been used in circuits connected to loudspeakers to create some of the sounds of electronic music. What inductance must be used with a 6.7 μF capacitor to produce a frequency of 10 kHz, which is near the middle of the audible range of frequencies?

10E. What capacitance would you connect across a 1.30 mH inductor to make the resulting oscillator resonate at 3.50 kHz?

11E. In an LC circuit with $L = 50$ mH and $C = 4.0$ μF, the current is initially a maximum. How long will it take before the capacitor is fully charged for the first time?

12E. Consider the circuit shown in Fig. 33-22. With switch S_1 closed and the other two switches open, the circuit has a time constant τ_C (see Section 28-8). With switch S_2 closed and the other two switches open, the circuit has a time constant τ_L (see Section 31-9). With switch S_3 closed and the other two switches open, the circuit oscillates with a period T. Show that $T = 2\pi\sqrt{\tau_C \tau_L}$.

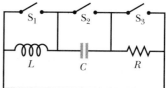

FIGURE 33-22 Exercise 12.

13E. Derive the differential equation for an LC circuit (Eq. 33-11) using the loop rule.

14E. A single loop consists of several inductors ($L_1, L_2, \ldots$), several capacitors ($C_1, C_2, \ldots$), and several resistors ($R_1, R_2, \ldots$) connected in series as shown, for example, in Fig. 33-23a. Show that regardless of the sequence of these circuit ele-

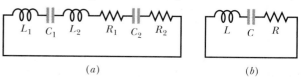

(a)   (b)

FIGURE 33-23 Exercise 14.

ments in the loop, the behavior of this circuit is identical to that of the simple *LC* circuit shown in Fig. 33-23*b*. (*Hint:* Consider the loop rule and see Problem 56 in Chapter 31)

15P. An oscillating *LC* circuit consisting of a 1.0 nF capacitor and a 3.0 mH coil has a maximum voltage of 3.0 V. (a) What is the maximum charge on the capacitor? (b) What is the maximum current through the circuit? (c) What is the maximum energy stored in the magnetic field of the coil?

16P. An oscillating *LC* circuit has an inductance of 3.00 mH and a capacitance of 10.0 μF. Calculate (a) the angular frequency and (b) the period of the oscillation. (c) At time $t = 0$ the capacitor is charged to 200 μC, and the current is zero. Sketch roughly the charge on the capacitor as a function of time.

17P. In an oscillating *LC* circuit in which $C = 4.00$ μF, the maximum potential difference across the capacitor during the oscillations is 1.50 V and the maximum current through the inductor is 50.0 mA. (a) What is the inductance *L*? (b) What is the frequency of the oscillations? (c) How much time does the charge on the capacitor take to rise from zero to its maximum value?

18P. In the circuit shown in Fig. 33-24 the switch has been in position *a* for a long time. It is now thrown to position *b*. (a) Calculate the frequency of the resulting oscillating current. (b) What is the amplitude of the current oscillations?

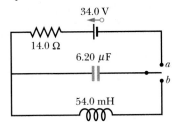

FIGURE 33-24 Problem 18.

19P. You are given a 10 mH inductor and two capacitors, of 5.0 μF and 2.0 μF capacitance. List the oscillation frequencies that can be generated by connecting these elements in various combinations.

20P. An *LC* circuit oscillates at a frequency of 10.4 kHz. (a) If the capacitance is 340 μF, what is the inductance? (b) If the maximum current is 7.20 mA, what is the total energy in the circuit? (c) What is the maximum charge on the capacitor?

21P. (a) In an oscillating *LC* circuit, in terms of the maximum charge on the capacitor, what is the charge there when the energy in the electric field is 50.0% of that in the magnetic field? (b) What fraction of a period must elapse following the time the capacitor is fully charged for this condition to arise?

22P. At some instant in an oscillating *LC* circuit, 75.0% of the total energy is stored in the magnetic field of the inductor. (a) In terms of the maximum charge on the capacitor, what is the charge there at this instant? (b) In terms of the maximum current in the inductor, what is the current there at this instant?

23P. An inductor is connected across a capacitor whose capacitance can be varied by turning a knob. We wish to make the frequency of the *LC* oscillations vary linearly with the angle of rotation of the knob, going from 2×10^5 to 4×10^5 Hz as the knob turns through 180°. If $L = 1.0$ mH, plot the required capacitance *C* as a function of the angle of rotation of the knob.

24P. A variable capacitor with a range from 10 to 365 pF is used with a coil to form a variable-frequency *LC* circuit to tune the input to a radio. (a) What ratio of maximum to minimum frequencies may be obtained with such a capacitor? (b) If this circuit is to obtain frequencies from 0.54 MHz to 1.60 MHz, the ratio computed in (a) is too large. By adding a capacitor in parallel to the variable capacitor, this range may be adjusted. How large should this added capacitor be, and what inductance should be chosen in order to obtain the desired range of frequencies?

25P. In an oscillating *LC* circuit, $L = 25.0$ mH and $C = 7.80$ μF. At time $t = 0$ the current is 9.20 mA, the charge on the capacitor is 3.80 μC, and the capacitor is charging. (a) What is the total energy in the circuit? (b) What is the maximum charge on the capacitor? (c) What is the maximum current? (d) If the charge on the capacitor is given by $q = Q\cos(\omega t + \phi)$, what is the phase angle ϕ? (e) Suppose the data are the same, except that the capacitor is discharging at $t = 0$. What then is ϕ?

26P. In an oscillating *LC* circuit, $L = 3.00$ mH and $C = 2.70$ μF. At $t = 0$ the charge on the capacitor is zero and the current is 2.00 A. (a) What is the maximum charge that will appear on the capacitor? (b) In terms of the period *T* of oscillation, how much time will elapse after $t = 0$ until the energy stored in the capacitor will be increasing at its greatest rate? (c) What is this greatest rate at which energy is transferred to the capacitor?

27P. Three identical inductors *L* and two identical capacitors *C* are connected in a two-loop circuit as shown in Fig. 33-25. (a) Suppose the currents are as shown in Fig. 33-25*a*. What is the current in the middle inductor? Write the loop equations and show that they are satisfied if the current oscillates with angular frequency $\omega = 1/\sqrt{LC}$. (b) Now suppose the currents are as shown in Fig. 33-25*b*. What is the current in the middle inductor? Write the loop equations and show that they are satisfied if the current oscillates with angular frequency $\omega = 1/\sqrt{3LC}$. Because the circuit can oscillate at two different frequencies, we cannot find an equivalent single-loop *LC* circuit to replace it.

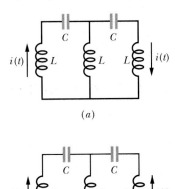

FIGURE 33-25
Problem 27.

28P. A series circuit containing inductance L_1 and capacitance C_1 oscillates at angular frequency ω. A second series circuit, containing inductance L_2 and capacitance C_2, oscillates at the same angular frequency. In terms of ω, what is the angular frequency of oscillation of a series circuit containing all four of these elements? Neglect resistance. (*Hint:* Use the formulas for equivalent capacitance and equivalent inductance; see Section 26-4 and Problem 56 in Chapter 31.

29P. In an oscillating LC circuit with $C = 64.0\ \mu$F, the current as a function of time is given by $i = (1.60)\sin(2500t + 0.680)$, where t is in seconds, i in amperes, and the phase angle in radians. (a) How soon after $t = 0$ will the current reach its maximum value? What are (b) the inductance L and (c) the total energy?

30P*. In Fig. 33-26 the 900 μF capacitor is initially charged to 100 V and the 100 μF capacitor is uncharged. Describe in detail how one might charge the 100 μF capacitor to 300 V by manipulating switches S_1 and S_2.

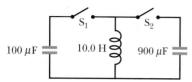

FIGURE 33-26 Problem 30.

SECTION 33-5 Damped Oscillations in an *RLC* Circuit

31E. What resistance R should be connected in series with an inductance $L = 220$ mH and capacitance $C = 12.0\ \mu$F for the maximum charge on the capacitor to decay to 99.0% of its initial value in 50.0 cycles? (Assume $\omega' \approx \omega$.)

32E. Consider a damped LC circuit. (a) Show that the damping term $e^{-Rt/2L}$ (which involves L but not C) can be rewritten in a more symmetric manner (involving L and C) as $e^{-\pi R\sqrt{C/L}(t/T)}$. Here T is the period of oscillation (neglecting resistance). (b) Using (a), show that the SI unit of $\sqrt{L/C}$ is the ohm. (c) Using (a), show that the condition that the fractional energy loss per cycle be small is $R \ll \sqrt{L/C}$.

33P. In an oscillating series RLC circuit, find the time required for the maximum energy present in the capacitor during an oscillation to fall to half its initial value. Assume $q = Q$ at $t = 0$.

34P. A single-loop circuit consists of a 7.20 Ω resistor, a 12.0 H inductor, and a 3.20 μF capacitor. Initially the capacitor has a charge of 6.20 μC and the current is zero. Calculate the charge on the capacitor N complete cycles later for $N = 5$, 10, and 100.

35P. At time $t = 0$ there is no charge on the capacitor of an RLC circuit but there is current I through the inductor. (a) Find the phase constant ϕ in Eq. 33-21 for the circuit. (b) Write an expression for the charge q on the capacitor as a function of time t and in terms of the current amplitude and angular frequency ω' of the oscillations.

36P. (a) By direct substitution of Eq. 33-21 into Eq. 33-20, show that $\omega' = \sqrt{(1/LC) - (R/2L)^2}$. (b) By what fraction does the fre-

quency of oscillation shift when the resistance is increased from 0 to 100 Ω in a circuit with $L = 4.40$ H and $C = 7.30\ \mu$F?

37P*. In an oscillating RLC circuit, show that the fraction of the energy lost per cycle of oscillation, $\Delta U/U$, is given to a close approximation by $2\pi R/\omega L$. The quantity $\omega L/R$ is often called the Q of the circuit (for "quality"). A high-Q circuit has low resistance and a low fractional energy loss ($= 2\pi/Q$) per cycle.

SECTION 33-8 Three Simple Circuits

38E. A 1.50 μF capacitor is connected as in Fig. 33-9a to an ac generator with $\mathscr{E}_m = 30.0$ V. What is the amplitude of the resulting alternating current if the frequency of the emf is (a) 1.00 kHz and (b) 8.00 kHz?

39E. A 50.0 mH inductor is connected as in Fig. 33-10a to an ac generator with $\mathscr{E}_m = 30.0$ V. What is the amplitude of the resulting alternating current if the frequency of the emf is (a) 1.00 kHz and (b) 8.00 kHz?

40E. A 50 Ω resistor is connected as in Fig. 33-8a to an ac generator with $\mathscr{E}_m = 30.0$ V. What is the amplitude of the resulting alternating current if the frequency of the emf is (a) 1.00 kHz and (b) 8.00 kHz?

41E. A 45.0 mH inductor has a reactance of 1.30 kΩ. (a) What is its operating frequency? (b) What is the capacitance of a capacitor with the same reactance at that frequency? (c) If the frequency is doubled, what are the reactances of the inductor and capacitor?

42E. A 1.50 μF capacitor has a capacitive reactance of 12.0 Ω. (a) What must be its operating frequency? (b) What will be the capacitive reactance if the frequency is doubled?

43E. (a) At what frequency would a 6.0 mH inductor and a 10 μF capacitor have the same reactance? (b) What would the reactance be? (c) Show that this frequency would be the natural frequency of an oscillating circuit with the same L and C.

44P. An ac generator emf is $\mathscr{E} = \mathscr{E}_m \sin \omega_d t$, with $\mathscr{E}_m = 25.0$ V and $\omega_d = 377$ rad/s. It is connected to a 12.7 H inductor. (a) What is the maximum value of the current? (b) When the current is a maximum, what is the emf of the generator? (c) When the emf of the generator is -12.5 V and increasing in magnitude, what is the current?

45P. The ac generator of Problem 44 is connected to a 4.15 μF capacitor. (a) What is the maximum value of the current? (b) When the current is a maximum, what is the emf of the generator? (c) When the emf of the generator is -12.5 V and increasing in magnitude, what is the current?

46P. An ac generator emf is $\mathscr{E} = \mathscr{E}_m \sin(\omega_d t - \pi/4)$, where $\mathscr{E}_m = 30.0$ V and $\omega_d = 350$ rad/s. The current produced in a connected circuit is $i(t) = I \sin(\omega_d t - 3\pi/4)$, where $I = 620$ mA. (a) At what time after $t = 0$ does the generator emf first reach a maximum? (b) At what time after $t = 0$ does the current first reach a maximum? (c) The circuit contains a single element other than the generator. Is it a capacitor, an inductor, or a resistor? Justify your answer. (d) What is the value of the capacitance, inductance, or resistance, as the case may be?

47P. An ac generator emf is $\mathscr{E} = \mathscr{E}_m \sin(\omega_d t - \pi/4)$, where

$\mathscr{E}_m$ = 30.0 V and ω_d = 350 rad/s. The current is given by $i(t)$ = $I \sin(\omega_d t + \pi/4)$, where I = 620 mA. (a) At what time after $t = 0$ does the generator emf first reach a maximum? (b) At what time after $t = 0$ does the current first reach a maximum? (c) The circuit contains a single element other than the generator. Is it a capacitor, an inductor, or a resistor? Justify your answer. (d) What is the value of the capacitance, inductance, or resistance, as the case may be?

48P. A three-phase generator G produces electrical power that is transmitted by means of three wires as shown in Fig. 33-27. The electric potentials (relative to a common reference level) of these wires are $V_1 = A \sin \omega_d t$, $V_2 = A \sin(\omega_d t - 120°)$, and $V_3 = A \sin(\omega_d t - 240°)$. Some types of heavy industrial equipment (for example, motors) have three terminals and are designed to be connected directly to these three wires. To use a more conventional two-terminal device (for example, a lightbulb), one connects it to any two of the three wires. Show that the potential difference between *any two* of the wires (a) oscillates sinusoidally with angular frequency ω_d and (b) has an amplitude of $A\sqrt{3}$.

Three-wire transmission line
FIGURE 33-27 Problem 48.

SECTION 33-9 The Series *RLC* Circuit

49E. (a) Find Z, ϕ, and I for the situation of Sample Problem 33-5 with the capacitor removed from the circuit, all other parameters remaining unchanged. (b) Draw to scale a phasor diagram like that of Fig. 33-11d for this new situation.

50E. (a) Find Z, ϕ, and I for the situation of Sample Problem 33-5 with the inductor removed from the circuit, all other parameters remaining unchanged. (b) Draw to scale a phasor diagram like that of Fig. 33-11d for this new situation.

51E. (a) Find Z, ϕ, and I for the situation of Sample Problem 33-5 with C = 70.0 μF, the other parameters remaining unchanged. (b) Draw a phasor diagram like that of Fig. 33-11d for this new situation and compare the two diagrams closely.

52E. A generator with an adjustable frequency of oscillation is wired in series to an inductor of L = 2.50 mH and a capacitor of C = 3.00 μF. At what frequency does the generator produce the largest possible current amplitude in the circuit?

53P. In Fig. 33-28, a generator with an adjustable frequency of oscillation is connected to a variable resistance R, a capacitor of C = 5.50 μF, and an inductor of inductance L. With R = 100 Ω, the amplitude of the current produced in the circuit by the genera-

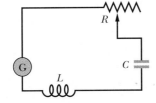

FIGURE 33-28 Problem 53.

tor is at half-maximum level when the generator's oscillations are at 1.30 and 1.50 kHz. (a) What is L? (b) If R is increased, what happens to the frequencies at which the current amplitude is at half-maximum level?

54P. Verify mathematically that the following geometric construction correctly gives both the impedance Z and the phase constant ϕ. Referring to Fig. 33-29, (i) draw an arrow in the positive y direction of magnitude X_C; (ii) draw a second arrow in the negative y direction of magnitude X_L; (iii) draw a third arrow of magnitude R in the positive x direction. Then the magnitude of the "resultant" of these arrows is Z, and the angle (measured clockwise from the positive x direction) of this resultant is ϕ.

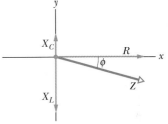

FIGURE 33-29 Problem 54.

55P. Can the amplitude of the voltage across an inductor be greater than the amplitude of the generator emf in an *RLC* circuit? Consider an *RLC* circuit with $\mathscr{E}_m$ = 10 V, R = 10 Ω, L = 1.0 H, and C = 1.0 μF. Find the amplitude of the voltage across the inductor at resonance.

56P. A coil of inductance 88 mH and unknown resistance and a 0.94 μF capacitor are connected in series with an alternating emf of frequency 930 Hz. If the phase constant between the applied voltage and the current is 75°, what is the resistance of the coil?

57P. When the generator emf in Sample Problem 33-5 is a maximum, what is the voltage across (a) the generator, (b) the resistor, (c) the capacitor, and (d) the inductor? (e) By summing these with appropriate signs, verify that the loop rule is satisfied.

58P. An ac generator of $\mathscr{E}_m$ = 220 V operating at 400 Hz causes oscillations in a series *RLC* circuit having R = 220 Ω, L = 150 mH, and C = 24.0 μF. Find (a) the capacitive reactance X_C, (b) the impedance Z, and (c) the current amplitude I. A second capacitor of the same capacitance is then connected in series with the other components. Determine whether the values of (d) X_C, (e) Z, and (f) I increase, decrease, or remain the same.

59P. An *RLC* circuit such as that of Fig. 33-7 has R = 5.00 Ω, C = 20.0 μF, L = 1.00 H, and $\mathscr{E}_m$ = 30.0 V. (a) At what angular frequency ω_d will the current amplitude have its maximum value, as in the resonance curves of Fig. 33-13? (b) What is this maximum value? (c) At what two angular frequencies ω_{d1} and ω_{d2} will the current amplitude be half this maximum value? (d) What is the fractional half-width [$= (\omega_{d1} - \omega_{d2})/\omega$] of the resonance curve for this circuit?

60P. For a certain series *RLC* circuit, the maximum generator emf is 125 V and the maximum current is 3.20 A. If the current leads the generator emf by 0.982 rad, what are (a) the impedance and (b) the resistance of the circuit? (c) Is the circuit predominantly capacitive or inductive?

61P. A series RLC circuit has a resonant frequency of 6.00 kHz. When it is driven at 8.00 kHz, it has an impedance of 1.00 kΩ and a phase constant of 45°. What are the values of (a) R, (b) L, and (c) C for this circuit?

62P. In a certain series RLC circuit operating at a frequency of 60.0 Hz, the maximum voltage across the inductor is 2.00 times the maximum voltage across the resistor and 2.00 times the maximum voltage across the capacitor. (a) By what phase angle does the current lag the generator emf? (b) If the maximum generator emf is 30.0 V, what should be the resistance of the circuit to obtain a maximum current of 300 mA?

63P. The circuit of Sample Problem 33-5 is not in resonance. (a) How can you tell? (b) What capacitor would you combine in parallel with the capacitor already in the circuit to bring resonance about? (c) What would the current amplitude then be?

64P. A generator is to be connected in series with an inductor of $L = 2.00$ mH and a capacitance C. You are to produce C by using capacitors of capacitances $C_1 = 4.00$ μF and $C_2 = 6.00$ μF, either singly or together. What resonant frequencies can the circuit have depending on the value of C?

65P. In Fig. 33-30, a generator with an adjustable frequency of oscillation is connected to resistance $R = 100$ Ω, inductances $L_1 = 1.70$ mH and $L_2 = 2.30$ mH, and capacitances $C_1 = 4.00$ μF, $C_2 = 2.50$ μF, and $C_3 = 3.50$ μF. (a) What is the resonant frequency of the circuit? (*Hint:* See Problem 56 in Chapter 31.) What happens to the resonant frequency if (b) the value of R is increased, (c) the value of L_1 is increased, and (d) capacitance C_3 is removed from the circuit?

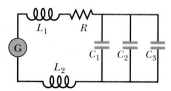

FIGURE 33-30 Problem 65.

66P. A series circuit with resistor–inductor–capacitor combination R_1, L_1, C_1 has the same resonant frequency as a second circuit with a different combination R_2, L_2, C_2. You now connect the two combinations in series. Show that this new circuit has the same resonant frequency as the separate circuits.

67P. Show that the fractional half-width (see Problem 59) of a resonance curve is given by

$$\frac{\Delta \omega_d}{\omega} = \sqrt{\frac{3C}{L}}\, R,$$

in which ω_d is the angular frequency at resonance and $\Delta \omega_d$ is the width of the resonance curve at half-amplitude. Note that $\Delta \omega_d / \omega$ decreases with R, as Fig. 33-13 shows. Use this formula to check the answer to Problem 59d.

68P*. The ac generator in Fig. 33-31 supplies 120 V at 60.0 Hz. With the switch open as in the diagram, the current leads the generator emf by 20.0°. With the switch in position 1 the current lags the generator emf by 10.0°. When the switch is in position 2, the current is 2.00 A. Find the values of R, L, and C.

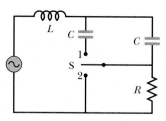

FIGURE 33-31 Problem 68.

SECTION 33-10   Power in Alternating-Current Circuits

69E. What is the maximum value of an ac voltage whose rms value is 100 V?

70E. An ac voltmeter with large impedance is connected in turn across the inductor, the capacitor, and the resistor in a series circuit having an alternating emf of 100 V (rms); it gives the same reading in volts in each case. What is this reading?

71E. (a) For the conditions in Problem 44c, is the generator supplying energy to or taking energy from the rest of the circuit? (b) Repeat for the conditions of Problem 45c.

72E. What direct current will produce the same amount of heat, in a particular resistor, as an alternating current that has a maximum value of 2.60 A?

73E. Calculate the average rate of energy dissipation in the circuits of Exercises 39, 40, 49, and 50.

74E. Show that the average rate of energy supplied to the circuit of Fig. 33-7 can also be written as $P_{av} = \mathcal{E}_{rms}^2 R / Z^2$. Show that this expression for average power gives reasonable results for a purely resistive circuit, for an RLC circuit at resonance, for a purely capacitive circuit, and for a purely inductive circuit.

75E. An electric motor connected to a 120 V, 60.0 Hz ac outlet does mechanical work at the rate of 0.100 hp (1 hp = 746 W). If it draws an rms current of 0.650 A, what is its effective resistance, in terms of power transfer? Is this the same as the resistance of its coils, as measured with an ohmmeter with the motor disconnected from the outlet?

76E. An air conditioner connected to a 120 V rms ac line is equivalent to a 12.0 Ω resistance and a 1.30 Ω inductive reactance in series. (a) Calculate the impedance of the air conditioner. (b) Find the average rate at which power is supplied to the appliance.

77E. An electric motor has an effective resistance of 32.0 Ω and an inductive reactance of 45.0 Ω when working under load. The rms voltage across the alternating source is 420 V. Calculate the rms current.

78P. Show mathematically, rather than graphically as in Fig. 33-14b, that the average value of $\sin^2(\omega t - \phi)$ over an integral number of half-cycles is $\frac{1}{2}$.

79P. For a sinusoidally driven series RLC circuit, show that over one complete cycle with period T (a) the energy stored in the capacitor does not change; (b) the energy stored in the inductor does not change; (c) the driving emf device supplies energy $(\frac{1}{2}T)\mathcal{E}_m I \cos \phi$; and (d) the resistor dissipates energy $(\frac{1}{2}T)RI^2$. (e) Show that the quantities found in (c) and (d) are equal.

80P. In a series oscillating *RLC* circuit, $R = 16.0\ \Omega$, $C = 31.2\ \mu F$, $L = 9.20$ mH, and $\mathscr{E} = \mathscr{E}_m \sin \omega_d t$ with $\mathscr{E}_m = 45.0$ V and $\omega_d = 3000$ rad/s. For time $t = 0.442$ ms find (a) the rate at which energy is being supplied by the generator, (b) the rate at which the energy in the capacitor is changing, (c) the rate at which the energy in the inductor is changing, and (d) the rate at which energy is being dissipated in the resistor. (e) What is the meaning of a negative result for any of (a), (b), and (c)? (f) Show that the results of (b), (c), and (d) sum to the result of (a).

81P. In Fig. 33-32 show that the average rate at which energy is dissipated in resistance R is a maximum when R is equal to the internal resistance r of the ac generator. (In the text we have tacitly assumed, up to this point, that $r = 0$.)

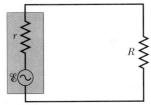

FIGURE 33-32 Problems 81 and 90.

82P. Figure 33-33 shows an ac generator connected to a "black box" through a pair of terminals. The box contains an *RLC* circuit, possibly even a multiloop circuit, whose elements and connections we do not know. Measurements outside the box reveal that

$$\mathscr{E}(t) = (75.0\ \text{V})\sin \omega_d t$$

and

$$i(t) = (1.20\ \text{A})\sin(\omega_d t + 42.0°).$$

(a) What is the power factor? (b) Does the current lead or lag the emf? (c) Is the circuit in the box largely inductive or largely capacitive? (d) Is the circuit in the box in resonance? (e) Must there be a capacitor in the box? An inductor? A resistor? (f) At what average rate is energy delivered to the box by the generator? (g) Why don't you need to know the angular frequency ω_d to answer all these questions?

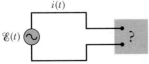

FIGURE 33-33 Problem 82.

83P. In an *RLC* circuit such as that of Fig. 33-7 assume that $R = 5.00\ \Omega$, $L = 60.0$ mH, $f_d = 60.0$ Hz, and $\mathscr{E}_m = 30.0$ V. For what values of the capacitance would the average rate at which energy is dissipated in the resistor be (a) a maximum and (b) a minimum? (c) What are these maximum and minimum energy dissipation rates? What are (d) the corresponding phase angles and (e) the corresponding power factors?

84P. A typical "light dimmer" used to dim the stage lights in a theater consists of a variable inductor L (whose inductance is adjustable between zero and L_{max}) connected in series with the lightbulb B as shown in Fig. 33-34. The electrical supply is 120 V (rms) at 60.0 Hz; the lightbulb is rated as "120 V, 1000 W." (a)

What L_{max} is required if the rate of energy dissipation in the lightbulb is to be varied by a factor of 5 from its upper limit of 1000 W? Assume that the resistance of the lightbulb is independent of its temperature. (b) Could one use a variable resistor (adjustable between zero and R_{max}) instead of an inductor? If so, what R_{max} is required? Why isn't this done?

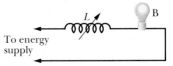

FIGURE 33-34 Problem 84.

85P. In Fig. 33-35, $R = 15.0\ \Omega$, $C = 4.70\ \mu F$, and $L = 25.0$ mH. The generator provides a sinusoidal voltage of 75.0 V (rms) and frequency $f = 550$ Hz. (a) Calculate the rms current. (b) Find the rms voltages V_{ab}, V_{bc}, V_{cd}, V_{bd}, V_{ad}. (c) At what average rate is energy dissipated by each of the three circuit elements?

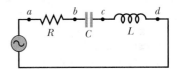

FIGURE 33-35 Problem 85.

SECTION 33-11 Transformers

86E. A generator supplies 100 V to the primary coil of a transformer of 50 turns. If the secondary coil has 500 turns, what is the secondary voltage?

87E. A transformer has 500 primary turns and 10 secondary turns. (a) If V_p is 120 V (rms), what is V_s, assuming an open circuit? (b) If the secondary is now connected to a resistive load of 15 Ω, what are the currents in the primary and secondary?

88E. Figure 33-36 shows an "autotransformer." It consists of a single coil (with an iron core). Three taps T_i are provided. Between taps T_1 and T_2 there are 200 turns, and between taps T_2 and T_3 there are 800 turns. Any two taps can be considered the "primary terminals" and any two taps can be considered the "secondary terminals." List all the ratios by which the primary voltage may be changed to a secondary voltage.

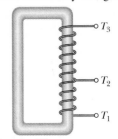

FIGURE 33-36 Exercise 88.

89P. An ac generator provides emf to a resistive load in a remote factory over a two-cable transmission line. At the factory a step-down transformer reduces the voltage from its (rms) transmission value V_t to a much lower value, safe and convenient for use in the factory. The transmission line resistance is 0.30 Ω/cable, and the power of the generator is 250 kW. Calculate the voltage drop along the transmission line and the rate at which energy is dissipated in the line as thermal energy if (a) $V_t = 80$ kV, (b) $V_t = 8.0$ kV, and (c) $V_t = 0.80$ kV. Comment on the acceptability of each choice.

90P. In Fig. 33-32 let the rectangular box on the left represent the (high-impedance) output of an audio amplifier, with $r = 1000\ \Omega$. Let $R = 10\ \Omega$ represent the (low-impedance) coil of a loudspeaker. For maximum transfer of energy to the load R we must have $R = r$, and that is not true in this case. However, a transformer can be used to ''transform'' resistances, making them behave electrically as if they were larger or smaller than they actually are. Sketch the primary and secondary coils of a transformer that can be introduced between the amplifier and the speaker in Fig. 33-32 to match the impedances. What must be the turns ratio?

Electronic Computation

91. A 45.0 μF capacitor and a 200 Ω resistor are connected in series with an ac source with a voltage V_s of 100 V. The frequency f of the source can be varied from 0 to 100 Hz. (a) Write an equation for the capacitive reactance X_C. (b) Simultaneously plot the resistance R, the capacitive reactance X_C, and the impedance Z versus f for the range $0 < f < 100$ Hz. (c) From the plots, determine the value of f for which $X_C = R$.

92. (a) For the situation of Problem 91, simultaneously plot the voltage V_C across the capacitor, the voltage V_R across the resistor, and the (constant) voltage V_s across the source versus f for the range $0 < f < 100$ Hz. (b) From the plots, determine the value of f for which $V_C = V_R$. (c) What is V_R at that frequency? (d) Determine the value of f for which $V_R = 0.50V_s$. (e) What is V_C at that frequency? (f) Determine the value of f for which $V_C = 0.50V_s$. (g) What is V_R at that frequency?

93. A 40.0 mH inductor and a 200 Ω resistor are connected in series with an ac source with a voltage V_s of 100 V. The frequency f of the source can be varied from 0 to 2500 Hz. (a) Write an equation for the inductive reactance X_L. (b) Simultaneously plot the resistance R, the inductive reactance X_L, and the impedance Z versus f for the range $0 < f < 2500$ Hz. (c) From the plots, determine the value of f for which $X_L = R$.

94. (a) For the situation of Problem 93, simultaneously plot the voltage V_L across the inductor, the voltage V_R across the resistor, and the (constant) voltage V_s across the source versus f for the range $0 < f < 2500$ Hz. (b) From the plots, determine the value of f for which $V_L = V_R$. (c) What is V_R at that frequency? (d) Determine the value of f for which $V_R = V_s/3$. (e) What is V_L at that frequency? (f) Determine the value of f for which $V_L = V_s/3$. (g) What is V_R at that frequency?

95. A 150.0 mH inductor, a 45.0 μF capacitor, and a 90.0 Ω resistor are connected in series with an ac source with a voltage V_s of 100 V. The frequency f of the source can be varied from 0 to 1000 Hz. (a) Simultaneously plot the capacitive reactance X_C and the inductive reactance X_L versus f for the range $0 < f < 200$ Hz. (b) From the plots, determine f at which $X_C = X_L$. (c) Plot the impedance Z of the circuit versus f for the range $0 < f < 188$ Hz and, from the plot, determine the minimum value of Z and the value of f at which it occurs.

Appendix **A**
*The International System of Units (SI)**

1. THE SI BASE UNITS

QUANTITY	NAME	SYMBOL	DEFINITION
length	meter	m	". . . the length of the path traveled by light in vacuum in 1/299,792,458 of a second." (1983)
mass	kilogram	kg	". . . this prototype [a certain platinum–iridium cylinder] shall henceforth be considered to be the unit of mass." (1889)
time	second	s	". . . the duration of 9,192,631,770 periods of the radiation corresponding to the transition between the two hyperfine levels of the ground state of the cesium-133 atom." (1967)
electric current	ampere	A	". . . that constant current which, if maintained in two straight parallel conductors of infinite length, of negligible circular cross section, and placed 1 meter apart in vacuum, would produce between these conductors a force equal to 2×10^{-7} newton per meter of length." (1946)
thermodynamic temperature	kelvin	K	". . . the fraction 1/273.16 of the thermodynamic temperature of the triple point of water." (1967)
amount of substance	mole	mol	". . . the amount of substance of a system which contains as many elementary entities as there are atoms in 0.012 kilogram of carbon-12." (1971)
luminous intensity	candela	cd	". . . the luminous intensity, in the perpendicular direction, of a surface of 1/600,000 square meter of a blackbody at the temperature of freezing platinum under a pressure of 101.325 newtons per square meter." (1967)

*Adapted from "The International System of Units (SI)," National Bureau of Standards Special Publication 330, 1972 edition. The definitions above were adopted by the General Conference of Weights and Measures, an international body, on the dates shown. In this book we do not use the candela.

A1

2. SOME SI DERIVED UNITS

QUANTITY	NAME OF UNIT	SYMBOL	
area	square meter	m^2	
volume	cubic meter	m^3	
frequency	hertz	Hz	s^{-1}
mass density (density)	kilogram per cubic meter	kg/m^3	
speed, velocity	meter per second	m/s	
angular velocity	radian per second	rad/s	
acceleration	meter per second per second	m/s^2	
angular acceleration	radian per second per second	rad/s^2	
force	newton	N	$kg \cdot m/s^2$
pressure	pascal	Pa	N/m^2
work, energy, quantity of heat	joule	J	$N \cdot m$
power	watt	W	J/s
quantity of electric charge	coulomb	C	$A \cdot s$
potential difference, electromotive force	volt	V	W/A
electric field strength	volt per meter (or newton per coulomb)	V/m	N/C
electric resistance	ohm	Ω	V/A
capacitance	farad	F	$A \cdot s/V$
magnetic flux	weber	Wb	$V \cdot s$
inductance	henry	H	$V \cdot s/A$
magnetic flux density	tesla	T	Wb/m^2
magnetic field strength	ampere per meter	A/m	
entropy	joule per kelvin	J/K	
specific heat	joule per kilogram kelvin	$J/(kg \cdot K)$	
thermal conductivity	watt per meter kelvin	$W/(m \cdot K)$	
radiant intensity	watt per steradian	W/sr	

3. THE SI SUPPLEMENTARY UNITS

QUANTITY	NAME OF UNIT	SYMBOL
plane angle	radian	rad
solid angle	steradian	sr

Appendix **B**
Some Fundamental Constants of Physics*

CONSTANT	SYMBOL	COMPUTATIONAL VALUE	BEST (1986) VALUE	
			VALUE[a]	UNCERTAINTY[b]
Speed of light in a vacuum	c	3.00×10^8 m/s	2.99792458	exact
Elementary charge	e	1.60×10^{-19} C	1.60217733	0.30
Gravitational constant	G	6.67×10^{-11} m^3/s$^2 \cdot$kg	6.67259	128
Universal gas constant	R	8.31 J/mol$\cdot$K	8.314510	8.4
Avogadro constant	N_A	6.02×10^{23} mol^{-1}	6.0221367	0.59
Boltzmann constant	k	1.38×10^{-23} J/K	1.380658	8.5
Stefan-Boltzmann constant	σ	5.67×10^{-8} W/m$^2 \cdot$K^4	5.67051	34
Molar volume of ideal gas at STP[d]	V_m	2.24×10^{-2} m^3/mol	2.241409	8.4
Permittivity constant	ϵ_0	8.85×10^{-12} F/m	8.85418781762	exact
Permeability constant	μ_0	1.26×10^{-6} H/m	1.25663706143	exact
Planck constant	h	6.63×10^{-34} J$\cdot$s	6.6260755	0.60
Electron mass[c]	m_e	9.11×10^{-31} kg	9.1093897	0.59
		5.49×10^{-4} u	5.48579903	0.023
Proton mass[c]	m_p	1.67×10^{-27} kg	1.6726231	0.59
		1.0073 u	1.0072764660	0.005
Ratio of proton mass to electron mass	m_p/m_e	1840	1836.152701	0.020
Electron charge-to-mass ratio	e/m_e	1.76×10^{11} C/kg	1.75881961	0.30
Neutron mass[c]	m_n	1.68×10^{-27} kg	1.6749286	0.59
		1.0087 u	1.0086649235	0.0023
Hydrogen atom mass[c]	m_{1_H}	1.0078 u	1.0078250316	0.0005
Deuterium atom mass[c]	m_{2_H}	2.0141 u	2.0141017779	0.0005
Helium atom mass[c]	$m_{4_{He}}$	4.0026 u	4.0026032	0.067
Muon mass	m_μ	1.88×10^{-28} kg	1.8835326	0.61
Electron magnetic moment	μ_e	9.28×10^{-24} J/T	9.2847701	0.34
Proton magnetic moment	μ_p	1.41×10^{-26} J/T	1.41060761	0.34
Bohr magneton	μ_B	9.27×10^{-24} J/T	9.2740154	0.34
Nuclear magneton	μ_N	5.05×10^{-27} J/T	5.0507866	0.34
Bohr radius	r_B	5.29×10^{-11} m	5.29177249	0.045
Rydberg constant	R	1.10×10^7 m^{-1}	1.0973731534	0.0012
Electron Compton wavelength	λ_C	2.43×10^{-12} m	2.42631058	0.089

[a]Values given in this column should be given the same unit and power of 10 as the computational value. [b]Parts per million. [c]Masses given in u are in unified atomic mass units, where 1 u = $1.6605402 \times 10^{-27}$ kg. [d]STP means standard temperature and pressure: 0°C and 1.0 atm (0.1 MPa).

*The values in this table were largely selected from a longer list in *Symbols, Units and Nomenclature in Physics* (IUPAP), prepared by E. Richard Cohen and Pierre Giacomo, 1986.

Appendix C
Some Astronomical Data

SOME DISTANCES FROM THE EARTH

To the moon*	3.82×10^8 m
To the sun*	1.50×10^{11} m
To the nearest star (Proxima Centauri)	4.04×10^{16} m
To the center of our galaxy	2.2×10^{20} m
To the Andromeda Galaxy	2.1×10^{22} m
To the edge of the observable universe	$\sim 10^{26}$ m

*Mean distance.

THE SUN, THE EARTH, AND THE MOON

PROPERTY	UNIT	SUN	EARTH	MOON
Mass	kg	1.99×10^{30}	5.98×10^{24}	7.36×10^{22}
Mean radius	m	6.96×10^8	6.37×10^6	1.74×10^6
Mean density	kg/m^3	1410	5520	3340
Free-fall acceleration at the surface	m/s^2	274	9.81	1.67
Escape velocity	km/s	618	11.2	2.38
Period of rotation[a]	—	37 d at poles[b] 26 d at equator[b]	23 h 56 min	27.3 d
Radiation power[c]	W	3.90×10^{26}		

[a]Measured with respect to the distant stars.

[b]The sun, a ball of gas, does not rotate as a rigid body.

[c]Just outside the Earth's atmosphere solar energy is received, assuming normal incidence, at the rate of 1340 W/m^2.

SOME PROPERTIES OF THE PLANETS

	MERCURY	VENUS	EARTH	MARS	JUPITER	SATURN	URANUS	NEPTUNE	PLUTO
Mean distance from sun, 10^6 km	57.9	108	150	228	778	1430	2870	4500	5900
Period of revolution, y	0.241	0.615	1.00	1.88	11.9	29.5	84.0	165	248
Period of rotation,[a] d	58.7	-243^b	0.997	1.03	0.409	0.426	-0.451^b	0.658	6.39
Orbital speed, km/s	47.9	35.0	29.8	24.1	13.1	9.64	6.81	5.43	4.74
Inclination of axis to orbit	<28°	≈3°	23.4°	25.0°	3.08°	26.7°	97.9°	29.6°	57.5°
Inclination of orbit to Earth's orbit	7.00°	3.39°		1.85°	1.30°	2.49°	0.77°	1.77°	17.2°
Eccentricity of orbit	0.206	0.0068	0.0167	0.0934	0.0485	0.0556	0.0472	0.0086	0.250
Equatorial diameter, km	4880	12,100	12,800	6790	143,000	120,000	51,800	49,500	2300
Mass (Earth = 1)	0.0558	0.815	1.000	0.107	318	95.1	14.5	17.2	0.002
Density (water = 1)	5.60	5.20	5.52	3.95	1.31	0.704	1.21	1.67	2.03
Surface value of g,[c] m/s^2	3.78	8.60	9.78	3.72	22.9	9.05	7.77	11.0	0.5
Escape velocity,[c] km/s	4.3	10.3	11.2	5.0	59.5	35.6	21.2	23.6	1.1
Known satellites	0	0	1	2	16 + ring	18 + rings	15 + rings	8 + rings	1

[a]Measured with respect to the distant stars.

[b]Venus and Uranus rotate opposite their orbital motion.

[c]Gravitational acceleration measured at the planet's equator.

Appendix **D**
Conversion Factors

Conversion factors may be read directly from these tables. For example, 1 degree = 2.778×10^{-3} revolutions, so $16.7° = 16.7 \times 2.778 \times 10^{-3}$ rev. The SI quantities are fully capitalized.

Adapted in part from G. Shortley and D. Williams, *Elements of Physics*, Prentice-Hall, Englewood Cliffs, NJ, 1971.

PLANE ANGLE

	°	′	″	RADIAN	rev
1 degree =	1	60	3600	1.745×10^{-2}	2.778×10^{-3}
1 minute =	1.667×10^{-2}	1	60	2.909×10^{-4}	4.630×10^{-5}
1 second =	2.778×10^{-4}	1.667×10^{-2}	1	4.848×10^{-6}	7.716×10^{-7}
1 RADIAN =	57.30	3438	2.063×10^5	1	0.1592
1 revolution =	360	2.16×10^4	1.296×10^6	6.283	1

SOLID ANGLE

1 sphere = 4π steradians = 12.57 steradians

LENGTH

	cm	METER	km	in.	ft	mi
1 centimeter =	1	10^{-2}	10^{-5}	0.3937	3.281×10^{-2}	6.214×10^{-6}
1 METER =	100	1	10^{-3}	39.37	3.281	6.214×10^{-4}
1 kilometer =	10^5	1000	1	3.937×10^4	3281	0.6214
1 inch =	2.540	2.540×10^{-2}	2.540×10^{-5}	1	8.333×10^{-2}	1.578×10^{-5}
1 foot =	30.48	0.3048	3.048×10^{-4}	12	1	1.894×10^{-4}
1 mile =	1.609×10^5	1609	1.609	6.336×10^4	5280	1

1 angström = 10^{-10} m

1 nautical mile = 1852 m
  = 1.151 miles = 6076 ft

1 fermi = 10^{-15} m

1 light-year = 9.460×10^{12} km

1 parsec = 3.084×10^{13} km

1 fathom = 6 ft

1 Bohr radius = 5.292×10^{-11} m

1 yard = 3 ft

1 rod = 16.5 ft

1 mil = 10^{-3} in.

1 nm = 10^{-9} m

AREA

	METER^2	cm^2	ft^2	in.^{2}
1 SQUARE METER =	1	10^4	10.76	1550
1 square centimeter =	10^{-4}	1	1.076×10^{-3}	0.1550
1 square foot =	9.290×10^{-2}	929.0	1	144
1 square inch =	6.452×10^{-4}	6.452	6.944×10^{-3}	1

1 square mile = 2.788×10^7 ft^2
  = 640 acres

1 barn = 10^{-28} m^2

1 acre = 43,560 ft^2

1 hectare = 10^4 m^2 = 2.471 acres

VOLUME

	METER^3	cm^3	L	ft^3	in.^{3}
1 CUBIC METER = 1	10^6	1000	35.31	6.102 × 10^4
1 cubic centimeter = 10^{-6}	1	1.000 × 10^{-3}	3.531 × 10^{-5}	6.102 × 10^{-2}
1 liter = 1.000 × 10^{-3}	1000	1	3.531 × 10^{-2}	61.02
1 cubic foot = 2.832 × 10^{-2}	2.832 × 10^4	28.32	1	1728
1 cubic inch = 1.639 × 10^{-5}	16.39	1.639 × 10^{-2}	5.787 × 10^{-4}	1

1 U.S. fluid gallon = 4 U.S. fluid quarts = 8 U.S. pints = 128 U.S. fluid ounces = 231 in.^{3}

1 British imperial gallon = 277.4 in.^{3} = 1.201 U.S. fluid gallons

MASS

Quantities in the colored areas are not mass units but are often used as such. When we write, for example, 1 kg "=" 2.205 lb, this means that a kilogram is a *mass* that *weighs* 2.205 pounds at a location where g has the standard value of 9.80665 m/s^2.

	g	KILOGRAM	slug	u	oz	lb	ton
1 gram = 1	0.001	6.852 × 10^{-5}	6.022 × 10^{23}	3.527 × 10^{-2}	2.205 × 10^{-3}	1.102 × 10^{-6}
1 KILOGRAM = 1000	1	6.852 × 10^{-2}	6.022 × 10^{26}	35.27	2.205	1.102 × 10^{-3}
1 slug = 1.459 × 10^4	14.59	1	8.786 × 10^{27}	514.8	32.17	1.609 × 10^{-2}
1 atomic mass unit = 1.661 × 10^{-24}	1.661 × 10^{-27}	1.138 × 10^{-28}	1	5.857 × 10^{-26}	3.662 × 10^{-27}	1.830 × 10^{-30}
1 ounce = 28.35	2.835 × 10^{-2}	1.943 × 10^{-3}	1.718 × 10^{25}	1	6.250 × 10^{-2}	3.125 × 10^{-5}
1 pound = 453.6	0.4536	3.108 × 10^{-2}	2.732 × 10^{26}	16	1	0.0005
1 ton = 9.072 × 10^5	907.2	62.16	5.463 × 10^{29}	3.2 × 10^4	2000	1

1 metric ton = 1000 kg

DENSITY

Quantities in the colored areas are weight densities and, as such, are dimensionally different from mass densities. See note for mass table.

	slug/ft^3	KILOGRAM/ METER^3	g/cm^3	lb/ft^3	lb/in.^{3}
1 slug per foot^3 = 1	515.4	0.5154	32.17	1.862 × 10^{-2}
1 KILOGRAM per METER^3 = 1.940 × 10^{-3}	1	0.001	6.243 × 10^{-2}	3.613 × 10^{-5}
1 gram per centimeter^3 = 1.940	1000	1	62.43	3.613 × 10^{-2}
1 pound per foot^3 = 3.108 × 10^{-2}	16.02	1.602 × 10^{-2}	1	5.787 × 10^{-4}
1 pound per inch^3 = 53.71	2.768 × 10^4	27.68	1728	1

TIME

	y	d	h	min	SECOND
1 year = 1	365.25	8.766 × 10^3	5.259 × 10^5	3.156 × 10^7
1 day = 2.738 × 10^{-3}	1	24	1440	8.640 × 10^4
1 hour = 1.141 × 10^{-4}	4.167 × 10^{-2}	1	60	3600
1 minute = 1.901 × 10^{-6}	6.944 × 10^{-4}	1.667 × 10^{-2}	1	60
1 SECOND = 3.169 × 10^{-8}	1.157 × 10^{-5}	2.778 × 10^{-4}	1.667 × 10^{-2}	1

SPEED

	ft/s	km/h	METER/ SECOND	mi/h	cm/s
1 foot per second = 1	1.097	0.3048	0.6818	30.48
1 kilometer per hour = 0.9113	1	0.2778	0.6214	27.78
1 METER per SECOND = 3.281	3.6	1	2.237	100
1 mile per hour = 1.467	1.609	0.4470	1	44.70
1 centimeter per second = 3.281×10^{-2}	3.6×10^{-2}	0.01	2.237×10^{-2}	1

1 knot = 1 nautical mi/h = 1.688 ft/s   1 mi/min = 88.00 ft/s = 60.00 mi/h

FORCE

Force units in the colored areas are now little used. To clarify: 1 gram-force (= 1 gf) is the force of gravity that would act on an object whose mass is 1 gram at a location where g has the standard value of 9.80665 m/s^2.

	dyne	NEWTON	lb	pdl	gf	kgf
1 dyne = 1	10^{-5}	2.248×10^{-6}	7.233×10^{-5}	1.020×10^{-3}	1.020×10^{-6}
1 NEWTON = 10^5	1	0.2248	7.233	102.0	0.1020
1 pound = 4.448×10^5	4.448	1	32.17	453.6	0.4536
1 poundal = 1.383×10^4	0.1383	3.108×10^{-2}	1	14.10	1.410×10^{-2}
1 gram-force = 980.7	9.807×10^{-3}	2.205×10^{-3}	7.093×10^{-2}	1	0.001
1 kilogram-force = 9.807×10^5	9.807	2.205	70.93	1000	1

PRESSURE

	atm	dyne/cm^2	inch of water	cm Hg	PASCAL	lb/in.^{2}	lb/ft^2
1 atmosphere = 1	1.013×10^6	406.8	76	1.013×10^5	14.70	2116
1 dyne per centimeter^2 = 9.869×10^{-7}	1	4.015×10^{-4}	7.501×10^{-5}	0.1	1.405×10^{-5}	2.089×10^{-3}
1 inch of water[a] at 4°C = 2.458×10^{-3}	2491	1	0.1868	249.1	3.613×10^{-2}	5.202
1 centimeter of mercury[a] at 0°C = 1.316×10^{-2}	1.333×10^4	5.353	1	1333	0.1934	27.85
1 PASCAL = 9.869×10^{-6}	10	4.015×10^{-3}	7.501×10^{-4}	1	1.450×10^{-4}	2.089×10^{-2}
1 pound per inch^2 = 6.805×10^{-2}	6.895×10^4	27.68	5.171	6.895×10^3	1	144
1 pound per foot^2 = 4.725×10^{-4}	478.8	0.1922	3.591×10^{-2}	47.88	6.944×10^{-3}	1

[a] Where the acceleration of gravity has the standard value of 9.80665 m/s^2.

1 bar = 10^6 dyne/cm^2 = 0.1 MPa   1 millibar = 10^3 dyne/cm^2 = 10^2 Pa   1 torr = 1 mm Hg

ENERGY, WORK, HEAT

Quantities in the colored areas are not properly energy units but are included for convenience. They arise from the relativistic mass–energy equivalence formula $E = mc^2$ and represent the energy released if a kilogram or unified atomic mass unit (u) is completely converted to energy (bottom two rows) or the mass that would be completely converted to one unit of energy (rightmost two columns).

	Btu	erg	ft·lb	hp·h	JOULE	cal	kW·h	eV	MeV	kg	u
1 British thermal unit =	1	1.055×10^{10}	777.9	3.929×10^{-4}	1055	252.0	2.930×10^{-4}	6.585×10^{21}	6.585×10^{15}	1.174×10^{-14}	7.070×10^{12}
1 erg =	9.481×10^{-11}	1	7.376×10^{-8}	3.725×10^{-14}	10^{-7}	2.389×10^{-8}	2.778×10^{-14}	6.242×10^{11}	6.242×10^{5}	1.113×10^{-24}	670.2
1 foot-pound =	1.285×10^{-3}	1.356×10^{7}	1	5.051×10^{-7}	1.356	0.3238	3.766×10^{-7}	8.464×10^{18}	8.464×10^{12}	1.509×10^{-17}	9.037×10^{9}
1 horsepower-hour =	2545	2.685×10^{13}	1.980×10^{6}	1	2.685×10^{6}	6.413×10^{5}	0.7457	1.676×10^{25}	1.676×10^{19}	2.988×10^{-11}	1.799×10^{16}
1 JOULE =	9.481×10^{-4}	10^{7}	0.7376	3.725×10^{-7}	1	0.2389	2.778×10^{-7}	6.242×10^{18}	6.242×10^{12}	1.113×10^{-17}	6.702×10^{9}
1 calorie =	3.969×10^{-3}	4.186×10^{7}	3.088	1.560×10^{-6}	4.186	1	1.163×10^{-6}	2.613×10^{19}	2.613×10^{13}	4.660×10^{-17}	2.806×10^{10}
1 kilowatt-hour =	3413	3.600×10^{13}	2.655×10^{6}	1.341	3.600×10^{6}	8.600×10^{5}	1	2.247×10^{25}	2.247×10^{19}	4.007×10^{-11}	2.413×10^{16}
1 electron-volt =	1.519×10^{-22}	1.602×10^{-12}	1.182×10^{-19}	5.967×10^{-26}	1.602×10^{-19}	3.827×10^{-20}	4.450×10^{-26}	1	10^{-6}	1.783×10^{-36}	1.074×10^{-9}
1 million electron-volts =	1.519×10^{-16}	1.602×10^{-6}	1.182×10^{-13}	5.967×10^{-20}	1.602×10^{-13}	3.827×10^{-14}	4.450×10^{-20}	10^{-6}	1	1.783×10^{-30}	1.074×10^{-3}
1 kilogram =	8.521×10^{13}	8.987×10^{23}	6.629×10^{16}	3.348×10^{10}	8.987×10^{16}	2.146×10^{16}	2.497×10^{10}	5.610×10^{35}	5.610×10^{29}	1	6.022×10^{26}
1 unified atomic mass unit =	1.415×10^{-13}	1.492×10^{-3}	1.101×10^{-10}	5.559×10^{-17}	1.492×10^{-10}	3.564×10^{-11}	4.146×10^{-17}	9.320×10^{8}	932.0	1.661×10^{-27}	1

POWER

	Btu/h	ft·lb/s	hp	cal/s	kW	WATT
1 British thermal unit per hour =	1	0.2161	3.929×10^{-4}	6.998×10^{-2}	2.930×10^{-4}	0.2930
1 foot-pound per second =	4.628	1	1.818×10^{-3}	0.3239	1.356×10^{-3}	1.356
1 horsepower =	2545	550	1	178.1	0.7457	745.7
1 calorie per second =	14.29	3.088	5.615×10^{-3}	1	4.186×10^{-3}	4.186
1 kilowatt =	3413	737.6	1.341	238.9	1	1000
1 WATT =	3.413	0.7376	1.341×10^{-3}	0.2389	0.001	1

MAGNETIC FIELD

	gauss	TESLA	milligauss
1 gauss =	1	10^{-4}	1000
1 TESLA =	10^{4}	1	10^{7}
1 milligauss =	0.001	10^{-7}	1

1 tesla = 1 weber/meter^2

MAGNETIC FLUX

	maxwell	WEBER
1 maxwell =	1	10^{-8}
1 WEBER =	10^{8}	1

GEOMETRY

Circle of radius r: circumference $= 2\pi r$; area $= \pi r^2$.

Sphere of radius r: area $= 4\pi r^2$; volume $= \frac{4}{3}\pi r^3$.

Right circular cylinder of radius r and height h:
area $= 2\pi r^2 + 2\pi rh$; volume $= \pi r^2 h$.

Triangle of base a and altitude h: area $= \frac{1}{2}ah$.

QUADRATIC FORMULA

If $ax^2 + bx + c = 0$, then $x = \dfrac{-b \pm \sqrt{b^2 - 4ac}}{2a}$.

TRIGONOMETRIC FUNCTIONS OF ANGLE θ

$$\sin\theta = \frac{y}{r} \quad \cos\theta = \frac{x}{r}$$

$$\tan\theta = \frac{y}{x} \quad \cot\theta = \frac{x}{y}$$

$$\sec\theta = \frac{r}{x} \quad \csc\theta = \frac{r}{y}$$

PYTHAGOREAN THEOREM

In this right triangle,
$$a^2 + b^2 = c^2$$

TRIANGLES

Angles are A, B, C

Opposite sides are a, b, c

Angles $A + B + C = 180°$

$$\frac{\sin A}{a} = \frac{\sin B}{b} = \frac{\sin C}{c}$$

$$c^2 = a^2 + b^2 - 2ab \cos C$$

Exterior angle $D = A + C$

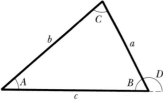

MATHEMATICAL SIGNS AND SYMBOLS

$=$ equals

$\approx$ equals approximately

$\sim$ is the order of magnitude of

$\neq$ is not equal to

$\equiv$ is identical to, is defined as

$>$ is greater than ($\gg$ is much greater than)

$<$ is less than ($\ll$ is much less than)

$\geq$ is greater than or equal to (or, is no less than)

$\leq$ is less than or equal to (or, is no more than)

$\pm$ plus or minus

$\propto$ is proportional to

Σ the sum of

$\bar{x}$ the average value of x

TRIGONOMETRIC IDENTITIES

$\sin(90° - \theta) = \cos\theta$

$\cos(90° - \theta) = \sin\theta$

$\sin\theta/\cos\theta = \tan\theta$

$\sin^2\theta + \cos^2\theta = 1$

$\sec^2\theta - \tan^2\theta = 1$

$\csc^2\theta - \cot^2\theta = 1$

$\sin 2\theta = 2\sin\theta\cos\theta$

$\cos 2\theta = \cos^2\theta - \sin^2\theta = 2\cos^2\theta - 1 = 1 - 2\sin^2\theta$

$\sin(\alpha \pm \beta) = \sin\alpha\cos\beta \pm \cos\alpha\sin\beta$

$\cos(\alpha \pm \beta) = \cos\alpha\cos\beta \mp \sin\alpha\sin\beta$

$\tan(\alpha \pm \beta) = \dfrac{\tan\alpha \pm \tan\beta}{1 \mp \tan\alpha\tan\beta}$

$\sin\alpha \pm \sin\beta = 2\sin\frac{1}{2}(\alpha \pm \beta)\cos\frac{1}{2}(\alpha \mp \beta)$

$\cos\alpha + \cos\beta = 2\cos\frac{1}{2}(\alpha + \beta)\cos\frac{1}{2}(\alpha - \beta)$

$\cos\alpha - \cos\beta = -2\sin\frac{1}{2}(\alpha + \beta)\sin\frac{1}{2}(\alpha - \beta)$

BINOMIAL THEOREM

$$(1 + x)^n = 1 + \frac{nx}{1!} + \frac{n(n-1)x^2}{2!} + \cdots \qquad (x^2 < 1)$$

EXPONENTIAL EXPANSION

$$e^x = 1 + x + \frac{x^2}{2!} + \frac{x^3}{3!} + \cdots$$

LOGARITHMIC EXPANSION

$$\ln(1 + x) = x - \tfrac{1}{2}x^2 + \tfrac{1}{3}x^3 - \cdots \qquad (|x| < 1)$$

TRIGONOMETRIC EXPANSIONS
(θ in radians)

$$\sin \theta = \theta - \frac{\theta^3}{3!} + \frac{\theta^5}{5!} - \cdots$$

$$\cos \theta = 1 - \frac{\theta^2}{2!} + \frac{\theta^4}{4!} - \cdots$$

$$\tan \theta = \theta + \frac{\theta^3}{3} + \frac{2\theta^5}{15} + \cdots$$

CRAMER'S RULE

Two simultaneous equations in unknowns x and y,

$$a_1 x + b_1 y = c_1 \qquad \text{and} \qquad a_2 x + b_2 y = c_2,$$

have the solutions

$$x = \frac{\begin{vmatrix} c_1 & b_1 \\ c_2 & b_2 \end{vmatrix}}{\begin{vmatrix} a_1 & b_1 \\ a_2 & b_2 \end{vmatrix}} = \frac{c_1 b_2 - c_2 b_1}{a_1 b_2 - a_2 b_1}$$

and

$$y = \frac{\begin{vmatrix} a_1 & c_1 \\ a_2 & c_2 \end{vmatrix}}{\begin{vmatrix} a_1 & b_1 \\ a_2 & b_2 \end{vmatrix}} = \frac{a_1 c_2 - a_2 c_1}{a_1 b_2 - a_2 b_1}.$$

PRODUCTS OF VECTORS

Let $\mathbf{i}$, $\mathbf{j}$, and $\mathbf{k}$ be unit vectors in the x, y, and z directions. Then

$$\mathbf{i} \cdot \mathbf{i} = \mathbf{j} \cdot \mathbf{j} = \mathbf{k} \cdot \mathbf{k} = 1, \qquad \mathbf{i} \cdot \mathbf{j} = \mathbf{j} \cdot \mathbf{k} = \mathbf{k} \cdot \mathbf{i} = 0,$$

$$\mathbf{i} \times \mathbf{i} = \mathbf{j} \times \mathbf{j} = \mathbf{k} \times \mathbf{k} = 0,$$

$$\mathbf{i} \times \mathbf{j} = \mathbf{k}, \qquad \mathbf{j} \times \mathbf{k} = \mathbf{i}, \qquad \mathbf{k} \times \mathbf{i} = \mathbf{j},$$

Any vector $\mathbf{a}$ with components a_x, a_y, and a_z along the x, y, and z axes can be written

$$\mathbf{a} = a_x \mathbf{i} + a_y \mathbf{j} + a_z \mathbf{k}.$$

Let $\mathbf{a}$, $\mathbf{b}$, and $\mathbf{c}$ be arbitrary vectors with magnitudes a, b, and c. Then

$$\mathbf{a} \times (\mathbf{b} + \mathbf{c}) = (\mathbf{a} \times \mathbf{b}) + (\mathbf{a} \times \mathbf{c})$$

$$(s\mathbf{a}) \times \mathbf{b} = \mathbf{a} \times (s\mathbf{b}) = s(\mathbf{a} \times \mathbf{b}) \qquad (s = \text{a scalar}).$$

Let θ be the smaller of the two angles between $\mathbf{a}$ and $\mathbf{b}$. Then

$$\mathbf{a} \cdot \mathbf{b} = \mathbf{b} \cdot \mathbf{a} = a_x b_x + a_y b_y + a_z b_z = ab \cos \theta$$

$$\mathbf{a} \times \mathbf{b} = -\mathbf{b} \times \mathbf{a} = \begin{vmatrix} \mathbf{i} & \mathbf{j} & \mathbf{k} \\ a_x & a_y & a_z \\ b_x & b_y & b_z \end{vmatrix}$$

$$= \mathbf{i} \begin{vmatrix} a_y & a_z \\ b_y & b_z \end{vmatrix} - \mathbf{j} \begin{vmatrix} a_x & a_z \\ b_x & b_z \end{vmatrix} + \mathbf{k} \begin{vmatrix} a_x & a_y \\ b_x & b_y \end{vmatrix}$$

$$= (a_y b_z - b_y a_z)\mathbf{i}$$
$$+ (a_z b_x - b_z a_x)\mathbf{j} + (a_x b_y - b_x a_y)\mathbf{k}$$

$$|\mathbf{a} \times \mathbf{b}| = ab \sin \theta$$

$$\mathbf{a} \cdot (\mathbf{b} \times \mathbf{c}) = \mathbf{b} \cdot (\mathbf{c} \times \mathbf{a}) = \mathbf{c} \cdot (\mathbf{a} \times \mathbf{b})$$

$$\mathbf{a} \times (\mathbf{b} \times \mathbf{c}) = (\mathbf{a} \cdot \mathbf{c})\mathbf{b} - (\mathbf{a} \cdot \mathbf{b})\mathbf{c}$$

DERIVATIVES AND INTEGRALS

In what follows, the letters u and v stand for any functions of x, and a and m are constants. To each of the indefinite integrals should be added an arbitrary constant of integration. The *Handbook of Chemistry and Physics* (CRC Press Inc.) gives a more extensive tabulation.

1. $\dfrac{dx}{dx} = 1$

2. $\dfrac{d}{dx}(au) = a\dfrac{du}{dx}$

3. $\dfrac{d}{dx}(u + v) = \dfrac{du}{dx} + \dfrac{dv}{dx}$

4. $\dfrac{d}{dx}x^m = mx^{m-1}$

5. $\dfrac{d}{dx}\ln x = \dfrac{1}{x}$

6. $\dfrac{d}{dx}(uv) = u\dfrac{dv}{dx} + v\dfrac{du}{dx}$

7. $\dfrac{d}{dx}e^x = e^x$

8. $\dfrac{d}{dx}\sin x = \cos x$

9. $\dfrac{d}{dx}\cos x = -\sin x$

10. $\dfrac{d}{dx}\tan x = \sec^2 x$

11. $\dfrac{d}{dx}\cot x = -\csc^2 x$

12. $\dfrac{d}{dx}\sec x = \tan x \sec x$

13. $\dfrac{d}{dx}\csc x = -\cot x \csc x$

14. $\dfrac{d}{dx}e^u = e^u\dfrac{du}{dx}$

15. $\dfrac{d}{dx}\sin u = \cos u\dfrac{du}{dx}$

16. $\dfrac{d}{dx}\cos u = -\sin u\dfrac{du}{dx}$

1. $\displaystyle\int dx = x$

2. $\displaystyle\int au\,dx = a\int u\,dx$

3. $\displaystyle\int (u + v)\,dx = \int u\,dx + \int v\,dx$

4. $\displaystyle\int x^m\,dx = \dfrac{x^{m+1}}{m+1} \qquad (m \neq -1)$

5. $\displaystyle\int \dfrac{dx}{x} = \ln |x|$

6. $\displaystyle\int u\dfrac{dv}{dx}\,dx = uv - \int v\dfrac{du}{dx}\,dx$

7. $\displaystyle\int e^x\,dx = e^x$

8. $\displaystyle\int \sin x\,dx = -\cos x$

9. $\displaystyle\int \cos x\,dx = \sin x$

10. $\displaystyle\int \tan x\,dx = \ln |\sec x|$

11. $\displaystyle\int \sin^2 x\,dx = \tfrac{1}{2}x - \tfrac{1}{4}\sin 2x$

12. $\displaystyle\int e^{-ax}\,dx = -\dfrac{1}{a}e^{-ax}$

13. $\displaystyle\int xe^{-ax}\,dx = -\dfrac{1}{a^2}(ax + 1)e^{-ax}$

14. $\displaystyle\int x^2 e^{-ax}\,dx = -\dfrac{1}{a^3}(a^2x^2 + 2ax + 2)e^{-ax}$

15. $\displaystyle\int_0^\infty x^n e^{-ax}\,dx = \dfrac{n!}{a^{n+1}}$

16. $\displaystyle\int_0^\infty x^{2n}e^{-ax^2}\,dx = \dfrac{1 \cdot 3 \cdot 5 \cdots (2n - 1)}{2^{n+1}a^n}\sqrt{\dfrac{\pi}{a}}$

17. $\displaystyle\int \dfrac{dx}{\sqrt{x^2 + a^2}} = \ln(x + \sqrt{x^2 + a^2})$

18. $\displaystyle\int \dfrac{x\,dx}{(x^2 + a^2)^{3/2}} = -\dfrac{1}{(x^2 + a^2)^{1/2}}$

19. $\displaystyle\int \dfrac{dx}{(x^2 + a^2)^{3/2}} = \dfrac{x}{a^2(x^2 + a^2)^{1/2}}$

Appendix F
Properties of the Elements

All physical properties are for a pressure of 1 atm unless otherwise specified.

ELEMENT	SYMBOL	ATOMIC NUMBER, Z	MOLAR MASS, g/mol	DENSITY, g/cm³ AT 20°C	MELTING POINT, °C	BOILING POINT, °C	SPECIFIC HEAT, J/(g·°C) AT 25°C
Actinium	Ac	89	(227)	10.06	1323	(3473)	0.092
Aluminum	Al	13	26.9815	2.699	660	2450	0.900
Americium	Am	95	(243)	13.67	1541	—	—
Antimony	Sb	51	121.75	6.691	630.5	1380	0.205
Argon	Ar	18	39.948	1.6626×10^{-3}	−189.4	−185.8	0.523
Arsenic	As	33	74.9216	5.78	817 (28 atm)	613	0.331
Astatine	At	85	(210)	—	(302)	—	—
Barium	Ba	56	137.34	3.594	729	1640	0.205
Berkelium	Bk	97	(247)	14.79	—	—	—
Beryllium	Be	4	9.0122	1.848	1287	2770	1.83
Bismuth	Bi	83	208.980	9.747	271.37	1560	0.122
Boron	B	5	10.811	2.34	2030	—	1.11
Bromine	Br	35	79.909	3.12 (liquid)	−7.2	58	0.293
Cadmium	Cd	48	112.40	8.65	321.03	765	0.226
Calcium	Ca	20	40.08	1.55	838	1440	0.624
Californium	Cf	98	(251)	—	—	—	—
Carbon	C	6	12.01115	2.26	3727	4830	0.691
Cerium	Ce	58	140.12	6.768	804	3470	0.188
Cesium	Cs	55	132.905	1.873	28.40	690	0.243
Chlorine	Cl	17	35.453	3.214×10^{-3} (0°C)	−101	−34.7	0.486
Chromium	Cr	24	51.996	7.19	1857	2665	0.448
Cobalt	Co	27	58.9332	8.85	1495	2900	0.423
Copper	Cu	29	63.54	8.96	1083.40	2595	0.385
Curium	Cm	96	(247)	13.3	—	—	—
Dysprosium	Dy	66	162.50	8.55	1409	2330	0.172
Einsteinium	Es	99	(254)	—	—	—	—
Erbium	Er	68	167.26	9.15	1522	2630	0.167
Europium	Eu	63	151.96	5.243	817	1490	0.163
Fermium	Fm	100	(237)	—	—	—	—
Fluorine	F	9	18.9984	1.696×10^{-3} (0°C)	−219.6	−188.2	0.753
Francium	Fr	87	(223)	—	(27)	—	—
Gadolinium	Gd	64	157.25	7.90	1312	2730	0.234
Gallium	Ga	31	69.72	5.907	29.75	2237	0.377
Germanium	Ge	32	72.59	5.323	937.25	2830	0.322
Gold	Au	79	196.967	19.32	1064.43	2970	0.131
Hafnium	Hf	72	178.49	13.31	2227	5400	0.144
Hahnium	Ha	105	—	—	—	—	—
Hassium	Hs	108	—	—	—	—	—

continued on next page

ELEMENT	SYMBOL	ATOMIC NUMBER, Z	MOLAR MASS, g/mol	DENSITY, g/cm³ AT 20°C	MELTING POINT, °C	BOILING POINT, °C	SPECIFIC HEAT, J/(g·°C) AT 25°C
Helium	He	2	4.0026	0.1664×10^{-3}	−269.7	−268.9	5.23
Holmium	Ho	67	164.930	8.79	1470	2330	0.165
Hydrogen	H	1	1.00797	0.08375×10^{-3}	−259.19	−252.7	14.4
Indium	In	49	114.82	7.31	156.634	2000	0.233
Iodine	I	53	126.9044	4.93	113.7	183	0.218
Iridium	Ir	77	192.2	22.5	2447	(5300)	0.130
Iron	Fe	26	55.847	7.874	1536.5	3000	0.447
Krypton	Kr	36	83.80	3.488×10^{-3}	−157.37	−152	0.247
Lanthanum	La	57	138.91	6.189	920	3470	0.195
Lawrencium	Lr	103	(257)	—	—	—	—
Lead	Pb	82	207.19	11.35	327.45	1725	0.129
Lithium	Li	3	6.939	0.534	180.55	1300	3.58
Lutetium	Lu	71	174.97	9.849	1663	1930	0.155
Magnesium	Mg	12	24.312	1.738	650	1107	1.03
Manganese	Mn	25	54.9380	7.44	1244	2150	0.481
Meitnerium	Mt	109	—	—	—	—	—
Mendelevium	Md	101	(256)	—	—	—	—
Mercury	Hg	80	200.59	13.55	−38.87	357	0.138
Molybdenum	Mo	42	95.94	10.22	2617	5560	0.251
Neodymium	Nd	60	144.24	7.007	1016	3180	0.188
Neon	Ne	10	20.183	0.8387×10^{-3}	−248.597	−246.0	1.03
Neptunium	Np	93	(237)	20.25	637	—	1.26
Nickel	Ni	28	58.71	8.902	1453	2730	0.444
Nielsbohrium	Ns	107	—	—	—	—	—
Niobium	Nb	41	92.906	8.57	2468	4927	0.264
Nitrogen	N	7	14.0067	1.1649×10^{-3}	−210	−195.8	1.03
Nobelium	No	102	(255)	—	—	—	—
Osmium	Os	76	190.2	22.59	3027	5500	0.130
Oxygen	O	8	15.9994	1.3318×10^{-3}	−218.80	−183.0	0.913
Palladium	Pd	46	106.4	12.02	1552	3980	0.243
Phosphorus	P	15	30.9738	1.83	44.25	280	0.741
Platinum	Pt	78	195.09	21.45	1769	4530	0.134
Plutonium	Pu	94	(244)	19.8	640	3235	0.130
Polonium	Po	84	(210)	9.32	254	—	—
Potassium	K	19	39.102	0.862	63.20	760	0.758
Praseodymium	Pr	59	140.907	6.773	931	3020	0.197
Promethium	Pm	61	(145)	7.22	(1027)	—	—
Protactinium	Pa	91	(231)	15.37 (estimated)	(1230)	—	—
Radium	Ra	88	(226)	5.0	700	—	—
Radon	Rn	86	(222)	9.96×10^{-3} (0°C)	(−71)	−61.8	0.092
Rhenium	Re	75	186.2	21.02	3180	5900	0.134
Rhodium	Rh	45	102.905	12.41	1963	4500	0.243
Rubidium	Rb	37	85.47	1.532	39.49	688	0.364
Ruthenium	Ru	44	101.107	12.37	2250	4900	0.239
Rutherfordium	Rf	104	—	—	—	—	—
Samarium	Sm	62	150.35	7.52	1072	1630	0.197

continued on next page

ELEMENT	SYMBOL	ATOMIC NUMBER, Z	MOLAR MASS, g/mol	DENSITY, g/cm³ AT 20°C	MELTING POINT, °C	BOILING POINT, °C	SPECIFIC HEAT, J/(g·°C) AT 25°C
Scandium	Sc	21	44.956	2.99	1539	2730	0.569
Seaborgium	Sg	106	—	—	—	—	—
Selenium	Se	34	78.96	4.79	221	685	0.318
Silicon	Si	14	28.086	2.33	1412	2680	0.712
Silver	Ag	47	107.870	10.49	960.8	2210	0.234
Sodium	Na	11	22.9898	0.9712	97.85	892	1.23
Strontium	Sr	38	87.62	2.54	768	1380	0.737
Sulfur	S	16	32.064	2.07	119.0	444.6	0.707
Tantalum	Ta	73	180.948	16.6	3014	5425	0.138
Technetium	Tc	43	(99)	11.46	2200	—	0.209
Tellurium	Te	52	127.60	6.24	449.5	990	0.201
Terbium	Tb	65	158.924	8.229	1357	2530	0.180
Thallium	Tl	81	204.37	11.85	304	1457	0.130
Thorium	Th	90	(232)	11.72	1755	(3850)	0.117
Thulium	Tm	69	168.934	9.32	1545	1720	0.159
Tin	Sn	50	118.69	7.2984	231.868	2270	0.226
Titanium	Ti	22	47.90	4.54	1670	3260	0.523
Tungsten	W	74	183.85	19.3	3380	5930	0.134
Uranium	U	92	(238)	18.95	1132	3818	0.117
Vanadium	V	23	50.942	6.11	1902	3400	0.490
Xenon	Xe	54	131.30	5.495×10^{-3}	−111.79	−108	0.159
Ytterbium	Yb	70	173.04	6.965	824	1530	0.155
Yttrium	Y	39	88.905	4.469	1526	3030	0.297
Zinc	Zn	30	65.37	7.133	419.58	906	0.389
Zirconium	Zr	40	91.22	6.506	1852	3580	0.276

The values in parentheses in the column of molar masses are the mass numbers of the longest-lived isotopes of those elements that are radioactive. Melting points and boiling points in parentheses are uncertain.

The data for gases are valid only when these are in their usual molecular state, such as H_2, He, O_2, Ne, etc. The specific heats of the gases are the values at constant pressure.

Source: Adapted from Wehr, Richards, Adair, *Physics of the Atom,* 4th ed., Addison-Wesley, Reading, MA, 1984, and from J. Emsley, *The Elements,* 2nd ed., Clarendon Press, Oxford, 1991.

Appendix G
Periodic Table of the Elements

| | | | Metals |
| Metalloids |
| Nonmetals |

Alkali metals IA

Noble gases 0

Transition metals

THE HORIZONTAL PERIODS

1	H (1)	IIA												IIIA	IVA	VA	VIA	VIIA	He (2)
2	Li (3)	Be (4)												B (5)	C (6)	N (7)	O (8)	F (9)	Ne (10)
3	Na (11)	Mg (12)	IIIB	IVB	VB	VIB	VIIB	VIIIB			IB	IIB		Al (13)	Si (14)	P (15)	S (16)	Cl (17)	Ar (18)
4	K (19)	Ca (20)	Sc (21)	Ti (22)	V (23)	Cr (24)	Mn (25)	Fe (26)	Co (27)	Ni (28)	Cu (29)	Zn (30)	Ga (31)	Ge (32)	As (33)	Se (34)	Br (35)	Kr (36)
5	Rb (37)	Sr (38)	Y (39)	Zr (40)	Nb (41)	Mo (42)	Tc (43)	Ru (44)	Rh (45)	Pd (46)	Ag (47)	Cd (48)	In (49)	Sn (50)	Sb (51)	Te (52)	I (53)	Xe (54)
6	Cs (55)	Ba (56)	* (57-71)	Hf (72)	Ta (73)	W (74)	Re (75)	Os (76)	Ir (77)	Pt (78)	Au (79)	Hg (80)	Tl (81)	Pb (82)	Bi (83)	Po (84)	At (85)	Rn (86)
7	Fr (87)	Ra (88)	† (89-103)	Rf (104)	Ha (105)	Sg (106)	Ns (107)	Hs (108)	Mt (109)	(110)	(111)	(112)						

Inner transition metals

Lanthanide series *

La (57)	Ce (58)	Pr (59)	Nd (60)	Pm (61)	Sm (62)	Eu (63)	Gd (64)	Tb (65)	Dy (66)	Ho (67)	Er (68)	Tm (69)	Yb (70)	Lu (71)

Actinide series †

Ac (89)	Th (90)	Pa (91)	U (92)	Np (93)	Pu (94)	Am (95)	Cm (96)	Bk (97)	Cf (98)	Es (99)	Fm (100)	Md (101)	No (102)	Lr (103)

The names for elements 104–109 (Rutherfordium, Hahnium, Seaborgium, Nielsbohrium, Hassium, and Meitnerium, respectively) are those recommended by the American Chemical Society Nomenclature Committee. As of 1996, the names and symbols for elements 104–108 have not yet been approved by the appropriate international body. Elements 110, 111 and 112 have been discovered but, as of 1996, have not been provisionally named.

Appendix H
Winners of the Nobel Prize in Physics*

1901 Wilhelm Konrad Röntgen *(1845–1923)* for the discovery of x rays

1902 Hendrik Antoon Lorentz *(1853–1928)* and Pieter Zeeman *(1865–1943)* for their researches into the influence of magnetism upon radiation phenomena

1903 Antoine Henri Becquerel *(1852–1908)* for his discovery of spontaneous radioactivity

Pierre Curie *(1859–1906)* and Marie Sklowdowska-Curie *(1867–1934)* for their joint researches on the radiation phenomena discovered by Becquerel

1904 Lord Rayleigh (John William Strutt) *(1842–1919)* for his investigations of the densities of the most important gases and for his discovery of argon

1905 Philipp Eduard Anton von Lenard *(1862–1947)* for his work on cathode rays

1906 Joseph John Thomson *(1856–1940)* for his theoretical and experimental investigations on the conduction of electricity by gases

1907 Albert Abraham Michelson *(1852–1931)* for his optical precision instruments and metrological investigations carried out with their aid

1908 Gabriel Lippmann *(1845–1921)* for his method of reproducing colors photographically based on the phenomena of interference

1909 Guglielmo Marconi *(1874–1937)* and Carl Ferdinand Braun *(1850–1918)* for their contributions to the development of wireless telegraphy

1910 Johannes Diderik van der Waals *(1837–1923)* for his work on the equation of state for gases and liquids

1911 Wilhelm Wien *(1864–1928)* for his discoveries regarding the laws governing the radiation of heat

1912 Nils Gustaf Dalén *(1869–1937)* for his invention of automatic regulators for use in conjunction with gas accumulators for illuminating lighthouses and buoys

1913 Heike Kamerlingh Onnes *(1853–1926)* for his investigations of the properties of matter at low temperatures which led, among other things, to the production of liquid helium

1914 Max von Laue *(1879–1960)* for his discovery of the diffraction of Röntgen rays by crystals

1915 William Henry Bragg *(1862–1942)* and William Lawrence Bragg *(1890–1971)* for their services in the analysis of crystal structure by means of x rays

1917 Charles Glover Barkla *(1877–1944)* for his discovery of the characteristic x rays of the elements

1918 Max Planck *(1858–1947)* for his discovery of energy quanta

1919 Johannes Stark *(1874–1957)* for his discovery of the Doppler effect in canal rays and the splitting of spectral lines in electric fields

1920 Charles-Édouard Guillaume *(1861–1938)* for the service he rendered to precision measurements in physics by his discovery of anomalies in nickel steel alloys

1921 Albert Einstein *(1879–1955)* for his services to theoretical physics, and especially for his discovery of the law of the photoelectric effect

1922 Niels Bohr *(1885–1962)* for the investigation of the structure of atoms, and of the radiation emanating from them

1923 Robert Andrews Millikan *(1868–1953)* for his work on the elementary charge of electricity and on the photoelectric effect

1924 Karl Manne Georg Siegbahn *(1886–1978)* for his discoveries and research in the field of x-ray spectroscopy

1925 James Franck *(1882–1964)* and Gustav Hertz *(1887–1975)* for their discovery of the laws governing the impact of an electron upon an atom

1926 Jean Baptiste Perrin *(1870–1942)* for his work on the discontinuous structure of matter, and especially for his discovery of sedimentation equilibrium

1927 Arthur Holly Compton *(1892–1962)* for his discovery of the effect named after him

Charles Thomson Rees Wilson (1869–1959) for his method of making the paths of electrically charged particles visible by condensation of vapor

1928 Owen Willans Richardson *(1879–1959)* for his work on the thermionic phenomenon and especially for the discovery of the law named after him

1929 Prince Louis Victor de Broglie *(1892–1987)* for his discovery of the wave nature of electrons

*See *Nobel Lectures, Physics,* 1901–1970, Elsevier Publishing Company, for biographies of the awardees and for lectures given by them on receiving the prize.

1930   Sir Chandrasekhara Venkata Raman *(1888–1970)* for his work on the scattering of light and for the discovery of the effect named after him

1932   Werner Heisenberg *(1901–1976)* for the creation of quantum mechanics, the application of which has, among other things, led to the discovery of the allotropic forms of hydrogen

1933   Erwin Schrödinger *(1887–1961)* and Paul Adrien Maurice Dirac *(1902–1984)* for the discovery of new productive forms of atomic theory

1935   James Chadwick *(1891–1974)* for his discovery of the neutron

1936   Victor Franz Hess *(1883–1964)* for the discovery of cosmic radiation

Carl David Anderson *(1905–1991)* for his discovery of the positron

1937   Clinton Joseph Davisson *(1881–1958)* and George Paget Thomson *(1892–1975)* for their experimental discovery of the diffraction of electrons by crystals

1938   Enrico Fermi *(1901–1954)* for his demonstrations of the existence of new radioactive elements produced by neutron irradiation, and for his related discovery of nuclear reactions brought about by slow neutrons

1939   Ernest Orlando Lawrence *(1901–1958)* for the invention and development of the cyclotron and for results obtained with it, especially for artificial radioactive elements

1943   Otto Stern *(1888–1969)* for his contribution to the development of the molecular-ray method and his discovery of the magnetic moment of the proton

1944   Isidor Isaac Rabi *(1898–1988)* for his resonance method for recording the magnetic properties of atomic nuclei

1945   Wolfgang Pauli *(1900–1958)* for the discovery of the Exclusion Principle (also called Pauli Principle)

1946   Percy Williams Bridgman *(1882–1961)* for the invention of an apparatus to produce extremely high pressures and for the discoveries he made therewith in the field of high-pressure physics

1947   Sir Edward Victor Appleton *(1892–1965)* for his investigations of the physics of the upper atmosphere, especially for the discovery of the so-called Appleton layer

1948   Patrick Maynard Stuart Blackett *(1897–1974)* for his development of the Wilson cloud-chamber method, and his discoveries therewith in nuclear physics and cosmic radiation

1949   Hideki Yukawa *(1907–1981)* for his prediction of the existence of mesons on the basis of theoretical work on nuclear forces

1950   Cecil Frank Powell *(1903–1969)* for his development of the photographic method of studying nuclear processes and his discoveries regarding mesons made with this method

1951   Sir John Douglas Cockcroft *(1897–1967)* and Ernest Thomas Sinton Walton *(1903–1995)* for their pioneer work on the transmutation of atomic nuclei by artificially accelerated atomic particles

1952   Felix Bloch *(1905–1983)* and Edward Mills Purcell *(1912–   )* for their development of new nuclear-magnetic precision methods and discoveries in connection therewith

1953   Frits Zernike *(1888–1966)* for his demonstration of the phase-contrast method, especially for his invention of the phase-contrast microscope

1954   Max Born *(1882–1970)* for his fundamental research in quantum mechanics, especially for his statistical interpretation of the wave function

Walther Bothe *(1891–1957)* for the coincidence method and his discoveries made therewith

1955   Willis Eugene Lamb *(1913–   )* for his discoveries concerning the fine structure of the hydrogen spectrum

Polykarp Kusch *(1911–1993)* for his precision determination of the magnetic moment of the electron

1956   William Shockley *(1910–1989)*, John Bardeen *(1908–1991)* and Walter Houser Brattain *(1902–1987)* for their researches on semiconductors and their discovery of the transistor effect

1957   Chen Ning Yang *(1922–   )* and Tsung Dao Lee *(1926–   )* for their penetrating investigation of the parity laws which has led to important discoveries regarding the elementary particles

1958   Pavel Aleksejevič Čerenkov *(1904–   )*, Il' ja Michajlovič Frank *(1908–1990)* and Igor' Evgen' evič Tamm *(1895–1971)* for the discovery and interpretation of the Cerenkov effect

1959   Emilio Gino Segrè *(1905–1989)* and Owen Chamberlain *(1920–   )* for their discovery of the antiproton

1960   Donald Arthur Glaser *(1926–   )* for the invention of the bubble chamber

1961   Robert Hofstadter *(1915–1990)* for his pioneering studies of electron scattering in atomic nuclei and for his thereby achieved discoveries concerning the structure of the nucleons

Rudolf Ludwig Mössbauer *(1929–   )* for his researches concerning the resonance absorption of γ rays and his discovery in this connection of the effect which bears his name

1962   Lev Davidovič Landau *(1908–1968)* for his pioneering theories of condensed matter, especially liquid helium

1963   Eugene P. Wigner *(1902–1995)* for his contributions to the theory of the atomic nucleus and the elementary particles, particularly through the discovery and application of fundamental symmetry principles

Maria Goeppert Mayer *(1906–1972)* and J. Hans D. Jensen *(1907–1973)* for their discoveries concerning nuclear shell structure

1964   Charles H. Townes *(1915–   )*, Nikolai G. Basov *(1922–   )* and Alexander M. Prochorov *(1916–   )* for fundamental work in the field of quantum electronics which has led to the construction of oscillators and amplifiers based on the maser–laser principle

1965 Sin-itiro Tomonaga *(1906–1979)*, Julian Schwinger *(1918–1994)* and Richard P. Feynman *(1918–1988)* for their fundamental work in quantum electrodynamics, with deep-ploughing consequences for the physics of elementary particles

1966 Alfred Kastler *(1902–1984)* for the discovery and development of optical methods for studying Hertzian resonance in atoms

1967 Hans Albrecht Bethe *(1906–   )* for his contributions to the theory of nuclear reactions, especially his discoveries concerning the energy production in stars

1968 Luis W. Alvarez *(1911–1988)* for his decisive contribution to elementary particle physics, in particular the discovery of a large number of resonance states, made possible through his development of the techniques of using the hydrogen bubble chamber and its data analysis

1969 Murray Gell-Mann *(1929–   )* for his contributions and discoveries concerning the classification of elementary particles and their interactions

1970 Hannes Alfvén *(1908–1995)* for fundamental work and discoveries in magneto-hydrodynamics with fruitful applications in different parts of plasma physics

Louis Néel *(1904–   )* for fundamental work and discoveries concerning antiferromagnetism and ferrimagnetism which have led to important applications in solid state physics

1971 Dennis Gabor *(1900–1979)* for his discovery of the principles of holography

1972 John Bardeen *(1908–1991)*, Leon N. Cooper *(1930–   )* and J. Robert Schrieffer *(1931–   )* for their development of a theory of superconductivity

1973 Leo Esaki *(1925–   )* for his discovery of tunneling in semiconductors

Ivar Giaever *(1929–   )* for his discovery of tunneling in superconductors

Brian D. Josephson *(1940–   )* for his theoretical prediction of the properties of a supercurrent through a tunnel barrier

1974 Antony Hewish *(1924–   )* for the discovery of pulsars

Sir Martin Ryle *(1918–1984)* for his pioneering work in radioastronomy

1975 Aage Bohr *(1922–   )*, Ben Mottelson *(1926–   )* and James Rainwater *(1917–1986)* for the discovery of the connection between collective motion and particle motion and the development of the theory of the structure of the atomic nucleus based on this connection

1976 Burton Richter *(1931–   )* and Samuel Chao Chung Ting *(1936–   )* for their (independent) discovery of an important fundamental particle

1977 Philip Warren Anderson *(1923–   )*, Nevill Francis Mott *(1905–1996)* and John Hasbrouck Van Vleck *(1899–1980)* for their fundamental theoretical investigations of the electronic structure of magnetic and disordered systems

1978 Peter L. Kapitza *(1894–1984)* for his basic inventions and discoveries in low-temperature physics

Arno A. Penzias *(1933–   )* and Robert Woodrow Wilson *(1936–   )* for their discovery of cosmic microwave background radiation

1979 Sheldon Lee Glashow *(1932–   )*, Abdus Salam *(1926–1996)*, and Steven Weinberg *(1933–   )* for their unified model of the action of the weak and electromagnetic forces and for their prediction of the existence of neutral currents

1980 James W. Cronin *(1931–   )* and Val L. Fitch *(1923–   )* for the discovery of violations of fundamental symmetry principles in the decay of neutral K mesons

1981 Nicolaas Bloembergen *(1920–   )* and Arthur Leonard Schawlow *(1921–   )* for their contribution to the development of laser spectroscopy

Kai M. Siegbahn *(1918–   )* for his contribution to high-resolution electron spectroscopy

1982 Kenneth Geddes Wilson *(1936–   )* for his method of analyzing the critical phenomena inherent in the changes of matter under the influence of pressure and temperature

1983 Subrehmanyan Chandrasekhar *(1910–1995)* for his theoretical studies of the structure and evolution of stars

William A. Fowler *(1911–1995)* for his studies of the formation of the chemical elements in the universe

1984 Carlo Rubbia *(1934–   )* and Simon van der Meer *(1925–   )* for their decisive contributions to the Large Project, which led to the discovery of the field particles W and Z, communicators of the weak interaction

1985 Klaus von Klitzing *(1943–   )* for his discovery of the quantized Hall resistance

1986 Ernst Ruska *(1906–1988)* for his invention of the electron microscope

Gerd Binnig *(1947–   )*, Heinrich Rohrer *(1933–   )* for their invention of the scanning tunneling microscope

1987 Karl Alex Müller *(1927–   )* and J. George Bednorz *(1950–   )* for their discovery of a new class of superconductors

1988 Leon M. Lederman *(1922–   )*, Melvin Schwartz *(1932–   )* and Jack Steinberger *(1921–   )* for the first use of a neutrino beam and the discovery of the muon neutrino

1989 Norman Ramsey *(1915–   )*, Hans Dehmelt *(1922–   )* and Wolfgang Paul *(1913–1993)* for their work that led to the development of atomic clocks and precision timing

1990 Jerome I. Friedman *(1930–   )*, Henry W. Kendall *(1926–   )* and Richard E. Taylor *(1929–   )* for demonstrating that protons and neutrons consist of quarks

1991 Pierre de Gennes *(1932–   )* for studies of order phenomena, such as in liquid crystals and polymers

1992 George Charpak *(1924–   )* for his invention of fast electronic detectors for high-energy particles

1993 Joseph H. Taylor *(1941–   )* and Russell A. Hulse *(1950–   )* for the discovery and interpretation of the first binary pulsar.

1994 Bertram N. Brockhouse *(1918–   )* and Clifford G. Shull *(1915–   )* for the development of neutron scattering techniques

1995 Martin L. Perl *(1927–   )* for the discovery of the tau lepton

Frederick Reines *(1918–   )* for the detection of the neutrino

1996 David M. Lee *(1931–   )*, Robert C. Richardson *(1937–   )* and Douglas D. Oscheroff *(1944–   )* for the discovery of superfluidity in helium-3

Answers to Checkpoints, Odd-Numbered Questions, Exercises, and Problems

Chapter 1

EP **3.** (a) 186 mi; (b) 3.0×10^8 mm **5.** (a) 10^9; (b) 10^{-4}; (c) 9.1×10^5 **7.** 32.2 km **9.** 0.020 km^3 **11.** (a) 250 ft^2; (b) 23.3 m^2; (c) 3060 ft^3; (d) 86.6 m^3 **13.** 8×10^2 km **15.** (a) 11.3 m^2/L; (b) 1.13×10^4 m^{-1}; (c) 2.17×10^{-3} gal/ft^2 **17.** (a) $d_{Sun}/d_{Moon} = 400$; (b) $V_{Sun}/V_{Moon} = 6.4 \times 10^7$; (c) 3.5×10^3 km **19.** (a) 0.98 ft/ns; (b) 0.30 mm/ps **21.** 3.156×10^7 s **23.** 5.79×10^{12} days **25.** (a) 0.013; (b) 0.54; (c) 10.3; (d) 31 m/s **27.** 15° **29.** 3.3 ft **31.** 2 days 5 hours **33.** (a) 2.99×10^{-26} kg; (b) 4.68×10^{46} **35.** 1.3×10^9 kg **37.** (a) 10^3 kg/m^3; (b) 158 kg/s **39.** (a) 1.18×10^{-29} m^3; (b) 0.282 nm

Chapter 2

CP **1.** b and c **2.** zero **3.** (a) 1 and 4; (b) 2 and 3; (c) 3 **4.** (a) plus; (b) minus; (c) minus; (d) plus **5.** 1 and 4 **6.** (a) plus; (b) minus; (c) $a = -g = -9.8$ m/s^2 **Q** **1.** (a) yes; (b) no; (c) yes; (d) yes **3.** (a) 2, 3; (b) 1, 3; (c) 4 **5.** all tie (see Eq. 2-16) **7.** (a) $-g$; (b) 2 m/s upward **9.** same **11.** $x = t^2$ and $x = 8(t - 2) + (1.5)(t - 2)^2$ **13.** increase **EP** **1.** (a) Lewis: 10.0 m/s, Rodgers: 5.41 m/s; (b) 1 h 10 min **3.** 309 ft **5.** 2 cm/y **7.** 6.71×10^8 mi/h, 9.84×10^8 ft/s, 1.00 ly/y **9.** (a) 5.7 ft/s; (b) 7.0 ft/s **11.** (a) 45 mi/h (72 km/h); (b) 43 mi/h (69 km/h); (c) 44 mi/h (71 km/h); (d) 0 **13.** (a) 28.5 cm/s; (b) 18.0 cm/s; (c) 40.5 cm/s; (d) 28.1 cm/s; (e) 30.3 cm/s **15.** (a) mathematically, an infinite number; (b) 60 km **17.** (a) 4 s $> t >$ 2 s; (b) 3 s $> t >$ 0; (c) 7 s $> t >$ 3 s; (d) $t = 3$ s **19.** 100 m **23.** (a) The signs of v and a are: AB: +, −; BC: 0, 0; CD: +, +; DE: +, 0; (b) no; (c) no **25.** (e) situations (a), (b), and (d) **27.** (a) 80 m/s; (b) 110 m/s; (c) 20 m/s^2 **29.** (a) 1.10 m/s, 6.11 mm/s^2; (b) 1.47 m/s, 6.11 mm/s^2 **31.** (a) 2.00 s; (b) 12 cm from left edge of screen; (c) 9.00 cm/s^2, to the left; (d) to the right; (e) to the left; (f) 3.46 s **33.** 0.556 s **35.** each, 0.28 m/s^2 **37.** 2.8 m/s^2 **39.** 1.62×10^{15} m/s^2 **41.** $21g$ **43.** (a) $25g$; (b) 400 m **45.** 90 m **47.** (a) 5.0 m/s^2; (b) 4.0 s; (c) 6.0 s; (d) 90 m **49.** (a) 5.00 m/s; (b) 1.67 m/s^2; (c) 7.50 m **51.** (a) 0.74 s; (b) -20 ft/s^2 **53.** (a) 0.75 s; (b) 50 m **55.** (a) 34.7 ft; (b) 41.6 s **57.** (a) 3.26 ft/s^2 **61.** (a) 31 m/s; (b) 6.4 s **63.** (a) 48.5 m/s; (b) 4.95 s; (c) 34.3 m/s; (d) 3.50 s **65.** (a) 5.44 s; (b) 53.3 m/s; (d) 5.80 m **67.** (a) 3.2 s; (b) 1.3 s **69.** 4.0 m/s **71.** (a) 350 ms; (b) 82 ms (each is for ascent and descent through the 15 cm) **73.** 857 m/s^2, upward **75.** (a) 1.23 cm; (b) 4 times, 9 times, 16 times, 25 times **77.** (a) 8.85 m/s; (b) 1.00 m **79.** 22 cm and 89 cm below the nozzle **81.** (a) 3.41 s; (b) 57 m **83.** (a) 40.0 ft/s **85.** 1.5 s **87.** (a) 5.4 s; (b) 41 m/s **89.** 20.4 m

91. (a) $d = v_i^2/2a' + T_R v_i$; (b) 9.0 m/s^2; (c) 0.66 s. **93.** (a) $v_j^2 = 2a'd_0(j - 1) + v_1^2$; (c) 7.0 m/s^2; (d) 14 m.

Chapter 3

CP **1.** (a) 7 m; (b) 1 m **2.** c, d, f **3.** (a) +, +; (b) +, −; (c) +, + **4.** (a) 90°; (b) 0 (vectors are parallel); (c) 180° (vectors are antiparallel) **5.** (a) 0° or 180°; (b) 90° **Q** **1.** **A** and **B** **3.** No, but **a** and $-\mathbf{b}$ are commutative: $\mathbf{a} + (-\mathbf{b}) = (-\mathbf{b}) + \mathbf{a}$. **5.** (a) **a** and **b** are parallel; (b) $\mathbf{b} = 0$; (c) **a** and **b** are perpendicular **7.** (a)–(c) yes (example: 5i and $-2\mathbf{i}$) **9.** all but e **11.** (a) minus, minus; (b) minus, minus **13.** (a) **B** and **C**, **D** and **E**; (b) **D** and **E** **15.** no (their orientations can differ) **17.** (a) 0 (vectors are parallel); (b) 0 (vectors are antiparallel) **EP** **1.** The displacements should be (a) parallel, (b) antiparallel, (c) perpendicular **3.** (b) 3.2 km, 41° south of west **5.** $\mathbf{a} + \mathbf{b}$: 4.2, 40° east of north; $\mathbf{b} - \mathbf{a}$: 8.0, 24° north of west **7.** (a) 38 units at 320°; (b) 130 units at 1.2°; (c) 62 units at 130° **9.** $a_x = -2.5$, $a_y = -6.9$ **11.** $r_x = 13$ m, $r_y = 7.5$ m **13.** (a) 14 cm, 45° left of straight down; (b) 20 cm, vertically up; (c) zero **15.** 4.74 km **17.** 168 cm, 32.5° above the floor **19.** $r_x = 12$, $r_y = -5.8$, $r_z = -2.8$ **21.** (a) $8\mathbf{i} + 2\mathbf{j}$, 8.2, 14°; (b) $2\mathbf{i} - 6\mathbf{j}$, 6.3, $-72°$ relative to **i** **23.** (a) 5.0, $-37°$; (b) 10, 53°; (c) 11, 27°; (d) 11, 80°; (e) 11, 260°; the angles are relative to $+x$, the last two vectors are in opposite directions **25.** 4.1 **27.** (a) $r_x = 1.59$, $r_y = 12.1$; (b) 12.2; (c) 82.5° **29.** 3390 ft, horizontally **31.** (a) -2.83 m, -2.83 m, $+5.00$ m, 0 m, 3.00 m, 5.20 m; (b) 5.17 m, 2.37 m; (c) 5.69 m, 24.6° north of east; (d) 5.69 m, 24.6° south of west **35.** (a) $a_x = 9.51$ m, $a_y = 14.1$ m; (b) $a_x' = 13.4$ m, $a_y' = 10.5$ m **37.** (a) $+y$; (b) $-y$; (c) 0; (d) 0; (e) $+z$; (f) $-z$; (g) ab, both; (h) ab/d, $+z$ **39.** yes **41.** (a) up, unit magnitude; (b) zero; (c) south, unit magnitude; (d) 1.00; (e) 0 **43.** (a) -18.8; (b) 26.9, $+z$ direction **45.** (a) 12, out of page; (b) 12, into page; (c) 12, out of page **47.** (a) $11\mathbf{i} + 5\mathbf{j} - 7\mathbf{k}$; (b) 120° **51.** (a) 57°; (b) $c_x = \pm 2.2$, $c_y = \mp 4.5$ **53.** (a) -21; (b) -9; (c) $5\mathbf{i} - 11\mathbf{j} - 9\mathbf{k}$

Chapter 4

CP **1.** (a) $(8\mathbf{i} - 6\mathbf{j})$ m; (b) yes, the xy plane **2.** (a) first; (b) third **3.** (1) and (3) a_x and a_y are both constant and thus **a** is constant; (2) and (4) a_y is constant but a_x is not, thus **a** is not **4.** 4 m/s^3, -2 m/s, 3 m **5.** (a) v_x constant; (b) v_y initially positive, decreases to zero, and then becomes progressively more negative; (c) $a_x = 0$ throughout; (d) $a_y = -g$ throughout **6.** (a) $-(4 \text{ m/s})\mathbf{i}$; (b) $-(8 \text{ m/s}^2)\mathbf{j}$ **7.** (1) 0, distance not changing; (2) $+70$ km/h, distance increasing; (3) $+80$ km/h, distance decreasing **Q** **1.** (1) and (3) a_y is constant but a_x is not and thus **a** is not; (2) a_x is constant but a_y

is not and thus **a** is not; (4) a_x and a_y are both constant and thus **a** is constant; -2 m/s², 3 m/s   **3.** (a) highest point; (b) lowest point   **5.** (a) all tie; (b) 1 and 2 tie (the rocket is shot upward), then 3 and 4 tie (it is shot into the ground!)   **7.** $(2\mathbf{i} - 4\mathbf{j})$ m/s   **9.** (a) all tie; (b) all tie; (c) c, b, a; (d) c, b, a **11.** (a) no; (b) same   **13.** (a) in your hands; (b) behind you; (c) in front of you   **15.** (a) straight down; (b) curved; (c) more curved   **17.** (a) 3; (b) 4.   **EP**   **1.** (a) $(-5.0\mathbf{i} + 8.0\mathbf{j})$ m; (b) 9.4 m, 122° from $+x$; (d) $(8\mathbf{i} - 8\mathbf{j})$ m; (e) 11 m, $-45°$ from $+x$ **3.** (a) $(-7.0\mathbf{i} + 12\mathbf{j})$ m; (b) xy plane   **5.** (a) 671 mi, 63.4° south of east; (b) 298 mi/h, 63.4° south of east; (c) 400 mi/h   **7.** (a) 6.79 km/h; (b) 6.96°   **9.** (a) $(3\mathbf{i} - 8t\mathbf{j})$ m/s; (b) $(3\mathbf{i} - 16\mathbf{j})$ m/s; (c) 16 m/s, $-79°$ to $+x$ **11.** (a) $(8t\mathbf{j} + \mathbf{k})$ m/s; (b) $8\mathbf{j}$ m/s²   **13.** $(-2.10\mathbf{i} + 2.81\mathbf{j})$ m/s²   **15.** (a) $-1.5\mathbf{j}$ m/s; (b) $(4.5\mathbf{i} - 2.25\mathbf{j})$ m   **17.** 60.0°   **19.** (a) 63 ms; (b) 1.6×10^3 ft/s   **21.** (a) 2.0 ns; (b) 2.0 mm; (c) $(1.0 \times 10^9\mathbf{i} - 2.0 \times 10^8\mathbf{j})$ cm/s   **23.** (a) 3.03 s; (b) 758 m; (c) 29.7 m/s   **25.** (a) 16 m/s, 23° above the horizontal; (b) 27 m/s, 57° below the horizontal   **27.** (a) 32.4 m; (b) -37.7 m   **29.** (b) 76°   **31.** (a) 51.8 m; (b) 27.4 m/s; (c) 67.5 m   **33.** (a) 194 m/s; (b) 38°   **35.** 1.9 in.   **37.** (a) 11 m; (b) 23 m; (c) 17 m/s, 63° below horizontal   **41.** (a) 73 ft; (b) 7.6°; (c) 1.0 s   **43.** 23 ft/s   **45.** (a) 11 m; (b) 45 m/s   **47.** 30 m above the release point   **49.** 19 ft/s   **51.** (a) 202 m/s; (b) 806 m; (c) 161 m/s, -171 m/s   **53.** (a) 20 cm; (b) no, the ball hits the net only 4.4 cm above the ground   **55.** yes; its center passes about 4.1 ft above the fence   **57.** (a) 9.00×10^{22} m/s², toward the center; (b) 1.52×10^{-16} s   **59.** (a) 6.7×10^6 m/s; (b) 1.4×10^{-7} s   **61.** (a) 7.49 km/s; (b) 8.00 m/s²   **63.** (a) 0.94 m; (b) 19 m/s; (c) 2400 m/s², toward center; (d) 0.05 s   **65.** (a) 1.3×10^5 m/s²; (b) 7.9×10^5 m/s² or $(8.0 \times 10^4)g$, toward the center; (c) both answers increase   **67.** (a) 0.034 m/s²; (b) 84 min   **69.** 2.58 cm/s²   **71.** 160 m/s²   **73.** 36 s, no   **75.** 0.018 mi/s² from either frame   **77.** 130°   **79.** 60°   **81.** (a) 5.8 m/s; (b) 16.7 m; (c) 67°   **83.** 185 km/h, 22° south of west   **85.** (a) from 75° east of south; (b) 30° east of north; substitute west for east to get second solution   **87.** (a) 30° upstream; (b) 69 min; (c) 80 min; (d) 80 min; (e) perpendicular to the current, the shortest possible time is 60 min   **89.** $0.83c$ **91.** (a) $0.35c$; (b) $0.62c$ **93.** For launch angles from 5° to 70°, it always moves away from the launch site. For a 75° launch angle, it moves toward the site from 11.5 s to 18.5 s after launch. For an 80° launch angle, it moves toward the site from 10.5 s to 20.5 s after launch. For an 85° launch angle, it moves toward the site from 10.5 s to 20.5 s after launch. For a 90° launch angle, it moves toward the site from 10 s to 20.5 s after launch.   **95.** (a) 1.6 s; (b) no; (c) 14 m/s; (d) yes   **97.** (a) $\Delta\mathbf{D} = (1.0$ m$)\mathbf{i} - (2.0$ m$)\mathbf{j} + (1.0$ m$)\mathbf{k}$; (b) 2.4 m; (c) $\bar{\mathbf{v}} = (0.025$ m/s$)\mathbf{i} - (0.050$ m/s$)\mathbf{j} + (0.025$ m/s$)\mathbf{k}$ (d) cannot be determined without additional information

Chapter 5

CP   **1.** c, d, and e **2.** (a) and (b) 2 N, leftward (acceleration is zero in each situation)   **3.** (a) and (b) 1, 4, 3, 2   **4.** (a) equal; (b) greater (acceleration is upward, thus net force on body must be upward)   **5.** (a) equal; (b) greater; (c) less   **6.** (a) increase; (b) yes; (c) same; (d) yes   **7.** (a) $F \sin\theta$; (b) increase   **8.** 0   **Q**   **1.** (a) yes; (b) yes; (c) yes; (d) yes   **3.** (a) 2 and 4; (b) 2 and 4   **5.** (a) 50 N, upward; (b) 150 N, upward   **7.** (a) less; (b) greater   **9.** (a) no; (b) no; (c) no   **11.** (a) increases; (b) increases; (c) decreases; (d) decreases   **13.** (a) 20 kg; (b) 18 kg; (c) 10 kg; (d) all tie; (e) 3, 2, 1   **15.** d, c, a, b **EP**   **1.** (a) $F_x = 1.88$ N, $F_y = 0.684$ N; (b) $(1.88\mathbf{i} + 0.684\mathbf{j})$ N   **3.** (a) $(-6.26\mathbf{i} - 3.23\mathbf{j})$ N; (b) 7.0 N, 207° relative to $+x$ **5.** $(-2\mathbf{i} + 6\mathbf{j})$ N   **7.** (a) 0; (b) $+20$ N; (c) -20 N; (d) -40 N; (e) -60 N   **9.** (a) $(1\mathbf{i} - 1.3\mathbf{j})$ m/s²; (b) 1.6 m/s² at $-50°$ from $+x$ **11.** (a) $\mathbf{F}_2$ and $\mathbf{F}_3$ are in the $-x$ direction, $\mathbf{a} = 0$; (b) $\mathbf{F}_2$ and $\mathbf{F}_3$ are in the $-x$ direction, $\mathbf{a}$ is on the x axis, $a = 0.83$ m/s²; (c) $\mathbf{F}_2$ and $\mathbf{F}_3$ are at 34° from $-x$ direction; $\mathbf{a} = 0$ **13.** (a) 22 N, 2.3 kg; (b) 1100 N, 110 kg; (c) 1.6×10^4 N, 1.6×10^3 kg   **15.** (a) 11 N, 2.2 kg; (b) 0, 2.2 kg   **17.** (a) 44 N; (b) 78 N; (c) 54 N; (d) 152 N   **19.** 1.18×10^4 N   **21.** 1.2×10^5 N   **23.** 16 N   **25.** (a) 13 ft/s²; (b) 190 lb   **27.** (a) 42 N; (b) 72 N; (c) 4.9 m/s²   **29.** (a) 0.02 m/s²; (b) 8×10^4 km; (c) 2×10^3 m/s   **31.** (a) 1.1×10^{-15} N; (b) 8.9×10^{-30} N   **33.** (a) 5500 N; (b) 2.7 s; (c) 4 times as far; (d) twice the time   **35.** (a) 4.9×10^5 N; (b) 1.5×10^6 N   **37.** (a) 110 lb, up; (b) 110 lb, down   **39.** (a) 0.74 m/s²; (b) 7.3 m/s²   **41.** (a) $\cos\theta$; (b) $\sqrt{\cos\theta}$ **43.** 1.8×10^4 N   **45.** (a) 4.6×10^3 N; (b) 5.8×10^3 N   **47.** (a) 250 m/s²; (b) 2.0×10^4 N   **49.** 23 kg   **51.** (a) 620 N; (b) 580 N   **53.** 1.9×10^5 lb   **55.** (a) rope breaks; (b) 1.6 m/s²   **57.** 4.6 N   **59.** (a) allow a downward acceleration with magnitude ≥ 4.2 ft/s²; (b) 13 ft/s or greater   **61.** 195 N, up   **63.** (a) 566 N; (b) 1130 N   **65.** 18,000 N   **67.** (a) 1.4×10^4 N; (b) 1.1×10^4 N; (c) 2700 N, toward the counterweight   **69.** 6800 N, at 21° to the line of motion of the barge   **71.** (a) 4.6 m/s²; (b) 2.6 m/s²   **73.** (b) $Fl/(m + M)$; (c) $MFl/(m + M)$; (d) $F(m + 2M)/2(m + M)$ **75.** $T_1 = 13$ N, $T_2 = 20$ N, $a = 3.2$ m/s²

Chapter 6

CP   **1.** (a) zero (because there is no attempt at sliding); (b) 5 N; (c) no; (d) yes   **2.** (a) same (10 N); (b) decreases; (c) decreases   **3.** greater   **4.** (a) **a** downward; **N** upward; (b) **a** and **N** upward   **5.** (a) $4R_1$; (b) $4R_1$ **6.** (a) same; (b) increases; (c) increases   **Q**   **1.** They slide at the same angle for all orders.   **3.** (a) upward; (b) horizontal, toward you; (c) no change; (d) increases; (e) increases   **5.** The frictional force $\mathbf{f}_s$ is initially directed up the ramp, decreases in magnitude to zero, and then is directed down the ramp, increasing in magnitude until the magnitude reaches $f_{s,\text{max}}$; thereafter, the magnitude of the frictional force is f_k, which is a constant smaller value.   **7.** (a) decreases; (b) decreases; (c) increases; (d) increases   **9.** (a) zero; (b) infinite   **11.** 4, 3; then 1, 2, and 5 tie   **13.** (a) less; (b) greater   **EP**   **1.** (a) 200 N; (b) 120 N   **3.** 2°   **5.** 440 N

7. (a) 110 N; (b) 130 N; (c) no; (d) 46 N; (e) 17 N
9. (a) 90 N; (b) 70 N; (c) 0.89 m/s^2 **11.** (a) no; (b) $(-12\mathbf{i} + 5\mathbf{j})$ N **13.** 20° **15.** (a) 0.13 N; (b) 0.12 **17.** $\mu_s = 0.58$, $\mu_k = 0.54$ **19.** (a) 0.11 m/s^2, 0.23 m/s^2; (b) 0.041, 0.029 **21.** 36 m **23.** (a) 300 N; (b) 1.3 m/s^2 **25.** (a) 66 N; (b) 2.3 m/s^2 **27.** (a) $\mu_k mg/(\sin\theta - \mu_k \cos\theta)$; (b) $\theta_0 = \tan^{-1}\mu_s$ **29.** (b) 3.0×10^7 N **31.** 100 N **33.** 3.3 kg **35.** (a) 11 ft/s^2; (b) 0.46 lb; (c) blocks move independently **37.** (a) 27 N; (b) 3.0 m/s^2 **39.** (a) 6.1 m/s^2, leftward; (b) 0.98 m/s^2, leftward **41.** (a) 3.0×10^5 N; (b) 1.2° **43.** 9.9 s **45.** 3.75 **47.** 12 cm **49.** 68 ft **51.** (a) 3210 N; (b) yes **53.** 0.078 **55.** (a) 0.72 m/s; (b) 2.1 m/s^2; (c) 0.50 N **57.** $\sqrt{Mgr/m}$ **59.** (a) 30 cm/s; (b) 180 cm/s^2, radially inward; (c) 3.6×10^{-3} N, radially inward; (d) 0.37 **61.** (a) 275 N; (b) 877 N **63.** 874 N **65.** (a) at the bottom of the circle; (b) 31 ft/s **67.** (a) 9.5 m/s; (b) 20 m **69.** 13° **71.** (a) 0.0338 N; (b) 9.77 N

Chapter 7

CP 1. (a) decrease; (b) same; (c) negative, zero **2.** d, c, b, a **3.** (a) same; (b) smaller **4.** (a) positive; (b) negative; (c) zero **5.** zero **Q 1.** all tie **3.** (a) increasing; (b) same; (c) same; (d) increasing **5.** (a) positive; (b) negative; (c) negative **7.** (a) positive; (b) zero; (c) negative; (d) negative; (e) zero; (f) positive **9.** all tie **11.** c, d, a and b tie; then f, e. **13.** (a) 3 m; (b) 3 m; (c) 0 and 6 m; (d) negative direction of x **15.** (a) A; (b) B **17.** twice **EP 1.** 1.8×10^{13} J **3.** (a) 3610 J; (b) 1900 J; (c) 1.1×10^{10} J **5.** (a) 1×10^5 megatons TNT; (b) 1×10^7 bombs **7.** father, 2.4 m/s; son, 4.8 m/s **9.** (a) 200 N; (b) 700 m; (c) -1.4×10^5 J; (d) 400 N, 350 m, -1.4×10^5 J **11.** 5000 J **13.** 47 keV **15.** 7.9 J **17.** 530 J **19.** -37 J **21.** (a) 314 J; (b) -155 J; (c) 0; (d) 158 J **23.** (a) 98 N; (b) 4.0 cm; (c) 3.9 J; (d) -3.9 J **25.** (a) $-3Mgd/4$; (b) Mgd; (c) $Mgd/4$ (d) $\sqrt{gd/2}$ **27.** 25 J **31.** -6 J **33.** (a) 12 J; (b) 4.0 m; (c) 18 J **35.** (a) -0.043 J; (b) -0.13 J **37.** (a) 6.6 m/s; (b) 4.7 m **39.** (a) up; (b) 5.0 cm; (c) 5.0 J **41.** 270 kW **43.** 235 kW **45.** 490 W **47.** (a) 100 J; (b) 67 W; (c) 33 W **49.** 0.99 hp **51.** (a) 0; (b) -350 W **53.** (a) 79.4 keV; (b) 3.12 MeV; (c) 10.9 MeV **55.** (a) 32 J; (b) 8 W; (c) 78°

Chapter 8

CP 1. no **2.** 3, 1, 2 **3.** (a) all tie; (b) all tie **4.** (a) CD, AB, BC (zero); (b) positive direction of x **5.** 2, 1, 3 **6.** decrease **7.** (a) seventh excited state, with energy E_7; (b) 1.3 eV **Q 1.** -40 J **3.** (c) and (d) tie; then (a) and (b) tie **5.** (a) all tie; (b) all tie **7.** (a) 3, 2, 1; (b) 1, 2, 3 **9.** less than (smaller decrease in potential energy) **11.** (a) $E < 3$ J, $K < 2$ J; (b) $E < 5$ J, $K < 4$ J **13.** (a) increasing; (b) decreasing; (c) decreasing; (d) constant in AB and BC, decreasing in CD **EP 1.** 15 J **3.** (a) 167 J; (b) -167 J; (c) 196 J; (d) 29 J **5.** (a) 0; (b) $mgh/2$; (c) mgh; (d) $mgh/2$; (e) mgh **7.** (a) -0.80 J; (b) -0.80 J; (c) $+1.1$ J **9.** (a) $mgL(1 - \cos\theta)$; (b) $-mgL(1 - \cos\theta)$; (c) $mgL(1 - \cos\theta)$ **11.** (a) 18 J;

(b) 0; (c) 30 J; (d) 0; (e) parts b and d **13.** (a) 2.08 m/s; (b) 2.08 m/s **15.** (a) $\sqrt{2gL}$; (b) $2\sqrt{gL}$; (c) $\sqrt{2gL}$ **17.** 830 ft **19.** (a) 6.75 J; (b) -6.75 J; (c) 6.75 J; (d) 6.75 J; (e) -6.75 J; (f) 0.459 m **21.** (a) 21.0 m/s; (b) 21.0 m/s **23.** (a) 0.98 J; (b) -0.98 J; (c) 3.1 N/cm **25.** (a) 39.2 J; (b) 39.2 J; (c) 4.00 m **27.** (a) 54 m/s; (b) 52 m/s; (c) 76 m, below **29.** (a) 39 ft/s; (b) 4.3 in. **31.** (a) 300 J; (b) 93.8 J; (c) 6.38 m **33.** (a) 4.8 m/s; (b) 2.4 m/s **35.** (a) $[v_0^2 + 2gL(1 - \cos\theta_0)]^{1/2}$; (b) $(2gL \cos\theta_0)^{1/2}$; (c) $[gL(3 + 2\cos\theta_0)]^{1/2}$ **37.** (a) $U(x) = -Gm_1m_2/x$; (b) $Gm_1m_2 d/x_1(x_1 + d)$ **39.** (a) $8mg$ leftward and mg downward; (b) $2.5R$ **43.** $mgL/32$ **47.** (a) $1.12(A/B)^{1/6}$; (b) repulsive; (c) attractive **49.** (a) turning point on left, none on right; molecule breaks apart; (b) turning points on both left and right; molecule does not break apart; (c) -1.2×10^{-19} J; (d) 2.2×10^{-19} J; (e) $\approx 1 \times 10^{-9}$ on each, directed toward the other; (f) $r < 0.2$ nm; (g) $r > 0.2$ nm; (h) $r = 0.2$ nm **51.** -25 J **53.** (a) 2200 J; (b) -1500 J; (c) 700 J **55.** 17 kW **57.** (a) -0.74 J; (b) -0.53 J **59.** -12 J **61.** 54% **63.** 880 MW **65.** (a) 39 kW; (b) 39 kW **67.** (a) 1.5 MJ; (b) 0.51 MJ; (c) 1.0 MJ; (d) 63 m/s **69.** (a) 67 J; (b) 67 J; (c) 46 cm **71.** Your force on the cabbage does work. **73.** (a) -0.90 J; (b) 0.46 J; (c) 1.0 m/s **75.** (a) 18 ft/s; (b) 18 ft **77.** 4.3 m **79.** (a) 31.0 J; (b) 5.35 m/s; (c) conservative **81.** 1.2 m **85.** in the center of the flat part **87.** (a) 24 ft/s; (b) 3.0 ft; (c) 9.0 ft; (d) 49 ft **89.** (a) 216 J; (b) 1180 N; (c) 432 J; (d) motor also supplies thermal energy to crate and belt **91.** (a) 1.1×10^{17} J; (b) 1.2 kg **93.** 7.28 MeV **95.** (a) release; (b) 17.6 MeV **97.** (a) 5.3 eV; (b) 0.9 eV **99.** (a) 7.2 J; (b) -7.2 J; (c) 86 cm; (d) 26 cm

Chapter 9

CP 1. (a) origin; (b) fourth quadrant; (c) on y axis below origin; (d) origin; (e) third quadrant; (f) origin **2.** (a) to (c) at the center of mass, still at the origin (their forces are internal to the system and cannot move the center of mass) **3.** (a) 1, 3, and then 2 and 4 tie (zero force); (b) 3 **4.** (a) 0; (b) no; (c) negative x **5.** (a) 500 km/h; (b) 2600 km/h; (c) 1600 km/h **6.** (a) yes; (b) no **Q 1.** point 4 **3.** (a) at the center of the sled; (b) $L/4$, to the right; (c) not at all (no net external force); (d) $L/4$, to the left; (e) L; (f) $L/2$; (g) $L/2$ **5.** (a) ac, cd, and bc; (b) bc; (c) bd and ad **7.** (a) 2 N, rightward; (b) 2 N, rightward; (c) greater than 2 N, rightward **9.** b, c, a **11.** (a) yes; (b) 6 kg·m/s in $-x$ direction; (c) can't tell **EP 1.** (a) 4600 km; (b) $0.73R_e$ **3.** (a) $x_{cm} = 1.1$ m, $y_{cm} = 1.3$ m; (b) shifts toward topmost particle **5.** $x_{cm} = -0.25$ m, $y_{cm} = 0$ **7.** in the iron, at midheight and midwidth, 2.7 cm from midlength **9.** $x_{cm} = y_{cm} = 20$ cm, $z_{cm} = 16$ cm **11.** (a) $H/2$; (b) $H/2$; (c) descends to lowest point and then ascends to $H/2$; (d) $(HM/m)(\sqrt{1 + m/M} - 1)$ **13.** 72 km/h **15.** (a) center of mass does not move; (b) 0.75 m **17.** 4.8 m/s **19.** (a) 22 m; (b) 9.3 m/s **21.** 53 m **23.** 13.6 ft **25.** (a) 52.0 km/h; (b) 28.8 km/h **27.** a proton **29.** (a) 30°; (b) $-0.572\mathbf{j}$ kg·m/s **31.** (a) $(-4.0 \times 10^4\ \mathbf{i})$ kg·m/s; (b) west; (c) 0 **33.** $0.707c$ **35.** 0.57 m/s, toward center of mass **37.** it increases by

4.4 m/s   **39.** (a) rocket case: 7290 m/s, payload: 8200 m/s; (b) before: 1.271×10^{10} J, after: 1.275×10^{10} J   **41.** (a) -1; (b) 1830; (c) 1830; (d) same   **43.** 14 m/s, 135° from the other pieces   **45.** 190 m/s
47. (a) $0.200v_{rel}$; (b) $0.210v_{rel}$; (c) $0.209v_{rel}$ **49.** (a) 1.57×10^6 N; (b) 1.35×10^5 kg; (c) 2.08 km/s   **51.** 108 m/s
53. 2.2×10^{-3} **57.** fast barge: 46 N more; slow barge: no change   **59.** (a) 7.8 MJ; (b) 6.2   **61.** 690 W
63. 5.5×10^6 N   **65.** 24 W   **67.** 100 m   **69.** (a) 860 N; (b) 2.4 m/s   **71.** (a) 3.0×10^5 J; (b) 10 kW; (c) 20 kW
73. (a) 2.1×10^6 kg; (b) $\sqrt{100 + 1.5t}$ m/s; (c) $(1.5 \times 10^6)/\sqrt{100 + 1.5t}$ N; (d) 6.7 km   **75.** $t = (3d/2)^{2/3}(m/2P)^{1/3}$

Chapter 10

CP **1.** (a) unchanged; (b) unchanged; (c) decreased
2. (a) zero; (b) positive; (c) positive direction of y **3.** (a) 4 kg·m/s; (b) 8 kg·m/s; (c) 3 J   **4.** (a) 0; (b) 4 kg·m/s
5. (a) 10 kg·m/s; (b) 14 kg·m/s; (c) 6 kg·m/s   **6.** (a) 2 kg·m/s; (b) 3 kg·m/s   **7.** (a) increases; (b) increases
Q **1.** all tie   **3.** b and c   **5.** (a) one stationary; (b) 2; (c) 5; (d) equal (pool player's result)   **7.** (a) 1 and 4 tie; then 2 and 3 tie; (b) 1; 3 and 4 tie; then 2   **9.** (a) rightward; (b) rightward; (c) smaller   **11.** positive direction of x axis
EP **1.** (a) 750 N; (b) 6.0 m/s   **3.** 6.2×10^4 N   **5.** 3000 N ($= 660$ lb)   **7.** 1.1 m   **9.** (a) 42 N·s; (b) 2100 N
11. (a) $(7.4 \times 10^3\,\mathbf{i} - 7.4 \times 10^3\,\mathbf{j})$ N·s; (b) $(-7.4 \times 10^3\,\mathbf{i})$ N·s; (c) 2.3×10^3 N; (d) 2.1×10^4 N; (e) $-45°$
13. (a) 1.0 kg·m/s; (b) 250 J; (c) 10 N; (d) 1700 N
15. 5 N   **17.** $2\mu v$ **19.** 990 N   **21.** (a) 1.8 N·s, upward; (b) 180 N, downward   **25.** 8 m/s   **27.** 38 km/s
29. 4.2 m/s   **31.** (a) 99 g; (b) 1.9 m/s; (c) 0.93 m/s
33. (a) 1.2 kg; (b) 2.5 m/s   **35.** 7.8 kg   **37.** (a) 1/3; (b) $4h$ **39.** 35 cm   **41.** 3.0 m/s   **43.** (a) $(10\mathbf{i} + 15\mathbf{j})$ m/s; (b) 500 J lost   **45.** (a) 2.7 m/s; (b) 1400 m/s   **47.** (a) A: 4.6 m/s, B: 3.9 m/s; (b) 7.5 m/s   **49.** 20 J for the heavy particle, 40 J for the light particle   **51.** $mv^2/6$ **53.** 13 tons
55. 25 cm   **57.** 0.975 m/s, 0.841 m/s   **59.** (a) 4.1 ft/s; (b) 1700 ft·lb; (c) $v_{24} = 5.3$ ft/s, $v_{32} = 3.3$ ft/s   **61.** (a) 30° from the incoming proton's direction; (b) 250 m/s and 430 m/s   **63.** (a) 41°; (b) 4.76 m/s; (c) no   **65.** $v = V/4$
67. (a) 117° from the final direction of B; (b) no   **69.** 120°
71. (a) 1.9 m/s, 30° to initial direction; (b) no   **73.** (a) 3.4 m/s, deflected by 17° to the right; (b) 0.95 MJ   **75.** (a) 117 MeV; (b) equal and opposite momenta; (c) π^-
77. (a) 4.94 MeV; (b) 0; (c) 4.85 MeV; (d) 0.09 MeV

Chapter 11

CP **1.** (b) and (c)   **2.** (a) and (d)   **3.** (a) yes; (b) no; (c) yes; (d) yes   **4.** all tie   **5.** 1, 2, 4, 3   **6.** (a) 1 and 3 tie, 4; then 2 and 5 tie (zero)   **7.** (a) downward in the figure; (b) less   **Q** **1.** (a) positive; (b) zero; (c) negative; (d) negative   **3.** (a) 2 and 3; (b) 1 and 3; (c) 4   **5.** (a) and (c)   **7.** (a) all tie; (b) 2, 3; then 1 and 4 tie   **9.** b, c, a
11. less   **13.** 90°; then 70° and 110° tie   **15.** Finite angular

displacements are not commutative.   **EP** **1.** (a) 1.50 rad; (b) 85.9°; (c) 1.49 m   **3.** (a) 0.105 rad/s; (b) 1.75×10^{-3} rad/s; (c) 1.45×10^{-4} rad/s   **5.** (a) $\omega(2) = 4.0$ rad/s, $\omega(4) = 28$ rad/s; (b) 12 rad/s²; (c) $\alpha(2) = 6.0$ rad/s², $\alpha(4) = 18$ rad/s²   **7.** (a) $\omega_0 + at^4 - bt^3$; (b) $\theta_0 + \omega_0 t + at^5/5 - bt^4/4$ **9.** 11 rad/s   **11.** (a) 9000 rev/min²; (b) 420 rev
13. (a) 30 s; (b) 1800 rad   **15.** 200 rev/min
17. (a) 2.0 rad/s²; (b) 5.0 rad/s; (c) 10 rad/s; (d) 75 rad
19. (a) 13.5 s; (b) 27.0 rad/s   **21.** (a) 340 s; (b) -4.5×10^{-3} rad/s²; (c) 98 s   **23.** (a) 1.0 rev/s²; (b) 4.8 s; (c) 9.6 s; (d) 48 rev   **25.** 6.1 ft/s² (1.8 m/s²), toward the center
27. 0.13 rad/s   **29.** 5.6 rad/s²   **31.** (a) 5.1 h; (b) 8.1 h
33. (a) 2.50×10^{-3} rad/s; (b) 20.2 m/s²; (c) 0   **35.** (a) -1.1 rev/min²; (b) 9900 rev; (c) -0.99 mm/s²; (d) 31 m/s²
37. (a) 310 m/s; (b) 340 m/s   **39.** (a) 1.94 m/s²; (b) 75.1°, toward the center of the track   **41.** 16 s   **43.** (a) 73 cm/s²; (b) 0.075; (c) 0.11   **45.** 12.3 kg·m²   **47.** first cylinder: 1100 J; second cylinder: 9700 J   **49.** (a) 221 kg·m²; (b) 1.10×10^4 J   **51.** (a) 6490 kg·m²; (b) 4.36 MJ
53. 0.097 kg·m²   **57.** (a) 1300 g·cm²; (b) 550 g·cm²; (c) 1900 g·cm²; (d) $A + B$ **59.** (a) 49 MJ; (b) 100 min
61. 4.6 N·m   **63.** (a) $r_1F_1 \sin\theta_1 - r_2F_2 \sin\theta_2$; (b) -3.8 N·m   **65.** 1.28 kg·m²   **67.** 9.7 rad/s², counterclockwise
69. (a) 155 kg·m²; (b) 64.4 kg   **71.** (a) 420 rad/s²; (b) 500 rad/s   **73.** small sphere: (a) 0.689 N·m and (b) 3.05 N; large sphere: (a) 9.84 N·m and (b) 11.5 N   **75.** 1.73 m/s²; 6.92 m/s²   **77.** (a) 1.4 m/s; (b) 1.4 m/s   **79.** (a) 19.8 kJ; (b) 1.32 kW   **81.** (a) 8.2×10^{28} N·m; (b) 2.6×10^{29} J; (c) 3.0×10^{21} kW   **83.** $\sqrt{9g/4\ell}$ **85.** (a) 4.8×10^5 N; (b) 1.1×10^4 N·m; (c) 1.3×10^6 J   **87.** (a) $3g(1 - \cos\theta)$; (b) $\frac{3}{2}g \sin\theta$; (c) 41.8°   **89.** (a) 5.6 rad/s²; (b) 3.1 rad/s
91. (a) 42.1 km/h; (b) 3.09 rad/s²; (c) 7.57 kW   **93.** (a) 3.4×10^5 g·cm²; (b) 2.9×10^5 g·cm²; (c) 6.3×10^5 g·cm²; (d) $(1.2$ cm$)$ $\mathbf{i}$ + $(5.9$ cm$)$ $\mathbf{j}$

Chapter 12

CP **1.** (a) same; (b) less   **2.** less   **3.** (a) $\pm z$; (b) $+y$; (c) $-x$ **4.** (a) 1 and 3 tie, then 2 and 4 tie, then 5 (zero); (b) 2 and 3   **5.** (a) 3, 1; then 2 and 4 tie (zero); (b) 3
6. (a) all tie (same τ, same t, thus same ΔL); (b) sphere, disk, hoop (reverse order of I)   **7.** (a) decreases; (b) same; (c) increases   **Q** **1.** (a) same; (b) block; (c) block
3. (a) greater; (b) same   **5.** (a) L; (b) $1.5L$ **7.** b, then c and d tie; then a and e tie (zero)   **9.** a, then b and c tie; then e, d (zero)   **11.** (a) same; (b) increases, because of decrease in rotational inertia   **13.** (a) 30 units clockwise; (b) 2 then 4, then the others; or 4 then 2, then the others   **15.** (a) spins in place; (b) rolls toward you; (c) rolls away from you
EP **1.** 1.00   **3.** (a) 59.3 rad/s; (b) -9.31 rad/s²; (c) 70.7 m   **5.** (a) -4.11 m/s²; (b) -16.4 rad/s²; (c) -2.54 N·m   **7.** (a) 8.0°; (b) $0.14g$ **9.** (a) 4.0 N, to the left; (b) 0.60 kg·m²   **11.** (a) $\frac{1}{2}mR^2$; (b) a solid circular cylinder
13. (a) $mg(R - r)$; (b) 2/7; (c) $(17/7)mg$ **15.** (a) $2.7R$; (b) $(50/7)mg$ **17.** (a) 13 cm/s²; (b) 4.4 s; (c) 55 cm/s; (d) 1.8×10^{-2} J; (e) 1.4 J; (f) 27 rev/s   **21.** (a) 24 N·m, in $+y$ direction; (b) 24 N·m, $-y$; (c) 12 N·m, $+y$;

(d) 12 N·m, $-y$ **23.** (a) $(-1.5\mathbf{i} - 4.0\mathbf{j} - \mathbf{k})$ N·m;
(b) $(-1.5\mathbf{i} - 4.0\mathbf{j} - \mathbf{k})$ N·m   **25.** $-2.0\mathbf{i}$ N·m   **27.** 9.8
kg·m²/s   **29.** (a) 12 kg·m²/s, out of page; (b) 3.0 N·m,
out of page   **31.** (a) 0; (b) $(8.0\mathbf{i} + 8.0\mathbf{k})$ N·m   **33.** (a) mvd;
(b) no; (c) 0, yes   **35.** (a) 3.15×10^{43} kg·m²/s; (b) 0.616
37. 4.5 N·m, parallel to xy plane at $-63°$ from $+x$
39. (a) 0; (b) 0; (c) $30t^3$ kg·m²/s, $90t^2$ N·m, both in $-z$
direction; (d) $30t^3$ kg·m²/s, $90t^2$ N·m, both in $+z$ direction
41. (a) $\frac{1}{2}mgt^2v_0\cos\theta_0$; (b) $mgtv_0\cos\theta_0$; (c) $mgtv_0\cos\theta_0$
43. (a) -1.47 N·m; (b) 20.4 rad; (c) -29.9 J; (d) 19.9 W
45. (a) 12.2 kg·m²; (b) 308 kg·m²/s, down   **47.** (a) 1/3;
(b) 1/9   **49.** $\omega_0 R_1 R_2 I_1/(I_1 R_2^2 + I_2 R_1^2)$ **51.** (a) 3.6 rev/s;
(b) 3.0; (c) work done by man in moving weights inward
53. (a) 267 rev/min; (b) 2/3   **55.** 3.0 min   **57.** 2.6 rad/s
59. (a) they revolve in a circle of 1.5 m radius at 0.93 rad/s;
(b) 8.4 rad/s; (c) $K_a = 98$ J, $K_b = 880$ J; (d) from the work
done in pulling inward   **61.** $ml/(M + m)(v/R)$
63. (a) $mvR/(I + MR^2)$; (b) $mvR^2/(I + MR^2)$ **65.** 1300 m/s
67. (a) 18 rad/s; (b) 0.92

69. $\theta = \cos^{-1}\left[1 - \dfrac{6m^2h}{\ell(2m + M)(3m + M)}\right]$

71. 5.28×10^{-35} J·s   **73.** Any three are spin up; the other is
spin down.   **75.** (a) The magnitude of the angular momentum
increases in proportion to t^2 and the magnitude of the torque
increases in proportion to t, in agreement with the second law
for rotation. (b) The magnitudes of the angular momentum and
torque again increase with time. But the change in the
magnitude of the angular momentum in any interval is less than
is predicted by proportionality to t^2 law and the change in the
torque is less than is predicted by proportionality to t. At any
position of the projectile the torque is less when drag is present
than when it is not.

Chapter 13

CP   **1.** c, e, f **2.** (a) no; (b) at site of $\mathbf{F}_1$, perpendicular to
plane of figure; (c) 45 N   **3.** (a) at C (to eliminate forces there
from a torque equation); (b) plus; (c) minus; (d) equal   **4.** d
5. (a) equal; (b) B; (c) B **Q**   **1.** (a) yes; (b) yes; (c) yes;
(d) no   **3.** b **5.** (a) yes; (b) no; (c) no (it could balance the
torques but the forces would then be unbalanced)   **7.** (a) a,
then b and c tie, then d **9.** (a) 20 N (the key is the pulley
with the 20 N weight); (b) 25 N   **11.** (a) $\sin\theta$; (b) same;
(c) larger   **13.** tie of A and B, then C **EP**   **1.** (a) two;
(b) seven   **3.** (a) 2.5 m; (b) 7.3°   **5.** 120°   **7.** 7920 N
9. (a) 840 N; (b) 530 N   **11.** 0.536 m   **13.** (a) 2770 N;
(b) 3890 N   **15.** (a) 1160 N, down; (b) 1740 N, up; (c) left,
stretched; (d) right, compressed   **17.** (a) 280 N; (b) 880 N,
71° above the horizontal   **19.** bars BC, CD, and DA are under
tension due to forces T, diagonals AC and BD are compressed
due to forces $\sqrt{2}T$ **21.** (a) 1800 lb; (b) 822 lb; (c) 1270 lb
23. (a) 49 N; (b) 28 N; (c) 57 N; (d) 29°   **25.** (a) 1900 N, up;
(b) 2100 N, down   **27.** (a) 340 N; (b) 0.88 m; (c) increases,
decreases   **29.** $W\sqrt{2rh - h^2}/(r - h)$ **31.** (a) $L/2$; (b) $L/4$;
(c) $L/6$; (d) $L/8$; (e) $25L/24$ **33.** (a) 6630 N; (b) $F_h = 5740$ N;
(c) $F_v = 5960$ N   **35.** 2.20 m   **37.** (a) 1.50 m; (b) 433 N;

(c) 250 N   **39.** (a) $a_1 = L/2$, $a_2 = 5L/8$, $h = 9L/8$;
(b) $b_1 = 2L/3$, $b_2 = L/2$, $h = 7L/6$ **41.** (a) 47 lb; (b) 120 lb;
(c) 72 lb   **43.** (a) 445 N; (b) 0.50; (c) 315 N
45. (a) 3.9 m/s²; (b) 2000 N on each rear wheel, 3500 N on
each front wheel; (c) 790 N on each rear wheel, 1410 N on
each front wheel   **47.** (a) 1.9×10^{-3}; (b) 1.3×10^7 N/m²;
(c) 6.9×10^9 N/m²   **49.** 3.1 cm   **51.** 2.4×10^9 N/m²
53. (a) 1.8×10^7 N; (b) 1.4×10^7 N; (c) 16   **55.** (a) 867 N;
(b) 143 N; (c) 0.165

Chapter 14

CP   **1.** all tie   **2.** (a) 1, tie of 2 and 4, then 3; (b) line d
3. negative y direction   **4.** (a) increase; (b) negative
5. (a) 2; (b) 1   **6.** (a) path 1 (decreased E (more negative)
gives decreased a); (b) less than (decreased a gives decreased T)
Q   **1.** (a) between, closer to less massive particle; (b) no;
(c) no (other than infinity)   **3.** $3GM^2/d^2$, leftward   **5.** b, tie
of a and c, then d **7.** b, a, c **9.** (a) negative; (b) negative;
(c) postive; (d) all tie   **11.** (a) all tie; (b) all tie
13. (a) same; (b) greater   **EP**   **1.** 19 m   **3.** 2.16
5. 1/2   **7.** 3.4×10^5 km   **9.** (a) 3.7×10^{-5} N, increasing y **11.** $M = m$ **13.** 3.2×10^{-7} N   **15.** $(GmM/d^2) \times$

$$\left[1 - \frac{1}{8(1 - R/2d)^2}\right]$$ **17.** 2.6×10^6 m   **19.** (a) $1.3 \times$

10^{12} m/s²; (b) 1.6×10^6 m/s   **21.** (a) 17 N; (b) 2.5
23. (b) 1.9 h   **27.** (a) $a_g = (3.03 \times 10^{43}$ kg·m/s²$)/M_h$;
(b) decrease; (c) 9.82 m/s²; (d) 7.30×10^{-15} m/s²; (e) no
29. 7.91 km/s   **31.** (a) $(3.0 \times 10^{-7}m)$ N; (b) $(3.3 \times 10^{-7}m)$
N; (c) $(6.7 \times 10^{-7}mr)$ N   **33.** (a) 9.83 m/s²; (b) 9.84 m/s²;
(c) 9.79 m/s²   **35.** (a) -1.4×10^{-4} J; (b) less; (c) positive;
(d) negative   **37.** (a) 0.74; (b) 3.7 m/s²; (c) 5.0 km/s
39. (a) 0.0451; (b) 28.5   **41.** $-Gm(M_E/R + M_M/r)$
43. (a) 5.0×10^{-11} J; (b) -5.0×10^{-11} J   **45.** (a) 1700 m/s;
(b) 250 km; (c) 1400 m/s   **47.** (a) 2.2×10^{-7} rad/s;
(b) 90 km/s   **51.** (a) -1.67×10^{-8} J; (b) 0.56×10^{-8} J
55. 6.5×10^{23} kg   **57.** 5×10^{10} **59.** (a) 7.82 km/s;
(b) 87.5 min   **61.** (a) 6640 km; (b) 0.0136   **63.** (a) 39.5
AU³/M_S·y²; (b) $T^2 = r^3/M$ **65.** (a) 1.9×10^{13} m;
(b) $3.5R_P$ **67.** south, at 35.4° above the horizon
71. $2\pi r^{3/2}/\sqrt{G(M + m/4)}$ **73.** $\sqrt{GM/L}$ **75.** (a) 2.8 y;
(b) 1.0×10^{-4} **77.** (a) 1/2; (b) 1/2; (c) B, by 1.1×10^8 J
79. (a) 54 km/s; (b) 960 m/s; (c) $R_p/R_a = v_a/v_p$ **81.** (a) $4.6 \times$
10^5 J; (b) 260   **83.** (a) 7.5 km/s; (b) 97 min; (c) 410 km;
(d) 7.7 km/s; (e) 92 min; (f) 3.2×10^{-3} N; (g) if the satellite–
Earth system is considered isolated, its $\mathbf{L}$ is conserved
85. (a) 5540 s; (b) 7.68 km/s; (c) 7.60 km/s; (d) 5.78×10^{10} J;
(e) -11.8×10^{10} J; (f) -6.02×10^{10} J; (g) $6.63 \times$
10^6 m; (h) 170 s, new orbit   **87.** (a) $(-7.0$ mm$)\mathbf{i} +$
$(3.0$ cm$)\mathbf{j}$; (b) $(-0.19$ m/s$)\mathbf{i} + (0.40$ m/s$)\mathbf{j}$ **89.** (a) $1.98 \times$
10^{30} kg; (b) 1.96×10^{30} kg

Chapter 15

CP   **1.** all tie   **2.** (a) all tie; (b) $0.95\rho_0$, ρ_0, $1.1\rho_0$
3. 13 cm³/s, outward   **4.** (a) all tie; (b) 1, then 2 and 3 tie, 4;

(c) 4, 3, 2, 1   **Q** **1.** e, then b and d tie, then a and c tie
3. (a) 1, 3, 2; (b) all tie; (c) no (you must consider the weight exerted on the scale via the walls)   **5.** 3, 4, 1, 2
7. (a) downward; (b) downward; (c) same   **9.** (a) same;
(b) same; (c) lower; (d) higher   **11.** (a) block 1, counterclockwise; block 2, clockwise; (b) block 1, tip more; block 2, right itself   **EP** **1.** 1000 kg/m³   **3.** 1.1×10^5 Pa or 1.1 atm   **5.** 2.9×10^4 N   **7.** 6.0 lb/in.²   **9.** 1.90×10^4 Pa   **11.** 5.4×10^4 Pa   **13.** 0.52 m   **15.** (a) 6.06×10^9 N; (b) 20 atm   **17.** 0.412 cm   **19.** $\frac{1}{4}\rho g A(h_2 - h_1)^2$
21. 44 km   **23.** (a) $\rho g W D^2/2$; (b) $\rho g W D^3/6$; (c) $D/3$
25. (a) 2.2; (b) 2.4   **27.** -3.9×10^{-3} atm   **29.** (a) fA/a;
(b) 20 lb   **31.** 1070 g   **33.** 1.5 g/cm³   **35.** 600 kg/m³
37. (a) 670 kg/m³; (b) 740 kg/m³   **39.** 390 kg
41. (a) 1.2 kg; (b) 1300 kg/m³   **43.** 0.126 m³   **45.** five
47. (a) 1.80 m³; (b) 4.75 m³   **49.** 2.79 g/cm³
51. (a) 9.4 N; (b) 1.6 N   **53.** 4.0 m   **55.** 28 ft/s   **57.** 43 cm/s   **59.** (a) 2.40 m/s; (b) 245 Pa   **61.** (a) 12 ft/s; (b) 13 lb/in.²   **63.** 0.72 ft·lb/ft³   **65.** (a) 2; (b) $R_1/R_2 = \frac{1}{2}$; (c) drain it until $h_2 = h_1/4$ **67.** 116 m/s   **69.** (a) 6.4 m³; (b) 5.4 m/s; (c) 9.8×10^4 Pa   **71.** (a) 560 Pa; (b) 5.0×10^4 N
73. 40 m/s   **75.** (b) $H - h$; (c) $H/2$ **77.** (b) 0.69 ft³/s
79. (b) 63.3 m/s

Chapter 16

CP **1.** (a) $-x_m$; (b) $+x_m$; (c) 0   **2.** a   **3.** (a) 5 J; (b) 2 J; (c) 5 J   **4.** all tie (in Eq. 16-32, m is included in I)
5. 1, 2, 3 (the ratio m/b matters; k does not)   **Q** **1.** c
3. (a) 0; (b) between 0 and $+x_m$; (c) between $-x_m$ and 0;
(d) between $-x_m$ and 0   **5.** (a) toward $-x_m$; (b) toward $+x_m$;
(c) between $-x_m$ and 0; (d) between $-x_m$ and 0; (e) decreasing;
(f) increasing   **7.** (a) 3, 2, 1; (b) all tie   **9.** 3, 2, 1
11. system with spring A **13.** b (infinite period; does not oscillate), c, a **15.** (a) same; (b) same; (c) same; (d) smaller; (e) smaller; (f) and (g) larger ($T = \infty$)   **EP** **1.** (a) 0.50 s;
(b) 2.0 Hz; (c) 18 cm   **3.** (a) 245 N/m; (b) 0.284 s
5. 708 N/m   **7.** $f > 500$ Hz   **9.** (a) 100 N/m; (b) 0.45 s
11. (a) 6.28×10^5 rad/s; (b) 1.59 mm   **13.** (a) 1.0 mm;
(b) 0.75 m/s; (c) 570 m/s²   **15.** (a) 1.29×10^5 N/m;
(b) 2.68 Hz   **17.** (a) 4.0 s; (b) $\pi/2$ rad/s; (c) 0.37 cm;
(d) (0.37 cm) cos $\frac{\pi}{2}t$; (e) (-0.58 cm/s) sin $\frac{\pi}{2}t$; (f) 0.58 cm/s;
(g) 0.91 cm/s²; (h) 0; (i) 0.58 cm/s   **19.** (b) 12.47 kg;
(c) 54.43 kg   **21.** 1.6 kg   **23.** (a) 1.6 Hz; (b) 1.0 m/s, 0;
(c) 10 m/s², ±10 cm; (d) $(-10$ N/m$)x$ **25.** 22 cm
27. (a) 25 cm; (b) 2.2 Hz   **29.** (a) 0.500 m; (b) -0.251 m;
(c) 3.06 m/s   **31.** (a) 0.183A; (b) same direction
37. (a) $k_1 = (n + 1)k/n$, $k_2 = (n + 1)k$; (b) $f_1 = \sqrt{(n + 1)/n}f$, $f_2 = \sqrt{n + 1}f$ **39.** (b) 42 min   **41.** (a) 200 N/m;
(b) 1.39 kg; (c) 1.91 Hz   **43.** (a) 130 N/m; (b) 0.62 s; (c) 1.6 Hz; (d) 5.0 cm; (e) 0.51 m/s   **45.** (a) 3/4; (b) 1/4; (c) $x_m/\sqrt{2}$
47. (a) 3.5 m; (b) 0.75 s   **49.** (a) 0.21 m; (b) 1.6 Hz;
(c) 0.10 m   **51.** (a) 0.0625 J; (b) 0.03125 J   **53.** 12 s
55. (a) 39.5 rad/s; (b) 34.2 rad/s; (c) 124 rad/s²   **57.** (a) 8.3 s;
(b) no   **59.** 9.47 m/s²   **61.** 8.77 s   **63.** 5.6 cm
65. $2\pi\sqrt{(R^2 + 2d^2)/2gd}$ **67.** (a) 0.205 kg·m²; (b) 47.7 cm;
(c) 1.50 s   **71.** (a) $2\pi\sqrt{(L^2 + 12x^2)/12gx}$; (b) 0.289 m

73. 9.78 m/s²   **75.** $2\pi\sqrt{m/3k}$ **77.** $(1/2\pi)(\sqrt{g^2 + v^4/R^2}/L)^{1/2}$
79. (b) smaller   **81.** (a) 2.0 s; (b) 18.5 N·m/rad
83. 0.29L **85.** 0.39   **87.** (a) 0.102 kg/s; (b) 0.137 J
89. $k = 490$ N/cm, $b = 1100$ kg/s   **91.** 1.9 in.
93. (a) $y_m = 8.8 \times 10^{-4}$ m, $T = 0.18$ s, $\omega = 35$ rad/s;
(b) $y_m = 5.6 \times 10^{-2}$ m, $T = 0.48$ s, $\omega = 13$ rad/s;
(c) $y_m = 3.3 \times 10^{-2}$ m, $T = 0.31$ s, $\omega = 20$ rad/s

Chapter 17

CP **1.** a, 2; b, 3; c, 1   **2.** (a) 2, 3, 1; (b) 3, then 1 and 2 tie
3. a **4.** 0.20 and 0.80 tie, then 0.60, 0.45   **5.** (a) 1; (b) 3;
(c) 2   **6.** (a) 75 Hz; (b) 525 Hz   **Q** **1.** $7d$ **3.** tie of A and B, then C, D **5.** intermediate (closer to fully destructive interference)   **7.** a and d tie, then b and c tie   **9.** (a) 8;
(b) antinode; (c) longer; (d) lower   **11.** (a) integer multiples of 3; (b) node; (c) node   **13.** string A **15.** decrease
EP **1.** (a) 75 Hz; (b) 13 ms   **3.** (a) 7.5×10^{14} to 4.3×10^{14} Hz; (b) 1.0 to 200 m; (c) 6.0×10^{16} to 3.0×10^{19} Hz
5. $y = 0.010 \sin \pi(3.33x + 1100t)$, with x and y in m and t in s   **11.** (a) $z = 3.0 \sin(60y - 10\pi t)$, with z in mm, y in cm, and t in s; (b) 9.4 cm/s   **13.** (a) $y = 2.0 \sin 2\pi(0.10x - 400t)$, with x and y in cm and t in s; (b) 50 m/s; (c) 40 m/s
15. (b) 2.0 cm/s; (c) $y = (4.0$ cm$) \sin(\pi x/10 - \pi t/5 + \pi)$, where x is in cm and t is in s; (d) -2.5 cm/s   **17.** 129 m/s
19. 135 N   **23.** (a) 15 m/s; (b) 0.036 N   **25.** $y = 0.12 \sin(141x + 628t)$, with y in mm, x in m, and t in s
27. (a) 5.0 cm; (b) 40 cm; (c) 12 m/s; (d) 0.033 s; (e) 9.4 m/s;
(f) $5.0 \sin(16x + 190t + 0.79)$, with x in m, y in cm, and t in s
29. (a) $v_1 = 28.6$ m/s, $v_2 = 22.1$ m/s; (b) $M_1 = 188$ g, $M_2 = 313$ g   **31.** (a) $\sqrt{k(\Delta l)(l + \Delta l)/m}$ **33.** (a) $P_2 = 2P_1$;
(b) $P_2 = P_1/4$ **35.** (a) 3.77 m/s; (b) 12.3 N; (c) zero; (d) 46.3 W; (e) zero; (f) zero; (g) ±0.50 cm   **37.** 82.8°, 1.45 rad, 0.23 wavelength   **39.** 5.0 cm   **41.** (a) 4.4 mm; (b) 112°
43. (a) $0.83y_1$; (b) 37°   **45.** (a) $2f_3$; (b) λ_3 **47.** 10 cm
49. (a) 82.0 m/s; (b) 16.8 m/s; (c) 4.88 Hz   **51.** 240 cm, 120 cm, 80 cm   **53.** 7.91 Hz, 15.8 Hz, 23.7 Hz   **55.** $f_{1A} = f_{4B}$, $f_{2A} = f_{8B}$ **57.** (a) 2.0 Hz, 200 cm, 400 cm/s; (b) $x = 50$ cm, 150 cm, 250 cm, etc.; (c) $x = 0$, 100 cm, 200 cm, etc.
63. (a) 1.3 m; (b) $y' = 0.002 \sin(9.4x) \cos(3800t)$, with x and y in m and t in s   **67.** (b) in the positive x direction; interchange the amplitudes of the original two traveling waves; (c) largest at $x = \lambda/4 = 6.26$ cm; smallest at $x = 0$ and $x = \lambda/2 = 12.5$ cm; (d) the largest amplitude is 4.0 mm, which is the sum of the amplitudes of the original traveling waves; the smallest amplitude is 1.0 mm, which is the difference of the amplitudes of the original traveling waves

Chapter 18

CP **1.** beginning to decrease (example: mentally move the curves of Fig. 18-6 rightward past the point at $x = 42$ m)
2. (a) fully constructive, 0; (b) fully constructive, 4   **3.** (a) 1 and 2 tie, then 3; (b) 3, then 1 and 2 tie   **4.** second
5. loosen   **6.** a, greater; b, less; c, can't tell; d, can't tell; e, greater; f, less   **7.** (a) 222 m/s; (b) $+20$ m/s   **Q** **1.** pulse

along path 2   **3.** (a) 2.0 wavelengths; (b) 1.5 wavelengths; (c) fully constructive, fully destructive   **5.** (a) exactly out of phase; (b) exactly out of phase   **7.** 70 dB   **9.** (a) two; (b) antinode   **11.** all odd harmonics   **13.** 501, 503, and 508 Hz; or 505, 507, and 508 Hz   **EP 1.** (a) $\approx 6\%$ **3.** the radio listener by about 0.85 s   **5.** 7.9×10^{10} Pa   **7.** If only the length is uncertain, it must be known to within 10^{-4} m. If only the time is imprecise, the uncertainty must be no more than one part in 10^8.   **9.** 43.5 m   **11.** 40.7 m   **13.** 100 kHz   **15.** (a) 2.29, 0.229, 22.9 kHz; (b) 1.14, 0.114, 11.4 kHz   **17.** (a) 6.0 m/s; (b) $y = 0.30 \sin(\pi x/12 + 50\pi t)$, with x and y in cm and t in s   **19.** 4.12 rad   **21.** (a) $343 \times (1 + 2m)$ Hz, with m being an integer from 0 to 28; (b) $686m$ Hz, with m being an integer from 1 to 29   **23.** (a) eight; (b) eight   **25.** 64.7 Hz, 129 Hz   **27.** (a) 0.080 W/m^2; (b) 0.013 W/m^2 **29.** 36.8 nm   **31.** (a) 1000; (b) 32   **33.** (a) 39.7 μW/m^2; (b) 171 nm; (c) 0.893 Pa   **35.** (a) 59.7; (b) 2.81×10^{-4} **37.** $s_m \propto r^{-1/2}$ **39.** (a) 5000; (b) 71; (c) 71   **41.** 171 m   **43.** 3.16 km   **45.** (a) 5200 Hz; (b) amplitude$_{SAD}$/amplitude$_{SBD}$ = 2   **47.** 20 kHz   **49.** by a factor of 4   **51.** water filled to a height of $\frac{7}{8}, \frac{5}{8}, \frac{3}{8}, \frac{1}{8}$ m   **53.** (a) 5.0 cm from one end; (b) 1.2; (c) 1.2   **55.** (a) 1130, 1500, and 1880 Hz   **57.** (a) 230 Hz; (b) higher   **59.** (a) node; (c) 22 s   **61.** 387 Hz   **63.** 0.02   **65.** 3.8 Hz   **67.** (a) 380 mi/h, away from owner; (b) 77 mi/h, away from owner   **69.** 15.1 ft/s   **71.** 2.6×10^8 m/s   **73.** (a) 77.6 Hz; (b) 77.0 Hz   **75.** 33.0 km   **79.** (a) 970 Hz; (b) 1030 Hz; (c) 60 Hz, no   **81.** (a) 1.02 kHz; (b) 1.04 kHz   **83.** 1540 m/s   **85.** 41 kHz   **87.** (a) 2.0 kHz; (b) 2.0 kHz   **89.** (a) 485.8 Hz; (b) 500.0 Hz; (c) 486.2 Hz; (d) 500.0 Hz   **91.** 1×10^6 m/s, receding   **93.** 0.13c

Chapter 19

CP 1. (a) all tie; (b) 50°X, 50°Y, 50°W   **2.** (a) 2 and 3 tie, then 1, then 4; (b) 3, 2, then 1 and 4 tie   **3.** A **4.** c and e **5.** (a) zero; (b) negative   **6.** b and d tie, then a, c **Q 1.** 25 S°, 25 U°, 25 R°   **3.** c, then the rest tie   **5.** B, then A and C tie   **7.** (a) both clockwise; (b) both clockwise   **9.** c, a, b **11.** upward (with liquid water on the exterior and at the bottom, $\Delta T = 0$ horizontally and downward)   **13.** at the temperature of your fingers   **15.** 3, 2, 1   **EP 1.** 2.71 K   **3.** 0.05 kPa, nitrogen   **5.** (a) 320°F; (b) -12.3°F   **7.** (a) -96°F; (b) 56.7°C   **9.** (a) -40°; (b) 575°; (c) Celsius and Kelvin cannot give the same reading   **11.** (a) Dimensions are inverse time   **13.** 4.4×10^{-3} cm   **15.** 0.038 in.   **17.** (a) 9.996 cm; (b) 68°C   **19.** 170 km   **21.** 0.32 cm^2 **23.** 29 cm^3 **25.** 0.432 cm^3 **27.** -157°C   **29.** 360°C   **35.** $+0.68$ s/h   **37.** (b) use 39.3 cm of steel and 13.1 cm of brass   **39.** (a) 523 J/kg·K; (b) 0.600; (c) 26.2 J/mol·K   **41.** 94.6 L   **43.** 109 g   **45.** 1.30 MJ   **47.** 1.9 times as great   **49.** (a) 33.9 Btu; (b) 172 F°   **51.** (a) 52 MJ; (b) 0°C   **53.** (a) 411 g; (b) 3.1¢   **55.** 0.41 kJ/kg·K   **57.** 3.0 min   **59.** 73 kW   **61.** 2.17 g   **63.** 33 m^2 **65.** 33 g   **67.** (a) 0°C; (b) 2.5°C   **69.** 2500 J/kg·K   **71.** A: 120 J, B: 75 J, C: 30 J   **73.** (a) -200 J; (b) -293 J;

(c) -93 J   **75.** -5.0 J   **77.** 33.3 kJ   **79.** 766°C   **81.** (a) 1.2 W/m·K, 0.70 Btu/ft·F°·h; (b) 0.030 ft^2·F°·h/Btu   **83.** 1660 J/s   **87.** arrangement b **89.** (a) 2.0 MW; (b) 220 W   **91.** (a) 17 kW/m^2; (b) 18 W/m^2 **93.** -6.1 nW   **95.** 0.40 cm/h   **97.** Cu-Al, 84.3°C; Al-brass, 57.6°C

Chapter 20

CP 1. all but c **2.** (a) all tie; (b) 3, 2, 1   **3.** gas A **4.** 5 (greatest change in T), then tie of 1, 2, 3, and 4   **5.** 1, 2, 3 ($Q_3 = 0$, Q_2 goes into work W_2, but Q_1 goes into greater work W_1 and increases gas temperature)   **Q 1.** increased but less than doubled   **3.** a, c, b   **5.** 1180 J   **7.** d, tie of a and b, then c   **9.** constant-volume process   **11.** (a) same; (b) increases; (c) decreases; (d) increases   **13.** -4 J   **15.** (a) 1, polyatomic; 2, diatomic; 3, monatomic; (b) more   **EP 1.** (a) 0.0127; (b) 7.65×10^{21} **3.** 6560   **5.** number of molecules in the ink $\approx 3 \times 10^{16}$; number of people $\approx 5 \times 10^{20}$; statement is wrong, by a factor of about 20,000   **7.** (a) 5.47×10^{-8} mol; (b) 3.29×10^{16} **9.** (a) 106; (b) 0.892 m^3 **11.** 27.0 lb/in.^{2} **13.** (a) 2.5×10^{25}; (b) 1.2 kg   **15.** 5600 J   **17.** 1/5   **19.** (a) -45 J; (b) 180 K   **21.** 100 cm^3 **23.** 198°F   **25.** 2.0×10^5 Pa   **27.** 180 m/s   **29.** 9.53×10^6 m/s   **31.** 313°C   **33.** 1.9 kPa   **35.** (a) 0.0353 eV, 0.0483 eV; (b) 3400 J, 4650 J   **37.** 9.1×10^{-6} **39.** (a) 6.75×10^{-20} J; (b) 10.7   **41.** 0.32 nm   **43.** 15 cm   **45.** (a) 3.27×10^{10}; (b) 172 m   **47.** (a) 22.5 L; (b) 2.25; (c) 8.4×10^{-5} cm; (d) same as (c)   **51.** (a) 3.2 cm/s; (b) 3.4 cm/s; (c) 4.0 cm/s   **53.** (a) v_P, v_{rms}, $\bar{v}$ (b) reverse ranking   **55.** (a) 1.0×10^4 K, 1.6×10^5 K; (b) 440 K, 7000 K   **57.** 4.7   **59.** (a) $2N/3v_0$; (b) $N/3$; (c) $1.22v_0$; (d) $1.31v_0$ **61.** $RT \ln(V_f/V_i)$ **63.** (a) 15.9 J; (b) 34.4 J/mol·K; (c) 26.1 J/mol·K   **65.** $(n_1C_1 + n_2C_2 + n_3C_3)/(n_1 + n_2 + n_3)$ **67.** (a) -5.0 kJ; (b) 2.0 kJ; (c) 5.0 kJ   **69.** (a) 0.375 mol; (b) 1090 J; (c) 0.714   **71.** (a) 14 atm; (b) 620 K   **79.** 0.63   **81.** (a) monatomic; (b) 2.7×10^4 K; (c) 4.5×10^4 mol; (d) 3.4 kJ, 340 kJ; (e) 0.01   **83.** 5 m^3 **85.** (a) in joules, in the order Q, ΔE_{int}, W: $1 \rightarrow 2$: 3740, 3740, 0; $2 \rightarrow 3$: 0, -1810, 1810; $3 \rightarrow 1$: -3220, -1930, -1290; cycle: 520, 0, 520; (b) $V_2 = 0.0246$ m^3, $p_2 = 2.00$ atm, $V_3 = 0.0373$ m^3, $p_3 = 1.00$ atm

Chapter 21

CP 1. a, b, c   **2.** smaller   **3.** c, b, a   **4.** a, d, c, b   **5.** b **Q 1.** increase   **3.** (a) increase; (b) same   **5.** equal   **7.** lower the temperature of the low temperature reservoir   **9.** (a) same; (b) increase; (c) decrease   **11.** (a) same; (b) increase; (c) decrease   **13.** more than the age of the universe   **EP 1.** 1.86×10^4 J   **3.** 2.75 mol   **7.** (a) 5.79×10^4 J; (b) 173 J/K   **9.** $+3.59$ J/K   **11.** (a) 14.6 J/K; (b) 30.2 J/K   **13.** (a) 4.45 J/K; (b) no   **15.** (a) 4500 J; (b) -5000 J; (c) 9500 J   **17.** (a) 57.0°C; (b) -22.1 J/K; (c) $+24.9$ J/K; (d) $+2.8$ J/K   **19.** (a) -710 mJ/K; (b) $+710$ mJ/K; (c) $+723$ mJ/K;

(d) -723 mJ/K; (e) $+13$ mJ/K; (f) 0   **23.** (a) (I) constant T, $Q = pV \ln 2$; constant V, $Q = 4.5pV$; (II) constant T, $Q = -pV \ln 2$; constant p, $Q = 7.5pV$; (b) (I) constant T, $W = pV \ln 2$; constant V, $W = 0$; (II) constant T, $W = -pV \ln 2$; constant p, $W = 3pV$; (c) $4.5pV$ for either case; (d) $4R \ln 2$ for either case   **25.** 0.75 J/K   **27.** (a) -943 J/K; (b) $+943$ J/K; (c) yes   **29.** (a) $3p_0V_0$; (b) $6RT_0$, $(3/2)R \ln 2$; (c) both are zero   **33.** (a) 31%; (b) 16 kJ   **35.** engine A, first; engine B, first and second; engine C, second; engine D, neither   **37.** 97 K   **39.** 99.99995%   **41.** 7.2 J/cycle   **43.** (a) 7200 J; (b) 960 J; (c) 13%   **45.** (a) 2270 J; (b) 14,800 J; (c) 15.4%; (d) 75.0%, greater   **49.** (a) 78%; (b) 81 kg/s   **55.** (a) 49 kJ; (b) 7.4 kJ   **57.** 21 J   **59.** (a) 0.071 J; (b) 0.50 J; (c) 2.0 J; (d) 5.0 J   **61.** 1.08 MJ   **63.** $[1 - (T_2/T_1)] \div [1 - (T_4/T_3)]$ **67.** (a) 1.27×10^{30}; (b) 7.9%; (c) 7.3%; (d) 7.3%; (e) 1.1%; (f) 0.0023%   **69.** (a) $W = N!/(n_1! \, n_2! \, n_3!)$; (b) $[(N/2)! \, (N/2)!]/[(N/3)! \, (N/3)! \, (N/3)!]$; (c) 4.2×10^{16}

Chapter 22

CP   **1.** C and D attract; B and D attract   **2.** (a) leftward; (b) leftward; (c) leftward   **3.** (a) a, c, b; (b) less than   **4.** $-15e$ (net charge of $-30e$ is equally shared)   **Q**   **1.** No, only for charged particles, charged particle-like objects, and spherical shells (including solid spheres) of uniform charge   **3.** a and b **5.** two points: one to the left of the particles and one between the protons   **7.** $6q^2/4\pi\epsilon_0 d^2$, leftward   **9.** (a) same; (b) less than; (c) cancel; (d) add; (e) the adding components; (f) positive direction of y; (g) negative direction of y; (h) positive direction of x; (i) negative direction of x **11.** (a) A, B, and D; (b) all four; (c) Connect A and D; disconnect them; then connect one of them to B. (There are two more solutions.)   **13.** (a) possibly; (b) definitely   **15.** same   **17.** D **EP**   **1.** 0.50 C   **3.** 2.81 N on each   **5.** (a) 4.9×10^{-7} kg; (b) 7.1×10^{-11} C   **7.** $3F/8$ **9.** (a) 1.60 N; (b) 2.77 N   **11.** (a) $q_1 = 9q_2$; (b) $q_1 = -25q_2$ **13.** either -1.00 μC and $+3.00$ μC or $+1.00$ μC and -3.00 μC   **15.** (a) 36 N, $-10°$ from the x axis; (b) $x = -8.3$ cm, $y = +2.7$ cm   **17.** (a) 5.7×10^{13} C, no; (b) 6.0×10^5 kg   **19.** (a) $Q = -2\sqrt{2}q$; (b) no   **21.** 3.1 cm   **23.** 2.89×10^{-9} N   **25.** -1.32×10^{13} C   **27.** (a) 3.2×10^{-19} C; (b) two   **29.** (a) 8.99×10^{-19} N; (b) 625   **31.** 5.1 m below the electron   **33.** 1.3 days   **35.** (a) 0; (b) 1.9×10^{-9} N   **37.** 10^{18} N   **39.** (a) ^{9}B; (b) ^{13}N; (c) ^{12}C   **41.** (a) $F = (Q^2/4\pi\epsilon_0 d^2)\alpha(1 - \alpha)$; (c) 0.5; (d) 0.15 and 0.85

Chapter 23

CP   **1.** (a) rightward; (b) leftward; (c) leftward; (d) rightward (p and e have same charge magnitude, and p is farther)   **2.** all tie   **3.** (a) toward positive y; (b) toward positive x; (c) toward negative y **4.** (a) leftward; (b) leftward; (c) decrease   **5.** (a) all tie; (b) 1 and 3 tie, then 2 and 4 tie   **Q**   **1.** (a) toward positive x; (b) downward and to the right;

(c) A **3.** two points: one to the left of the particles, the other between the protons   **5.** (a) yes; (b) toward; (c) no (the field vectors are not along the same line); (d) cancel; (e) add; (f) adding components; (g) toward negative y **7.** (a) 3, then 1 and 2 tie (zero); (b) all tie; (c) 1 and 2 tie, then 3   **9.** (a) rightward; (b) $+q_1$ and $-q_3$, increase; q_2, decrease; n, same   **11.** a, b, c **13.** (a) 4, 3, 1, 2; (b) 3, then 1 and 4 tie, then 2   **EP**   **1.** (a) 6.4×10^{-18} N; (b) 20 N/C   **3.** to the right in the figure   **7.** 56 pC   **9.** 3.07×10^{21} N/C, radially outward   **13.** (a) $q/8\pi\epsilon_0 d^2$, to the left; $3q/\pi\epsilon_0 d^2$, to the right; $7q/16\pi\epsilon_0 d^2$, to the left   **15.** 0   **17.** 9:30   **19.** $E = q/\pi\epsilon_0 a^2$, along bisector, away from triangle   **21.** $7.4q/4\pi\epsilon_0 d^2$, $28°$ counterclockwise to $+x$ **23.** 6.88×10^{-28} C$\cdot$m   **25.** $(1/4\pi\epsilon_0)(p/r^3)$, antiparallel to **p**   **29.** $R/\sqrt{2}$ **31.** $(1/4\pi\epsilon_0)(4q/\pi R^2)$, toward decreasing y **37.** (a) 0.10 μC; (b) 1.3×10^{17}; (c) 5.0×10^{-6} **39.** 3.51×10^{15} m/s^2 **41.** 6.6×10^{-15} N   **43.** 2.03×10^{-7} N/C, up   **45.** (a) -0.029 C; (b) repulsive forces would explode the sphere   **47.** (a) 1.92×10^{12} m/s^2; (b) 1.96×10^5 m/s   **49.** (a) 8.87×10^{-15} N; (b) 120   **51.** 1.64×10^{-19} C ($\approx 3\%$ high)   **53.** (a) 0.245 N, $11.3°$ clockwise from the $+x$ axis; (b) $x = 108$ m, $y = -21.6$ m   **55.** 27μm   **57.** (a) yes; (b) upper plate, 2.73 cm   **59.** (a) 0; (b) 8.5×10^{-22} N$\cdot$m; (c) 0   **61.** $(1/2\pi)\sqrt{pE/I}$ **63.** (a) $E = (2q/4\pi\epsilon_0 d^2)(\alpha/(1 + \alpha^2)^{3/2})$; (c) 0.707; (d) 0.21 and 1.9

Chapter 24

CP   **1.** (a) $+EA$; (b) $-EA$; (c) 0; (d) 0   **2.** (a) 2; (b) 3; (c) 1   **3.** (a) equal; (b) equal; (c) equal   **4.** (a) $+50e$; (b) $-150e$ **5.** 3 and 4 tie, then 2, 1   **Q**   **1.** (a) 8 N$\cdot$m^2/C; (b) 0   **3.** (a) all tie (zero); (b) all tie   **5.** $+13q/\epsilon_0$ **7.** all tie   **9.** all tie   **11.** 2σ, σ, 3σ; or 3σ, σ, 2σ **13.** (a) all tie ($E = 0$); (b) all tie   **15.** (a) same ($E = 0$); (b) decrease; (c) decrease (to zero); (d) same   **EP**   **1.** (a) 693 kg/s; (b) 693 kg/s; (c) 347 kg/s; (d) 347 kg/s; (e) 575 kg/s   **3.** (a) 0; (b) -3.92 N$\cdot$m^2/C; (c) 0; (d) 0 for each field   **5.** (a) enclose $2q$ and $-2q$, or enclose all four charges; (b) enclose $2q$ and q; (c) not possible   **7.** 2.0×10^5 N$\cdot$m^2/C   **9.** $q/6\epsilon_0$ **11.** (a) $-\pi R^2 E$; (b) $\pi R^2 E$ **13.** -4.2×10^{-10} C   **15.** 0 through each of the three faces meeting at q, $q/24\epsilon_0$ through each of the other faces   **17.** 2.0 μC/m^2 **19.** (a) 4.5×10^{-7} C/m^2; (b) 5.1×10^4 N/C   **21.** (a) -3.0×10^{-6} C; (b) $+1.3 \times 10^{-5}$ C   **23.** (a) 0.32 μC; (b) 0.14 μC   **27.** (a) $E = q/2\pi\epsilon_0 Lr$, radially inward; (b) $-q$ on both inner and outer surfaces; (c) $E = q/2\pi\epsilon_0 Lr$, radially outward   **29.** 3.6 nC   **31.** (b) $\rho R^2/2\epsilon_0 r$ **33.** (a) 5.3×10^7 N/C; (b) 60 N/C   **35.** 5.0 nC/m^2 **37.** 0.44 mm   **39.** (a) 4.9×10^{-22} C/m^2; (b) downward   **41.** (a) $\rho x/\epsilon_0$; (b) $\rho d/2\epsilon_0$ **43.** (a) -750 N$\cdot$m^2/C; (b) -6.64 nC   **45.** (a) 4.0×10^6 N/C; (b) 0   **47.** (a) 0; (b) $q_a/4\pi\epsilon_0 r^2$; (c) $(q_a + q_b)/4\pi\epsilon_0 r^2$ **51.** (a) $-q$; (b) $+q$; (c) $E = q/4\pi\epsilon_0 r^2$, radially outward; (d) $E = 0$; (e) $E = q/4\pi\epsilon_0 r^2$, radially outward; (f) 0; (g) $E = q/4\pi\epsilon_0 r^2$, radially outward; (h) yes, charge is induced; (i) no; (j) yes; (k) no; (l) no   **53.** (a) $E = (q/4\pi\epsilon_0 a^3)r$; (b) $E = q/4\pi\epsilon_0 r^2$; (c) 0; (d) 0; (e) inner, $-q$; outer, 0   **55.** $q/2\pi a^2$ **59.** $\alpha = 0.80$

Chapter 25

CP **1.** (a) negative; (b) increase **2.** (a) positive; (b) higher **3.** (a) rightward; (b) 1, 2, 3, 5: positive; 4, negative; (c) 3, then 1, 2, and 5 tie, then 4 **4.** all tie **5.** a, c (zero), b **6.** (a) 2, then 1 and 3 tie; (b) 3; (c) accelerate leftward **7.** closer (half of 9.23 fm) **Q** **1.** (a) higher; (b) positive; (c) negative; (d) all tie **3.** (a) 1 and 2; (b) none; (c) no; (d) 1 and 2 yes, 3 and 4 no **5.** b, then a, c, and d tie **7.** (a) negative; (b) zero **9.** (a) 1, then 2 and 3 tie; (b) 3 **11.** left **13.** a, b, c **15.** (a) c, b, a; (b) zero **17.** (a) positive; (b) positive; (c) negative; (d) all tie **19.** (a) no; (b) yes **21.** no (a particle at the intersection would have two different potential energies) **23.** (a)–(b) all tie; (c) C, B, A; (d) all tie **EP** **1.** 1.2 GeV **3.** (a) 3.0×10^{10} J; (b) 7.7 km/s; (c) 9.0×10^4 kg **7.** 2.90 kV **9.** 8.8 mm **11.** (a) 136 MV/m; (b) 8.82 kV/m **13.** (b) because $V = 0$ point is chosen differently; (c) $q/(8\pi\epsilon_0 R)$; (d) potential differences are independent of the choice for the $V = 0$ point **15.** (a) -4500 V; (b) -4500 V **17.** 843 V **19.** 2.8×10^5 **21.** $x = d/4$ and $x = -d/2$ **23.** none **25.** (a) 3.3 nC; (b) 12 nC/m^2 **27.** 6.4×10^8 V **29.** 190 MV **31.** (a) -4.8 nm; (b) 8.1 nm; (c) no **33.** 16.3 μV **35.** (a) $\dfrac{2\lambda}{4\pi\epsilon_0} \ln\left[\dfrac{L/2 + (L^2/4 + d^2)^{1/2}}{d}\right]$; (b) 0 **37.** (a) $-5Q/4\pi\epsilon_0 R$; (b) $-5Q/4\pi\epsilon_0(z^2 + R^2)^{1/2}$ **39.** $0.113\sigma R/\epsilon_0$ **41.** $(Q/4\pi\epsilon_0 L) \ln(1 + L/d)$ **43.** 670 V/m **45.** $p/2\pi\epsilon_0 r^3$ **47.** 39 V/m, $-x$ direction **51.** (a) $\dfrac{c}{4\pi\epsilon_0}[\sqrt{L^2 + y^2} - y]$; (b) $\dfrac{c}{4\pi\epsilon_0}\left[1 - \dfrac{y}{\sqrt{L^2 + y^2}}\right]$ **53.** (a) 2.5 MV; (b) 5.1 J; (c) 6.9 J **55.** (a) 2.72×10^{-14} J; (b) 3.02×10^{-31} kg, about $\frac{1}{3}$ of accepted value **57.** (a) 0.484 MeV; (b) 0 **59.** 2.1 d **61.** 0 **63.** (a) 27.2 V; (b) -27.2 eV; (c) 13.6 eV; (d) 13.6 eV **65.** 1.8×10^{-10} J **67.** 1.48×10^7 m/s **69.** $qQ/4\pi\epsilon_0 K$ **71.** 0.32 km/s **73.** 1.6×10^{-9} m **77.** (a) $V_1 = V_2$; (b) $q_1 = q/3$, $q_2 = 2q/3$; (c) 2 **79.** (a) -0.12 V; (b) 1.8×10^{-8} N/C, radially inward **81.** (a) 12,000 N/C; (b) 1800 V; (c) 5.8 cm **83.** (c) 4.24 V

Chapter 26

CP **1.** (a) same; (b) same **2.** (a) decreases; (b) increases; (c) decreases **3.** (a) V, $q/2$; (b) $V/2$, q **4.** (a) $q_0 = q_1 + q_{34}$; (b) equal (C_3 and C_4 are in series) **5.** (a) same; (b)–(d) increase; (e) same (same potential difference across same plate separation) **6.** (a) same; (b) decrease; (c) increase **Q** **1.** a, 2; b, 1; c, 3 **3.** (a) increase; (b) same **5.** (a) parallel; (b) series **7.** (a) $C/3$; (b) $3C$; (c) parallel **9.** (a) equal; (b) less **11.** (a)–(d) less **13.** (a) 2; (b) 3; (c) 1 **15.** Increase plate separation d, but also plate area A, keeping A/d constant. **EP** **1.** 7.5 pC **3.** 3.0 mC **5.** (a) 140 pF; (b) 17 nC **7.** (a) 84.5 pF; (b) 191 cm^2 **9.** (a) 11 cm^2; (b) 11 pF; (c) 1.2 V **13.** (b) 4.6×10^{-5}/K **15.** 7.33 μF **17.** 315 mC **19.** (a) 10.0 μF; (b) $q_2 = 0.800$ mC, $q_1 = 1.20$ mC; (c) 200 V for both **21.** (a) $d/3$; (b) $3d$ **25.** (a) five in series; (b) three

arrays as in (a) in parallel (and other possibilities) **27.** 43 pF **29.** (a) 50 V; (b) 5.0×10^{-5} C; (c) 1.5×10^{-4} C **31.** (a) $q_1 = 9.0$ μC, $q_2 = 16$ μC, $q_3 = 9.0$ μC, $q_4 = 16$ μC; (b) $q_1 = 8.4$ μC, $q_2 = 17$ μC, $q_3 = 11$ μC, $q_4 = 14$ μC **33.** 99.6 nJ **35.** 72 F **37.** 4.9% **39.** 0.27 J **41.** 0.11 J/m^3 **43.** (a) 2.0 J **45.** (a) $q_1 = 0.21$ mC, $q_2 = 0.11$ mC, $q_3 = 0.32$ mC; (b) $V_1 = V_2 = 21$ V, $V_3 = 79$ V; (c) $U_1 = 2.2$ mJ, $U_2 = 1.1$ mJ, $U_3 = 13$ mJ **47.** (a) $q_1 = q_2 = 0.33$ mC, $q_3 = 0.40$ mC; (b) $V_1 = 33$ V, $V_2 = 67$ V, $V_3 = 100$ V; (c) $U_1 = 5.6$ mJ, $U_2 = 11$ mJ, $U_3 = 20$ mJ **53.** Pyrex **55.** (a) 6.2 cm; (b) 280 pF **57.** 0.63 m^2 **59.** (a) 2.85 m^3; (b) 1.01×10^4 **61.** (a) $\epsilon_0 A/(d-b)$; (b) $d/(d-b)$; (c) $-q^2 b/2\epsilon_0 A$, sucked in **65.** $\dfrac{\epsilon_0 A}{4d}\left(\kappa_1 + \dfrac{2\kappa_2\kappa_3}{\kappa_2 + \kappa_3}\right)$ **67.** (a) 13.4 pF; (b) 1.15 nC; (c) 1.13×10^4 N/C; (d) 4.33×10^3 N/C **69.** (a) 7.1; (b) 0.77 μC **71.** (a) 0.606; (b) 0.394

Chapter 27

CP **1.** 8 A, rightward **2.** (a)–(c) rightward **3.** a and c tie, then b **4.** Device 2 **5.** (a) and (b) tie, then (d), then (c) **Q** **1.** a, b, and c tie, then d (zero) **3.** b, a, c **5.** tie of A, B, and C, then a tie of $A + B$ and $B + C$, then $A + B + C$ **7.** (a)–(c) 1 and 2 tie, then 3 **9.** C, A, B **11.** b, a, c **13.** (a) conductors: 1 and 4; semiconductors: 2 and 3; (b) 2 and 3; (c) all four **EP** **1.** 1.25×10^{15} **3.** 6.7 μC/m^2 **5.** 14-gauge **7.** (a) 2.4×10^{-5} A/m^2; (b) 1.8×10^{-15} m/s **9.** 0.67 A, toward the negative terminal **11.** (a) 0.654 μA/m^2; (b) 83.4 MA **13.** 13 min **15.** (a) $J_0 A/3$; (b) $2J_0 A/3$ **17.** 2.0×10^{-8} $\Omega \cdot$m **19.** 100 V **21.** (a) 1.53 kA; (b) 54.1 MA/m^2; (c) 10.6×10^{-8} $\Omega \cdot$m, platinum **23.** (a) 253°C; (b) yes **25.** (a) 0.38 mV; (b) negative; (c) 3 min 58 s **27.** 54 Ω **29.** 3.0 **31.** (a) 1.3 mΩ; (b) 4.6 mm **33.** (a) 6.00 mA; (b) 1.59×10^{-8} V; (c) 21.2 nΩ **35.** 2000 K **37.** (a) copper: 5.32×10^5 A/m^2, aluminum: 3.27×10^5 A/m^2; (b) copper: 1.01 kg/m, aluminum: 0.495 kg/m **39.** 0.40 Ω **41.** (a) $R = \rho L/\pi ab$ **43.** 14 kC **45.** 11.1 Ω **47.** (a) 1.0 kW; (b) 25¢ **49.** 0.135 W **51.** (a) 1.74 A; (b) 2.15 MA/m^2; (c) 36.3 mV/m; (d) 2.09 W **53.** (a) 1.3×10^5 A/m^2; (b) 94 mV **55.** (a) $4.46 for a 31-day month; (b) 144 Ω; (c) 0.833 A **57.** 660 W **59.** (a) 3.1×10^{11}; (b) 25 μA; (c) 1300 W, 25 MW **61.** 27 cm/s **63.** (a) 120 Ω; (b) 107 Ω; (c) 5.3×10^{-3}/C°; (d) 5.9×10^{-3}/C°; (e) 276 Ω

Chapter 28

CP **1.** (a) rightward; (b) all tie; (c) b, then a and c tie; (d) b, then a and c tie **2.** (a) all tie; (b) R_1, R_2, R_3 **3.** (a) less; (b) greater; (c) equal **4.** (a) $V/2$, i; (b) V, $i/2$ **5.** (a) 1, 2, 4, 3; (b) 4, tie of 1 and 2; then 3 **Q** **1.** 3, 4, 1, 2 **3.** (a) no; (b) yes; (c) all tie (the circuits are the same) **5.** parallel, R_2, R_1, series **7.** (a) equal; (b) more **9.** (a) less; (b) less; (c) more **11.** C_1, 15 V; C_2, 35 V; C_3, 20 V; C_4, 20 V; C_5, 30 V **13.** 60 μC **15.** c, b, a **17.** (a) all tie; (b) 1, 3, 2 **19.** 1, 3, and 4 tie (8 V on each resistor), then 2 and 5 tie (4 V on each resistor) **EP** **1.** (a) $320; (b) 4.8¢ **3.** 11 kJ

5. (a) counterclockwise; (b) battery 1; (c) B　**7.** (a) 80 J;
(b) 67 J; (c) 13 J converted to thermal energy within battery
9. (a) 14 V; (b) 100 W; (c) 600 W; (d) 10 V, 100 W
11. (a) 50 V; (b) 48 V; (c) B is connected to the negative
terminal　**13.** 2.5 V　**15.** (a) 6.9 km; (b) 20 Ω
17. 8.0 Ω　**19.** 10^{-6}　**21.** the cable　**23.** (a) 1000 Ω;
(b) 300 mV; (c) 2.3×10^{-3}　**25.** (a) 3.41 A or 0.586 A;
(b) 0.293 V or 1.71 V　**27.** 5.56 A　**29.** 4.0 Ω and 12 Ω
31. 4.50 Ω　**33.** 0.00, 2.00, 2.40, 2.86, 3.00, 3.60, 3.75,
3.94 A　**35.** $V_d - V_c = +0.25$ V, by all paths　**37.** three
39. (a) 2.50 Ω; (b) 3.13 Ω　**41.** nine　**43.** (a) left branch:
0.67 A down; center branch: 0.33 A up; right branch: 0.33 A
up; (b) 3.3 V　**47.** (a) 120 Ω; (b) $i_1 = 51$ mA, $i_2 = i_3 =$
19 mA, $i_4 = 13$ mA　**49.** (a) 19.5 Ω; (b) 0; (c) ∞; (d) 82.3 W,
57.6 W　**51.** (a) Cu: 1.11 A, Al: 0.893 A; (b) 126 m
53. (a) 13.5 kΩ; (b) 1500 Ω; (c) 167 Ω; (d) 1480 Ω
55. 0.45 A　**57.** (a) 12.5 V; (b) 50 A　**59.** -0.9%
65. (a) 0.41τ; (b) 1.1τ　**67.** 4.6　**69.** (a) 0.955 μC/s;
(b) 1.08 μW; (c) 2.74 μW; (d) 3.82 μW　**71.** 2.35 MΩ
73. 0.72 MΩ　**75.** 24.8 Ω to 14.9 kΩ　**77.** (a) at $t = 0$,
$i_1 = 1.1$ mA, $i_2 = i_3 = 0.55$ mA; at $t = \infty$, $i_1 = i_2 = 0.82$ mA,
$i_3 = 0$; (c) at $t = 0$, $V_2 = 400$ V; at $t = \infty$, $V_2 = 600$ V;
(d) after several time constants ($\tau = 7.1$ s) have elapsed
79. (a) $V_T = -ir + \mathscr{E}$; (b) 13.6 V; (c) 0.060 Ω
81. (a) 6.4 V; (b) 3.6 W; (c) 17 W; (d) -5.6 W; (e) a

Chapter 29

CP　**1.** $a, +z; b, -x; c, \mathbf{F}_B = 0$　**2.** 2, then tie of 1 and 3
(zero); (b) 4　**3.** (a) $+z$ and $-z$ tie, then $+y$ and $-y$ tie, then
$+x$ and $-x$ tie (zero); (b) $+y$　**4.** (a) electron;
(b) clockwise　**5.** $-y$　**6.** (a) all tie; (b) 1 and 4 tie, then 2
and 3 tie　Q　**1.** (a) all tie; (b) 1 and 2 (charge is negative)
3. a, no, $\mathbf{v}$ and $\mathbf{F}_B$ must be perpendicular; b, yes; c, no, $\mathbf{B}$ and
$\mathbf{F}_B$ must be perpendicular　**5.** (a) $\mathbf{F}_E$; (b) $\mathbf{F}_B$　**7.** (a) right;
(b) right　**9.** (a) negative; (b) equal; (c) equal; (d) half a
circle　**11.** (a) $\mathbf{B}_1$; (b) $\mathbf{B}_1$ into page; $\mathbf{B}_2$ out of page; (c) less
13. all　**15.** all tie　**17.** (a) positive; (b) (1) and (2) tie, then
(3) which is zero　EP　**1.** M/QT　**3.** (a) 9.56×10^{-14} N, 0;
(b) $0.267°$　**5.** (a) east; (b) 6.28×10^{14} m/s^2; (c) 2.98 mm
7. 0.75k T　**9.** (a) 3.4×10^{-4} T, horizontal and to the left
as viewed along $\mathbf{v}_0$; (b) yes, if velocity is the same as the
electron's velocity　**11.** $(-11.4\mathbf{i} - 6.00\mathbf{j} + 4.80\mathbf{k})$ V/m
13. 680 kV/m　**17.** (b) 2.84×10^{-3}　**19.** (a) 1.11×10^7 m/s;
(b) 0.316 mm　**21.** (a) 0.34 mm; (b) 2.6 keV
23. (a) 2.05×10^7 m/s; (b) 467 μT; (c) 13.1 MHz; (d) 76.3 ns
25. (a) 2.60×10^6 m/s; (b) 0.109 μs; (c) 0.140 MeV;
(d) 70 kV　**29.** (a) 1.0 MeV; (b) 0.5 MeV　**31.** $R_d = \sqrt{2}R_p$;
$R_\alpha = R_p$.　**33.** (a) $B\sqrt{mq/2V}\,\Delta x$; (b) 8.2 mm　**37.** (a) $-q$;
(b) $\pi m/qB$　**39.** $B_{\min} = \sqrt{mV/2ed^2}$　**41.** (a) 22 cm;
(b) 21 MHz　**43.** neutron moves tangent to original path,
proton moves in a circular orbit of radius 25 cm　**45.** 28.2 N,
horizontally west　**47.** 20.1 N　**49.** $Bitd/m$, away from
generator　**51.** -0.35k N　**53.** 0.10 T, at 31° from the
vertical　**55.** 4.3×10^{-3} N·m, negative y　**59.** $qvaB/2$
61. (a) 540 Ω, in series; (b) 2.52 Ω, in parallel
63. (a) 12.7 A; (b) 0.0805 N·m　**65.** (a) 0.184 A·m^2;

(b) 1.45 N·m　**67.** (a) 20 min; (b) 5.9×10^{-2} N·m
69. (a) $(8.0 \times 10^{-4}$ N·m$)(-1.2\mathbf{i} - 0.90\mathbf{j} + 1.0\mathbf{k})$;
(b) -6.0×10^{-4} J

Chapter 30

CP　**1.** a, c, b　**2.** b, c, a　**3.** d, tie of a and c, then b
4. d, a, tie of b and c (zero)　Q　**1.** c, d, then a and b tie
3. 2 and 4　**5.** a, b, c　**7.** b, d, c, a (zero)　**9.** (a) 1, $+x$;
2, $-y$; (b) 1, $+y$; 2, $+x$　**11.** outward　**13.** c and
d tie, then b, a　**15.** d, then tie of a and e, then b, c
17. 0 (dot product is zero)　EP　**1.** 32.1 A
3. (a) 3.3 μT; (b) yes　**5.** (a) $(0.24\mathbf{i})$ nT; (b) 0; (c) $(-43\mathbf{k})$ pT;
(d) $(0.14\mathbf{k})$ nT　**7.** (a) 16 A; (b) west to east　**9.** 0
11. (a) 0; (b) $\mu_0 i/4R$, into the page; (c) same as (b)
13. $\mu_0 i\theta (1/b - 1/a)/4\pi$, out of page　**15.** (a) 1.0 mT, out of
the figure; (b) 0.80 mT, out of the figure　**25.** 200 μT, into
page　**27.** (a) it is impossible to have other than $B = 0$
midway between them; (b) 30 A　**29.** 4.3 A, out of page
35. $0.338\mu_0 i^2/a$, toward the center of the square
37. (b) to the right　**39.** (b) 2.3 km/s　**41.** $+5\mu_0 i$
47. (a) $\mu_0 ir/2\pi c^2$; (b) $\mu_0 i/2\pi r$; (c) $\dfrac{\mu_0 i}{2\pi(a^2 - b^2)}\,\dfrac{a^2 - r^2}{r}$;
(d) 0　**49.** $3i/8$, into page　**53.** 0.30 mT　**55.** 108 m
61. 0.272 A　**63.** (a) 4; (b) 1/2　**65.** (a) 2.4 A·m^2;
(b) 46 cm　**67.** (a) $\mu_0 i(1/a + 1/b)/4$, into page; (b) $\frac{1}{2}i\pi(a^2 + b^2)$,
into page　**69.** (a) 79 μT; (b) 1.1×10^{-6} N·m
71. (b) $(0.060\,\mathbf{j})$ A·m^2; (c) $(9.6 \times 10^{-11}\mathbf{j})$ T, $(-4.8 \times 10^{-11}\mathbf{j})$ T　**73.** (a) B from sum: 7.069×10^{-5} T; $\mu_0 in =$
5.027×10^{-5} T; 40% difference; (b) B from sum: 1.043×10^{-4} T; $\mu_0 in = 1.005 \times 10^{-4}$ T; 4% difference; (c) B from
sum: 2.506×10^{-4} T; $\mu_0 in = 2.513 \times 10^{-4}$ T; 0.3%
difference　**75.** (a) $\mathbf{B} = (\mu_0/2\pi)\,[i_1/(x - a) + i_2/x]\mathbf{j}$;
(b) $\mathbf{B} = (\mu_0/2\pi)\,(i_1/a)\,(1 + b/2)\mathbf{j}$

Chapter 31

CP　**1.** b, then d and e tie, and then a and c tie (zero)　**2.** a
and b tie, then c (zero)　**3.** c and d tie, then a and b tie
4. b, out; c, out; d, into; e, into　**5.** d and e　**6.** (a) 2, 3, 1
(zero); (b) 2, 3, 1　**7.** a and b tie, then c　Q　**1.** (a) all tie
(zero); (b) all tie (nonzero); (c) 3, then tie of 1 and 2 (zero)
3. out　**5.** (a) into; (b) counterclockwise; (c) larger
7. (a) leftward; (b) rightward　**9.** c, a, b　**11.** (a) 1, 3, 2;
(b) 1 and 3 tie, then 2　**13.** a, tie of b and c　**15.** (a) more;
(b) same; (c) same; (d) same (zero)　**17.** a, 2; b, 4; c, 1; d, 3
EP　**1.** 57 μWb　**3.** 1.5 mV　**5.** (a) 31 mV; (b) right to
left　**7.** (a) 0.40 V; (b) 20 A　**9.** (b) 58 mA　**11.** 1.2 mV
13. 1.15 μWb　**15.** 51 mV, clockwise when viewed along the
direction of $\mathbf{B}$　**17.** (a) 1.26×10^{-4} T, 0, -1.26×10^{-4} T;
(b) 5.04×10^{-8} V　**19.** (b) no　**21.** 15.5 μC
23. (a) 24 μV; (b) from c to b　**25.** (b) design it so that
$Nab = (5/2\pi)$ m^2　**27.** (a) 0.598 μV; (b) counterclockwise
29. (a) $\dfrac{\mu_0 ia}{2\pi}\left(\dfrac{2r + b}{2r - b}\right)$; (b) $2\mu_0 iabv/\pi R(4r^2 - b^2)$
31. $A^2 B^2/R\Delta t$　**33.** (a) 48.1 mV; (b) 2.67 mA; (c) 0.128 mW
35. $v_t = mgR/B^2 L^2$　**37.** 268 W　**39.** (a) 240 μV; (b) 0.600

mA; (c) 0.144 μW; (d) 2.88×10^{-8} N; (e) same as (c)
41. 1, -1.07 mV; 2, -2.40 mV; 3, 1.33 mV   **43.** at a:
4.4×10^7 m/s^2, to the right; at b: 0; at c: 4.4×10^7 m/s^2, to the
left   **45.** 0.10 μWb   **47.** (a) 800; (b) 2.5×10^{-4} H/m
49. (a) $\mu_0 i/W$; (b) $\pi\mu_0 R^2/W$ **51.** (a) decreasing;
(b) 0.68 mH   **53.** (a) 0.10 H; (b) 1.3 V   **55.** (a) 16 kV;
(b) 3.1 kV; (c) 23 kV   **57.** (b) so that the changing magnetic
field of one does not induce current in the other;
(c) $1/L_{eq} = \sum\limits_{j=1}^{N} (1/L_j)$ **59.** 6.91   **61.** 1.54 s   **63.** (a) 8.45
ns; (b) 7.37 mA   **65.** $(42 + 20t)$ V   **67.** 12.0 A/s
69. (a) $i_1 = i_2 = 3.33$ A; (b) $i_1 = 4.55$ A, $i_2 = 2.73$ A;
(c) $i_1 = 0$, $i_2 = 1.82$ A; (d) $i_1 = i_2 = 0$ **71.** $\mathscr{E}L_1/R(L_1 + L_2)$
73. (a) $i(1 - e^{-Rt/L})$ **75.** $1.23\tau_L$ **77.** (a) 240 W;
(b) 150 W; (c) 390 W   **79.** (a) 97.9 H; (b) 0.196 mJ
81. (a) 10.5 mJ; (b) 14.1 mJ   **83.** (a) 34.2 J/m^3; (b) 49.4
mJ   **85.** 1.5×10^8 V/m   **87.** $(\mu_0 l/2\pi)\ln(b/a)$ **89.** (a) 1.3
mT; (b) 0.63 J/m^3 **91.** (a) 1.0 J/m^3; (b) 4.8×10^{-15} J/m^3
93. (a) 1.67 mH; (b) 6.00 mWb   **95.** 13 H   **99.** magnetic
field exists only within the cross section of solenoid 1

Chapter 32

CP 1. d, b, c, a (zero)   **2.** (a) 2; (b) 1   **3.** (a) away;
(b) away; (c) less   **4.** (a) toward; (b) toward; (c) less
5. a, c, b, d (zero)   **6.** tie of b, c, and d, then a
Q 1. (a) a, c, f; (b) bar gh **3.** supplied   **5.** (a) all down;
(b) 1 up, 2 down, 3 zero   **7.** (a) 1 up, 2 up, 3 down;
(b) 1 down, 2 up, 3 zero   **9.** (a) rightward; (b) leftward
11. (a) decreasing; (b) decreasing   **13.** (a) tie of a and b, then
c, d; (b) none (plate lacks circular symmetry, so **B** is not
tangent to a circular loop); (c) none   **15.** 1/4   **EP 1.** (b)
sign is minus; (c) no, compensating positive flux through open
end near magnet   **3.** 47 μWb, inward   **5.** 55 μT   **7.** (a)
600 MA; (b) yes; (c) no   **9.** (a) 31.0 μT, 0°; (b) 55.9 μT,
73.9°; (c) 62.0 μT, 90°   **11.** 4.6×10^{-24} J   **13.** (a) $5.3 \times$
10^{11} V/m; (b) 20 mT; (c) 660   **15.** (a) 7; (b) 7; (c) $3h/2\pi$, 0;
(d) $3eh/4\pi m$, 0; (e) $3.5h/2\pi$; (f) 8   **17.** (b) in the direction
of the angular momentum vector   **19.** $\Delta\mu = e^2 r^2 B/4m$
21. 20.8 mJ/T   **23.** yes   **25.** (a) 4 K; (b) 1 K
29. (a) 3.0 μT; (b) 5.6×10^{-10} eV   **31.** (a) 8.9 A·m^2;
(b) 13 N·m   **35.** (a) 0.14 A; (b) 79 μC   **37.** $2.4 \times$
10^{13} V/m·s   **39.** 1.9 pT   **41.** 7.5×10^5 V/s
43. 7.2×10^{12} V/m·s   **45.** (a) 2.1×10^{-8} A, downward;
(b) clockwise   **47.** (a) 0.63 μT; (b) 2.3×10^{12} V/m·s
49. (a) 2.0 A; (b) 2.3×10^{11} V/m·s; (c) 0.50 A;
(d) 0.63 μT·m   **51.** (a) 7.60 μA; (b) 859 kV·m/s;
(c) 3.39 mm; (d) 5.16 pT

Chapter 33

CP 1. (a) $T/2$, (b) T, (c) $T/2$, (d) $T/4$ **2.** (a) 5 V;
(b) 150 μJ   **3.** (a) 1; (b) 2   **4.** (a) C, B, A; (b) 1, A;
2, B; 3, S; 4, C; (c) A **5.** (a) increases; (b) decreases
6. (a) 1, lags; 2, leads; 3, in phase; (b) 3 ($\omega_d = \omega$ when
$X_L = X_C$)   **7.** (a) increase (circuit is mainly capacitive;
increase C to decrease X_C to be closer to resonance for

maximum P_{av}); (b) closer   **8.** step-up   **Q 1.** (a) $T/4$,
(b) $T/4$, (c) $T/2$ (see Fig. 33-2), (d) $T/2$ (see Eq. 31-40)
3. b, a, c **5.** (a) 3, 1, 2; (b) 2, tie of 1 and 3
7. slower   **9.** (a) 1 and 4; (b) 2 and 3   **11.** (a) 3,
then 1 and 2 tie; (b) 2, 1, 3   **13.** (a) negative; (b) lead
15. (a)–(c) rightward, increase   **EP 1.** 9.14 nF
3. 45.2 mA   **5.** (a) 6.00 μs; (b) 167 kHz; (c) 3.00 μs
7. (a) 89 rad/s; (b) 70 ms; (c) 25 μF   **9.** 38 μH
11. 7.0×10^{-4} s   **15.** (a) 3.0 nC; (b) 1.7 mA; (c) 4.5 nJ
17. (a) 3.60 mH; (b) 1.33 kHz; (c) 0.188 ms   **19.** 600, 710,
1100, 1300 Hz   **21.** (a) $Q/\sqrt{3}$; (b) 0.152   **25.** (a) 1.98 μJ;
(b) 5.56 μC; (c) 12.6 mA; (d) $-46.9°$; (e) $+46.9°$ **27.** (a) 0;
(b) $2i(t)$ **29.** (a) 356 μs; (b) 2.50 mH; (c) 3.20 mJ
31. 8.66 mΩ **33.** $(L/R)\ln 2$ **35.** (a) $\pi/2$ rad; (b) $q = $
$(I/\omega') e^{-Rt/2L} \sin \omega' t$ **39.** (a) 0.0955 A; (b) 0.0119 A
41. (a) 4.60 kHz; (b) 26.6 nF; (c) $X_L = 2.60$ kΩ, $X_C = $
0.650 kΩ **43.** (a) 0.65 kHz; (b) 24 Ω **45.** (a) 39.1 mA;
(b) 0; (c) 33.9 mA   **47.** (a) 6.73 ms; (b) 2.24 ms;
(c) capacitor; (d) 59.0 μF   **49.** (a) $X_C = 0$, $X_L = 86.7$ Ω,
$Z = 182$ Ω, $I = 198$ mA, $\phi = 28.5°$ **51.** (a) $X_C = 37.9$ Ω,
$X_L = 86.7$ Ω, $Z = 167$ Ω, $I = 216$ mA, $\phi = 17.1°$
53. (a) 2.35 mH; (b) they move away from 1.40 kHz
55. 1000V   **57.** (a) 36.0 V; (b) 27.3 V; (c) 17.0 V;
(d) -8.34 V   **59.** (a) 224 rad/s; (b) 6.00 A; (c) 228 rad/s,
219 rad/s; (d) 0.040   **61.** (a) 707 Ω; (b) 32.2 mH;
(c) 21.9 nF   **63.** (a) resonance at $f = 1/2\pi\sqrt{LC} = 85.7$ Hz;
(b) 15.6 μF; (c) 225 mA   **65.** (a) 796 Hz; (b) no change;
(c) decreased; (d) increased   **69.** 141 V   **71.** (a) taking;
(b) supplying   **73.** 0, 9.00 W, 3.14 W, 1.82 W
75. 177 Ω, no   **77.** 7.61 A   **83.** (a) 117 μF; (b) 0;
(c) 90.0 W, 0; (d) 0°, 90°; (e) 1, 0   **85.** (a) 2.59 A;
(b) 38.8 V, 159 V, 224 V, 64.2 V, 75.0 V; (c) 100 W for R,
0 for L and C.   **87.** (a) 2.4 V; (b) 3.2 mA, 0.16 A
89. (a) 1.9 V, 5.9 W; (b) 19 V, 590 W; (c) 0.19 kV, 59 kW
91. (a) $X_C = [(2\pi)(45 \times 10^{-6}$ F$)f]^{-1}$; (c) 17.7 Hz
93. (a) $X_L = (2\pi)(40 \times 10^{-3}$ H$)f$; (c) 796 Hz
95. (b) 61 Hz; (c) 90 Ω and 61 Hz

Chapter 34

CP 1. (a) (Use Fig. 34-5.) On right side of rectangle, **E** is in
negative y direction; on left side, $\mathbf{E} + d\mathbf{E}$ is greater and in
same direction; (b) **E** is downward. On right side, **B** is in
negative z direction; on left side, $\mathbf{B} + d\mathbf{B}$ is greater and in
same direction.   **2.** positive direction of x **3.** (a) same;
(b) decrease   **4.** a, d, b, c (zero)   **5.** a **6.** (a) yes; (b) no
Q 1. (a) positive direction of z; (b) x **3.** (a) same;
(b) increase; (c) decrease   **5.** both 20° clockwise from the y
axis   **7.** two   **9.** b, 30°; c, 60°; d, 60°; e, 30°; f, 60°
11. d, b, a, c **13.** (a) b; (b) blue; (c) c **15.** 1.5
EP 1. (a) 4.7×10^{-3} Hz; (b) 3 min 32 s   **3.** (a) 4.5×10^{24}
Hz; (b) 1.0×10^4 km or 1.6 Earth radii   **7.** it would steadily
increase; (b) the summed discrepancies between the apparent
time of eclipse and those observed from x; the radius of Earth's
orbit   **9.** 5.0×10^{-21} H   **11.** 1.07 pT   **17.** 4.8×10^{-29}
W/m^2 **19.** 4.51×10^{-10} **21.** 89 cm   **23.** 1.2 MW/m^2
25. 820 m   **27.** (a) 1.03 kV/m; 3.43 μT   **29.** (a) 1.4 $\times$

10^{-22} W; (b) 1.1×10^{15} W   **31.** (a) 87 mV/m; (b) 0.30 nT; (c) 13 kW   **33.** 3.3×10^{-8} Pa   **35.** (a) 4.7×10^{-6} Pa; (b) 2.1×10^{10} times smaller   **37.** 5.9×10^{-8} Pa   **39.** (a) 3.97 GW/m²; (b) 13.2 Pa; (c) 1.67×10^{-11} N; (d) 3.14×10^3 m/s²   **41.** $I(2 - frac)/c$ **43.** $p_{r\perp} \cos^2 \theta$ **45.** 1.9 mm/s   **47.** (b) 580 nm   **49.** (a) 1.9 V/m; (b) 1.7×10^{-11} Pa   **51.** 1/8   **53.** 3.1%   **55.** 20° or 70°   **57.** 19 W/m²   **59.** (a) 2 sheets; (b) 5 sheets   **61.** 180°   **63.** 1.26   **65.** 1.07 m   **69.** (a) 0; (b) 20°; (c) still 0 and 20°   **73.** 1.41   **75.** 1.22   **77.** 182 cm   **79.** (a) no; (b) yes; (c) about 43°   **81.** (a) 35.6°; (b) 53.1°   **83.** (b) 23.2°   **85.** (a) 53°; (b) yes   **87.** (a) 55.5°; (b) 55.8°

Chapter 35

CP Kaleidoscope answer: two mirrors that form a V with an angle of 60°   **1.** 0.2d, 1.8d, 2.2d   **2.** (a) real; (b) inverted; (c) same   **3.** (a) e; (b) virtual, same   **4.** virtual, same as object, diverging   **Q** **1.** c   **3.** (a) a; (b) c   **5.** (a) no; (b) yes (fourth is off mirror ed)   **7.** (a) from infinity to the focal point; (b) decrease continually   **9.** d (infinite), tie of a and b, then c   **11.** mirror, equal; lens, greater   **13.** (a) all but variation 2; (b) for 1, 3, and 4: right, inverted; for 5 and 6: left, same   **15.** (a) less; (b) less   **EP** **1.** (a) virtual; (b) same; (c) same; (d) $D + L$ **3.** 40 cm   **7.** (a) 7; (b) 5; (c) 1 to 3; (d) depends on the position of O and your perspective   **11.** new illumination is 10/9 of the old   **15.** 10.5 cm   **19.** (a) 2.00; (b) none   **23.** 1.14   **25.** (b) separate the lenses by a distance $f_2 - |f_1|$, where f_2 is the focal length of the converging lens   **27.** 45 mm, 90 mm   **29.** (a) $+40$ cm; (b) at infinity   **33.** (a) 40 cm, real; (b) 80 cm, real; (c) 240 cm, real; (d) -40 cm, virtual; (e) -80 cm, virtual; (f) -240 cm, virtual   **35.** same orientation, virtual, 30 cm to left of second lens, $m = 1$ **37.** (a) final image coincides in location with the object; it is real, inverted, and $m = -1.0$ **39.** (a) coincides in location with the original object and is enlarged 5.0 times; (c) virtual; (d) yes

45. $i = \dfrac{(2 - n)r}{2(n - 1)}$, to the right of the right side of the sphere

47. 2.1 mm   **49.** (b) when image is at near point   **51.** (b) farsighted   **53.** -125

Chapter 36

CP **1.** b (least n), c, a   **2.** (a) top; (b) bright intermediate illumination (phase difference is 2.1 wavelengths)   **3.** (a) 3λ, 3; (b) 2.5λ, 2.5   **4.** a and d tie (amplitude of resultant wave is $4E_0$), then b and c tie (amplitude of resultant wave is $2E_0$)   **5.** (a) 1 and 4; (b) 1 and 4   **Q** **1.** a, c, b   **3.** (a) 300 nm; (b) exactly out of phase   **5.** c   **7.** (a) increase; (b) 1λ   **9.** down   **11.** (a) maximum; (b) minimum; (c) alternates   **13.** d   **15.** (a) 0.5 wavelength; (b) 1 wavelength   **17.** bright   **19.** all   **EP** **1.** (a) 5.09×10^{14} Hz; (b) 388 nm; (c) 1.97×10^8 m/s   **5.** 2.1×10^8 m/s   **7.** the time is longer for the pipeline containing air, by about 1.55 ns   **9.** 22°, refraction reduces θ   **11.** (a) pulse 2; (b) $0.03L/c$ **13.** (a) 1.70 (or 0.70); (b) 1.70 (or 0.70); (c) 1.30

(or 0.30); (d) brightness is identical, close to fully destructive interference   **15.** (a) 0.833; (b) intermediate, closer to fully constructive interference   **17.** $(2m + 1)\pi$ **19.** 2.25 mm   **21.** 648 nm   **23.** 1.6 mm   **25.** 16   **27.** 0.072 mm   **29.** 8.75λ   **31.** 0.03%   **33.** 6.64 μm   **35.** $y = 17 \sin(\omega t + 13°)$ **39.** (a) 1.17 m, 3.00 m, 7.50 m; (b) no   **41.** $I = \frac{1}{9}I_m[1 + 8 \cos^2(\pi d \sin \theta/\lambda)]$, I_m = intensity of central maximum   **43.** $L = (m + \frac{1}{2})\lambda/2$, for $m = 0, 1, 2, \ldots$ **45.** 0.117 μm, 0.352 μm   **47.** λ/5   **49.** 70.0 nm   **51.** none   **53.** (a) 552 nm; (b) 442 nm   **55.** 338 nm   **59.** $2n_2L/\cos \theta_r = (m + \frac{1}{2})\lambda$, for $m = 0, 1, 2, \ldots$, where $\theta_r = \sin^{-1}[(\sin \theta_i)/n_2]$ **61.** intensity is diminished by 88% at 450 nm and by 94% at 650 nm   **63.** (a) dark; (b) blue end   **65.** 1.89 μm   **67.** 1.00025   **69.** (a) 34; (b) 46   **73.** 588 nm   **75.** 1.0003   **77.** $I = I_m \cos^2(2\pi x/\lambda)$

Chapter 37

CP **1.** (a) expand; (b) expand   **2.** (a) second side maximum; (b) 2.5   **3.** (a) red; (b) violet   **4.** diminish   **5.** (a) increase; (b) same   **6.** (a) left; (b) less   **Q** **1.** (a) contract; (b) contract   **3.** with megaphone (larger opening, less diffraction)   **5.** four   **7.** (a) larger; (b) red   **9.** (a) decrease; (b) same; (c) in place   **11.** (a) A; (b) left; (c) left; (d) right   **EP** **1.** 690 nm   **3.** 60.4 μm   **5.** (a) 2.5 mm; (b) 2.2×10^{-4} rad   **7.** (a) 70 cm; (b) 1.0 mm   **9.** 41.2 m from the central axis   **11.** 160°   **15.** (d) 53°, 10°, 5.1°   **19.** (a) 1.3×10^{-4} rad; (b) 10 km   **21.** 50 m   **23.** 30.5 μm   **25.** 1600 km   **27.** (a) 17.1 m; (b) 1.37×10^{-10} **29.** 27 cm   **31.** 4.7 cm   **33.** (a) 0.347°; (b) 0.97°   **35.** (a) red; (b) 130 μm   **37.** five   **41.** λD/d   **43.** (a) 5.05 μm; (b) 20.2 μm   **45.** (a) 3.33 μm; (b) 0, $\pm 10.2°$, $\pm 20.7°$, $\pm 32.0°$, $\pm 45.0°$, $\pm 62.2°$ **47.** all wavelengths shorter than 635 nm   **49.** 13,600   **51.** 500 nm   **53.** (a) three; (b) 0.051°   **55.** 523 nm   **61.** 470 nm to 560 nm   **63.** 491   **65.** 3650   **67.** (a) 1.0×10^4 nm; (b) 3.3 mm   **69.** (a) 0.032°/nm, 0.076°/nm, 0.24°/nm; (b) 40,000, 80,000, 120,000   **71.** (a) tan θ; (b) 0.89   **73.** 0.26 nm   **75.** 6.8°   **77.** (a) 170 pm; (b) 130 pm   **81.** 0.570 nm   **83.** 30.6°, 15.3° (clockwise); 3.08°, 37.8° (counterclockwise)

Chapter 38

CP **1.** (a) same (speed of light postulate); (b) no (the start and end of the flight are spatially separated); (c) no (again, because of the spatial separation)   **2.** (a) Sally's; (b) Sally's   **3.** a, negative; b, positive; c, negative   **4.** (a) right; (b) more   **5.** (a) equal; (b) less   **Q** **1.** all tie (pulse speed is c)   **3.** (a) C_1; (b) C_1 **5.** (a) 3, 2, 1; (b) 1 and 3 tie, then 2   **7.** (a) negative; (b) positive   **9.** c, then b and d tie, then a   **11.** (a) 3, tie of 1 and 2, then 4; (b) 4, tie of 1 and 2, then 3; (c) 1, 4, 2, 3   **13.** greater than f_1 **EP** **1.** (a) 3×10^{-18}; (b) 8.2×10^{-8}; (c) 1.1×10^{-6}; (d) 3.7×10^{-5}; (e) 0.10   **3.** 0.75c   **5.** 0.99c   **7.** 55 m   **9.** 1.32 m   **11.** 0.63 m   **13.** 6.4 cm   **15.** (a) 26 y; (b) 52 y; (c) 3.7 y   **17.** (b) 0.999 999 15c   **19.** (a) $x' = 0$, $t' = 2.29$ s; (b) $x' =$

6.55×10^8 m, $t' = 3.16$ s   **21.** (a) 25.8 μs; (b) small flash
23. (a) 1.25; (b) 0.800 μs   **25.** 2.40 μs   **27.** (a) 0.84c, in
the direction of increasing x; (b) 0.21c, in the direction of
increasing x; the classical predictions are 1.1c and 0.15c
29. (a) 0.35c; (b) 0.62c **31.** 1.2 μs   **33.** seven
35. 22.9 MHz   **37.** +2.97 nm   **39.** (a) $\tau_0/\sqrt{1 - v^2/c^2}$
41. (a) 0.134c; (b) 4.65 keV; (c) 1.1%   **43.** (a) 0.9988, 20.6;
(b) 0.145, 1.01; (c) 0.073, 1.0027   **45.** (a) 5.71 GeV,
6.65 GeV, 6.58 GeV/c; (b) 3.11 MeV, 3.62 MeV,
3.59 MeV/c **47.** 18 smu/y   **49.** (a) 0.943c; (b) 0.866c
51. (a) 256 kV; (b) 0.746c **53.** $\sqrt{8}mc$ **55.** 6.65×10^6 mi,
or 270 earth circumferences   **57.** 110 km   **59.** (a) 2.7 $\times$

10^{14} J; (b) 1.8×10^7 kg; (c) 6.0×10^6 **61.** 4.00 u, probably
a helium nucleus   **63.** 330 mT
65.

SIGNAL	TIME SENT (h)	TIME REPLY RECEIVED (h)	TIME REPORTED	DISTANCE (m)
1	6.0	400	11.8	2.10×10^{14}
2	12.0	800	23.6	4.19×10^{14}
3	18.0	1200	35.5	6.29×10^{14}
4	24.0	1600	47.3	8.38×10^{14}
5	30.0	2000	59.1	1.05×10^{15}

67. (a) $vt \sin \theta$; (b) $t[1 - (v/c) \cos \theta]$; (c) 3.24c

Index

Page references followed by lowercase roman t indicate material in tables. Page references followed by lowercase italic *n* indicate material in footnotes.

Photo Credits

Chapter 16
Page 372: Tom van Dyke/Sygma. Page 373: Kent Knudson/FPG International. Page 387: Bettmann Archive. Page 392: Courtesy NASA.

Chapter 17
Page 399: John Visser/Bruce Coleman, Inc. Page 415: Richard Megna/Fundamental Photographs. Page 416: Courtesy T.D. Rossing, Northern Illinois University.

Chapter 18
Page 425: Stephen Dalton/Animals Animals. Page 426: Howard Sochurak/The Stock Market. Page 433: Ben Rose/The Image Bank. Page 434: Bob Gruen/Star File. Page 435: John Eastcott/Yva Momativk/DRK Photo. Page 441: Philippe Plailly/Science Photo Library/Photo Researchers. Page 444: Courtesy NASA.

Chapter 19
Page 453: Tom Owen Edmunds/The Image Bank. Page 459: AP/Wide World Photos. Page 471: Peter Arnold/Peter Arnold, Inc. Page 472: Courtesy Daedalus Enterprises, Inc.

Chapter 20
Page 484: Bryan and Cherry Alexander Photography.

Chapter 21
Page 509: Steven Dalton/Photo Researchers. Page 517: Richard Ustinich/The Image Bank. Page 523 (left): Cary Wolinski/Stock, Boston. Page 523 (right): Courtesy of Professor Bernard Hallet, Quaternary Research Center, University of Washington, Seattle. Page 534: Jeff Werner.

Chapter 22
Page 537: Michael Watson. Page 538: Fundamental Photographs. Page 539: Courtesy Xerox. Page 540: Johann Gabriel Doppelmayr, Neuentdeckte Phaenomena von Bewünderswurdigen Würckungen der Natur, Nuremberg, 1744. Page 547: Courtesy Lawrence Berkeley Laboratory.

Chapter 23
Page 554: Quesada/Burke Studios. Page 565: Russ Kinne/Comstock, Inc. Page 566: Courtesy Environmental Elements Corporation.

Chapter 24
Page 579: E.R. Degginger/Bruce Coleman, Inc. Page 589 (top): Courtesy E. Philip Krider, Institute for Atmospheric Physics, University of Arizona, Tucson. Page 589 (bottom): C. Johnny Autery.

Chapter 25
Pages 601 and 605: Courtesy NOAA. Page 617: Courtesy Westinghouse Corporation.

Chapter 26
Page 628: Goivaux Communication/Phototake. Page 629: Paul Silvermann/Fundamental Photographs. Page 638: ©The Harold E. Edgerton 1992 Trust/Courtesy Palm Press, Inc. Page 639: Courtesy The Royal Institute, England.

Chapter 27
Page 651: UPI/Corbis-Bettmann. Page 657: The Image Works. Page 663: Laurie Rubie/Tony Stone Images/New York, Inc. Page 666 (left): Courtesy AT&T Bell Laboratories. Page 666 (right): Courtesy Shoji Tonaka/International Superconductivity Technology Center, Tokyo, Japan.

Chapter 28
Page 673: Norbert Wu. Page 674: Courtesy Southern California Edison Company.

Chapter 29
Page 700: Johnny Johnson/Earth Scenes/Animals Animals. Page 701: Schneps/The Image Bank. Page 703: Courtesy Lawrence Berkeley Laboratory, University of California. Page 704: Courtesy Dr. Richard Cannon, Southeast Missouri State University, Cape Girardeau. Page 708: Courtesy John Le P. Webb, Sussex University, England. Page 710: Courtesy Dr. L.A. Frank, University of Iowa. Page 712: Courtesy Fermi National Accelerator Laboratory.

Chapter 30
Page 728: Michael Brown/Florida Today/Gamma Liaison. Page 730: Courtesy Education Development Center.

Chapter 31
Page 752: Dan McCoy/Black Star. Page 756: Courtesy Fender Musical Instruments Corporation. Page 760: Courtesy Jenn-Air Co. Page 765: Courtesy The Royal Institute, England.

Chapter 32
Page 786: Bob Zehring. Page 787: Runk/Schoenberger/Grant Heilman Photography. Page 794: Peter Lerman. Page 797 (top): Courtesy Ralph W. DeBlois. Page 797 (bottom): R.E. Rosenweig, Research and Science Laboratory, courtesy Exxon Co. USA.

Chapter 33
Page 811: Courtesy Haverfield Helicopter Co. Page 814: Courtesy Hewlett Packard. Page 827 (left): Steve Kagan/Gamma Liaison. Page 827 (right): Ted Cowell/Black Star.

Chapter 34
Page 841: Courtesy Hansen Publications. Page 850: ©1992 Ben and Miriam Rose, from the collection of the Center for Creative Photography, Tucson. Page 854: Diane Schiumo/Fundamental Photographs. Page 855: PSSC Physics, 2nd edition; ©1975 D.C. Heath and Co. with Education Development Center, Newton, MA. Page 856: Courtesy Lockheed Advanced Development Company. Page 858 (top right): Tony Stone Images/New York, Inc. Page 858 (bottom left): Courtesy Bausch & Lomb. Page 861: Will and Deni McIntyre/Photo Researchers. Page 867: Cornell University.

Chapter 35
Page 872: Courtesy Courtauld Institute Galleries, London. Page 874 (left): Frans Lanting/Minden Pictures, Inc. Page 874 (right): Wayne Sorce. Page 882: Dr. Paul A. Zahl/Photo Researchers.